IMAGING THROUGH TURBULENCE

The CRC Press
Laser and Optical Science and Technology Series

Editor-in-Chief: Marvin J. Weber

Handbooks of Laser Science and Technology

Volume I: Lasers and Masers, Edited by Marvin J. Weber

Volume II: Gas Lasers, Edited by Marvin J. Weber

Volume III: Optical Materials, Part 1, Edited by Marvin J. Weber

Volume IV: Optical Materials, Part 2, Edited by Marvin J. Weber

Volume V: Optical Materials, Part 3, Edited by Marvin J. Weber

Supplement I: Lasers, Edited by Marvin J. Weber

Thermodynamic and Kinetic Aspects of the Vitreous State

Sergei V. Nemilov

Forthcoming Handbook Titles

Supplement II: Optical Materials, Edited by Marvin J. Weber

Handbook of Laser Wavelengths, Edited by Marvin J. Weber

IMAGING THROUGH TURBULENCE

Michael C. Roggemann
Byron Welsh

CRC Press
Boca Raton Boston New York Washington London

Library of Congress Cataloging-in-Publication Data

Roggemann, Michael C.
Imaging through turbulence / authors, Michael C. Roggemann, Bryon Welsh.
p. cm.
Includes bibliographical references and index.
ISBN 0-8493-3787-9 (alk. paper)
1. Imaging Systems. 2. Atmospheric turbulence. 3. Optics, Adaptive. 4. Speckle. I. Welsh, Bryon. II. Title.
TK8315.R64 1996
621.36′7--dc20

95-45367
CIP

Direct all inquiries to CRC Press LLC, 2000 Corporate Blvd., N.W., Boca Raton, Florida 33431.

International Standard Book Number 0-8493-3787-9
Library of Congress Card Number 95-45367
Printed in the United States of America 2 3 4 5 6 7 8 9 0
Printed on acid-free paper

PREFACE

The optical effects of atmospheric turbulence arise from random inhomogeneities in the temperature distribution of the atmosphere. A consequence of these temperature inhomogeneities is that the index of refraction distribution of the atmosphere is random. Plane waves striking the atmosphere from space objects acquire an aberration as they propagate through the atmosphere. The plane wave's surface of constant phase is no longer planar when intercepted by an astronomical telescope. The practical consequence of atmospheric turbulence is that resolution is generally limited by turbulence rather than by the optical design and quality of a telescope. Very large telescopes, on the order of four to eight meters in diameter, have been successfully built to conduct astronomy. These telescopes have the full light gathering capability of their large aperture, allowing extremely dim objects to be observed. However, at visible and near infrared wavelengths the angular resolution achieved is equivalent to that obtained with a much smaller telescope, on the order of a few tens of centimeters in diameter.

Since the effects of turbulence on imaging systems were widely recognized in the 1950's and 1960's a number of methods have been proposed to mitigate these effects. Three broad classes of techniques to mitigate turbulence effects exist: (1) pure post processing techniques, such as the so-called speckle imaging methods, which use specialized image measurements and image post processing; (2) adaptive optics techniques, which afford a mechanical means of sensing and correcting for turbulence effects as they occur; and (3) hybrid methods, which combine elements of the post processing techniques and the adaptive optics techniques. Within each of these broad classes there exists one or more imaging technique for overcoming turbulence effects. Each of these techniques has its own set of performance limits, and hardware and software requirements.

The goal of this book is to describe turbulence effects on imaging systems, and to present the essential concepts and limits of performance for various techniques for overcoming the effects of turbulence on imaging systems. This information is scattered across a wide body of technical literature, with the consequent difficulties arising from disparate viewpoints and non-uniform notation and analysis techniques. The discussion presented here integrates this large body of technical literature. Every effort has been made to accurately represent the state of the art as it exists at this writing. However, the area of imaging through turbulence is an active research area – future innovations will no doubt lead to greater advances.

The material presented in this book is written at the first or second year of graduate school level. Review material is presented in the areas of Fourier optics and statistical optics. However, this book will be most accessible to those who already have a firm grounding in these fields. Many graduate students and active workers in this field now share this material in their common background.

This book was prepared with the idea of providing two levels of understanding. At the first level, the goal is to provide a fundamental understanding of the workings of each technique. Meaningful questions include: "How does the method work?","What type of measurements are required?", and "What kind of data processing is required?". The second level of understanding pertains to the statistical performance of the imaging technique in the case of adaptive optics and of the estimators in the case of post processing techniques. The key questions here are: "What are the appropriate statistical measures of performance?",

"What are the appropriate statistical estimators?", "What is the limit of detection for a given method in the presence of all sources of randomness?", and "How do the various imaging methods compare in a signal-to-noise ratio sense?". Understanding at the first level applies primarily to the mechanical and mathematical operation of the method at hand, and will generally (though not always) be obtainable from knowledge of Fourier optics. The second level of understanding derives primarily from statistical concepts, requiring the proper treatment of all sources of randomness, including turbulence statistics, photon statistics, and detector noise statistics. Both levels of understanding are critical to a researcher seeking to choose the best imaging technique for a particular problem.

This book is organized as follows. In Chapter 1 an overview of the problem of imaging through turbulence, and techniques for overcoming turbulence effects is provided. Chapter 2 provides relevant background material from the fields of Fourier and statistical optics. In Chapter 3 statistical models for the index of refraction distribution of the atmosphere are developed, the effects of a random index of refraction distribution on wave propagation are presented, and the implications of turbulence effects on wave propagation for imaging systems are derived. Modal expansions for the turbulence–induced aberration using Zernike polynomials and a form of the Karhunen–Loève expansion are also provided in Chapter 3. Since simulations of atmospheric turbulence effects play a prominent role in analyzing the performance of techniques to overcome turbulence effects, techniques for simulating turbulence–induced wave front phase effects are also included in Chapter 3. Speckle imaging techniques, including the Labeyrie's speckle interferometry method for estimating the modulus of the Fourier transform of a space object, and the Knox–Thompson and bispectrum techniques for reconstructing the phase of the Fourier transform of a space object are presented in Chapter 4. Adaptive optics techniques for overcoming turbulence effects are presented in Chapter 5, including hardware components and models, performance metrics, and analysis techniques. Finally, in Chapter 6 hybrid imaging techniques are presented including deconvolution from wave front sensing, deconvolution of adaptive optics images, compensated speckle imaging, compensated deconvolution from wave front sensing, and frame selection. Wherever possible we provide a theoretical expression for the appropriate performance measures, and present evaluations of the performance in a graphical, tabular, or image form. In many cases a simulation is used to obtain the performance results presented. Where simulations have been used a description of the simulation technique is provided.

Finally, we wish to thank all of those who made direct or indirect contributions to this book. We most sincerely thank our families for showing great patience and providing loving support during this project. No project of this magnitude can be attacked without such support, and we are profoundly grateful. The technical reviewers, specifically Dr. Chuck Matson, Dr. Dave Sandler, and Dr. Peter McCullough, provided an enormous number of great suggestions for improving both the accuracy and the quality of the presentation in this book. The editors and staff of CRC Press, most notably Bob Stern, were always supportive and helpful, and contributed both the motivation for this book, and provided the "nuts and bolts" support needed to make a book. Thank you one and all for your help and support!

Michael C. Roggemann
Byron M. Welsh

Contents

1

Introduction

1.1 Overview of the problem area

The simple experiment of turning a telescope with a diameter of 20 cm or larger towards a bright star on a clear night and measuring the point spread function of the combined atmosphere-telescope system at visible wavelengths reveals a critical defect in the measured image. Even if the telescope is completely free of aberrations, the point spread function (PSF) measured over an exposure time on the order of several seconds will be much broader than the PSF predicted by diffraction alone. A short exposure image, with exposure time of a few tens of milliseconds, demonstrates a different manifestation of the same phenomenon – images which have a speckled appearance. Examples of these image defects are shown in Fig. 1.1. The physical origin of these image defects is known to be atmospheric turbulence. In the absence of compensating techniques, atmospheric turbulence imposes a fundamental limit on the angular resolution of many optical systems. Resolution limits imposed by turbulence profoundly limit the performance of imaging systems which must look through the atmosphere.

Atmospheric turbulence affects imaging systems by virtue of wave propagation through a medium with a nonuniform index of refraction. Light in a narrow spectral band approaching the atmosphere from a distant light source, such as a star, is well modeled by a plane wave. The planar nature of this wave remains unchanged as long as the wave propagates through free space, which has a uniform index of refraction. The atmosphere, however, does not have a uniform index of refraction. Rather, the atmosphere contains a multitude of randomly distributed regions of uniform index of refraction, referred to as turbulent eddies. The index of refraction varies from eddy to eddy. As a result, the light wave that reaches the surface of the Earth is not planar. Excursions of this wave from a plane are manifested as random aberrations in astronomical imaging systems. Anyone who wears glasses understands the general effects of aberrations on imaging systems; they generally broaden the point spread function of the image, lowering the resolution which can be achieved compared to an ideal system. Imagine now the situation in which these aberrations are random and evolve with time. In this situation glasses (i.e., fixed correcting optics) are no longer capable of correcting for the random aberrations. It is the randomness and time evolution of the random aberrations caused by atmospheric turbulence that make imaging through the Earth's atmosphere a difficult and challenging problem. These aberrations are the underlying causes of the image defects shown in Fig. 1.1.

The practical consequence of atmospheric turbulence is that resolution is generally limited by turbulence rather than by the optical design and optical quality of a telescope. Even at the best observatory sites, the angular resolution is limited to approximately 1 arcsec ($\sim 5\mu$ rad) at visible wavelengths regardless of the size of the telescope. Compare this resolution to the theoretically achievable resolution of 0.013 arcsec for a 8 meter telescope. The Hubble Space Telescope achieves a diffraction limited resolution of 0.05 arcsec due to the fact it is above the Earth's atmosphere. Very large telescopes, on the order of four to eight meters in diameter, have been successfully built for astronomical imaging. These telescopes have the full light gathering capability of their large aperture, allowing extremely dim objects to be observed. However, at visible and near infrared wavelengths, the angular resolution achieved is

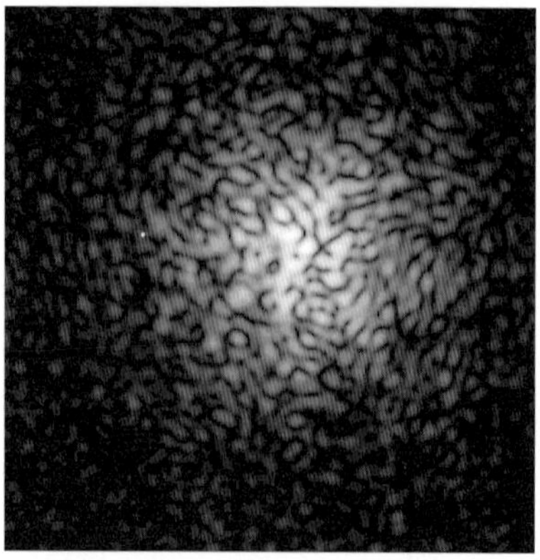

(a)

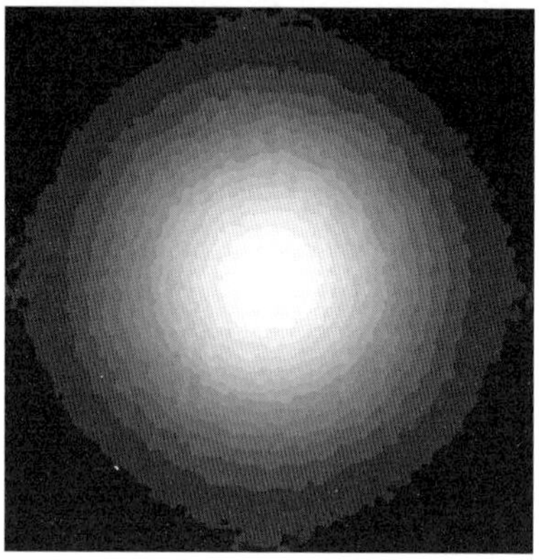

(b)

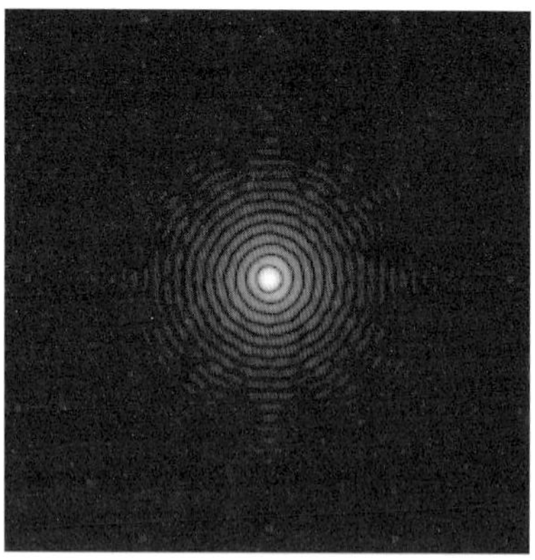

(c)

FIGURE 1.1
Simulated star images: (a) short exposure image; (b) long exposure image; (c) diffraction-limited image. For these results the telescope diameter is D = 1 meter, the atmospheric turbulence conditions represent average seeing at a good observatory location (i.e., the atmospheric coherence diameter r_0 = 10 centimeters) and the mean wavelength is λ = 550 nanometers.

equivalent to that obtained with a much smaller telescope, on the order of a few tens of centimeters in diameter.

Since the effects of turbulence on imaging systems were widely recognized in the 1950's and 1960's, a number of methods have been proposed to mitigate these effects. Three broad classes of techniques to mitigate turbulence effects exist: 1) pure post processing techniques, such as the so-called speckle imaging methods, which use specialized image measurements and image post processing; 2) adaptive optics techniques, which use mechanical means to sense and correct for turbulence effects as they occur; and 3) hybrid methods, which combine elements of the post processing techniques and the adaptive optics techniques. Within each of these broad classes there exists one or more imaging techniques for overcoming turbulence effects. Each of these techniques has its own set of performance limits, hardware requirements, and software requirements.

The goal of this book is to provide a single reference on how turbulence affects imaging systems, and the on various techniques for overcoming of the effects of turbulence on imaging systems. Even though we focus on applications involving ground-based systems looking up through the Earth's atmosphere, most of the theoretical wave propagation and imaging system results can also be applied to applications involving horizontal propagation. Presently, this information is scattered across a wide body of technical literature. The discussion presented here integrates this large body of technical literature, and provides information regarding the tradeoffs between the various imaging methods. Every effort has been made to accurately represent the state of the art as it exists at this writing. However, the area of imaging through turbulence is an active research area – future innovations will no doubt lead to greater advances.

1.2 Historical overview of imaging through turbulence

This section provides a brief discussion of the history of understanding turbulence effects on imaging systems, and the efforts to overcome the limits imposed by atmospheric turbulence. The literature in this area is far too extensive to cite all contributions. A topical survey of key results is provided below.

1.2.1 Recognition of turbulence effects

Issac Newton was aware that, in the absence of any correction, it is impossible to attain diffraction limited performance at visible wavelengths with a ground-based telescope bigger than a few tens of centimeters in diameter [1], regardless of the design and optical quality of the telescope. In Newton's day some of the optical consequences of atmospheric turbulence were known. The twinkling of the stars was well known, and it had also been noted that the planets did not twinkle. Further, by Newton's time it was known that the point spread function of a telescope obtained by looking at a star was significantly broader than the point spread function which could be observed under laboratory conditions. Newton correctly attributed these effects to "tremors" in the atmosphere [1, page 423]:

> *"If the theory of making Telescopes could at length be fully brought into Practice, yet would there be certain Bounds beyond which Telescopes could not perform. For the air through which we look upon the Stars, is in perpetual Tremor; as may be seen by the tremulous Motion of Shadows cast from high Towers, and by the twinkling of the fix'd stars."*

Newton was also able to explain qualitatively why stars twinkle when viewed with the naked eye, but do not twinkle when viewed with telescopes:

> *"But these Stars do not twinkle when viewed through Telescopes which have large apertures. For the Rays of Light which pass through divers parts of the aperture, tremble each of them apart, and by means of their various and sometimes contrary Tremors, fall at one and the*

> *same time upon different points at the bottom of the Eye, and their trembling Motions are too quick and confused to be perceived severally."*

Though we would use more modern terms to describe this phenomenon today, Newton's insight that atmospheric turbulence was the cause of this effect was correct. Newton also noted that the point spread function of a telescope looking through turbulence is broader than would be expected in the absence of the atmosphere. As a result, large telescopes could be used to measure dim objects by virtue of the light gathering capability of a large aperture, but a large telescope alone could not overcome the effects of atmospheric turbulence:

> *"And all these illuminated Points constitute one broad lucid Point, composed of those many trembling Points confusedly and insensibly mixed with one another by very short and swift Tremors, and thereby cause the Star to appear broader than it is, and without any trembling of the whole. Long Telescopes may cause Objects to appear brighter and larger than short ones can do, but they cannot be so formed as to take away the confusion of the Rays which arises from the Tremors of the Atmosphere."*

Newton's suggestion that observatories be placed atop high mountains to partially mitigate the effects of atmospheric turbulence remains the standard wisdom for choosing observatory sites:

> *"The only Remedy is a most serene and quiet Air, such as may perhaps be found on the tops of the highest Mountains above the grosser Clouds."*

Understanding the origin of the optical effects of atmospheric turbulence did little to improve the state-of-the-art of astronomy until modern times. In Newton's day the only available light detector, and the only processor of optical signals was the human visual system. The invention of photographic film in the early 1800's eventually resulted in the ability to permanently record images measured through turbulence, but the combined effects of poor film sensitivity and an interest in viewing dim objects resulted in long exposure image measurements. These long exposure images contained the result of a very large number of realizations of the random turbulence effects averaged into a single measurement. The resulting images were similar in character to those shown in Fig. 1.1b – the images of stars were much broader than the images that would arise due to diffraction alone. By the 1950's, film systems had progressed to the point where it was possible to measure short exposure images of bright objects, essentially freezing the turbulence effects during the image measurement time. The first short exposure images were reported to look like a "bunch of grapes" [2], containing what are now called "speckles" (see Fig. 1.1a). The speckles were observed to be approximately diffraction limited in extent. These first short exposure images provided a hint that high resolution information was somehow encoded in short exposure image measurements. Further advances in turbulence understanding, light detection devices, and computerized signal processing were required to exploit this insight.

1.2.2 Understanding turbulence effects on wave propagation and imaging systems

Atmospheric turbulence arises from heating and cooling of the Earth's surface by the sun. Sunlight warms large land masses during daylight hours, and these warm land masses heat the air. During the night the Earth's surface gradually cools, and this heat is also coupled into the air. Heating the air in this manner results in large spatial scale motions. This air motion eventually becomes turbulent, with the result that the large spatial scale motions break up into progressively smaller scale motions, eventually giving rise to randomly sized and distributed pockets of air, each having a characteristic temperature. These pockets of air are the turbulent eddies referred to earlier. The index of refraction of air is sensitive to temperature, and hence, the atmosphere exhibits variations in the index of refraction. Plane waves propagating through the atmosphere are no longer planar when they arrive at the surface of the Earth.

The study of turbulent air motion is a problem in the field of fluid mechanics. During the 1940's, Kolmogorov [3] developed a model for how energy is transported from large scale turbulent eddies to small scale turbulent eddies. Kolmogorov's model provides a spatial power spectrum for the index of refraction fluctuations. Tatarskii applied Kolmogorov's model to solve the wave equation for propagation through regions of weak random index fluctuations [4]. Fried used Tatarskii's results to describe turbulence effects in terms of Zernike polynomials [5], and to derive a useful single parameter, r_0, referred to as the atmospheric coherence diameter, to describe the effects of turbulence on the resolution of imaging systems [6]. The seminal contributions of Kolmogorov, Tatarskii, and Fried form the theoretical basis of nearly all the work which has followed in atmospheric optics.

1.2.3 Speckle imaging

The first work to partially overcome the effects of turbulence relied on post detection computer processing of large numbers of short exposure images. Methods for processing speckled images and reconstructing an image of the object are generally referred to as speckle imaging techniques owing to the fact that the data consist of a set of short exposure, speckled images. In speckle imaging the object is generally estimated by first estimating its Fourier transform, and then inverting the Fourier data to obtain an estimate of the object. The Fourier transform of an object is, in general, a complex-valued quantity. Hence, both the modulus and the phase of the Fourier transform of the object must be estimated to obtain an estimate of the object. In speckle imaging the Fourier modulus and phase estimation processes are handled separately.

Labeyrie showed that the energy spectrum of an object, that is, the squared modulus of the Fourier transform of the object, could be estimated from a large data set of short exposure images similar to the one shown in Fig. 1.1a [7]. This basic method for estimating the energy spectrum of an object using Labeyrie's technique, referred to as speckle interferometry, is shown in the block diagram in Fig. 1.2. Two sets of data are required for speckle interferometry: a set of speckled images of the object of interest, and a second set of speckled images of a nearby reference star. Both sets of data are processed as shown in Fig. 1.2. The images are Fourier transformed, the squared modulus of the Fourier transform is computed, and the squared modulus data are averaged over both the object and reference data sets. The reference data are then used in a deconvolution procedure to remove the combined effects of the telescope and atmosphere from the estimate of the object's energy spectrum.

Estimating the energy spectrum of an object in this manner is possible because the second moment of the short exposure optical transfer function of the combined atmospheric-telescope system, referred to as the speckle transfer function, is non-zero out to the diffraction limited cutoff frequency of the telescope [8]. The energy spectrum of certain objects often contains unique information about the object, such as in the case of double stars, where the energy spectrum encodes both the separation and the relative brightness of the stars [2]. However, the phase of the Fourier transform of an object, referred to as the phase spectrum of the object, is required to create an actual image. Two methods of obtaining the phase spectrum of the object have been developed, the so-called Knox-Thompson or cross spectrum method [9], and the bispectrum method [10]. Both the cross spectrum and the bispectrum methods are based on the fact that certain higher order moments of the complex Fourier transform of speckled images contain information about the phase spectrum of the object. An additional step is required to reconstruct the phase spectrum from the cross spectrum or bispectrum.

Speckle imaging methods suffer from a poor signal-to-noise ratio (SNR) due to the degree of randomness in the atmospheric turbulence-imposed transfer function, and due to the statistics associated with photoelectric detection of light. In fact, at mid and high spatial frequencies the expected SNR of the energy spectrum for a single speckled frame is bounded from above by unity for an infinitely bright point object, such as a star, and is in general much smaller for extended dim objects [11, 12]. While the SNR for speckle imaging can be improved through averaging many independent frames of data, the practical consequence of these limits is that several hundred to a few thousand images may be needed to obtain an acceptable SNR at high spatial frequencies.

An alternative post detection processing method has also been used to obtain high resolution images through turbulence. This alternative method combines information from a wavefront sensor and a short

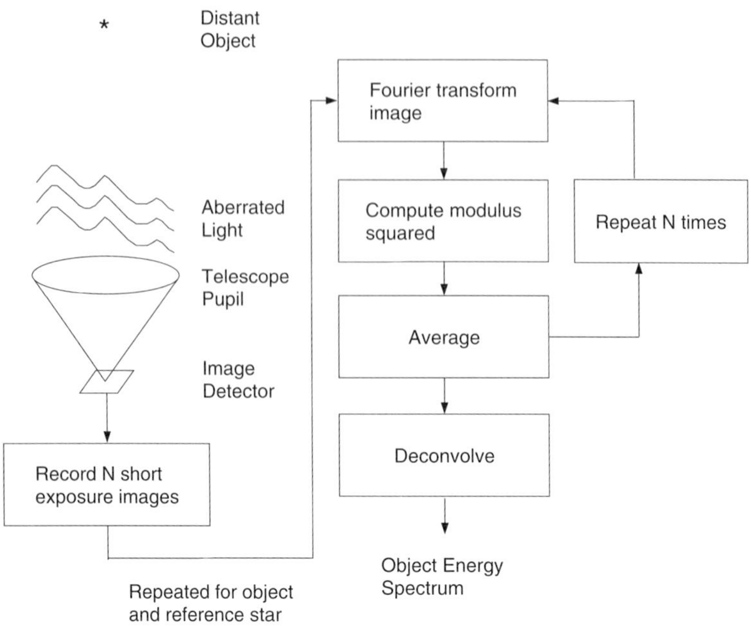

FIGURE 1.2
Block diagram for speckle imaging.

exposure image measurement [13, 14]. This method is referred to as deconvolution from wavefront sensing (DWFS), and is shown in block diagram form in Fig. 1.3. In DWFS simultaneous measurements are made of a short exposure image and the output of a wavefront sensor (WFS). The WFS data is processed to yield an estimate of the instantaneous optical transfer function which caused the short exposure image measurement, and this estimate is used in a post processing scheme designed to deconvolve the effects of the combined atmosphere-telescope system from the measured image. DWFS offers the advantage over speckle imaging of providing both the modulus and the phase of the Fourier transform of the object directly, but at the expense of measuring and processing WFS data.

1.2.4 Adaptive optics

The notion that aberrations induced by the atmosphere could be compensated by using mechanical means was first put forth by Babcock [15]. Babcock's conjecture arose from the realization that the physical origin of the turbulence-induced aberration is spatial variations in the optical path length between the object and the telescope. If these optical path length differences could be mechanically adjusted before the light is focused into an image, then the measured image will be superior to an uncompensated image. Systems which perform turbulence compensation using mechanical means now exist, and are generically referred to as adaptive optics systems [16, 17]. Adaptive optics provide a means of sensing the atmospheric turbulence-induced aberration and partially correcting for this aberration in real time. Image quality is improved by reducing the aberration caused by the atmosphere, resulting in a narrower point spread function and thus improved resolution. To illustrate the significant improvement in imaging performance that is possible using an adaptive optical telescope consider imaging a binary star pair through the Earth's atmosphere. Figure. 1.4(a) shows the image of the binary star pair assuming no atmospheric effects (i.e., diffraction limited) using a 3 meter telescope at visible wavelengths ($\lambda = 0.5\mu$m). The stars are separated by $2\lambda/D$ radians which corresponds to 0.33 μrad

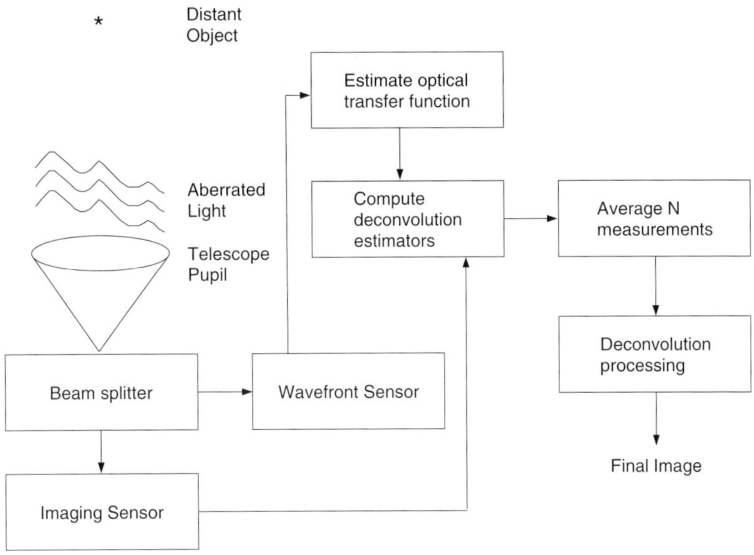

FIGURE 1.3
Block diagram for deconvolution from wavefront sensing.

(0.069 arcsec) for $\lambda = 0.5\ \mu$m. The brightness of the star on the right is one half that of the star on the left. The bottom two images shown in Fig. 1.4 illustrate the effects of atmospheric turbulence. These two images correspond to (b) a long exposure image for a conventional telescope and (c) a long exposure for a telescope employing adaptive optics. These images clearly demonstrate the severe performance degradation caused by the atmosphere, while at the same time illustrating the ability of adaptive optics to compensate for the effects of the turbulence and dramatically improve performance.

The functional elements of an adaptive optics system are shown Fig. 1.5. These include the deformable mirror (DM), the WFS, and an actuator command computer. The DM is a flexible mirror whose surface figure can be modified at high speed in response to applied electrical signals. A widely used method for constructing DMs is to bond piezo-electric actuators between a thin, continuous face sheet mirror and a rigid base plate [18]. Voltages applied to the actuators cause them to expand or contract, changing the figure of the mirror surface. The WFS is a device which measures gradients in the phase aberration across small regions in the telescope pupil, called subapertures. The DM, WFS, and actuator command computer are arranged in a closed loop system. Aberrated light entering the telescope is first reflected from the deformable mirror (DM). Some of this light is focused to form an image, and some is used to provide signals to the WFS. An actuator command computer processes WFS measurements and calculates the set of electrical signals to apply to the DM to achieve the desired DM figure. The entire process, from WFS measurement to DM update, must be performed at speeds commensurate with the rate of change of the turbulence-induced phase errors. In an ideal adaptive optics system, the surface of the DM is configured from instant to instant to present an approximation of the conjugate of the turbulence-induced phase error so that the wave reflected from the DM will more closely approximate a plane wave.

Since Babcock's proposal, much attention has been devoted to developing practical adaptive optics hardware and to understanding the limits of performance of adaptive optics techniques. In particular, good optical quality, high speed, and stable DMs have been developed, along with high speed, light efficient wavefront sensors, and reconstruction techniques to optimally map WFS measurements to actuator commands [16, 19]. Further, methods have been developed to analyze the average optical behavior of adaptive optics systems [20], and to understand the effects of the adaptive optics system

(a)

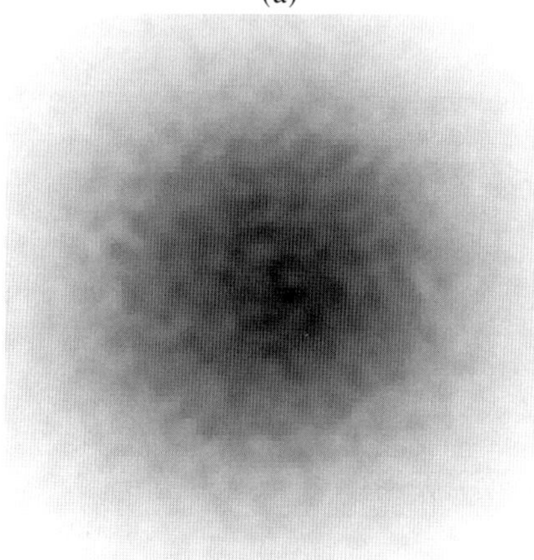

(b)

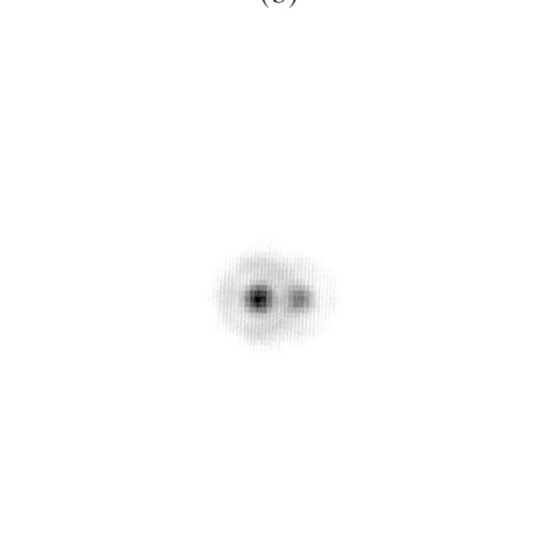

(c)

FIGURE 1.4
Simulated long exposure imagery of a binary star pair for (a) a diffraction limited telescope (i.e., no atmospheric turbulence effects), (b) a conventional telescope operating in the presence of atmospheric turbulence and, (c) a telescope using adaptive optics. The results were computed via simulation for the adaptive optical telescope described in Chapter 5. The stars are separated by 0.33 μrad (0.069 arcsec) and the imaging wavelength is 0.5 μm. The brightness of the star on the right is one half that of the star on the left.

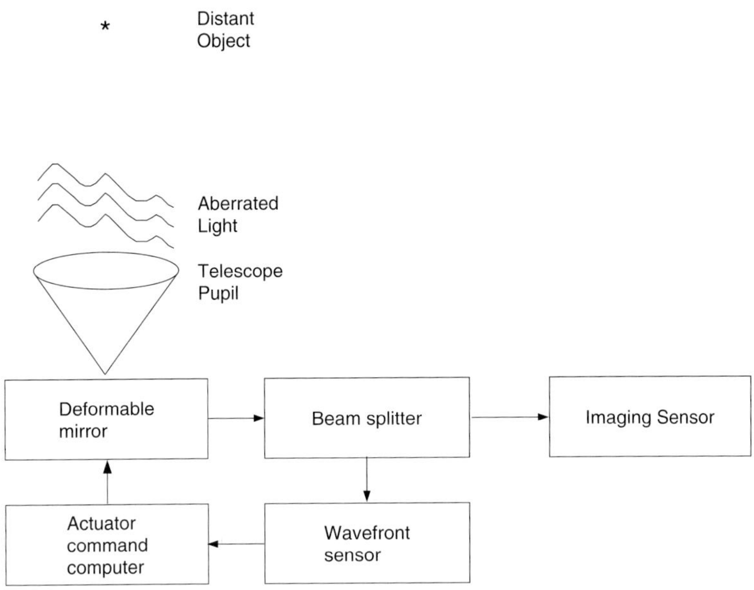

FIGURE 1.5
Block diagram of an adaptive optics system.

statistics on the statistics of measured images [21, 22]. Experiments have shown that adaptive optics systems can vastly improve the resolution of systems which must form images through the atmosphere [23, 24]. However, it is impossible for an adaptive optics system to fully restore the imaging performance of a ground-based telescope due to finite signal and measurement noise in the WFS, finite spatial sampling of the phase aberration by the WFS, finite number of degrees-of-freedom available in the DM, and time delays between sensing and correcting for the turbulence-induced aberration.

In most of the adaptive optics systems built to date some light from the object is used to provide signals to the WFS. This method of providing signals to the WFS is referred to as self-referenced operation. Finite signal availability to the wavefront sensor is a fundamental limitation to performance of an adaptive optics system since the rate of arrival of photons from the object generally cannot be controlled by the observer. In practice, WFS subapertures are small, typically on the order of 10 - 20 cm across, and the signal integration times quite short, on the order of a few milliseconds. As a result, there is a limit to how dim an object can be and still provide acceptable signal levels to the WFS.

A relatively new method of providing signals for WFS measurements uses laser light provided by a laser transmitter associated with the telescope [25, 26]. The transmitted laser beam is focused on a spot in the center of the field of view of the telescope at a finite altitude above the telescope. Within the region where the beam is focused some of the light is scattered from the atmospheric constituents, and some of the scattered light is returned in the direction of the telescope. This scattered light is provided to the WFS to measure the turbulence-induced aberration. The laser spot in this method is referred to as an artificial beacon. Use of artificial beacons removes the fundamental problem of low signal associated with self-referenced adaptive optics.

Another major limitation in the performance of adaptive optics systems is an effect referred to as anisoplanatism [27]. Anisoplanatism refers to the degradation caused by using a reference source for wavefront compensation that is separated or distinct from the object of interest. This separation causes the optical paths from the reference and the object to traverse different regions of the atmosphere, resulting in distinct wavefront perturbations for each wavefront. As the separation between the reference and the object increases, the correlation between the two wavefronts decreases. This decreased correlation results in a degradation of the ability of the adaptive optical system to correct the object wavefront. Another type of anisoplanatism must also be considered when using artificial guide stars. The finite altitude of an artificial beacon gives rise to an effect called "focus anisoplanatism" [28, 29, 30]. Again, the basic problem is that the optical paths from the artificial guide star and the object traverse different regions of the atmosphere. This path difference is a problem even when the artificial guide star is in the same direction as the object.

Adaptive optics systems are hardware intensive. Adaptive optics systems which have WFS subapertures with an area of approximately the square of the atmospheric coherence length, r_0, and DM actuators placed on a Cartesian grid with grid spacing of approximately r_0, are referred to as fully compensated systems. Fully compensated 4 meter diameter–class telescopes require WFSs with several hundred to a few thousand subapertures for operation at visible wavelengths, and roughly an equal number of DM degrees–of–freedom. Further, the adaptive optics system must be controlled at rates of several tens of Hz to a few hundred Hz, requiring very high speed data transfers and stable control. As a result of the complexity and expense of adaptive optics systems, effort has been given to exploring systems which are simpler than fully compensated systems by virtue of using fewer, larger WFS subapertures, and fewer DM degrees-of-freedom. These simpler systems are referred to as partially compensated adaptive optics systems [22, 31, 32, 33].

1.2.5 Hybrid approaches

Analysis of the statistical performance of partially compensated adaptive optics systems has shown that partially compensated systems do not provide the same quality of optical performance as fully compensated systems [32]. In particular, the average PSF of a partially compensated system is broader than the PSF of a fully compensated system. Images measured with partially compensated adaptive optics are more severely blurred than images measured with fully compensated systems, but high spatial frequency, and hence, high resolution information can be measured with good SNR using partially compensated systems. The high spatial frequency information in an image can be recovered by using post detection image reconstruction [24]. Imaging techniques which combine predetection correction using adaptive optics and post detection image reconstruction are called hybrid imaging systems.

Hybrid imaging techniques have been studied as an alternative to fully compensated adaptive optics for large telescopes, providing some correction for turbulence effects by means of adaptive optics, and some correction through image post processing. Partially compensated adaptive optics are much less complicated and expensive than fully compensated telescopes. For example, a fully compensated telescope with diameter of 4 m operating under seeing conditions of $r_0 = 10$ cm would require approximately 1250 DM actuators and WFS subapertures if the actuators were placed on a Cartesian grid with 10 cm spacing, and the subapertures were 10 cm × 10 cm. Doubling the separation of the actuators to 20 cm and doubling the edge length of the subapertures to 20 cm reduce the required number of subapertures and actuators to about 315, resulting in a partially compensated system which has much simpler hardware, but reduced optical performance. Hybrid imaging techniques have also been examined as a means to overcome some of the effects of finite control bandwidth in adaptive optics systems [34].

Hybrid imaging techniques are shown in block diagram form in Fig. 1.6. Adaptive optics hardware is used to partially correct for turbulence effects, and images are recorded at the imaging port of the system. Typically, images of both the object of interest and images of a reference star are recorded [24]. The reference star images are used as part of the image reconstruction process. WFS measurements may also be recorded for use in the image reconstruction [34].

The performance of hybrid imaging systems has been analyzed in the Fourier transform domain of the image. Analyses have treated the problem of estimating the Fourier transform, or spectrum of the object, and then recovering an intensity image of the object from the Fourier data. The simplest type of hybrid imaging, referred to as the linear method, corrects for short time scale tilt errors in the wavefront after correction by the adaptive optics, averages to improve the SNR of the data, and uses a deconvolution procedure to boost the amplitude of the spectral components of the image [24, 35]. Speckle imaging

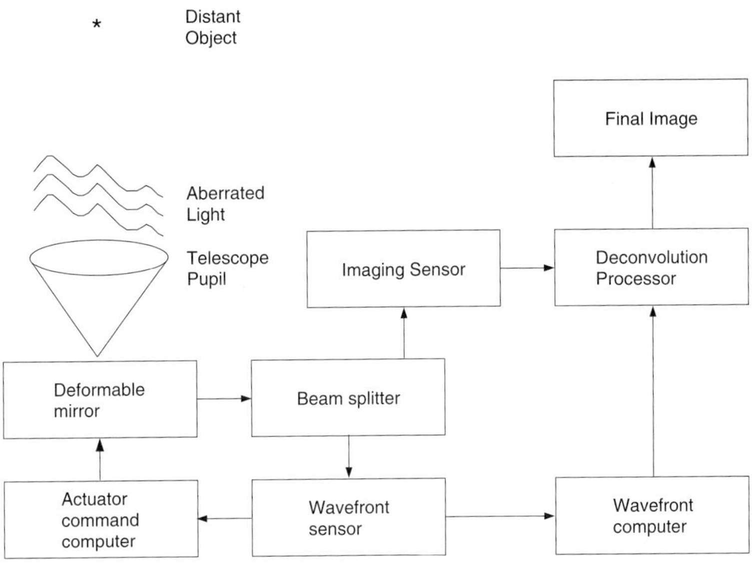

FIGURE 1.6
Block diagram of a hybrid imaging system.

techniques have also been applied to compensated imagery [31, 36], as have DWFS-based methods [34]. These studies have shown that a tradeoff exists between predetection compensation and post detection image processing. In particular, it is possible to improve the quality of imagery using specialized measurements and post processing.

1.3 Overview of the book

The remainder of this book is organized as follows. Chapter 2 contains a review of the relevant background material from the areas of Fourier optics and statistical optics. Key imaging concepts, such as the frequency domain analysis of imaging systems, temporal and spatial coherence of light, and the semi-classical model of light detection processes are presented in Chapter 2.

Chapter 3 provides a discussion of turbulence effects on imaging systems. Chapter 3 begins with a discussion of the spatial and spectral statistics of index of refraction fluctuations in the atmosphere. Models for turbulence strength are also presented. The statistics of wave propagation through random media are then presented, and used to introduce the first-order effects of turbulence on imaging systems. Modal expansions for the turbulence-induced phase aberration are then presented. Chapter 3 concludes with a discussion of techniques for creating random phase screens which simulate the effects of turbulence on the phase of a wave propagating through the atmosphere.

Chapter 4 presents a discussion of speckle imaging. This chapter begins with an overview of speckle imaging, explaining the basic measurement and data processing steps. Labeyrie's technique for estimating the modulus of the Fourier transform of an object, typically referred to as speckle interferometry, is then presented. The nature of the speckle transfer function and the signal-to-noise ratio imposed limits on the performance of speckle interferometry are emphasized. Techniques for estimating the phase of the Fourier transform of an object are then presented. Both the Knox-Thompson and bispectrum techniques for Fourier phase estimation are discussed. Image reconstruction for speckle imaging is then treated, and several examples of speckle imaging performance are provided. Important considerations for simulating speckle imaging are also discussed.

Adaptive optics systems are presented in Chapter 5. Chapter 5 begins with a discussion of factors which degrade adaptive optical system performance, such as finite signal to the wave front sensor and time delay effects. Hardware components and mathematical models for the components then presented. Emphasis is placed on wave front sensing and reconstruction concepts. Adaptive optics system performance modeling and performance metrics are then discussed. The two standard approaches to modeling performance are presented: numerical analysis; and simulation. A large body of performance results are presented in Chapter 5.

Hybrid imaging methods are presented in Chapter 6. Though DWFS is presented in this chapter as a post processing technique, implementing DWFS requires use of a wave front sensor. Hence, the presentation of DWFS is placed in Chapter 6 so that wave front sensing and reconstruction concepts developed in Chapter 5 may be used in the discussion. Next, hybrid imaging techniques are discussed, including deconvolution of compensated images, compensated speckle imaging, compensated DWFS, frame selection. Signal-to-noise ratio imposed limits on performance are discussed for each of these imaging techniques. Methods for simulating performance, and a representative set of performance results are provided in each case.

Bibliography

[1] I. Newton, "Optics," in *Great Books of the Western World* (R. M. Hutchins, ed.), vol. 34, Chicago: Encyclopedia of Brittanica, 1952.

[2] F. Roddier, "The effects of atmospheric turbulence in optical astronomy," in *Progress in Optics* (E. Wolf, ed.), vol. XIX, New York: North-Holland, 1981.

[3] A. N. Kolmogorov, "The local structure of turbulence in incompressible viscous fluids for very large reynolds' numbers," in *Turbulence, Classic Papers on Statistical Theory* (S. K. Friedlander and L. Topper, eds.), pp. 151–155, New York: Wiley-Interscience, 1961.

[4] V. I. Tatarskii, *Wave Propagation in a Turbulent Medium.* New York: Dover Publications, 1967.

[5] D. L. Fried, "Statistics of a geometric representation of wavefront distortion," *J. Opt. Soc. Am.*, vol. 55, pp. 1427–1435, 1965.

[6] D. L. Fried, "Optical resolution through a randomly inhomogeneous medium for very long and very short exposures," *J. Opt. Soc. Am.*, vol. 56, pp. 1372–1379, 1966.

[7] A. Labeyrie, "Attainment of diffraction limited resolution in large telescopes by fourier analyzing speckle patterns in star images," *Astron. Astrophys.*, vol. 6, p. 85, 1970.

[8] D. Korff, "Analysis of a method for obtaining near-diffraction-limited information in the presence of atmospheric turbulence," *J. Opt. Soc. Am.*, vol. 63, pp. 971–980, 1973.

[9] K. T. Knox and B. J. Thompson, "Recovery of images from atmospherically degraded short exposure images," *Astrophys. J.*, vol. 193, pp. L45–L48, 1974.

[10] A. W. Lohmann, G. Weigelt, and B. Wirnitzer, "Speckle masking in astronomy: triple correlation theory and applications," *Appl. Opt.*, vol. 22, pp. 4028–4037, 1983.

[11] J. W. Goodman, *Statistical Optics.* New York: John Wiley & Sons, 1985.

[12] J. W. Goodman and J. F. Belsher, "Photon limitations in imaging and image restoration," contract, Rome Air Development Center, Griffiss AFB, NY, 1977.

[13] D. L. Fried, "Post detection wavefront compensation," in *SPIE Proceedings on Digital Image Recovery and Synthesis*, vol. 828, pp. 127–133, 1987.

[14] J. Primot, G. Rousset, and J. C. Fontanella, "Deconvolution from wave-front sensing: a new technique for compensating turbulence-degraded images," *J. Opt. Soc. Am. A*, vol. 7, pp. 1589–1608, 1990.

[15] H. W. Babcock, "The possibility of compensating astronomical seeing," *Publ. Astron. Soc. Pac.*, vol. 65, pp. 229–236, 1953.

[16] J. W. Hardy, "Active optics: A new technology for the control of light," *Proc. IEEE*, vol. 66, pp. 651–697, 1978.

[17] J. W. Hardy, "Adaptive optics - a progress review," in *Proc. SPIE on Active and Adaptive Optical Systems*, vol. 1542, pp. 2–17, 1991.

[18] M. A. Ealey and J. A. Wellman, "Deformable mirrors: design fundamentals, key performance specifications, and parametric trades," in *Proc. SPIE on Active and Adaptive Optical Components*, vol. 1543, pp. 36–51, 1991.

[19] E. P. Wallner, "Optimal wave front correction using slope measurements," *J. Opt. Soc. Am.*, vol. 73, pp. 1771–1776, 1983.

[20] B. M. Welsh and C. S. Gardner, "Performance analysis of adaptive optics systems using slope sensors," *J. Opt. Soc. Am. A*, vol. 6, pp. 1913–1923, 1989.

[21] M. C. Roggemann, "Limited degree-of-freedom adaptive optics and image reconstruction," *Appl. Opt.*, vol. 30, pp. 4227–4233, 1991.

[22] M. C. Roggemann, "Optical performance of fully and partially compensated adaptive optics systems using least-squares and minimum variance phase reconstruction," *Comp. & Elec. Eng.*, vol. 18, pp. 451–466, 1992.

[23] R. Q. Fugate, B. L. Ellerbroek, C. H. Higgins, M. P. Jelonek, W. J. Lange, A. C. Slavin, W. J. Wild, D. M. Winker, J. M. Wynia, J. M. Spinhirne, B. R. Boeke, R. E. Ruane, J. F. Moroney, M. D. Oliker, D. W. Sindle, and R. A. Cleis, "Two generations of laser-guide-star adaptive-optics experiments at the starfire optical range," *J. Opt. Soc. Am. A*, vol. 11, pp. 310–314, 1994.

[24] M. C. Roggemann, D. W. Tyler, and M. F. Bilmont, "Linear reconstruction of compensated images: theory and experimental results," *Appl. Opt.*, vol. 31, pp. 7429–7441, 1992.

[25] R. Q. Fugate, D. L. Fried, G. A. Ameer, B. R. Boeke, S. L. Browne, P. H. Roberts, R. E. Ruane, G. A. Tyler, and L. M. Wopat, "Measurement of atmospheric wavefront distortion using scattered light from a laser guide star," *Nature*, vol. 353, pp. 144–146, 1991.

[26] C. S. Gardner, B. M. Welsh, and L. A. Thompson, "Design and performance analysis of adaptive optical telescopes using laser guide stars," *Proc. IEEE*, vol. 78, pp. 1721–1743, 1990.

[27] D. L. Fried, "Anisoplanatism in adaptive optics," *J. Opt. Soc. Am.*, vol. 72, pp. 52–61, 1982.

[28] B. M. Welsh and C. S. Gardner, "Effects of turbulence induced anisoplanatism on the imaging performance of adaptive astronomical telescopes using laser guide stars," *J. Opt. Soc. Am. A*, vol. 8, pp. 69–80, 1991.

[29] D. L. Fried, "Analysis of fundamental limits to artificial guide star adaptive optics system performance for astronomical imaging," *J. Opt. Soc. Am. A*, vol. 11, pp. 277–287, 1994.

[30] B. M. Welsh, "Imaging performance analysis of adaptive telescopes using laser guide stars," *Appl. Opt.*, vol. 30, pp. 5021–5030, 1991.

[31] P. Nisenson and R. Barakat, "Partial atmospheric correction with adaptive optics," *J. Opt. Soc. Am. A*, vol. 4, pp. 2249–2253, 1987.

[32] M. C. Roggemann and C. L. Matson, "Power spectrum and fourier phase spectrum estimation by using fully and partially compensating adaptive optics and bispectrum postprocessing," *J. Opt. Soc. Am. A*, vol. 9, pp. 1525–1535, 1992.

[33] F. Roddier, M. Northcott, and J. E. Graves, "A simple low-order adaptive optics system for near-infrared applications," *Pub. Astron. Soc. Pac.*, vol. 103, pp. 131–149, 1991.

[34] M. C. Roggemann and J. A. Meinhardt, "Image reconstruction by means of wave-front sensor measurements in closed-loop adaptive-optics systems," *J. Opt. Soc. Am. A*, vol. 10, pp. 1996–2007, 1993.

[35] B. L. McGlamery, "Restoration of turbulence-degraded images," *J. Opt. Soc. Am.*, vol. 57, pp. 293–297, 1966.

[36] M. C. Roggemann, E. L. Caudill, D. W. Tyler, M. J. Fox, M. A. Von Bokern, and C. L. Matson, "Compensated speckle imaging: theory and experimental results," *Appl. Opt.*, vol. 33, pp. 3099–3110, 1994.

2

Background: Fourier and Statistical Optics

2.1 Introduction

As in the study of any new field it is necessary to present some basic background material essential for developing a well founded understanding of the area. In the study of optical imaging and atmospheric optics, it quickly becomes apparent that an understanding of Fourier and statistical optics is critical. Fourier optics allows problems associated with diffraction and imaging to be cast into an easily understood "linear systems" framework. This framework is particularly convenient and easy to understand for those trained in classical linear systems theory. A Fourier optics background allows us to characterize the deterministic optical problem. In many situations this deterministic approach is adequate. In dealing with optical wave propagation through the atmosphere though, it is important to realize that the inherent randomness of propagation affects an optical system in a way that can only be described in a statistical manner. The detection of light also gives rise to random effects that fundamentally affect the performance of an optical system. In order to understand these statistical effects, a background in statistical optics is also required.

The Fourier optics materials presented in this chapter are the key results needed for more detailed discussions in later chapters. Excellent introductory texts covering more detailed discussions of Fourier optics include those of references [1] and [2]. Fourier optics allow us to understand optical phenomena that are not handled by geometrical optics. Even without the degradations caused by imaging through atmospheric turbulence, an optical imaging system suffers a fundamental performance limitation not predicted by geometrical optics. Understanding this fundamental performance limitation is important in developing an understanding of the degradations caused by wave propagation through the atmosphere. Propagation of the optical field through the components of an imaging system gives rise to an effect called diffraction. Diffraction causes the system to depart from the ideal performance predicted by geometrical optics by causing light, that geometrical optics predicts would be focused to a point, to be spread over a small region around the point. The amount of spreading is a function of the optical wavelength and the size and shape of the optical components. The phrase "diffraction limited performance" refers to this fundamental effect that can only be improved upon by increasing the size of the optics or decreasing the optical wavelength. The study of Fourier optics allows us to characterize these diffraction effects in a linear systems framework, which in turn facilitates an intuitive understanding of their effects. In this brief introduction to Fourier optics we first present the basic formula for calculating diffraction effects in wave propagation problems: the Rayleigh-Sommerfeld diffraction formula. The Rayleigh-Sommerfeld diffraction formula allows us to derive both the Fresnel and Fraunhofer diffraction formulas, which are special cases of the more general formula. These diffraction formulas are then applied to wave propagation through lenses, which in turn sets up the discussion of imaging.

The examples of coherent and incoherent imaging offer two opportunities to illustrate the importance of understanding statistical optics. In the Fourier optics discussion of imaging given in Section 2.2 we discuss imaging for coherent and incoherent optical fields. These two cases are extremes of the more general situation in which the coherence of the optical field is somewhere between fully coherent and

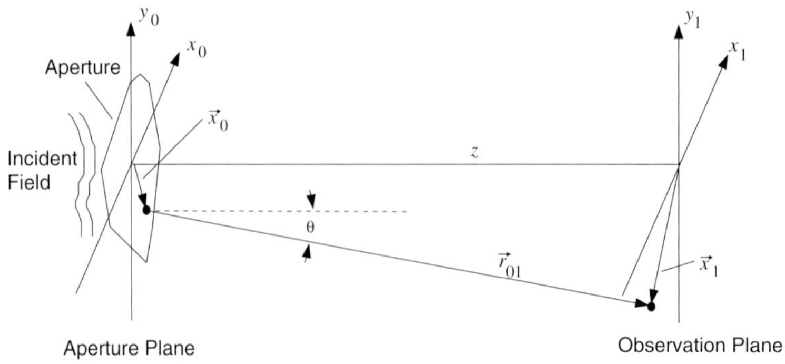

FIGURE 2.1
Diffraction geometry.

fully incoherent. The coherence properties of an optical field can be discussed from both temporal and spatial viewpoints. In either case, these properties qualitatively tell us how well the optical field interferes with delayed and/or spatially shifted versions of itself. In some imaging scenarios, particularly for imaging through the atmosphere, the spatial coherence properties of the optical field determine the performance limit of the system. In fact, it is natural to discuss the atmospherically induced degradations in terms of the effects on the coherence properties of the optical field. In cases in which the coherence properties establish the performance limitations, diffraction limited performance is not achieved. Besides the inherent coherence properties of the optical field, the very process of detecting an optical irradiance distribution using a detector induces a random noise effect referred to as shot noise, or photon noise. Shot noise and other noise sources, such as detector read noise, serve to degrade performance in ways that represent fundamental limitations. Again, these limitations can only be discussed in a statistical sense. Reference [3] provides an excellent introduction and reference for the beginning student in statistical optics. Other outstanding references include [4], [5], and [6]. In summary, the coherence properties of an optical field and the randomness inherent in the light detection process are key concepts in developing an understanding of the issues associated with imaging through atmospheric turbulence.

Fourier optics concepts are introduced in the next section. This section includes discussions of scalar diffraction theory, lenses, and imaging. Section 2.3 presents the introductory statistical optics concepts relevant to the imaging material presented in later chapters. Included in this section are discussions of optical coherence, imaging as an interferometric process, and an introduction to the photoelectric light detection process.

2.2 Fourier optics

In this section we present the basic formulation for calculating the optical field at an observation point due to a field in a diffracting aperture some distance away. With this formulation, calculations involving optical propagation through an optical system simply become repeated applications of the diffraction formula. The diffraction formula forms the basis for all subsequent results.

2.2.1 Scalar diffraction theory

Figure 2.1 illustrates the basic geometry of the diffraction calculation. On the left a field is incident upon an aperture in the x_0y_0 plane. The x_1y_1 plane defines the observation plane and the vector $\vec{r}_{01}$ is pointing from a source point $\vec{x}_0$ to the observation point $\vec{x}_1$. The problem of interest is the calculation of the optical field at the observation point given knowledge of the field in the aperture plane. Before presenting the diffraction formula, it is important to present the underlying assumptions used to obtain the results. The diffraction formula is a result of scalar diffraction theory. In scalar diffraction theory the optical field quantities are treated as scalar quantities and their vector nature ignored. Scalar diffraction theory has been shown to be accurate if the diffracting aperture is large with respect to the optical wavelength and the observation plane is many wavelengths away from the diffracting plane [1]. These two conditions are easily satisfied for most situations of interest involving optical fields and are certainly satisfied for any application of interest presented in this book. Monochromatic optical fields are also assumed unless explicitly stated otherwise.

Rayleigh-Sommerfeld diffraction formula

The Rayleigh-Sommerfeld diffraction formula is a rigorous solution to the scalar diffraction problem shown in Fig. 2.1 and represents the most general solution given the assumptions stated above. This formula is a result of rigorous application of the Helmholtz wave equation to the problem. The wave equation is solved by use of Green's theorem and a clever choice of a Green's function. The Rayleigh-Sommerfeld diffraction formula is given by [1]

$$u_1(\vec{x}_1) = \frac{1}{j\lambda}\int d\vec{x}_0 u_0(\vec{x}_0)\frac{\exp(jkr_{01})}{|\vec{r}_{01}|}\cos(\theta), \tag{2.1}$$

where $u(\vec{x})$ is the complex phasor representation of the optical field at the point $\vec{x}$ in the plane of interest, λ is the optical wavelength, k is the optical wave number $(2\pi/\lambda)$, $r_{01} = |\vec{r}_{01}|$, and the notation $\int d\vec{x}$ designates integration over the entire xy plane. The factor $\cos(\theta)$ is called the obliquity factor with θ defined in Fig. 2.1. Unless stated otherwise, the subscript on the field quantity u will indicate the plane in which the field is being specified. In Eq. (2.1), $u_1(\vec{x}_1)$ is the field in the observation plane and $u_0(\vec{x}_0)$ is the field in the aperture plane. It should also be clear from Fig. 2.1 that the quantities $\vec{r}_{01}$ and θ are functions of $\vec{x}_0$ and $\vec{x}_1$. The interpretation of Eq. (2.1) is appealing in that it reinforces the Huygens-Fresnel principle. The Huygens-Fresnel principle states that a field from a diffracting aperture can be described as a superposition of spherical waves from an infinite number of point sources within the aperture. Equation (2.1) is a direct mathematical statement of the Huygens-Fresnel principle in that the integral represents an infinite sum of expanding spherical waves $(\exp(jk\vec{r}_{01})/|\vec{r}_{01}|)$ for each point in the aperture, weighted by the incident field $u_0(\vec{x}_0)$ and the obliquity factor.

It is important to recognize that Eq. (2.1) is a superposition integral and can thus be written in the following way

$$u_1(\vec{x}_1) = \int d\vec{x}_0 u_0(\vec{x}_0)h_d(\vec{x}_1, \vec{x}_0), \tag{2.2}$$

where $h_d(\vec{x}_1, \vec{x}_0)$ is the free space propagation impulse response given by

$$h_d(\vec{x}_1, \vec{x}_0) = \frac{1}{j\lambda}\frac{\exp(jkr_{01})}{|\vec{r}_{01}|}\cos(\theta). \tag{2.3}$$

The subscript d is used to indicate that the impulse response is associated with the diffraction geometry in Fig. 2.1. Due to the linearity of wave propagation, it is not surprising that the diffraction calculation can be cast in the superposition formulation of Eq. (2.2). Later in this section we will see that imaging is also a linear operation. We further note that $\vec{r}_{01}$ and $\cos(\theta)$ can be written in terms of $\vec{x}_1 - \vec{x}_0$ which implies the superposition integral in Eq. (2.2) reduces to a simpler convolutional form. In other words we can write

$$u_1(\vec{x}_1) = \int \mathrm{d}\vec{x}_0 u_0(\vec{x}_0) h_d(\vec{x}_1 - \vec{x}_0), \\ = u_0(\vec{x}_1) * h_d(\vec{x}_1), \tag{2.4}$$

where the notation $*$ designates 2-dimensional convolution. The convolutional form has extremely useful consequences in that the rich set of Fourier analysis tools applicable to linear, shift-invariant systems is now available to solve diffraction problems. By considering the 2-dimensional Fourier transforms of $u_0(\vec{x}_0)$ and $u_1(\vec{x}_1)$, we can recast the Rayleigh-Sommerfeld diffraction formula in the spatial frequency domain.

Before presenting the frequency domain equivalent of Eq. (2.4), the mathematical convention for the Fourier transform is introduced. The Fourier transform of the 2-dimensional function $g(\vec{x})$ is defined in the following way:

$$G(\vec{f}) = \mathcal{F}[g(\vec{x})] = \int \mathrm{d}\vec{x} g(\vec{x}) \exp[-j2\pi \vec{f} \cdot \vec{x}], \tag{2.5}$$

where $G(\vec{f})$ is the Fourier transform of $g(\vec{x})$ and $\vec{f}$ is the spatial frequency vector having units of m^{-1}. In most instances the lower case and capital letter representations of a function are used to indicate a Fourier transform pair. The inverse Fourier transform is defined as

$$g(\vec{x}) = \mathcal{F}^{-1}[G(\vec{f})] = \int \mathrm{d}\vec{f} G(\vec{f}) \exp[j2\pi \vec{f} \cdot \vec{x}]. \tag{2.6}$$

The frequency domain representation of the convolution of Eq. (2.4) can be written

$$U_1(\vec{f}) = U_0(\vec{f}) H_d(\vec{f}), \tag{2.7}$$

where $U_1(\vec{f})$, $U_0(\vec{f})$ and $H_d(\vec{f})$ are the Fourier transforms of $u_1(\vec{x}_1)$, $u_0(\vec{x}_0)$ and $h_d(\vec{x})$, respectively. The quantities $U_1(\vec{f})$ and $U_0(\vec{f})$ can be interpreted as the angular distribution or spectrum of plane waves comprising the fields at the aperture and observation planes, respectively. In this interpretation the spatial frequency vector $\vec{f}$ is the x and y-directed components of a 3-dimensional vector that points in the direction of propagation of a single plane wave. Due to the fact the plane must satisfy the wave equation, the z component of this propagation vector must satisfy $f_z = \sqrt{(\lambda^{-1} - f_x^2 - f_y^2)}$ where f_x and f_y are the x and y-directed components of $\vec{f}$. The function $H_d(\vec{f})$ is the transfer function associated with free space propagation from the aperture plane to the observation plane. This transfer function is found by Fourier transforming Eq. (2.3) and is given by [1]

$$H_d(\vec{f}) = \begin{cases} \exp\left(j2\pi z\sqrt{\lambda^{-2} - |\vec{f}|^2}\right) & |\vec{f}| < \lambda^{-1} \\ 0 & \text{elsewhere} \end{cases}. \tag{2.8}$$

We next consider two special cases of the Rayleigh-Sommerfeld diffraction formula: the Fresnel and Fraunhofer diffraction formulas.

Fresnel diffraction formula

Even though the Rayleigh-Sommerfeld formula is the most general of the diffraction formulas, this formulation is difficult to use. This difficulty arises because of the term r_{01} in the exponent of Eq. (2.3). In its exact form, r_{01} is given by (see Fig. 2.1)

$$r_{01} = \sqrt{z^2 + (x_0 - x_1)^2 + (y_0 - y_1)^2}. \tag{2.9}$$

where the scalar notation x and y designates the x- and y-directed components of the vector $\vec{x}$. The variable z is the separation of the aperture and observation planes. It is the square root in Eq. (2.9) that makes working with the Rayleigh-Sommerfeld diffraction difficult.

The Fresnel approximation involves assuming z is much larger than the transverse coordinates x_0, y_0, x_1 and y_1 and using the binomial expansion of r_{01}. If z is large enough, paraxial conditions are said to exist, and only the lowest order terms of the binomial expansion need to be kept. Using the binomial expansion and keeping the lowest order terms results in the Fresnel free space propagation impulse response [1]:

$$h_d(\vec{x}_1 - \vec{x}_0) = \frac{\exp(jkz)}{j\lambda} \frac{\exp\left(j\frac{k|\vec{x}_0 - \vec{x}_1|^2}{2z}\right)}{|\vec{r}_{01}|} \cos(\theta). \tag{2.10}$$

The main point to notice concerning Eq. (2.10) is the quadratic form of the second exponential term. The Fresnel approximation in essence is simply an approximation of the spherical phase front arising in the Rayleigh-Sommerfeld diffraction formula with a field having a quadratic phase front. The quadratic phase front is an accurate approximation for points very near the optical axis (z axis in Fig. 2.1). As in the Rayleigh-Sommerfeld formulation, the Fresnel diffraction formula is cast in a linear, shift-invariant systems framework. The corresponding Fresnel transfer function is obtained by Fourier transforming Eq. (2.10):

$$H_d(\vec{f}) = \begin{cases} \exp(jkz)\exp(-j\pi\lambda z|\vec{f}|^2) & |\vec{f}| < \lambda^{-1} \\ 0 & \text{elsewhere} \end{cases} . \tag{2.11}$$

The Fresnel diffraction formula is often written with the quadratic phase term in an expanded form. Substituting Eq. (2.10) into Eq. (2.4) and expanding the quadratic phase term gives

$$\begin{aligned} u_1(\vec{x}_1) = {} & \frac{\exp(jkz)}{j\lambda z} \exp\left[j\frac{k}{2z}|\vec{x}_1|^2\right] \\ & \times \int d\vec{x}_0 u_0(\vec{x}_0) \exp\left[j\frac{k}{2z}|\vec{x}_0|^2\right] \exp\left[j\frac{2\pi}{\lambda z}\vec{x}_0 \cdot \vec{x}_1\right]. \end{aligned} \tag{2.12}$$

Using the Fourier transform operator notation, $\mathcal{F}[\cdot]$, Eq. (2.12) can also be written as

$$u_1(\vec{x}_1) = \frac{\exp(jkz)}{j\lambda z} \exp\left[j\frac{k}{2z}|\vec{x}_1|^2\right] \mathcal{F}\left[u_0(\vec{x}_0)\exp\left[j\frac{k}{2z}|\vec{x}_0|^2\right]\right]\Bigg|_{\vec{f}=\vec{x}_1/\lambda z} , \tag{2.13}$$

where the subscript indicates that the Fourier transform is to be evaluated at the spatial frequency $\vec{f} = \vec{x}_0/\lambda z$. A final approximation, made in the next section, will further simplify the diffraction equation.

Fraunhofer diffraction formula

In many cases the propagation distance between the aperture and observation planes is so large that we can make another approximation known as the Fraunhofer approximation. If the propagation distance, z, is large enough, then the quadratic phase term inside the Fourier transform of Eq. (2.13) can be ignored. The requirement on the propagation distance can easily be shown to be

$$z \gg \frac{k|\vec{x}_1|^2}{2}, \tag{2.14}$$

for all source points $\vec{x}_1$. If this condition is true, Eq. (2.13) can be rewritten ignoring the second quadratic

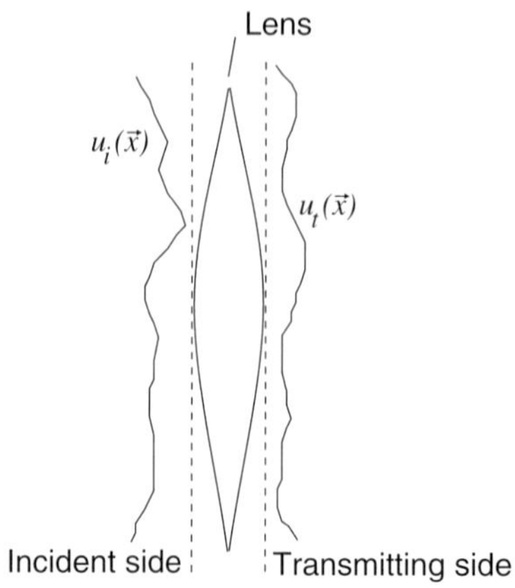

FIGURE 2.2
Lens geometry.

exponential term:

$$u_1(\vec{x}_1) = \frac{\exp(jkz)}{j\lambda z} \exp\left[j\frac{k}{2z}|\vec{x}_1|^2\right] \mathcal{F}\left[u_0(\vec{x}_0)\right]\Big|_{\vec{f}=\vec{x}_1/\lambda z}. \tag{2.15}$$

The noteworthy feature of Eq. (2.15) is that the observation field is simply a Fourier transform of the aperture field multiplied by quadratic and range dependent exponential terms. This calculation is often simple to perform and nearly always quick and easy using computers and the Fast Fourier Transform (FFT) algorithm.

The diffraction formulas given in the previous three subsections form the foundation of Fourier optics. Since we are ultimately interested in imaging, we now turn our attention to the application of the diffraction formulas to optical systems containing lenses.

2.2.2 Lenses

In order to discuss imaging, it is necessary to introduce a model of the effect a lens has on an incident optical field. The basic problem is to develop a model that characterizes how a field is changed by propagating from the incident side of a lens to the transmitting side as shown in Fig. 2.2. The following relationship between the incident and transmitted fields is assumed:

$$u_t(\vec{x}) = u_i(\vec{x})t_l(\vec{x}), \tag{2.16}$$

where $u_i(\vec{x})$ and $u_t(\vec{x})$ are the incident and transmitted fields, respectively, and $t_l(\vec{x})$ is a transparency function describing the effect of the lens. Using "thin lens" assumptions, the transparency function takes the form [1]:

$$t_l(\vec{x}) = \exp\left[-j\frac{k}{2f}|\vec{x}|^2\right], \tag{2.17}$$

where f is the focal length of the lens. This lens model along with the diffraction formulas of the previous sections form the foundation for understanding imaging.

2.2.3 Coherent image formation

As stated in the introduction to this chapter, an imaging system experiences fundamental performance

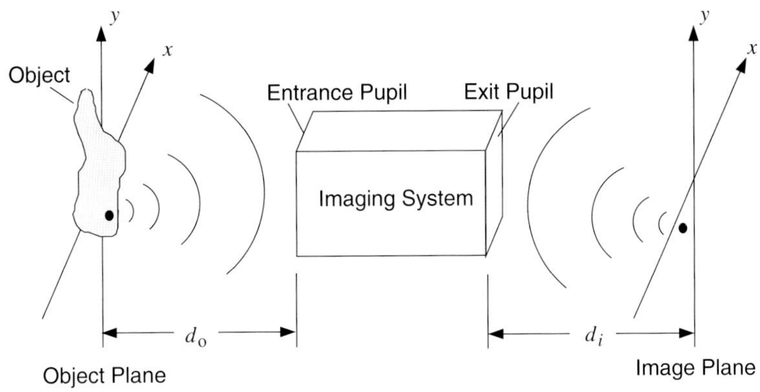

FIGURE 2.3
Generalized imaging geometry.

limitations not predicted by geometrical optics. These limitations arise from the wave nature of light and can be predicted by Fourier optics techniques. In particular, we use the Fresnel diffraction formula and the associated assumptions to study how the optical field propagates through an imaging system. Here we dwell on the key results of the analysis. The reader is referred to the detailed presentation in Chapter 5 of reference [1]. The word "coherent" in the title of this section refers to the spatial coherence of the illuminating optical field. Coherent illumination implies the optical field is perfectly correlated or deterministic.

Consider the generalized imaging system shown in Fig. 2.3. The object being imaged is located in the object plane on the left. The image field is located in the image plane on the right. For our purposes, the imaging system is a collection of one or more optical elements that transform incident diverging spherical waves into converging spherical waves (as illustrated in Fig. 2.3). Without loss of generality, we also assume the system results in an image having unity magnification. The unity magnification assumption is important for the linear systems framework that is presented below. Imaging systems having non-unity magnification can be cast into the linear systems framework simply by analyzing the system as if it gives unity magnification and then scaling the size of the image to account for any magnification. We also ignore any constants in the relationship between the image and object fields. As far as the performance of an imaging system is concerned, constant scaling terms are of no consequence and thus can be ignored.

For the case of coherent, monochromatic illumination and object and image sizes sufficiently small to satisfy the Fresnel paraxial conditions, we obtain the following simple relationship between the object field, $u_o(\vec{x})$, and the image field, $u_i(\vec{x})$:

$$u_i(\vec{x}) = u_o(\vec{x}) * h(\vec{x}), \tag{2.18}$$

where $h(\vec{x})$ is the imaging system impulse response. A straightforward application of the Fresnel diffraction analysis of Section 2 along with the lens transparency model of Eq. (2.17) shows that the imaging system impulse response is given by [1]

$$\begin{aligned} h(\vec{x}) &= \int \mathrm{d}\vec{f}\, W(\vec{f}\lambda d_i) \exp\left[-j2\pi\vec{f}\cdot\vec{x}\right], \\ &= \mathcal{F}\left[W(\vec{f}\lambda d_i)\right], \end{aligned} \tag{2.19}$$

where $W(\vec{x})$ is the pupil function of the imaging system, λ is the optical wavelength, and d_i is the distance between the exit pupil and the image plane. The pupil function takes on a constant value for locations within the pupil, and is zero outside the pupil. In order to obtain the seemingly simple result given in Eq. (2.19), two quadratic phase terms must be eliminated from the diffraction analysis result. Reference [1] gives a detailed argument for eliminating these two exponential terms as well as a number of mathematical manipulations to obtain the *space invariant* impulse response of Eq. (2.19).

Equation (2.19) shows there exists a very simple relationship between the pupil function of a coherent imaging system and its impulse response. This simple relationship is the Fourier transform! Later, when we introduce the effects of aberrations on the imaging system, we find that the pupil function, $W(\vec{x})$, will take on a more generalized role. In the case of pupil plane aberrations, the generalized pupil function describes both the entrance pupil of the system as well as the aberrations. The most important fact to draw from the results presented in Eqs. (2.18) and (2.19) is that imaging can be cast into a linear, shift-invariant system framework. The convolution of Eq. (2.18) leads directly to a frequency domain analysis of coherent imaging, which is the subject of the next section.

2.2.4 Frequency domain analysis of coherent imaging

The convolution of Eq. (2.18) indicates that imaging problems can be cast into the familiar framework of linear, shift-invariant systems. This framework allows us to characterize and solve imaging problems in the frequency domain. Taking the Fourier transform of both sides of Eq. (2.18), and using the Fourier transform relationship between convolution in the space domain and multiplication in the frequency domain gives

$$U_i(\vec{f}) = U_o(\vec{f})H(\vec{f}), \tag{2.20}$$

where $U_i(\vec{f})$, $U_o(\vec{f})$ and $H(\vec{f})$ are the Fourier transforms of optical field quantities $u_i(\vec{x})$ and $u_o(\vec{x})$ and the impulse response $h(\vec{x})$, respectively. The quantity $H(\vec{f})$ is called the coherent transfer function (CTF). Taking the Fourier transform of the impulse response given in Eq. (2.19) gives the result

$$H(\vec{f}) = W(\vec{f}\lambda d_i). \tag{2.21}$$

Recall the pupil function defines the physical extent of the system pupil. We also see that the pupil function describes the system transfer function. In other words, the pupil function indicates which spatial frequencies are passed by the system. Those spatial frequencies corresponding to locations within the pupil are passed while those corresponding to locations outside the pupil are not.

Before moving to incoherent imaging we present two coherent imaging examples. Consider first a system having a rectangular pupil described by

$$W(\vec{x}) = \text{rect}\left(\frac{x}{D_x}\right)\text{rect}\left(\frac{y}{D_y}\right), \tag{2.22}$$

where the rect$(\cdot)$ is defined as

$$\text{rect}(x) = \begin{cases} 1 & |x| < 1/2 \\ 0 & \text{elsewhere} \end{cases}, \tag{2.23}$$

and D_x and D_y are the x- and y-directed dimensions of the rectangular aperture. The CTF for the imaging system is found by substituting Eq. (2.22) into Eq. (2.21):

$$H(f_x, f_y) = \text{rect}\left(\frac{f_x\lambda d_i}{D_x}\right)\text{rect}\left(\frac{f_y\lambda d_i}{D_y}\right), \tag{2.24}$$

where f_x and f_y are the x- and y-directed components of the spatial frequency vector $\vec{f}$. A plot of a cross-section of the CTF is shown in Fig. 2.4. One of the interesting points to notice from Fig. 2.4 is that the cutoff frequency of the CTF is proportional to the size of the entrance pupil and inversely proportional to the optical wavelength. The cutoff frequency is given by $D_x/2\lambda d_i$ in the x-direction and $D_y/2\lambda d_i$ in the y-direction.

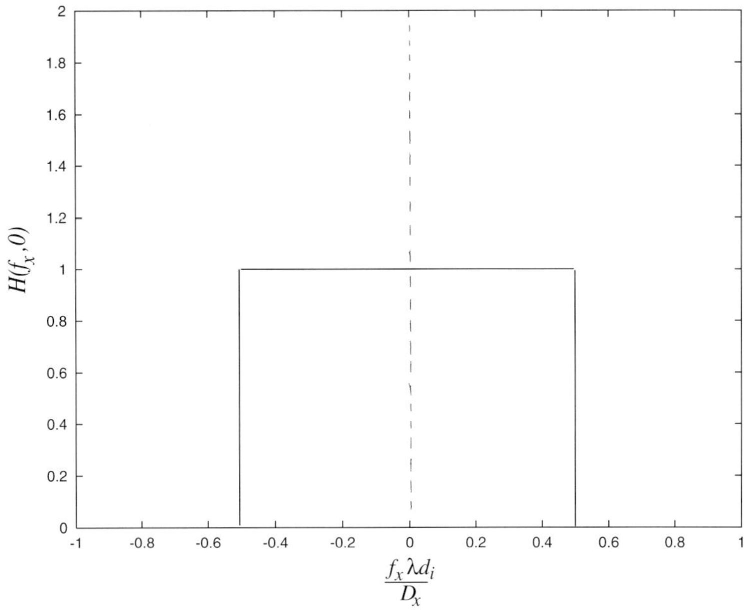

FIGURE 2.4
Cross-section of the CTF $H(f_x, f_y)$ plotted versus f_x for the case of a rectangular pupil.

The impulse response of the imaging system is the inverse Fourier transform of Eq. (2.24) and is given by

$$h(x, y) = \frac{D_x D_y}{(\lambda d_i)^2} \operatorname{sinc}\left(\frac{x D_x}{\lambda d_i}\right) \operatorname{sinc}\left(\frac{y D_y}{\lambda d_i}\right), \tag{2.25}$$

where $\operatorname{sinc}(x) = \sin(\pi x)/\pi x$. A plot of the x–axis slice of the impulse response is shown in Fig. 2.5. The spatial resolution of the imaging system can be equated to the distance from the origin to the first zero crossing of the impulse response. This criterion is referred to as the Rayleigh resolution criterion and is $\lambda d_i / D_x$ in the x-direction and $\lambda d_i / D_y$ in the y-direction for the rectangular aperture.

For the second example consider an imaging system with a circular pupil:

$$W(\vec{x}) = \operatorname{circ}\left(\frac{|\vec{x}|}{D/2}\right), \tag{2.26}$$

where the circ$(\cdot)$ is defined as

$$\operatorname{circ}(x) = \begin{cases} 1 & |\vec{x}| < 1 \\ 0 & \text{elsewhere} \end{cases}, \tag{2.27}$$

and D is the diameter of the pupil. The CTF is again found by substituting Eq. (2.26) into Eq. (2.21):

$$H(\vec{f}) = \operatorname{circ}\left(\frac{|\vec{f}|\lambda d_i}{D/2}\right). \tag{2.28}$$

A plot of a center cross-section of the CTF is the same as the cross-section shown for the rectangular aperture in Fig. 2.4 with D equated to D_x.

The impulse response of the circular imaging system is the inverse Fourier transform of Eq. (2.28). Due to the circular symmetry of $H(\vec{f})$, the inverse Fourier transform can be converted into a Fourier-Bessel transform [1]. The Fourier-Bessel transform of a circularly symmetric function $g(r)$ is given

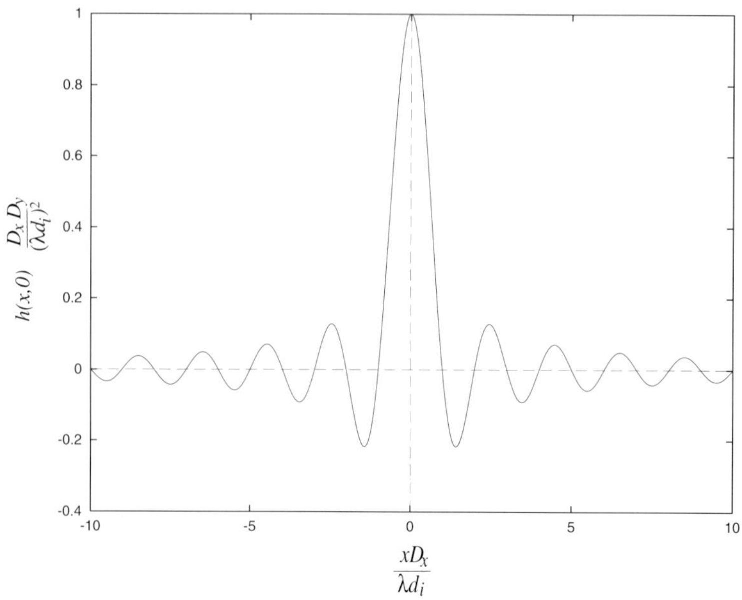

FIGURE 2.5
Cross-section of the coherent impulse response $h(x, 0)$ plotted versus x for the case of a rectangular pupil.

by

$$G(\rho) = 2\pi \int_0^\infty \mathrm{d}r \; r g(r) J_o(2\pi r \rho), \tag{2.29}$$

where $r = |\vec{x}|$ is the radial spatial coordinate, $\rho = |\vec{f}|$ is the radial spatial frequency coordinate and $J_o(\cdot)$ is a Bessel function of the first kind, zero order. The inverse Fourier-Bessel transform is identical to the forward transform with the $g(r)$ and $G(\rho)$ swapped:

$$g(r) = 2\pi \int_0^\infty \mathrm{d}\rho \; \rho G(\rho) J_o(2\pi r \rho). \tag{2.30}$$

Substituting Eq. (2.28) into Eq. (2.30) and evaluating the integral gives the impulse response for the circular pupil:

$$h(r) = \frac{D^2}{(2\lambda d_i)^2} \frac{J_1\left(2\pi \dfrac{rD}{2\lambda d_i}\right)}{\dfrac{rD}{2\lambda d_i}}, \tag{2.31}$$

where $J_1(\cdot)$ is a Bessel function of the first kind, order one. A plot of a cross-section of the impulse response is shown in Fig. 2.6. Again we can equate the spatial resolution to the distance to the first zero crossing of the impulse response. This criterion corresponds to a spatial resolution of $1.22\lambda d_i/D$.

As seen from Eqs. (2.19) and (2.21), the coherent imaging system response depends on the distance d_i. In many situations it is useful to represent the impulse response and CTF in a way that is independent of d_i. This goal is easily accomplished by considering the angular frequency response of the system. Consider the following transformations relating linear space to angle space:

$$\vec{\alpha} = \frac{\vec{x}}{d_i},$$

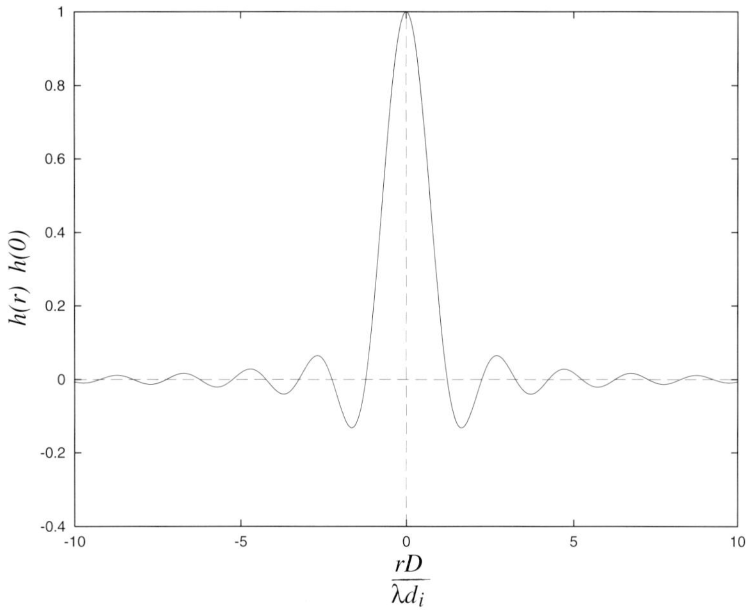

FIGURE 2.6
Cross-section of the coherent impulse response $h(r)$ plotted versus radial coordinate r for the case of a circular pupil.

$$\vec{\beta} = \vec{f} d_i, \tag{2.32}$$

where $\vec{\alpha}$ and $\vec{\beta}$ are angle equivalents to the space variables $\vec{x}$ and $\vec{f}$. Using these new angular coordinates we can rewrite Eqs. (2.19) and (2.21) in the following way:

$$\begin{aligned} h(\vec{\alpha}) &= \int \mathrm{d}\vec{\beta} W(\vec{\beta}\lambda) \exp\left[-j2\pi\vec{\beta}\cdot\vec{\alpha}\right], \\ &= \mathcal{F}\left[W(\vec{\beta}\lambda)\right], \end{aligned} \tag{2.33}$$

$$H(\vec{\beta}) = W(\vec{\beta}\lambda), \tag{2.34}$$

where we have again ignored any resulting constant factors. Note that to be consistent with the earlier notation we should have defined new functions for the angular impulse response and CTF in Eqs. (2.33) and (2.34). For convenience the same notation is used and the arguments of $h(\cdot)$ and $H(\cdot)$ will guide the reader in the proper interpretation.

2.2.5 Incoherent image formation

The results given in the previous section are applicable for the case of coherent, monochromatic illumination of the object. In this section the coherent imaging results of the previous section will be extended to the case of incoherent illumination. The phrase "incoherent illumination" represents the opposite extreme of "coherent illumination". The conventional model of incoherent illumination is that each field point on the object is completely independent of all other field points. In reality we can never have an optical field that behaves as this model would suggest. We find that real optical fields will have some relationship for two points if the two points are close enough (less than λ apart, for example). In most situations this relationship will not invalidate the results of our idealized model if the imaging system under consideration is unable to resolve the small distances involved [1].

We begin our discussion of incoherent imaging by considering the image irradiance obtained from a coherent imaging system. The irradiance observed in the image plane of a coherent imaging system, $i(\vec{x})$, is given by the squared modulus of Eq. (2.18):

$$i(\vec{x}) = |u_i(\vec{x})|^2 = |u_o(\vec{x}) * h_i(\vec{x})|^2. \tag{2.35}$$

Expanding Eq. (2.35) gives:

$$i(\vec{x}) = \int \mathrm{d}\vec{x}' \int \mathrm{d}\vec{x}'' u_o(\vec{x}')u_o^*(\vec{x}'')h_i(\vec{x} - \vec{x}')h_i^*(\vec{x} - \vec{x}''), \tag{2.36}$$

where the superscript $*$ designates complex conjugate. Equation (2.36) describes the irradiance observed for a coherent imaging system and in its current form can not be further simplified. Let us now consider the consequence of assuming that $u_o(\vec{x})$ is an incoherent optical field.

In the case of incoherent illumination, the object field $u_o(\vec{x})$ is inherently a random quantity, which in turn implies the irradiance $i(\vec{x})$ is also random. The average irradiance observed in the imaging plane becomes the quantity of interest. Taking an average[1] of both sides of Eq. (2.36) gives

$$\begin{aligned}\langle i(\vec{x})\rangle &= \left\langle \int \mathrm{d}\vec{x}' \int \mathrm{d}\vec{x}'' u_o(\vec{x}')u_o^*(\vec{x}'')h_i(\vec{x} - \vec{x}')h_i^*(\vec{x} - \vec{x}'')\right\rangle, \\ &= \int \mathrm{d}\vec{x}' \int \mathrm{d}\vec{x}'' \langle u_o(\vec{x}')u_o^*(\vec{x}'')\rangle h_i(\vec{x} - \vec{x}')h_i^*(\vec{x} - \vec{x}''),\end{aligned} \tag{2.37}$$

where $\langle\cdot\rangle$ is the ensemble average operator, and it has been moved inside the integral in the second line of Eq. (2.37). The coherence properties of the incoherent field can now be used to simplify Eq. (2.37). The conventional incoherent model for the correlation, $\langle u_o(\vec{x}')u_o^*(\vec{x}'')\rangle$, is given by [3]

$$\langle u_o(\vec{x}')u_o^*(\vec{x}'')\rangle = \kappa\langle o(\vec{x}')\rangle\delta(\vec{x}' - \vec{x}''), \tag{2.38}$$

where $\langle o(\vec{x})\rangle$ is the average irradiance of the object field and $\delta(\vec{x})$ is the Dirac-delta function. For notational simplicity the angle brackets enclosing $i(\vec{x})$ and $o(\vec{x})$ are dropped and henceforth, irradiance quantities are assumed average values. The constant κ is simply required to give the equality in Eq. (2.38) [3] and will be dropped in the subsequent results. Substituting Eq. (2.38) into Eq. (2.37) and performing the integration over $\vec{x}''$ gives

$$\begin{aligned}i(\vec{x}) &= \int \mathrm{d}\vec{x}' o(\vec{x}')|h_i(\vec{x} - \vec{x}')|^2, \\ &= o(\vec{x}) * |h_i(\vec{x})|^2.\end{aligned} \tag{2.39}$$

The remarkable fact to notice from Eq. (2.39) is that there exists a convolution relationship between the image and object irradiance distributions. The impulse response characterizing the input/output relationship between the two irrandiances is $|h_i(\vec{x})|^2$. By convention this impulse response is called the point spread function (PSF) and is designated with the following notation:

$$s(\vec{x}) = |h_i(\vec{x})|^2. \tag{2.40}$$

It is interesting to note that, in the case of incoherent illumination, the imaging system responds linearly to the object irradiance distribution; whereas for coherent illumination, the system responds

[1] The average can be an ensemble or time average due to the assumption of ergodicity for $u_o(\vec{x})$. A time average can be interpreted as the integration process that any real detector requires to detect the irradiance of an optical field.

linearly to the object field distribution. Just as we saw for coherent imaging, the convolution relationship of Eq. (2.39) leads directly to a frequency domain analysis, which is the subject of the next section.

2.2.6 Frequency domain analysis of incoherent imaging

Taking the Fourier transform of both sides of Eq. (2.39) gives the frequency domain equivalent of the incoherent imaging result of the previous section:

$$I(\vec{f}) = O(\vec{f})\mathcal{H}(\vec{f}), \tag{2.41}$$

where $I(\vec{f})$ and $O(\vec{f})$ are Fourier transforms of $i(\vec{x})$ and $o(\vec{x})$, respectively, and $\mathcal{H}(\vec{f})$ is the Fourier transform of the PSF, $s(\vec{x})$. The frequency domain quantities I and O are referred to as the object and image spectra and the transfer function $\mathcal{H}(\vec{f})$ is called the optical transfer function (OTF). By convention, the OTF is normalized to unity at $\vec{f} = 0$. This normalization is partly for mathematical convenience and partly due to the fact that image irradiances are always positive and the normalization equalizes the dc component of the object and image irradiances. The dc component represents the overall irradiance in the image and the normalization forces the overall irradiance in the image and the object to be equal.

The relationship between $\mathcal{H}(\vec{f})$ and the coherent impulse response $h(\vec{x})$ is found by Fourier transforming Eq. (2.40):

$$\begin{aligned}\mathcal{H}(\vec{f}) &= \frac{\mathcal{F}[s(\vec{x})]}{\mathcal{F}[s(\vec{x})]|_{\vec{f}=0}} = \frac{\mathcal{F}[h(\vec{x})h^*(\vec{x})]}{\mathcal{F}[h(\vec{x})h^*(\vec{x})]|_{\vec{f}=0}} \\ &= \frac{\mathcal{F}[h(\vec{x})] \star \mathcal{F}[h(\vec{x})]}{[\mathcal{F}[h(\vec{x})] \star \mathcal{F}[h(\vec{x})]]|_{\vec{f}=0}},\end{aligned} \tag{2.42}$$

where the denominator term is required in order to normalize the OTF to unity at $\vec{f} = 0$. The notation $\star$ is the correlation operator and is defined in the following way for two arbitrary functions $f(\vec{x})$ and $g(\vec{x})$:

$$f(\vec{x}) \star g(\vec{x}) = \int d\vec{x}' f(\vec{x}' - \vec{x})g^*(\vec{x}'). \tag{2.43}$$

The OTF, $\mathcal{H}(\vec{f})$, can also be related to the system pupil function $W(\vec{x})$ by substituting the Fourier transform of $h_i(\vec{x})$, found in Eq. (2.21), into Eq. (2.42):

$$\mathcal{H}(\vec{f}) = \frac{W(\vec{f}\lambda d_i) \star W(\vec{f}\lambda d_i)}{W(0) \star W(0)}. \tag{2.44}$$

Aberrations in the pupil of the imaging system can be also be incorporated within the linear systems framework discussed above. In this case the definition of the pupil function, $W(\vec{x})$, is generalized to account for both the finite pupil size and the pupil plane aberrations. The generalized pupil function may now be complex and can be interpreted as the product of a physical pupil function and a function describing the aberration:

$$W(\vec{x}) = W_p(\vec{x})e^{j\phi(\vec{x})}, \tag{2.45}$$

where $W_p(\vec{x})$ is a real valued function describing the physical pupil and $\phi(\vec{x})$ describes the pupil plane phase aberration. The OTF of the aberrated imaging system can be found by substituting this generalized pupil function into the OTF expression given in Eq. (2.44).

The section is concluded with a continuation of the examples given at the end of the coherent imaging section (Section 2.2.4). First consider the rectangular pupil function given in Eq. (2.22). Substituting

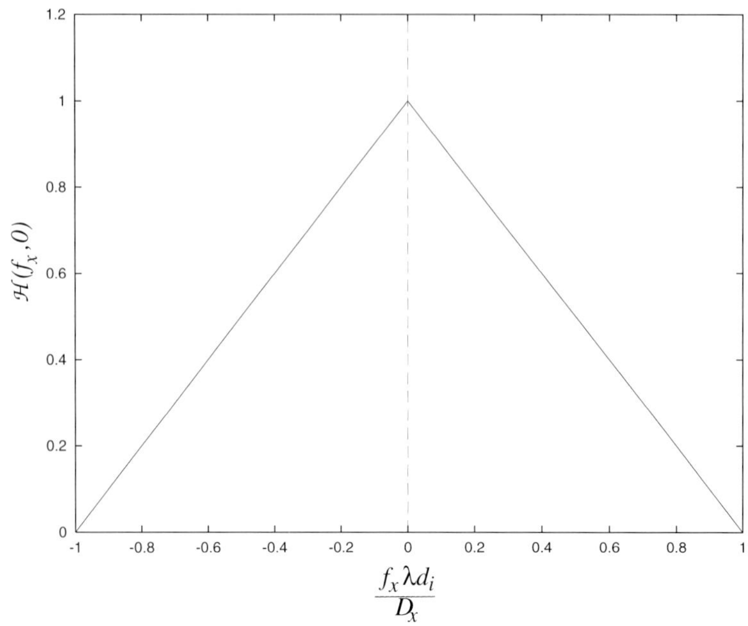

FIGURE 2.7
Cross-section of the OTF $\mathcal{H}(f_x, f_y)$ plotted versus f_x for the case of a rectangular pupil.

the pupil function in Eq. (2.44) gives the following expression for the OTF [1]:

$$\mathcal{H}(f_x, f_y) = \frac{\left[\text{rect}\left(\frac{f_x \lambda d_i}{D_x}\right)\text{rect}\left(\frac{f_y \lambda d_i}{D_y}\right)\right] \star \left[\text{rect}\left(\frac{f_x \lambda d_i}{D_x}\right)\text{rect}\left(\frac{f_y \lambda d_i}{D_y}\right)\right]}{[\text{rect}(0)\text{rect}(0)] \star [\text{rect}(0)\text{rect}(0)]}$$
$$= \text{tri}\left(\frac{f_x \lambda d_i}{D_x}\right)\text{tri}\left(\frac{f_y \lambda d_i}{D_y}\right), \tag{2.46}$$

where tri$(\cdot)$ is the triangle function and is defined as follows:

$$\text{tri}(x) = \begin{cases} 1 - |x| & |x| < 1 \\ 0 & \text{elsewhere} \end{cases}. \tag{2.47}$$

Fig. 2.7 illustrates a cross-section of the OTF. Note that the cutoff frequency is $f_c = D_x/\lambda d_i$ along the x-direction and is $f_c = D_y/\lambda d_i$ along the y-direction. The corresponding PSF is found by Fourier transforming Eq. (2.46). Performing this transformation gives

$$s(x, y) = \frac{D_x D_y}{(\lambda d_i)^2}\text{sinc}^2\left(\frac{x D_x}{\lambda d_i}\right)\text{sinc}^2\left(\frac{y D_y}{\lambda d_i}\right). \tag{2.48}$$

Fig. 2.8 illustrates a cross section of the PSF.

Next, consider the circular pupil function given in Eq. (2.26). Substituting this pupil function in Eq. (2.44) gives the following expression for the OTF:

$$\mathcal{H}(\rho) = \begin{cases} \frac{2}{\pi}\left[\cos^{-1}\left(\frac{\rho \lambda d_i}{D}\right) - \frac{\rho \lambda d_i}{D}\sqrt{1 - \left(\frac{\rho \lambda d_i}{D}\right)^2}\right] & \rho \leq D/\lambda d_i \\ 0 & \text{otherwise} \end{cases}, \tag{2.49}$$

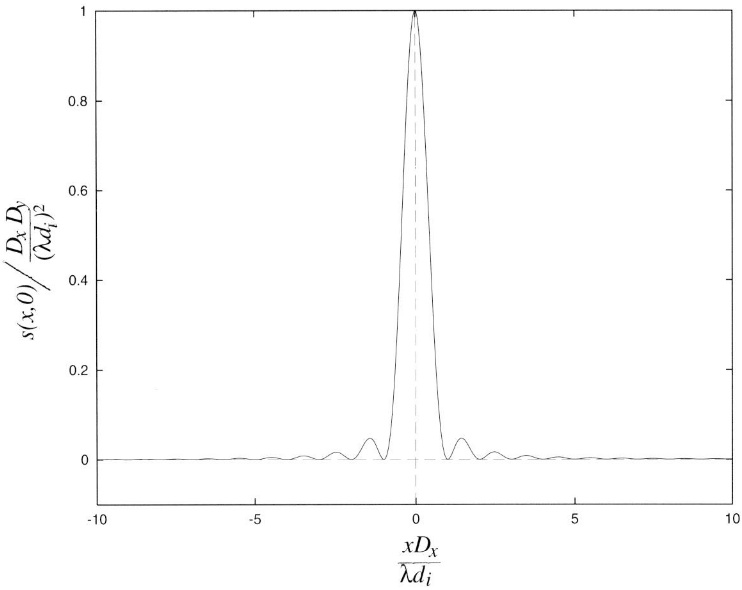

FIGURE 2.8
Cross-section of the PSF $s(x, y)$ plotted versus x for the case of a rectangular pupil.

Fig. 2.9 illustrates a cross-section of the OTF. Note that the cutoff frequency is $f_c = D/\lambda d_i$. The corresponding PSF is found by employing the inverse Fourier-Bessel transform on Eq. (2.49). Performing this transformation gives

$$\frac{s(r)}{s(0)} = \frac{J_1^2\left(\frac{rD}{2\lambda d_i}\right)}{\left(\frac{rD}{2\lambda d_i}\right)^2}, \tag{2.50}$$

Fig. 2.10 illustrates a cross section of the PSF.

2.3 Statistical optics

As stated in the introduction of this chapter, statistical optics plays a key role in understanding incoherent imaging as well as imaging in the presence of a random medium such as the atmosphere. Both the temporal and spatial coherence properties of the optical field affect the performance of an imaging system. As an example, the temporal coherence properties of an optical field determine the contrast of the speckled images used in the post processing image reconstruction techniques discussed in Chapter 1 and later in Chapter 4. These speckled images represent interference effects in the image plane. The degree to which the optical field is able to interfere with itself is described by its temporal coherence.

The spatial coherence of an optical field plays an even larger role in characterizing the performance of imaging systems. Later in this section we introduce an interferometric view of imaging that builds on the concept of spatial coherence. The interferometric view of imaging and an understanding of spatial coherence allow us to understand the effects of imaging through a random medium, such as the atmosphere. As shown later in this section, the spatial coherence of an optical field is inextricably

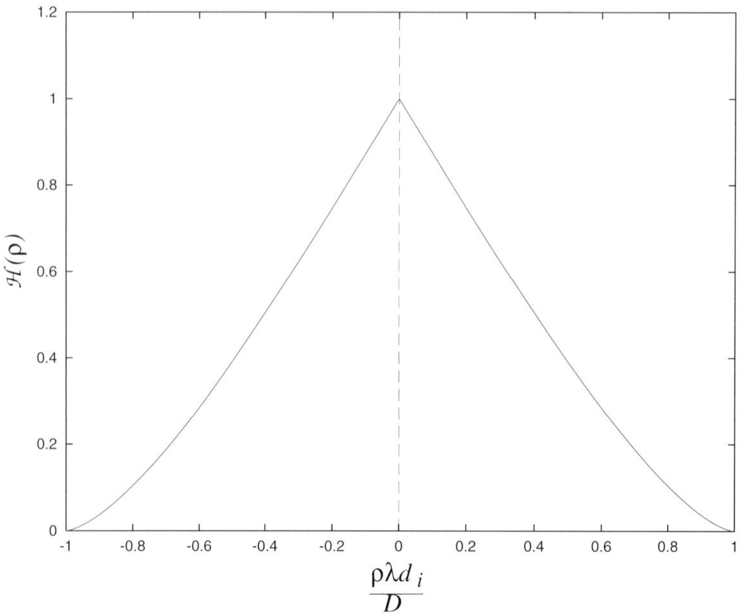

FIGURE 2.9
OTF $\mathcal{H}(\rho)$ plotted versus radial spatial frequency ρ for the case of a circular pupil.

related to the spatial structure of the source giving rise to the field. This dependence can be viewed as the means by which an imaging system is capable of producing an image. The imaging system is essentially decoding the source shape and structure from the coherence properties of the pupil plane optical field. In the case of imaging through the atmosphere, the randomness of the medium distorts the coherence properties in a way that makes it impossible for the imaging system to recover a diffraction limited image.

A complete understanding of the random degradations of an optical system requires that noise in the light detection process be properly treated. A fundamental source of noise in light detection is due to the random arrival times and locations of photoevents in a detector material. The physical origin of this noise is attributed to the quantum nature of light-matter interactions and is referred to interchangeably as shot noise, photon noise, and Poisson noise. Photon noise is a form of signal-dependent noise, requiring use of special analysis techniques. An additional source of noise in the light detection process is additive noise attributed to the readout electronics of some detectors. Additive noise is typically independent of the signal, allowing use of standard random process analysis techniques.

In the next subsection we present an introduction to the temporal and spatial coherence properties of optical fields. In Section 2.3.2 the semi-classical model for photoelectric light detection is presented.

2.3.1 Temporal and spatial coherence

The temporal and spatial coherence properties of an optical field can be mathematically expressed through correlation functions. Knowledge of these correlation functions tells all there is to know about the second order coherence properties of the optical field. The coherence properties are related to the degree to which an optical field interferes with itself and other fields. It is through this relationship to interference that we can experimentally investigate the coherence properties. The Michelson interferometer is an experiment that is capable of measuring the temporal coherence of an optical field, while the Young's

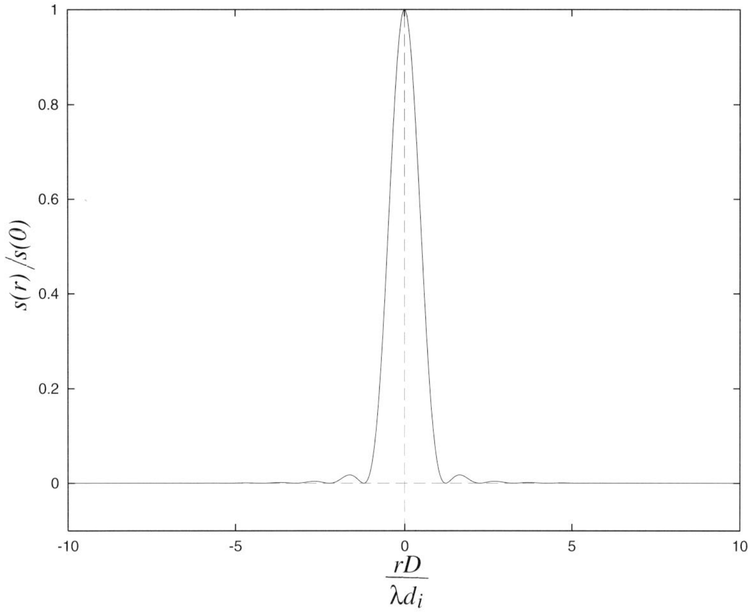

FIGURE 2.10
PSF $s(r)$ plotted versus radial coordinate r for the case of a circular pupil.

experiment is designed to measure the spatial coherence [3]. These two experiments will be described in the following subsections along with an introduction to the effect of propagation on the coherence of an optical field. We end this section with the introduction of an interferometric view of imaging that ties the spatial coherence of an optical field to the statistical performance characteristics of an imaging system. This interferometric view of imaging will be particularly useful in our discussion of atmospheric turbulence effects on imaging in Chapter 3.

Correlation functions

In the previous discussion of Fourier optics the notation $u(\vec{x})$ was used to designate the complex phasor representation of an optical field. By not explicitly showing a time dependence, monochromatic conditions were assumed. For the discussion to follow we relax this monochromatic condition by allowing the field to have a finite bandwidth. To reflect the finite bandwidth, $u(\vec{x}, t)$ is adopted as the complex representation of the optical field at position $\vec{x}$ and time t. The space-time correlation function for $u(\vec{x}, t)$ is defined mathematically in the following way:

$$\Gamma_u(\vec{x}, t; \vec{x}', t') = \mathrm{E}\{u(\vec{x}, t)u^*(\vec{x}', t')\}, \tag{2.51}$$

where the operator $\mathrm{E}\{\cdot\}$ is the ensemble average operator. Equation (2.51) is a second order, space-time correlation function that contains information about both the spatial and temporal coherence properties of the field. It is important to note that Γ_u is, in general, complex. Later we find it useful to write Γ_u in a polar representation (magnitude and phase):

$$\Gamma_u(\vec{x}, t; \vec{x}', t') = |\Gamma_u(\vec{x}, t; \vec{x}', t')| \exp(j2\pi\overline{\nu}(t - t') + j\psi(\vec{x}, t; \vec{x}', t')), \tag{2.52}$$

where $\overline{\nu}$ is the mean temporal frequency of $u(\vec{x}, t)$, and $\psi(\vec{x}, t; \vec{x}', t')$ is a phase term. In writing Eq. (2.52), we have anticipated the $2\pi\overline{\nu}(t - t')$ dependence in the exponent. This particular dependence will be shown later for wide sense stationary processes.

The expectation operator in Eq. (2.51) can also be interpreted as a time average. In the time average case, the operator $\mathrm{E}\{\cdot\}$ is usually replaced with the notation $\langle\cdot\rangle$. Equating the ensemble and time average is valid if the random process $u(\vec{x}, t)$ is ergodic. A random process is ergodic if any average calculated along a particular realization is equal to the same average calculated across the ensemble of realizations. In the case of a temporal random process, the average calculated along a particular realization is equivalent to a time average. Temporal ergodicity is a well accepted assumption for thermal optical sources. The time average interpretation is physically appealing since it can be closely related to the integration or exposure time of an optical detector. The assumption of ergodicity is discussed in more detail in reference [3].

The correlation function given in Eq. (2.51) is important because in the detection of the irradiance of an optical field statistical correlations play a major role in the outcome. As an example, consider an optical field that has been split into two identical beams. Assume the two beams are laterally shifted and temporally delayed from one another. Superimposing the beams and detecting the resulting irradiance forms an interference experiment. This type of interference arises in both Michelson's and Young's interferometers as well as many other interference related phenomena (such as imaging). The signal from the optical detector will be proportional to

$$\langle i(\vec{x}, t)\rangle = \langle |u(\vec{x}, t) + u(\vec{x} + \Delta\vec{x}, t + \tau)|^2\rangle, \tag{2.53}$$

where $i(\vec{x}, t)$ is the instantaneous irradiance and $\Delta\vec{x}$ and τ are the spatial and temporal shifts of the beams, respectively. The time average brackets are used since the detector effectively performs a time average over many temporal periods of the optical field. In general we will drop the angle brackets and the explicit time dependence on the left hand side of Eq. (2.53) and interpret $i(\vec{x})$ as the time average quantity. The interesting thing to notice from this experiment is that a number of second order correlation functions result from expanding Eq. (2.53):

$$\begin{aligned} i(\vec{x}) &= \langle |u(\vec{x}, t)|^2\rangle + \langle |u(\vec{x} + \Delta\vec{x}, t + \tau)|^2\rangle \\ &\quad + \langle u(\vec{x}, t) u^*(\vec{x} + \Delta\vec{x}, t + \tau)\rangle + \langle u^*(\vec{x}, t) u(\vec{x} + \Delta\vec{x}, t + \tau)\rangle. \end{aligned} \tag{2.54}$$

Substituting the definition of the correlation function found in Eq. (2.51) into Eq. (2.54) gives the detected irradiance explicitly as a function of space-time correlation functions:

$$\begin{aligned} i(\vec{x}) &= \Gamma_u(\vec{x}, t; \vec{x}, t) + \Gamma_u(\vec{x} + \Delta\vec{x}, t + \tau; \vec{x} + \Delta\vec{x}, t + \tau) \\ &\quad + \Gamma_u(\vec{x}, t; \vec{x} + \Delta\vec{x}, t + \tau) + \Gamma_u(\vec{x} + \Delta\vec{x}, t + \tau; \vec{x}, t). \end{aligned} \tag{2.55}$$

This irradiance expression shows that the detected signal is explicitly dependent on the second order coherence properties of the optical field. The situation under consideration may seem a bit contrived, but any optical problem involving light detection ultimately depends on the optical correlation functions as seen above. In particular, we will see later that imaging can be viewed entirely from an interferometric viewpoint.

It is useful to continue with the example above to illustrate how the detected irradiance $i(\vec{x})$ can be used to measure the coherence properties of the optical field. In fact the form of the irradiance given in Eq. (2.55) is closely related to the irradiance seen in two well known interferometers: the Michelson's and Young's interferometers. Both of these experiments are used to measure the coherence properties of an optical field. We first note that the self coherence terms, $\Gamma_u(\vec{x}, t; \vec{x}, t)$ and $\Gamma_u(\vec{x} + \Delta\vec{x}, t + \tau; \vec{x} + \Delta\vec{x}, t + \tau)$, are constant with respect to time. This fact stems from the assumption of temporal ergodicity. We also note that the self coherence terms represent the irradiance of the two beams before being superimposed onto the detector, $i_u(\vec{x})$ and $i_u(\vec{x} + \Delta\vec{x})$. Temporal ergodicity also implies that $u(\vec{x}, t)$ is stationary which allows us to write the correlation functions as solely a function of the difference of the time coordinates. In other words we can write

$$\Gamma_u(\vec{x}, t; \vec{x}', t') = \Gamma_u(\vec{x}, \vec{x}', t - t'). \tag{2.56}$$

Noting that the two cross terms of Eq. (2.55) are complex conjugates of each other and using the new facts given above, results in an interference signal of the following form:

$$i(\vec{x}) = i_u(\vec{x}) + i_u(\vec{x} + \Delta\vec{x}) + 2\mathrm{Re}\{\Gamma_u(\vec{x}, \vec{x} + \Delta\vec{x}, \tau)\}, \tag{2.57}$$

where the $\mathrm{Re}\{\cdot\}$ operator extracts the real part of the argument. In order to make any further progress, a mathematical form for the optical field $u(\vec{x}, t)$ must be assumed. The assumed form will not result in a loss of generality, but will aid in the interpretation of the irradiance expression given above. It is reasonable to assume the bandwidth of the optical field is much smaller than the mean temporal frequency. In the field of communications it is common to express narrowband signals with a complex phasor having a slowly varying complex amplitude or envelope. The phrase "slowly varying" is a reference to the rate of change of the complex amplitude of the phasor relative to the temporal rate of rotation of the phasor. Using this time varying phasor representation, $u(\vec{x}, t)$ is written in the following way

$$u(\vec{x}, t) = a(\vec{x}, t)e^{j2\pi\bar{\nu}t}, \tag{2.58}$$

where $a(\vec{x}, t)$ is the slowly varying complex amplitude. The relationship between the correlation function Γ_u and the correlation function associated with the complex amplitude $a(\vec{x}, t)$ is

$$\begin{aligned}
\Gamma_u(\vec{x}, \vec{x}', t - t') &= \mathrm{E}\{u(\vec{x}, t)u^*(\vec{x}', t')\}, \\
&= \mathrm{E}\{a(\vec{x}, t)e^{j2\pi\bar{\nu}t}a^*(\vec{x}', t')e^{-j2\pi\bar{\nu}t'}\}, \\
&= \mathrm{E}\{a(\vec{x}, t)a^*(\vec{x}', t')\}e^{j2\pi\bar{\nu}(t-t')}, \\
&= \Gamma_a(\vec{x}, \vec{x}', t - t')e^{j2\pi\bar{\nu}(t-t')},
\end{aligned} \tag{2.59}$$

where Γ_a is the correlation function of the complex envelope $a(\vec{x}, t)$. Substituting Eq. (2.59) into Eq. (2.57) while also using the polar representation of Γ_u given in Eq. (2.52) results in

$$\begin{aligned}
i(\vec{x}) &= i_u(\vec{x}) + i_u(\vec{x} + \Delta\vec{x}) \\
&\quad + 2|\Gamma_a(\vec{x}, \vec{x} + \Delta\vec{x}, \tau)| \cos(2\pi\bar{\nu}\tau + \psi(\vec{x}, \vec{x} + \Delta\vec{x}, \tau)).
\end{aligned} \tag{2.60}$$

From this expression, note that the coherence properties of the optical field are encoded in the measured irradiance $i(\vec{x})$. The first two terms of Eq. (2.60) encode the irradiance of the two optical beams, $i_u(\vec{x})$ and $i_u(\vec{x} + \Delta\vec{x})$. The last term encodes the space-time cross coherence properties of the field. The space-time cross coherence is modulated onto a sinusoidal signal of frequency $\bar{\nu}$. Both the amplitude and the phase of the sinusoid contain information about the complex coherence function Γ_a.

This simple experiment results in measured irradiance values that encode both the irradiance and cross coherence properties of the field. Understanding the coherence properties of light is important since any optical experiment ultimately involves detecting irradiance, and this irradiance is, in turn, inextricably related to the coherence of the field. For the case of incoherent imaging, the coherence of the field in the imaging system pupil plays the key role of determining performance. Before discussing the statistical aspects of imaging we first examine the information that can be obtained from an experiment like the one described above. First, consider the temporal coherence properties of the optical field by letting the spatial shift, $\Delta\vec{x}$, go to zero. Letting $\Delta\vec{x} = 0$ reduces Eq. (2.60) to the expression that describes the irradiance measured in a Michelson interferometer [3].

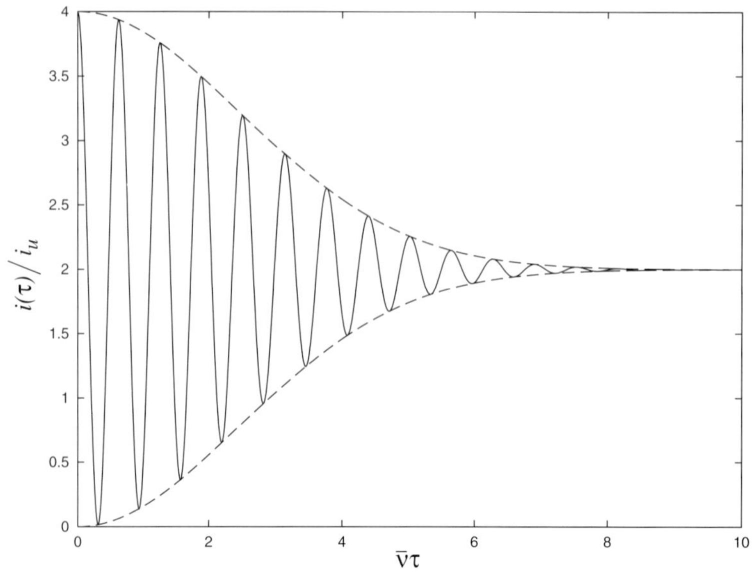

FIGURE 2.11
Representative interferogram from a Michelson's interferometer.

Temporal coherence experiment

In a Michelson interferometer the optical field is split into two beams. One of the beams is temporally delayed from the other. The two beams are superimposed on a detector and the resulting irradiance is measured. The measured irradiance is an interference phenomenon that is a function of the time delay, τ, between the two beams. Letting $\Delta\vec{x} = 0$ in equation (2.60) gives the expression that describes the irradiance in a Michelson interferometer:

$$i(\tau) = 2i_u + 2|\Gamma_a(\tau)|\cos(2\pi\bar{\nu}\tau + \psi(\tau)), \tag{2.61}$$

where we have, for convenience, dropped the dependence on $\vec{x}$, but have explicitly shown the dependence on the time delay τ. From this expression we see the measured irradiance is a constant plus an oscillating term. Fig. 2.11 illustrates a representative plot of the irradiance as a function of τ. This plot is called an interferogram. The oscillating part of the signal is called a fringe and the constant portion is twice the magnitude of the irradiance of each beam. The magnitude of the fringes are controlled by the absolute value of the correlation function, $|\Gamma_a(\tau)|$.

Consider two cases: large and small values of τ. In the case of a time delay large enough to cause the correlation function Γ_a to be nearly zero, the resulting interferogram is a constant irradiance having a value of twice the irradiance of the individual beams, as seen in the right hand side of Fig. 2.11. This case represents a situation in which the beams do not interfere with each other and fringes are not observed. If, on the other hand, τ is small enough that the optical beams are fully correlated, the interferogram will show an irradiance pattern that oscillates between zero and four times the irradiance of the individual beams. In this case the fringes are maximally visible since the contrast is maximum, as seen in the left hand side of Fig. 2.11.

The visibility of the fringes is the most commonly measured quantity in an experiment like this. As the name implies, the visibility tells us something about how visible the fringes are. The visibility is mathematically defined as follows:

$$\mathcal{V} = \frac{i_{max} - i_{min}}{i_{max} + i_{min}}, \tag{2.62}$$

where i_{max} and i_{min} are the maximum and minimum values of $i(\tau)$. In Fig. 2.11 the maximum and minimum irradiances are taken from the envelope shown by the dashed line. The visibility is useful since it is a quantitative measure of how well the two beams interfere. It is easy to see that for the zero correlation case discussed above (right hand side of Fig. 2.11), the visibility will also be zero. In the case of the fully correlated beams (left hand side of Fig. 2.11), it is also easy to see that the visibility will take on the maximum value of one.

The visibility is closely related to the correlation function, Γ_a. Solving Eq. (2.61) for the maximum and minimum values of the measured irradiance (by considering the maximum and minimum values of the cosine term) results in the following relationship between $\mathcal{V}$ and Γ_a:

$$\begin{aligned} \mathcal{V} &= \frac{4|\Gamma_a(\tau)|}{4\Gamma_a(0)}, \\ &= \frac{|\Gamma_a(\tau)|}{\Gamma_a(0)}, \\ &= |\gamma_a(\tau)|, \end{aligned} \tag{2.63}$$

where $\gamma_a(\tau)$ is the normalized temporal correlation coefficient of the complex envelope of the optical field. The magnitude of γ_a ranges between 0 and 1, and as shown in Eq. (2.63) is equal to the visibility, $\mathcal{V}$. Only when the two optical beams are of equal irradiance, as we have assumed here, does the simple relationship in Eq. (2.63) remain valid. Reference [3] presents a more general relationship between $\mathcal{V}$ and γ_a when the two beams have unequal irradiances.

The interference observed in a Michelson interferometer can range from no interference (zero visibility) to maximum interference (unity visibility). The characteristics of the two beams that control the degree to which they interfere have not been discussed. Since the optical field $u(\vec{x}, t)$ is wide sense stationary (WSS), the Wiener-Khinchin theorem [3] states that the temporal spectrum of $u(\vec{x}, t)$ is a Fourier transform of the temporal correlation function $\Gamma_a(\tau)$:

$$\mathcal{G}(\nu - \overline{\nu}) = \frac{1}{4}\mathcal{F}[\Gamma_a(\tau)], \tag{2.64}$$

where $\mathcal{G}(\nu)$ is the temporal power spectrum of the complex random process $u(\vec{x}, t)$ and ν is temporal frequency (Hz). Equation (2.64) is significant in that it makes the connection between the interferogram and the power spectrum of the optical field. By conducting an experiment using a Michelson interferometer, we can measure the temporal power spectrum of the optical source. The interferogram contains all that is required to fully characterize Γ_a, which in turn can be Fourier transformed to obtain $\mathcal{G}(\nu)$.

Equation (2.64) is also significant from the point of view that it qualitatively indicates what characteristics cause fields to interfere with delayed versions of themselves. A broad bandwidth optical field will have a narrow temporal correlation function. A narrow correlation function implies that the time delay τ must be small in order for Γ_a to have a non-zero value. In other words, the optical field will only interfere with itself for small time delays. In this case the optical source is said to have a short correlation time. If the optical field has a very narrow bandwidth, the temporal correlation function will be correspondingly wide. In this case, the time delay can be large while still providing high visibility interference fringes. This source is said to have a long correlation time.

As an example of the importance of temporal coherence, consider the post processing image reconstruction techniques discussed in Chapter 1. The temporal spectrum of the optical field determines the contrast of the irradiance peaks in the detected image. These irradiance peaks are referred to as

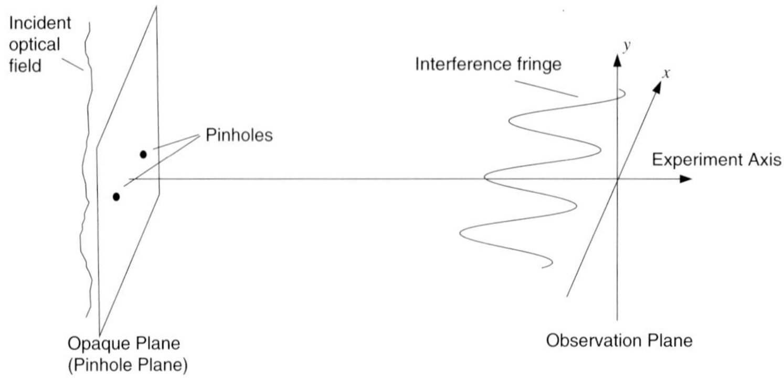

FIGURE 2.12
Geometry of Young's experiment.

speckle and they are formed by the interference of the optical fields arising from different points on the pupil. The interfering optical fields arriving at a point in the image plane have each incurred slightly different time delays in the propagation from the pupil plane. This time delay controls the degree to which the optical fields are able to interfere and thus form speckles. The goal in many of these image reconstruction techniques is to form high contrast (high visibility), short exposure speckle patterns. A high speckle contrast results in a better signal-to-noise ratio. In order to achieve these high contrast speckles, the bandwidth of the optical field is usually restricted to be as small as possible. The small bandwidth increases the temporal coherence, thereby enhancing the contrast of the speckles.

In the next section we consider the case in which the time delay, τ, is set to zero and only consider a spatial shift of the two optical beams. An experiment of this type is similar to Young's experiment in which the spatial coherence properties of the optical field are measured.

Spatial coherence experiment

In Young's experiment [3] an opaque plane contains two pin holes. An incident optical field on one side of the opaque plane causes two fields to arise on the other side of the plane. The geometry is shown in Fig. 2.12. An observation plane is placed some distance behind the pinhole plane and is used to measure the interference of the light from the two holes. In this experiment the two pinholes act as sampling points and the interference seen in the observation plane can be used to infer something about the coherence of the field sampled at the two pinholes. Figure 2.12 shows a cross-section of the two dimensional fringe pattern seen in the observation plane. Unfortunately, due to the geometry shown in Fig. 2.12, the interference pattern seen in the observation plane has a mix of both temporal and spatial coherence effects. The temporal effects arise for observation points away from the axis of the system due to the differing path lengths. The spatial coherence effects are also part of the measurement since the optical field is being sampled at two spatially distinct points. It is only on the axis that the visibility of the fringe pattern is solely a function of the spatial coherence of the optical field.

In this section we are only interested in spatial coherence. Consequently, we are not interested in presenting the detailed geometrical considerations inherent in Young's experiment. With this in mind, consider the irradiance expression given in Eq. (2.60) for the experiment introduced in the beginning of this section. By letting $\tau = 0$ in Eq. (2.60), the expression describes two beams that are spatially shifted from one another with a zero time delay. The detected irradiance reduces to

$$i(\Delta\vec{x}) = i_u(\vec{x}) + i_u(\vec{x}+\Delta\vec{x}) + 2|\Gamma_a(\vec{x},\vec{x}+\Delta\vec{x})|\cos(\psi(\vec{x},\vec{x}+\Delta\vec{x})), \tag{2.65}$$

where we have dropped the time delay dependence since it is zero and are explicitly showing the dependence on the spatial shift $\Delta\vec{x}$. The first two terms in Eq. (2.65) represent the irradiance of the optical field at the sample points $\vec{x}$ and $\vec{x}+\Delta\vec{x}$. The second term contains information about the cross-coherence of the two beams at the sample points $\vec{x}$ and $\vec{x}+\Delta\vec{x}$.

Whereas temporal coherence can be extracted in the Michelson interferometer, the spatial coherence properties can be obtained by extracting $\Gamma_a(\vec{x},\vec{x}+\Delta\vec{x})$ from the measured irradiance given in Eq. (2.65). The spatial coherence is measured for various sampling points by varying both $\vec{x}$ and $\Delta\vec{x}$. Varying these quantities is equivalent to moving the pinholes around in Young's experiment. Just as we saw for the temporal correlation function, the visibility of the measured fringe pattern is directly related to the spatial correlation function $\Gamma_a(\vec{x},\vec{x}+\Delta\vec{x})$. The relationship between visibility and Γ_a is established in reference [3] and is given by

$$\begin{aligned}\mathcal{V} &= \frac{|\Gamma_a(\vec{x},\vec{x}+\Delta\vec{x})|}{\sqrt{\Gamma_a(\vec{x},\vec{x})\Gamma_a(\vec{x}+\Delta\vec{x},\vec{x}+\Delta\vec{x})}},\\ &= |\gamma_a(\vec{x},\vec{x}+\Delta\vec{x})|,\end{aligned} \tag{2.66}$$

where we are assuming the beams have equal irradiance at the two sample points (i.e., $i_u(\vec{x}) = i_u(\vec{x}+\Delta\vec{x})$.)

In the case of temporal coherence, the motivation for measuring $\Gamma_a(\tau)$ was established by presenting the connection between $\Gamma_a(\tau)$ and the temporal power spectrum of the optical field. To establish a similar motivation for studying spatial coherence we turn to imaging. Specifically, an interferometric view of imaging is presented that explicitly makes use of the knowledge of the spatial coherence of the optical field in the imaging system pupil. This view of imaging is particularly useful for understanding how atmospheric turbulence degrades imaging performance. Before presenting this view of imaging, a fundamental spatial coherence propagation result is presented.

Van Cittert-Zernike theorem

Young's experiment can be used to measure the spatial coherence of an optical field sampled at two points by a pair of pinholes. The question of how the experiment will change if the pinhole locations are fixed, but the distance of the optical source is allowed to vary, is the subject of this section. In this section we show that the coherence properties change with propagation and present the basic theoretical result that describes how.

For the following discussion assume the optical field is quasimonochromatic. The quasimonochromatic assumption insures that the bandwidth of the optical field is finite, but small enough so that temporal coherence effects can be ignored. In other words, the temporal coherence length of the field is much longer than any temporal delays that might arise in the experiment. Consider the propagation geometry shown in Fig. 2.13. An optical field propagates from the source plane on the left to the observation plane on the right. By straightforward application of the Fresnel diffraction formula of Eq. (2.10), it is easy to show that the spatial coherence function $\Gamma_a(\vec{x}_1,\vec{x}_2)$ obeys the following propagation law:

$$\Gamma_a(\vec{x}_1,\vec{x}_2) = \frac{1}{\overline{\lambda}^2}\int d\vec{x}'\int d\vec{x}''\,\Gamma_a(\vec{x}',\vec{x}'')\,\frac{\exp\left[-j\frac{2\pi}{\overline{\lambda}}(\vec{r}_2-\vec{r}_2)\right]}{r_1 r_2}\cos(\theta_1)\cos(\theta_2), \tag{2.67}$$

where $\vec{r}_1$, $\vec{r}_2$, θ_1 and θ_2 are shown in Fig. 2.13, the integrations are carried out over the source plane coordinates, and the mean wavelength of the optical field is designated by $\overline{\lambda}$. Unfortunately, it is difficult

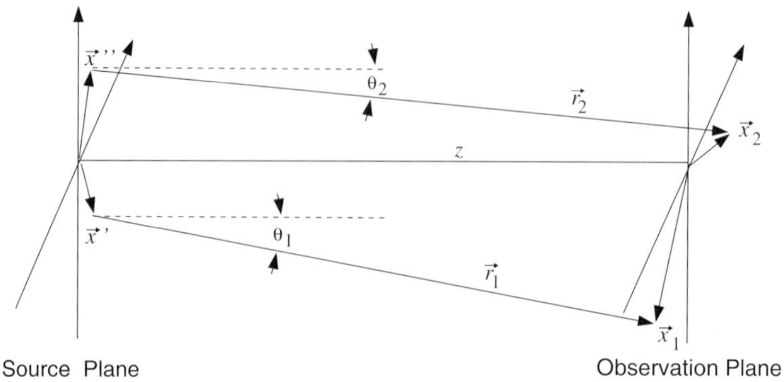

FIGURE 2.13
Geometry for derivation of the propagation law for the spatial correlation function Γ_a.

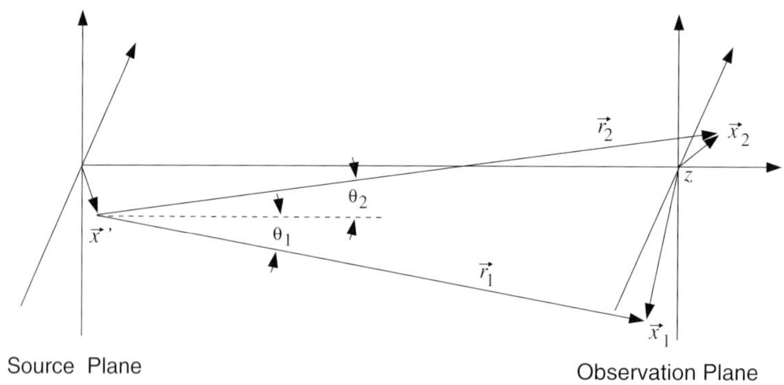

FIGURE 2.14
Geometry applicable for propagation of Γ_a for an incoherent source.

to gain any insight from Eq. (2.67). By considering a specific set of conditions, we can derive one of the most significant theorems associated with incoherent imaging.

Start by assuming the optical field in the source plane is spatially incoherent. The model for an incoherent field was presented in Eq. (2.38) and is repeated here:

$$\Gamma_a(\vec{x}', \vec{x}'') = \kappa i_s(\vec{x}')\delta(\vec{x}' - \vec{x}''), \tag{2.68}$$

where $i_s(\vec{x}')$ is the irradiance distribution of the optical field in the source plane. Substituting Eq. (2.68) into Eq. (2.67) and performing the integration over $\vec{x}''$ gives

$$\Gamma_a(\vec{x}_1, \vec{x}_2) = \frac{\kappa}{\overline{\lambda}^2} \int d\vec{x}'\, i_s(\vec{x}') \frac{\exp\left[-j\frac{2\pi}{\overline{\lambda}}(\vec{r}_2 - \vec{r}_2)\right]}{r_1 r_2} \cos(\theta_1)\cos(\theta_2). \tag{2.69}$$

The geometry in Fig. 2.13 must be modified to reflect the fact the vectors $\vec{r}_1$ and $\vec{r}_2$ originate from the same source point for Eq. (2.69). Figure 2.14 shows the new geometry. To simplify Eq. (2.69) further we need to make a number of assumptions. The assumptions are similar to the paraxial assumptions used to obtain the Fresnel diffraction formula in Section 2.2.1. First, the source and observation points are assumed to be sufficiently close to the z-axis so that r_1 and r_2 in the denominator can be approximated with z, where z is the perpendicular distance between the source and observation plane. Second, using the same assumption as presented above, the cosine terms can be replaced with their small argument approximations (i.e., unity). Finally, the binomial expansion of the exponential argument in Eq. (2.69) is used to replace r_1 and r_2 with the following approximations

$$r_1 \cong z + \frac{(x_1 - x')^2 + (y_1 - y')^2}{2z},$$
$$r_2 \cong z + \frac{(x_2 - x')^2 + (y_2 - y')^2}{2z}, \tag{2.70}$$

where x_1, y_1, x_2, and y_2 are the x- and y-directed components of observation coordinates $\vec{x}_1$ and $\vec{x}_2$, and x' and y' are the x- and y-directed components of the source coordinate $\vec{x}'$. Substituting Eq. (2.70) into Eq. (2.69) and invoking the assumptions presented above gives

$$\Gamma_a(\vec{x}_1, \vec{x}_2) = \frac{\kappa e^{-j\psi}}{(\bar{\lambda} z)^2} \int d\vec{x}'\, i_s(\vec{x}') \exp\left[-j\frac{2\pi}{\bar{\lambda} z}\vec{x}' \cdot (\vec{x}_2 - \vec{x}_1)\right],$$
$$= \frac{\kappa e^{-j\psi}}{(\bar{\lambda} z)^2} \mathcal{F}\left[i_s(\vec{x}')\right]\Big|_{\vec{f}=(\vec{x}_2-\vec{x}_1)/\bar{\lambda} z}, \tag{2.71}$$

where ψ is a quadratic phase term given by

$$\psi = \frac{\pi}{\bar{\lambda} z}(|\vec{x}_2|^2 - |\vec{x}_1|^2). \tag{2.72}$$

Equation (2.71) is known as the Van Cittert-Zernike theorem and is an extremely significant result for imaging of incoherent sources. The Van Cittert-Zernike theorem indicates that the spatial coherence properties of the optical field some distance away from the source are related by a Fourier transform to the irradiance distribution of the source. This result indicates that imaging can be accomplished by measuring the spatial coherence function $\Gamma_a(\vec{x}_1, \vec{x}_2)$ rather than forming a conventional image with a lens. The process of measuring $\Gamma_a(\vec{x}_1, \vec{x}_2)$ is called interferometry. As discussed earlier in this section, it is possible to measure $\Gamma_a(\vec{x}_1, \vec{x}_2)$ by performing an experiment like Young's experiment. In practice, experiments to measure Γ_a are more sophisticated, but the basic concept remains the same. Once $\Gamma_a(\vec{x}_1, \vec{x}_2)$ is measured, the image of the object can be constructed via an inversion of Eq. (2.71). The Van Cittert-Zernike theorem forms the basis of all incoherent, interferometric imaging, including astronomical radio and optical interferometry.

The most significant advantage of conducting interferometry over conventional imaging is the resolution advantage associated with interferometry. In conventional imaging, resolution is tied to the size of the system pupil. Beyond a few meters in diameter, telescopes become very difficult and expensive to manufacture. The resolution of interferometry is only limited by the distance between the two optical field sample points. In astronomical optical interferometry, the optical field is sampled with a pair of telescopes separated by a given distance, referred to as a baseline. The baseline between the two telescopes determines the angular resolution of the measurement, not the telescope diameters. A major disadvantage associated with interferometry is the requirement to completely sample all possible baselines, $\vec{x}_2 - \vec{x}_2$, for the inversion process. The tolerances involved in bringing the two sampled optical fields together to interfere on an optical detector also makes this type of imaging difficult [7].

In the next section we present an interferometric view of conventional imaging that makes use of the Van Cittert-Zernike theorem.

Interferometric view of imaging

When considering the Fourier transform of a two dimensional function, the transformation can be interpreted as a decomposition of the function into a set of two dimensional sinusoidal signals of varying amplitudes and phases. Each component of this decomposition has a complex magnitude with a corresponding spatial frequency $\vec{f}$. The complex magnitude specifies the amplitude and phase of the sinusoid. The direction of the spatial frequency vector $\vec{f}$ specifies the direction of the sinusoidal oscillation and the magnitude of $\vec{f}$ specifies the spatial rate of oscillation. We now shift our attention to the image $i(\vec{x})$ captured by an incoherent imaging system. Just as discussed above, the Fourier spectrum of $i(\vec{x})$ can be interpreted as a decomposition of the image into a set of sinusoidal signals. Any one of these sinusoidal signals can also be interpreted as the fringe observed in a Young's experiment [3]. In particular, the sinusoidal components of an image can be interpreted as arising from a pair of pinholes placed in the pupil plane of the imaging system (see Fig. 2.12). The separation of the pinhole pair is related to the spatial frequency of the observed fringe by the relationship $\Delta\vec{x} = \overline{\lambda} d_i \vec{f}$, where $\Delta\vec{x}$ is the vector separation of the pinholes and d_i is the distance between the pupil and image planes.

The concept just introduced indicates that pairs of points in the pupil of an imaging system can be interpreted as forming sinusoidal fringes in the imaging plane. The sinusoidal fringes arise because of a Young's-like phenomenon: the fields from the pair of pupil points interfere in the image plane and form a fringe pattern just as we see in Fig. 2.12. The fringes from all possible pairs of points in the pupil each give rise to a fringe which all add together to form the total image $i(\vec{x})$. The degree to which pairs of pupil plane points form fringes is controlled by the spatial coherence of the fields at each pair of points. These sinusoidal fringes are entirely analogous to the Fourier components in the Fourier decomposition discussed above. With this view of imaging we can see that imaging can be interpreted as an interferometric process. The lens of the imaging system is simply a means of causing the optical fields to interfere with the proper phase in the image plane.

Reference [3] presents a mathematical framework for the interferometric view of imaging that reinforces the concepts presented above. In particular, the following result for the Fourier spectrum of an image is presented:

$$I(\vec{f}) = \int \mathrm{d}\vec{x}\, W(\vec{x})W(\vec{x} - \overline{\lambda} d_i \vec{f})\Gamma_a(\vec{x}, \vec{x} - \overline{\lambda} d_i \vec{f}), \tag{2.73}$$

where $I(\vec{f})$ is the Fourier spectrum of the image $i(\vec{x})$, $W(\vec{x})$ is the pupil function of the imaging system, and $\Gamma_a(\vec{x}, \vec{x}')$ is the spatial correlation function of the optical field in the pupil. Equation (2.73) shows that the Fourier spectrum of the image is closely related to the spatial coherence of the optical field in the pupil. As discussed in the introduction to Section 2.3, the imaging system can be thought of as decoding the image from the spatial coherence of the pupil plane field. This decoding process is explained through Eq. (2.73). Equation (2.73) shows that the Fourier components of the image are a result of an infinite sum of interference effects from field points in the pupil separated by $\Delta\vec{x} = \overline{\lambda} d_i \vec{f}$. The integration operation in Eq. (2.73) essentially slides this pair of field sample points over all possible pairs in the pupil and sums each fringe contribution to obtain the total image Fourier component corresponding to spatial frequency $\vec{f}$.

Equation (2.73) is applicable for the case of imaging an object with arbitrary spatial correlation properties. In the case of an incoherent object, Eq. (2.73) simplifies to

$$I(\vec{f}) = \Gamma_a(\overline{\lambda} d_i \vec{f}) \int \mathrm{d}\vec{x} W(\vec{x})W(\vec{x} - \overline{\lambda} d_i \vec{f}), \tag{2.74}$$

where we have written Γ_a as a function of only the separation of the two points in the pupil. This simplification follows from the Van Cittert-Zernike theorem given in Eq. (2.71). The use of the lens in the imaging system cancels out the quadratic phase term ψ in this theorem which allows Γ_a to be written as only a function of the field point separations, $\Delta\vec{x} = \overline{\lambda} d_i \vec{f}$.

An interesting observation to make first about Eq. (2.74) is that the integral term on the right hand side of the equation is essentially the OTF of the incoherent imaging system. Another interesting fact is that all pairs of points separated by $\overline{\lambda} d_i \vec{f}$ contribute to a particular sinusoidal fringe in an "in phase" manner since the OTF portion of Eq. (2.74) is real and positive (for an unaberrated system). With these two facts in mind it is not surprising that the OTF takes the form that it does. OTF's are maximum at $\vec{f} = 0$ and taper away to zero at the cutoff frequency. Consider two pinhole separations, $\overline{\lambda d_i \vec{f_1}}$ and $\overline{\lambda d_i \vec{f_2}}$. These pinhole pairs are shifted over the aperture and each position of the pair results in a contribution to the image Fourier component corresponding to the spatial frequencies $\vec{f_1}$ and $\vec{f_2}$. In an integral sense, there will be a relatively larger number of pinhole pairs for the smaller separation than there will be for the larger separation. As a result, smaller spatial frequencies will be attenuated less than higher spatial frequencies. In the limit as the pinhole separation becomes as large as the system pupil diameter, there will be fewer and fewer pairs of pinholes that contribute to the spatial frequencies in the image plane. This limiting separation corresponds to the cutoff frequency of the pupil.

In the next chapter we examine the effects of a random phase and amplitude perturbation in the pupil plane of the system. These random perturbations affect how the fringes add together in the image plane and as a result determine how the overall image is degraded. It is this degradation and the characterization of the perturbations that are the subjects of the next chapter.

2.3.2 Semi-classical model for the photoelectric light detection process

Up to now we have treated the problem of detecting the image plane irradiance as if it were a deterministic process. However, the deterministic model for light detection is incomplete in that it fails to account for noise observed in an image when a finite light level is available for measuring the image. The physical origin of this noise is the quantized, random nature of photon-matter interactions. Random effects arising from photon-matter interactions are presented in this subsection in the context of the imaging problem. The ratio of the mean to the standard deviation of the measured image spectrum, referred to as the image spectrum signal-to-noise ratio (SNR), will be used as the metric for evaluating the effects of noise on the image spectrum. The image spectrum SNR is a measure of the relative strengths of the image spectrum signal and the random variability of the signal, providing a useful means of evaluating the effects of various sources of noise on the quality of the measurement. High SNR indicates good estimation of an image spectral component, while low SNR indicates poor estimation. The image spectrum SNR and other related SNRs are closely related to image quality and will be of fundamental interest to understanding the limits of performance of the imaging techniques discussed in Chapters 4–6. We shall see that the random nature of the light detection process imposes a limit to detection for imaging systems which is not apparent from the earlier treatment of wave propagation and image formation.

Analysis of the effects of noise on an imaging system in the spatial frequency domain is motivated by the expression for the OTF given in Eq. (2.44), and the expression for the generalized pupil function given in Eq. (2.45). These two expressions provide a linear relationship between the spatial-domain aberration, $\phi(\vec{x})$, and the spatial frequency-domain OTF. This key result ultimately allows knowledge of the spatial statistics of the turbulence-induced aberration to be used to calculate the statistics of the OTF. Additional motivation is provided by the spectral properties of the detected image model. As we shall see, this model consists of a set of Dirac delta functions coinciding with the locations of photon detection sites. These delta functions are highly localized and discontinuous in space, causing analytic difficulties. However, the Fourier transform of the detected image is continuous everywhere and can be easily analyzed. The primary focus of the later chapters is to derive the limits of estimating the Fourier spectrum components of images for various imaging techniques. Spatial frequency domain SNR expressions are required to understand the performance limits of these techniques. In this section the statistics of the image spectrum are analyzed and used to derive the image spectrum SNR.

Photon-matter interactions which occur in light detectors are fundamentally random. Specifically, the exact location and time of a photon absorption event in a detector cannot be predicted beforehand. This fundamental source of randomness gives rise to noise in imaging systems which is often called photon noise, or shot noise. Photon noise imposes limitations on the performance of imaging systems which are generally more severe than the limitations imposed by diffraction effects alone, particularly under low light levels. Photon noise is fundamental to all photoelectric detectors and is a form of signal dependent noise. That is, the strength of photon noise depends on the number and distribution of the detected photoevents. Special methods have been developed to treat the signal dependent nature of photon noise. These methods are presented in the following subsections. Photon-limited detectors, that is, detectors capable of recording the time and location of individual photoevents, can be closely approximated in practice. Other common detectors provide signals which are corrupted by additive, signal independent noise. Additive noise is treated here in a very general sense, allowing the physical origin of the noise to be ignored while providing a detailed analysis of its effects.

The image degradation caused by detecting a finite number of photoevents is illustrated in Figure 2.15, where negative images are shown for clarity. Figure 2.15a is a computer graphics rendering of a satellite which serves as the object in this example. Figure 2.15b shows the PSF of a diffraction-limited imaging system which was convolved with Fig. 2.15a to obtain the diffraction-limited image shown in Fig. 2.15c. Figure 2.15c contains degradations due only to the wave nature of light. The effects of a finite number of photoevents on the image are shown in Figs. 2.15d–2.15f, where there are approximately 10^6, 10^5, and 10^4 photoevents per image, respectively. Inspection of these figures clearly demonstrates the strong effect that photon noise has on image quality.

Photoelectric light detection occurs when a photon interacts with a detector to create an electrical signal which is then sensed electronically. Proper treatment of noise in the photoelectric light detection process requires that the discrete, or quantized nature of energy transfer in light-matter interactions be taken into account. We adopt the semi-classical model for photoelectric light detection in an imaging detector. The semi-classical model has the key characteristic that all propagation-related phenomena are treated with classical electro-magnetic fields techniques until the light-matter interaction occurs at the detector surface [3]. The required classical electro-magnetic fields results have been reviewed in the Fourier optics discussion of Section 2.2.

The semi-classical model rests on three assumptions about the statistical nature of the photoevents. These assumptions are

1. The probability of a single photoevent occurring in an area, $\mathrm{d}A$, which is small compared to the coherence area of the light, during a time interval, $\mathrm{d}t$, which is short compared to the coherence time of the light, is

$$P(1, \mathrm{d}t, \mathrm{d}A) = \eta \, \mathrm{d}t \, \mathrm{d}A \, i(\vec{x}, t), \tag{2.75}$$

   where $\vec{x}$ is a two-dimensional spatial coordinate on the detector surface, $i(\vec{x}, t)$ is the classical image irradiance, and η is the quantum efficiency defined below;

2. the probability of more than one photoevent occurring in $\mathrm{d}A$ during $\mathrm{d}t$ is negligibly small compared to the probability that zero or one photoevent occurs; and

3. the numbers of photoevents, K, occurring during non-overlapping space or time intervals are statistically independent.

These assumptions are sufficient to show that the random variable K obeys Poisson statistics [3]. Predictions related to the photoelectric detection of light based on the semi-classical theory have been confirmed repeatedly by experiment [8].

Light energy is detected by converting photons into electrical signals, generally by creating a free electron. Energy levels in light are quantized, with photons having energy, E, given by

$$E = \frac{hc}{\lambda} = h\nu, \tag{2.76}$$

where h is Plank's constant, $h = 6.626 \times 10^{-34}$ Joule-sec, c is the speed of light, λ is the wavelength of the light, and ν is the temporal frequency of the light. Some fraction, η, of the photons falling on the detector are absorbed by the detector material. The quantity η is referred to as the quantum efficiency of the detector, and is generally wavelength dependent. Photon absorption is accomplished by a change in the energy state of an electron in the detector material. In the case of photo-cathodes, this change in energy level is sufficient to free an electron from the surface of the detector, providing a signal which can

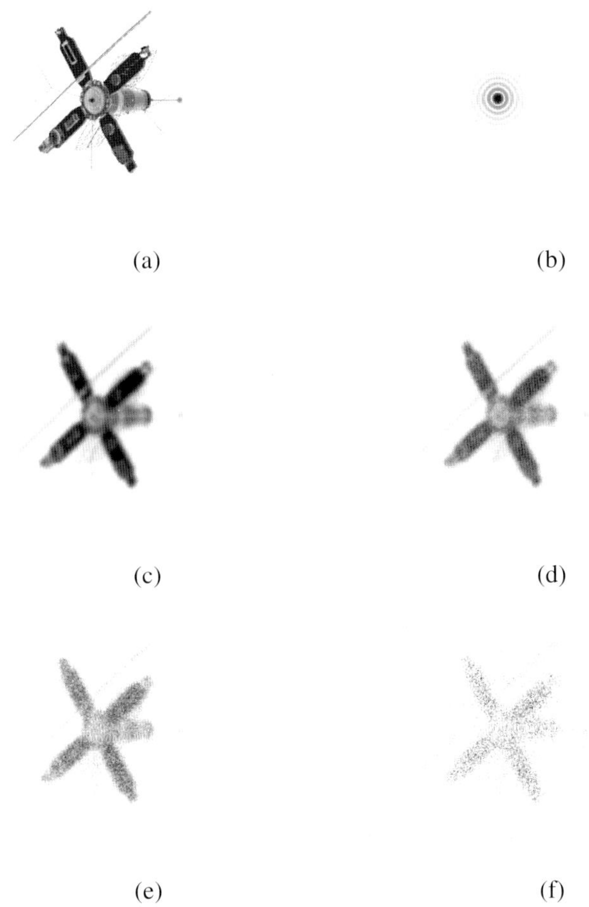

FIGURE 2.15
Simulated satellite images: (a) original object; (b) logarithm of PSF; (c) diffraction-limited image, infinite signal; (d) image with $K \approx 10^6$ photoevents; (e) image with $K \approx 10^5$ photoevents; (f) image with $K \approx 10^4$ photoevents. Negative images are displayed for clarity.

be subsequently amplified and measured by electrical means. In the case of semiconductor detectors, this change in energy level moves a valence electron into the conduction band of the semiconductor. Conduction band electrons drift in the presence of an electric field, creating a signal which can be amplified and measured by electrical means. In either case, the free electrons created are called photoelectrons, and the act of detecting a photon is referred to as a photoevent.

Photoelectric detectors convert an incident flux of photons into a discrete number, K, of photoelectrons during the measurement time, τ. The mean number of photoevents, $\overline{K}$, arising during an integration time, τ, in some area, $\mathrm{d}A$, of the detector is related to the classical irradiance, $i(t)$ (power/area), by

$$\overline{K} = \frac{\eta \mathrm{d}A}{h\nu} \int_{t}^{t+\tau} \mathrm{d}t\, i(t), \tag{2.77}$$

where t represents the independent time variable. The best possible detector of light allows for the direct measurement of K, and the location, $\vec{x}_k$, of each photoevent. Such detectors are referred to as photon-counting, or photon-limited detectors. As we shall show, even photon-limited detectors provide a noise-corrupted signal. The physical origins of this noise are the random arrival times and locations of photoevents in photoelectric detectors.

The random variable, K, strictly obeys Poisson statistics only for the case where the image irradiance is deterministic [9]. For example, photoevents in a detector caused by a fixed optical system forming an image of a deterministic object under constant illumination obey Poisson statistics. Photoevents arising from an image irradiance having random variations in space or time are not strictly Poisson random processes, but rather are doubly stochastic Poisson random processes [3]. Under these conditions the random variable, K, is not Poisson distributed, but is conditionally Poisson distributed, where the conditioning is upon the instantaneous irradiance distribution during the integration time, τ. Photoevents arising from an image formed with light which has passed through the atmosphere is one example of a doubly stochastic Poisson random process. The atmosphere causes the PSF to be random, thereby causing the image irradiance to be random. It is not, in general, possible to derive the probability density function for K when there are random fluctuations in the irradiance distribution. However, it is possible to obtain the moments of K under doubly stochastic conditions [10], and this will be sufficient for the SNR analyses which follow.

In the following subsections we review Poisson random processes. A mathematical model for the detected image is introduced, and this model is used to compute the image spectrum SNR of a deterministic imaging system forming an image of a fixed object. As previously mentioned, the image spectrum SNR provides a measure of the relative strengths of the signal, as characterized by its mean value, and the fluctuations of the signal due to noise, as characterized by the standard deviation of the signal. High values of the image spectrum SNR for some spectral component indicate good estimation of that spectral component, while low values of the image spectrum SNR indicate poor estimation of that spectral component. We then introduce the mathematical techniques used to analyze doubly stochastic Poisson random processes and establish the method for calculating the means and variances of quantities governed by doubly stochastic Poisson random processes required to obtain SNR expressions.

Poisson random processes and imaging

In this section we introduce the mathematical properties of Poisson random processes, present a model for the detected image, obtain the Fourier transform of the detected image, and compute the mean and variance of the spectral components of the detected image for the case of a deterministic OTF. A random process is formed from an underlying random experiment with outcomes in the set $\{Q\}$, and some measure of the probability of each possible outcome. The random process, $f(Q; x)$, with independent variable, x, is formed by associating a function with each outcome in $\{Q\}$, along with the measure of probability of the outcome [11]. The set of functions $\{f(Q; x)\}$ are called sample functions of the random process $f(Q; x)$. Often the dependence upon Q is dropped so that a sample function is represented simply by $f(x)$.

Poisson random processes have sample functions which consist of collections of Dirac delta functions with the following defining characteristics [3]:

1. the probability that K delta functions fall within the area A , $P(K; A)$, is given by

$$P(K; A) = \frac{1}{K!}\left(\int_A \mathrm{d}\vec{x}\, \lambda(\vec{x})\right)^K \exp\left\{-\int_A \mathrm{d}\vec{x}\, \lambda(\vec{x})\right\}, \tag{2.78}$$

   where $\lambda(\vec{x}) \geq 0$, is referred to as the rate function of the random process, the notation $\int_A \mathrm{d}\vec{x}$ represents integration over the values of $\vec{x}$ contained in A, and

2. the number of delta functions falling in non-overlapping areas are statistically independent.

The rate function has the physical interpretation that the mean number of delta functions occurring in an area $\mathrm{d}\vec{x}$ is given by $\lambda(\vec{x})\mathrm{d}\vec{x}$. The rate function is proportional to the classical image irradiance in an imaging system. The probability density function of the location of the nth event, $p(\vec{x}_n)$, is given by

$$p(\vec{x}_n) = \frac{\lambda(\vec{x}_n)}{\int_A \mathrm{d}\vec{x}\, \lambda(\vec{x})}. \tag{2.79}$$

The mean, $\overline{K}$, the second moment, $\overline{K^2}$, and the variance, σ_K^2, of the number of delta functions occurring in the area A are given by

$$\overline{K} = \int_A \mathrm{d}\vec{x}\, \lambda(\vec{x}), \tag{2.80}$$

$$\overline{K^2} = \overline{K} + \left(\overline{K}\right)^2, \tag{2.81}$$

$$\sigma_K^2 = \overline{K^2} - \left(\overline{K}\right)^2 = \overline{K}, \tag{2.82}$$

where the overline is used to indicate the statistical expectation operator. If the rate function, $\lambda(\vec{x})$, is deterministic then all randomness in K can be attributed to the stochastic fluctuations in the number of photoevents. An example where $\lambda(\vec{x})$ is constant arises when a fixed object under constant illumination is imaged onto a detector by an deterministic optical system.

We now apply the Poisson random process concepts presented above to the problem of computing the first and second moments of the Fourier spectrum of a photon-limited image. The detected photon-limited image is modeled as the sum of delta functions whose locations correspond to the locations of the photoevents contributing to the image. Mathematically, the detected image, $d(\vec{x})$, is defined

$$d(\vec{x}) = \sum_{n=1}^{K} \delta(\vec{x} - \vec{x}_n), \tag{2.83}$$

where $\vec{x}_n$ is the location of the nth photoevent in the image plane, and K is the total number of photoevents composing the image. The Fourier transform of $d(\vec{x})$, $D(\vec{f})$, is defined as

$$D(\vec{f}) = \int \mathrm{d}\vec{x}\, d(\vec{x}) \exp\{-j2\pi\vec{f}\cdot\vec{x}\}, \tag{2.84}$$

where the integration is over all space. Equation (2.84) can be evaluated as

$$\begin{aligned} D(\vec{f}) &= \int \mathrm{d}\vec{x} \sum_{n=1}^{K} \delta(\vec{x} - \vec{x}_n) \exp\{-j2\pi \vec{f} \cdot \vec{x}\} \\ &= \sum_{n=1}^{K} \exp\{-j2\pi \vec{f} \cdot \vec{x}_n\}. \end{aligned} \tag{2.85}$$

We now evaluate the first and second moments of $D(\vec{f})$ for the case of a fixed image irradiance, $i(\vec{x})$.

We shall require the following definitions. The nth statistical moment of a continuous random process $f(\vec{y})$ is defined as

$$E\{f^n(\vec{y})\} = \int \mathrm{d}\vec{y}\, f^n(\vec{y}) p_y(\vec{y}), \tag{2.86}$$

where $p_y(\vec{y})$ is the probability density function of the random variable $\vec{y}$, and the integration is taken over all space. The notation $E\{\bullet\}$ will be used interchangeably with the overbar to indicate the expected value of the quantity of interest. The nth moment of a discrete random process, $g(\vec{y}_m)$, is given by

$$E\{g^n(\vec{y}_m)\} = \sum_{m=-\infty}^{\infty} g^n(\vec{y}_m) P(\vec{y}_m), \tag{2.87}$$

where $\{\vec{y}_m\}$ is a discrete set of points, and $P(\vec{y}_m)$ is the probability of the occurrence of $\vec{y}_m$.

The random quantities in Eq. (2.85) are K, the total number of photoevents, and $\vec{x}_n$, the locations of the photoevents. Thus, to compute the first two moments of $D(\vec{f})$, expectations over the joint distribution of $\vec{x}_n$ and K must be computed. Using Bayes rule, the joint probability density function of $\vec{x}_n$ and K, $p(\vec{x}_n, K)$ may be written:

$$p(\vec{x}_n, K) = p(\vec{x}_n|K) P(K), \tag{2.88}$$

where $p(\vec{x}_n|K)$ is a conditional probability density function of $\vec{x}_n$ conditioned upon knowledge of K, and $P(K)$ is the probability that K photoevents occurred in the image.

The required expectation for the first moment of $D(\vec{f})$ may thus be written

$$\begin{aligned} E\{D(\vec{f})\} &= E\left\{\sum_{n=1}^{K} \exp\{-j2\pi \vec{f} \cdot \vec{x}_n\}\right\} \\ &= \sum_{K=0}^{\infty} \left\{\int \mathrm{d}\vec{x}_n \left(\sum_{n=1}^{K} \exp\{-j2\pi \vec{f} \cdot \vec{x}_n\}\right) p(\vec{x}_n|K)\right\} P(K). \end{aligned} \tag{2.89}$$

The inner integral is the expected value of $D(\vec{f})$ over all possible $\vec{x}_n$, conditioned upon K. We adopt the shorthand $E_{x|K}\{\bullet\}$ for this conditional expectation. The outer summation is the expected value of the inner expectation over all possible values of K. We adopt the shorthand $E_K\{\bullet\}$ for this expectation. Using this notation Eq. (2.89) can be written as:

$$E\{D(\vec{f})\} = E_K\left\{E_{x|K}\{D(\vec{f})\}\right\}. \tag{2.90}$$

The expectation of Eq. (2.90) is now evaluated. The inner expectation is

$$E_{x|K}\{D(\vec{f})\} = \int \mathrm{d}\vec{x}_n \left(\sum_{n=1}^{K} \exp\{-j2\pi \vec{f} \cdot \vec{x}_n\}\right) p(\vec{x}_n|K). \tag{2.91}$$

Since the rate function is deterministic, and hence, independent of K, Eq. (2.79) can be substituted for

$p(\vec{x}_n|K)$. Interchanging the orders of integration and summation yields

$$\mathrm{E}_{x|K}\{D(\vec{f})\} = \frac{\sum_{n=1}^{K} \int \mathrm{d}\vec{x}_n\, \lambda(\vec{x}_n) \exp\{-j2\pi \vec{f} \cdot \vec{x}_n\}}{\int \mathrm{d}\vec{x}_n\, \lambda(\vec{x}_n)} . \tag{2.92}$$

From Eq. (2.80) it is evident that the denominator of Eq. (2.92) is the mean number of photoevents in the image, $\overline{K}$. Further, the integral in the numerator of Eq. (2.92) is the Fourier transform of the rate function. Letting $\mathcal{F}\{\lambda(\vec{x}_n)\} = \Lambda(\vec{f})$, Eq. (2.92) becomes

$$\mathrm{E}_{x|K}\{D(\vec{f})\} = \sum_{n=1}^{K} \frac{\Lambda(\vec{f})}{\overline{K}} = K\Lambda_n(\vec{f}) , \tag{2.93}$$

where $\Lambda_n(\vec{f})$ is a normalized version of $\Lambda(\vec{f})$ defined as:

$$\Lambda_n(\vec{f}) = \frac{\Lambda(\vec{f})}{\overline{K}} . \tag{2.94}$$

Note that

$$\Lambda(\vec{0}) = \int \mathrm{d}\vec{x}_n\, \lambda(\vec{x}_n) = \overline{K}, \tag{2.95}$$

so that $\Lambda_n(\vec{0}) = 1$. The outer expectation in Eq. (2.90) is given by

$$\mathrm{E}_K\{\mathrm{E}_{x|K}\{D(\vec{f})\}\} = \mathrm{E}_K\{K\Lambda_n(\vec{f})\} = \overline{K}\Lambda_n(\vec{f}). \tag{2.96}$$

In an imaging system the relationship between $\Lambda_n(\vec{f})$ and the spectrum of the classical image irradiance, $I(\vec{f})$, is

$$\Lambda_n(\vec{f}) = \frac{I(\vec{f})}{I(\vec{0})} = \frac{\mathcal{H}(\vec{f})O(\vec{f})}{O(\vec{0})} , \tag{2.97}$$

where $O(\vec{f})$ is the spectrum of the object irradiance distribution. We define the normalized object spectrum, $O_n(\vec{f})$, as

$$O_n(\vec{f}) = \frac{O(\vec{f})}{O(\vec{0})} . \tag{2.98}$$

Normalizing $\Lambda(\vec{f})$ in this way allows effects due to the relative strengths of the spectral components of the classical image and effects due to the finite number of photoevents to be isolated in the analysis.

Using this normalization, the final form for $\mathrm{E}\{D(\vec{f})\}$ is

$$\mathrm{E}\{D(\vec{f})\} = \overline{K}\mathcal{H}(\vec{f})O_n(\vec{f}), \tag{2.99}$$

where the term $\overline{K}$ arises from the finite number of photoevents available for image formation, the term $\mathcal{H}(\vec{f})$ is due to the optics, and the term $O_n(\vec{f})$ is due the object. The dependence of $\mathrm{E}\{D(\vec{f})\}$ upon the object is fundamental, arising from the relationship in Eq. (2.97). The second moment of $D(\vec{f})$, $\mathrm{E}\{|D(\vec{f})|^2\}$, is now evaluated.

The second moment $\mathrm{E}\{|D(\vec{f})|^2\}$ is evaluated using the method established in Eqs. (2.90) to (2.96):

$$\mathrm{E}\{|D(\vec{f})|^2\} = \mathrm{E}_K\left\{\mathrm{E}_{x|K}\{|D(\vec{f})|^2\}\right\} . \tag{2.100}$$

The inner expectation is expanded as

$$\begin{aligned}
\mathrm{E}_{x|K}\{|D(\vec{f})|^2\} &= \mathrm{E}_{x|K}\left\{\left(\sum_{n=1}^{K}\exp\{-j2\pi\vec{f}\cdot\vec{x}_n\}\right)\left(\sum_{m=1}^{K}\exp\{+j2\pi\vec{f}\cdot\vec{x}_m\}\right)\right\} \\
&= \mathrm{E}_{x|K}\left\{\sum_{n=1}^{K}\sum_{m=1}^{K}\exp\{-j2\pi\vec{f}\cdot(\vec{x}_n-\vec{x}_m)\}\right\}. \qquad (2.101)
\end{aligned}$$

There are two types of terms in Eq. (2.101): 1) K terms for which $n = m$; and 2) $(K^2 - K)$ terms for which $n \neq m$. Terms for which $n = m$ are trivially evaluated, and contribute a total of K to the double summation in Eq. (2.101). Terms for which $n \neq m$ are evaluated using the statistical independence of photoevents falling in non-overlapping locations. Mathematically, if $n \neq m$, then the joint probability density function of $\vec{x}_n$ and $\vec{x}_m$ is written

$$p(\vec{x}_n, \vec{x}_m) = p(\vec{x}_n)p(\vec{x}_m). \qquad (2.102)$$

Thus, the inner expectation of a single term of Eq. (2.101) for which $n \neq m$ is written

$$\begin{aligned}
&\mathrm{E}_{x|K}\left\{\exp\{-j2\pi\vec{f}\cdot(\vec{x}_n-\vec{x}_m)\}\right\} \\
&= \mathrm{E}_{x|K}\left\{\exp\{-j2\pi\vec{f}\cdot\vec{x}_n\}\right\}\mathrm{E}_{x|K}\left\{\exp\{j2\pi\vec{f}\cdot\vec{x}_m\}\right\} \\
&= \frac{\int \mathrm{d}\vec{x}_n\,\lambda(\vec{x}_n)\exp\{-j2\pi\vec{f}\cdot\vec{x}_n\}}{\int \mathrm{d}\vec{x}_n\,\lambda(\vec{x}_n)} \times \frac{\int \mathrm{d}\vec{x}_m\,\lambda(\vec{x}_m)\exp\{j2\pi\vec{f}\cdot\vec{x}_m\}}{\int \mathrm{d}\vec{x}_m\,\lambda(\vec{x}_m)} \\
&= \frac{|\Lambda(\vec{f})|^2}{(\overline{K})^2} \\
&= |\Lambda_n(\vec{f})|^2 \\
&= |\mathcal{H}(\vec{f})O_n(\vec{f})|^2. \qquad (2.103)
\end{aligned}$$

Since there are $K^2 - K$ terms for which $n \neq m$, the total contribution of these terms to $\mathrm{E}_{x|K}\{|D(\vec{f})|^2\}$ is $(K^2 - K)$ times the contribution of one such term. The outer expectation for the second moment of $D(\vec{f})$ for the $n \neq m$ terms is

$$\begin{aligned}
\mathrm{E}_{n\neq m}\{|D(\vec{f})|^2\} &= \mathrm{E}_K\left\{(K^2-K)|\mathcal{H}(\vec{f})O_n(\vec{f})|^2\right\} \\
&= \mathrm{E}_K\{(K^2-K)\}|\mathcal{H}(\vec{f})O_n(\vec{f})|^2 \\
&= (\overline{K})^2|\mathcal{H}(\vec{f})O_n(\vec{f})|^2, \qquad (2.104)
\end{aligned}$$

where Eq. (2.81) was used to evaluate $\mathrm{E}_K\{K^2 - K\}$, and the subscript $n \neq m$ is used to indicate that this is the contribution to the expectation from the $n \neq m$ terms. The final form of the second moment of $D(\vec{f})$ is obtained by summing the contributions of the terms for which $n = m$ and $n \neq m$:

$$\mathrm{E}\{|D(\vec{f})|^2\} = \overline{K} + (\overline{K})^2|\mathcal{H}(\vec{f})O_n(\vec{f})|^2. \qquad (2.105)$$

In Eq. (2.105) the terms $\overline{K}$ arise from the finite number of photoevents, the term $\mathcal{H}(\vec{f})$ is due to the optics, and the term $O_n(\vec{f})$ is due to the object.

The variance of $D(\vec{f})$, $\mathrm{var}\{D(\vec{f})\}$, is obtained from its first and second moments by

$$\begin{aligned}
\mathrm{var}\{D(\vec{f})\} &= \mathrm{E}\{|D(\vec{f})|^2\} - |\mathrm{E}\{D(\vec{f})\}|^2 \\
&= \overline{K} + (\overline{K})^2|\mathcal{H}(\vec{f})O_n(\vec{f})|^2 - (\overline{K})^2|\mathcal{H}(\vec{f})O_n(\vec{f})|^2 \\
&= \overline{K}. \qquad (2.106)
\end{aligned}$$

Note that while the first and second moments of $D(\vec{f})$ depend upon $\mathcal{H}(\vec{f})$ and $O_n(\vec{f})$, the variance of $D(\vec{f})$ depends only upon $\overline{K}$ for the case of a deterministic OTF and a fixed object.

Spatial frequency domain signal-to-noise ratio of an image

In this subsection the results obtained for the mean and variance of the Fourier transform of a detected image are used to define and derive the signal-to-noise ratio (SNR) of $D(\vec{f})$. A model for additive noise is also introduced, and the effects of additive noise on the SNR of $D(\vec{f})$ are analyzed. The SNR of $D(\vec{f})$ is defined [12]

$$\mathrm{SNR}_D(\vec{f}) = \frac{|\mathrm{E}\{D(\vec{f})\}|}{[\mathrm{var}\{D(\vec{f})\}]^{1/2}}, \tag{2.107}$$

and is referred to as the image spectrum SNR. The image reconstruction techniques described in the subsequent chapters use spatial frequency domain methods to obtain estimates of object irradiance distributions from blurred and noisy images. The definition of SNR given in Eq. (2.107) is quite general and will be widely applied to other random quantities in succeeding chapters. We dwell on deriving the SNR of $D(\vec{f})$ in this section to both illustrate the analysis technique and to obtain a useful result. $\mathrm{SNR}_D(\vec{f})$ is now derived for the case of a fixed OTF and photon-limited detection.

Using the results obtained in Eqs. (2.99) and (2.106) the SNR of an image obtained with a deterministic OTF and a photon-limited image detection system is given by

$$\mathrm{SNR}_D(\vec{f}) = \overline{K}^{1/2} |\mathcal{H}(\vec{f}) O_n(\vec{f})|. \tag{2.108}$$

In Eq. (2.108) $\overline{K}^{1/2}$ accounts for the effects of a finite number of photoevents, $\mathcal{H}(\vec{f})$ accounts for the effects of the optics, and $O_n(\vec{f})$ accounts for the effects of the object. The following observations can be made regarding Eq. (2.108):

1. Increasing the mean number of photoevents always improves $\mathrm{SNR}_D(\vec{f})$.
2. $\mathrm{SNR}_D(\vec{f})$ goes to zero at a spatial frequency no higher than the diffraction-limited cutoff of the optics, since $\mathcal{H}(\vec{f})$ goes to zero there.
3. Aberrations, which generally broaden the point spread function of the optics and narrow $\mathcal{H}(\vec{f})$, have the effect of lowering $\mathrm{SNR}_D(\vec{f})$.
4. For irradiance images, $\mathrm{SNR}_D(\vec{f})$ takes its maximum value at $\vec{f} = 0$ and tends to decrease with increasing spatial frequency because the modulus of the OTF, $|\mathcal{H}(\vec{f})|$, and the modulus of the normalized object spectrum, $|O_n(\vec{f})|$, tend to decrease with increasing $|\vec{f}|$.

Example image spectrum SNRs are provided in Fig. 2.16, where the image spectrum SNRs for the images in Figs. 2.15d–2.15f are presented. The image spectrum SNRs presented in Fig. 2.16 are shown as a function of a single spatial frequency variable which is normalized to the diffraction-limited cutoff frequency of the telescope. To obtain these results two-dimensional image spectrum SNRs were computed using Eq. (2.108). The resulting values of $\mathrm{SNR}_D(\vec{f})$ lying on circles of constant radii were averaged and plotted. Note that the image spectrum SNR plots are consistent with the photoevent levels in Figs. 2.15d–2.15f. The image with 10^6 photoevents, Fig. 2.15d, has the highest image spectrum SNR, while the image with 10^4 photoevents, Fig. 2.15f, has the lowest image spectrum SNR. The image spectrum SNR plots are also consistent with a subjective assessment of the quality of the images in Figs. 2.15d–2.15f. The image corresponding to the highest image spectrum SNR, Fig. 2.15d, has the highest quality, while the image corresponding to the lowest image spectrum SNR, Fig. 2.15f, has the lowest quality. This relationship between image spectrum SNR and image quality has been previously noted [12], however, there is not a general theory relating the image spectrum SNR to any quantitative measure of image quality.

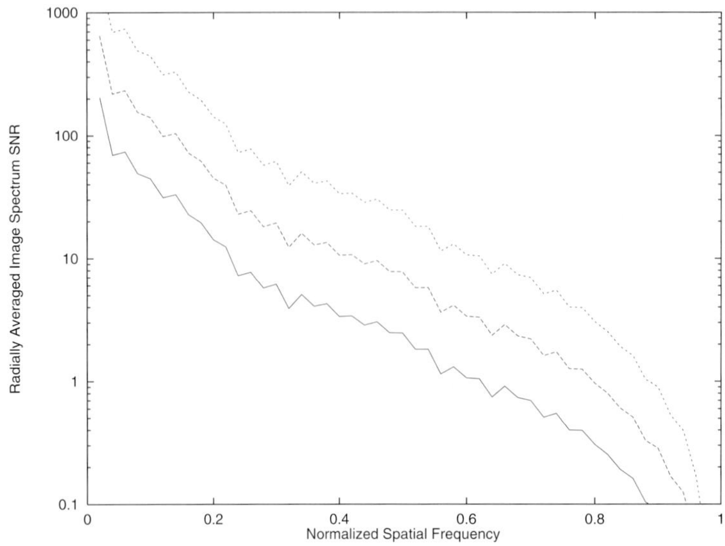

FIGURE 2.16
Radially averaged image spectrum SNRs for the images shown in Figs. (2.15d)–(2.15f). Top curve is for $\overline{K} = 10^6$, middle curve is for $\overline{K} = 10^5$, and bottom curve is for $\overline{K} = 10^4$.

Additive noise is present in many imaging detectors used in practice and has the effect of reducing the image spectrum SNR. Additive noise is distinct from photon noise in that additive noise is statistically independent of the signal, while photon noise is signal dependent. Further, the physical origin of additive noise is generally associated with the detector material and the readout electronics, while the source of photon noise is quantized nature of light-matter interactions. We now introduce the model for the detected image containing additive noise, and analyze the effects of this noise on $\text{SNR}_D(\vec{f})$. The model for a detected image with additive noise is given by

$$d(\vec{x}) = \sum_{n=1}^{K} \delta(\vec{x} - \vec{x}_n) + \sum_{p=1}^{P} n_p \delta(\vec{x} - \vec{x}_p), \tag{2.109}$$

where $\vec{x}_p$ is the location of the pth pixel in the image, n_p is a random variable representing the amount of additive noise at the pth pixel in units of photoevents, and P is the total number of pixels in the image. The first term in Eq. (2.109) describes the photon-limited image, and the second term describes the additive noise. The random variable, n_p, is assumed to have the following properties:

1. $\text{E}\{n_p\} = 0$: n_p is assumed to be zero mean,
2. $\text{E}\{n_p n_{p'}\} = 0$ if $p \neq p'$: additive noise is assumed to be spatially uncorrelated,
3. $\text{E}\{n_p n_{p'}\} = \sigma_n^2$ if $p = p'$: a single variance describes the additive noise at each pixel, with units of photoelectrons squared per pixel, and
4. n_p is statistically independent of K and $\vec{x}_n$.

For example, the output of widely used charge-coupled device (CCD) detectors is corrupted with additive noise commonly referred to as read noise. Read noise in CCDs is specified in root mean square (RMS) electrons per pixel per read out which is exactly the standard deviation of n_p, σ_n. The effect of additive noise on the mean and variance of the Fourier transform of $d(\vec{x})$ is now analyzed.

The Fourier transform of the detected image can be evaluated using Eq. (2.85), and is given by:

$$D(\vec{f}) = \sum_{n=1}^{K} \exp\{-j2\pi\vec{f}\cdot\vec{x}_n\} + \sum_{p=1}^{P} n_p \exp\{-j2\pi\vec{f}\cdot\vec{x}_p\}. \quad (2.110)$$

We can now evaluate the mean and variance of the image spectrum in the presence of additive noise to obtain a new expression for the image spectrum SNR for images corrupted by additive noise. Since n_p is zero-mean, the expected value of $D(\vec{f})$ is given by

$$\begin{aligned} \mathrm{E}\{D(\vec{f})\} &= \mathrm{E}\left\{\sum_{n=1}^{K} \exp\{-j2\pi\vec{f}\cdot\vec{x}_n\} + \sum_{p=1}^{P} n_p \exp\{-j2\pi\vec{f}\cdot\vec{x}_p\}\right\} \\ &= \overline{K}\mathcal{H}(\vec{f})O_n(\vec{f}), \end{aligned} \quad (2.111)$$

where the mean of $D(\vec{f})$ in the presence of additive is seen to be equal to the photon-limited case of Eq. (2.99). The additive noise terms in $D_n(\vec{f})$ are zero mean, uncorrelated, and statistically independent of the photon-limited detection term. Hence, additive noise increases the second moment of $D(\vec{f})$ by the amount

$$\mathrm{E}\left\{\left|\sum_{p=1}^{P} n_p \exp\{-j2\pi\vec{f}\cdot\vec{x}_p\}\right|^2\right\} = P\sigma_n^2. \quad (2.112)$$

The variance of $D(\vec{f})$ is obtained with reference to Eq. (2.106) and is given by

$$\mathrm{var}\{D(\vec{f})\} = \overline{K} + P\sigma_n^2. \quad (2.113)$$

Finally, the image spectrum SNR of an image corrupted with additive noise is given by

$$\mathrm{SNR}_D(\vec{f}) = \frac{\overline{K}|\mathcal{H}(\vec{f})O_n(\vec{f})|}{(\overline{K} + P\sigma_n^2)^{1/2}}. \quad (2.114)$$

Eq. (2.114) shows that additive noise always reduces the image spectrum SNR, however, this SNR-reducing effect can, in principle, be overcome by increasing $\overline{K}$. In the limits of no read noise, $\sigma_n^2 = 0$ in Eq. (2.114), or infinite signal, $\overline{K} \rightarrow \infty$, the photon-limited noise image spectrum SNR is obtained. Finally, Eq. (2.114) indicates that the minimum number of pixels, P, should be used to maximize the image spectrum SNR when additive noise is present.

Doubly stochastic Poisson random processes

Up to now we have discussed only the effects on the detected image spectrum of randomness arising from the random locations of photoevents, the random number of photoevents, and additive noise in the image measurement. An additional source of randomness arises from atmospheric turbulence – specifically, the rate function, $\lambda(\vec{x})$, is random due to atmospheric turbulence effects. As a consequence of this random aberration, the point spread function, and hence the OTF, of the imaging system are random, causing the classical image irradiance to be random. Poisson random processes which have random rate functions are referred to as doubly stochastic Poisson random processes. Detailed treatment of the statistics of the OTF is deferred to Chapters 3 and 5. In this section we analyze the effects of a random OTF on the mean, second moment, and variance of $D(\vec{f})$.

Let the OTF of an imaging system, $\mathcal{H}(\vec{f})$, be random. The first and second moments of $\mathcal{H}(\vec{f})$ are dependent on $\vec{f}$, and are represented by:

1. the mean of $\mathcal{H}(\vec{f})$ is denoted $\overline{\mathcal{H}}(\vec{f})$,
2. the second moment of $\mathcal{H}(\vec{f})$ is denoted $\overline{|\mathcal{H}(\vec{f})|^2}$, and
3. the variance of $\mathcal{H}(\vec{f})$ is denoted $\text{var}\{\mathcal{H}(\vec{f})\}$.

To evaluate the mean of $D(\vec{f})$, the method of Eq. (2.90) is extended to account for the randomness of $\mathcal{H}(\vec{f})$:

$$\mathrm{E}\{D(\vec{f})\} = \mathrm{E}_{\mathcal{H}}\left\{\mathrm{E}_{K|\mathcal{H}}\left\{\mathrm{E}_{x|K,\mathcal{H}}\{D(\vec{f})\}\right\}\right\}. \tag{2.115}$$

The inner expectation is interpreted as the expectation over the photoevent locations, $\vec{x}_n$, conditioned upon K and a specific realization of $\mathcal{H}(\vec{f})$. The middle expectation is over the number of photoevents, K, conditioned upon $\mathcal{H}(\vec{f})$. Finally, the outer expectation is the expectation over all possible realizations of $\mathcal{H}(\vec{f})$, or equivalently, over all corresponding rate functions, $\lambda(\vec{x})$. The inner expectation is expanded as

$$\begin{aligned}
\mathrm{E}_{x|K,\mathcal{H}}\{D(\vec{f})\} &= \mathrm{E}_{x|K,\mathcal{H}}\left\{\sum_{n=1}^{K}\exp\{-j2\pi\vec{f}\cdot\vec{x}_n\}\right\} \\
&= \int \mathrm{d}\vec{x}_n\left(\sum_{n=1}^{K}\exp\{-j2\pi\vec{f}\cdot\vec{x}_n\}\right)p(\vec{x}_n|K,\mathcal{H}) \\
&= \frac{\sum_{n=1}^{K}\int \mathrm{d}\vec{x}_n\,\lambda(\vec{x}_n)\exp\{-j2\pi\vec{f}\cdot\vec{x}_n\}}{\int \mathrm{d}\vec{x}_n\,\lambda(\vec{x}_n)} \\
&= K\Lambda_n(\vec{f}) \\
&= K\mathcal{H}(\vec{f})O_n(\vec{f}),
\end{aligned} \tag{2.116}$$

where, in writing the conditional probability density function of x_n, we have used Eq. (2.79) since the conditioning is upon a specific rate function. The middle expectation is evaluated as:

$$\begin{aligned}
\mathrm{E}_{K|\mathcal{H}}\left\{\mathrm{E}_{x|K,\mathcal{H}}\{D(\vec{f})\}\right\} &= \mathrm{E}_{K|\mathcal{H}}\{K\mathcal{H}(\vec{f})O_n(\vec{f})\} \\
&= \overline{K}\mathcal{H}(\vec{f})O_n(\vec{f}).
\end{aligned} \tag{2.117}$$

The outer expectation is given by

$$\begin{aligned}
\mathrm{E}_{\mathcal{H}}\left\{\mathrm{E}_{K|\mathcal{H}}\left\{\mathrm{E}_{x|K,\mathcal{H}}\{D(\vec{f})\}\right\}\right\} &= \mathrm{E}_{\mathcal{H}}\{\overline{K}\mathcal{H}(\vec{f})O_n(\vec{f})\} \\
&= \overline{K}\,\overline{\mathcal{H}}(\vec{f})O_n(\vec{f}),
\end{aligned} \tag{2.118}$$

which is the final form of the mean of $D(\vec{f})$ for the case of a random OTF.

The second moment of $D(\vec{f})$ is also evaluated using the method of Eq. (2.115):

$$\begin{aligned}
&\mathrm{E}\{|D(\vec{f})|^2\} = \mathrm{E}_{\mathcal{H}}\left\{\mathrm{E}_{K|\mathcal{H}}\left\{\mathrm{E}_{x|K,\mathcal{H}}\{|D(\vec{f})|^2\}\right\}\right\} \\
&= \mathrm{E}_{\mathcal{H}}\left\{\mathrm{E}_{K|\mathcal{H}}\left\{\mathrm{E}_{x|K,\mathcal{H}}\left\{\left(\sum_{n=1}^{K}\exp\{-j2\pi\vec{f}\cdot\vec{x}_n\}\right)\times\left(\sum_{m=1}^{K}\exp\{+j2\pi\vec{f}\cdot\vec{x}_m\}\right)\right\}\right\}\right\} \\
&= \mathrm{E}_{\mathcal{H}}\left\{\mathrm{E}_{K|\mathcal{H}}\left\{\mathrm{E}_{x|K,\mathcal{H}}\left\{\sum_{n=1}^{K}\sum_{m=1}^{K}\exp\{-j2\pi\vec{f}\cdot(\vec{x}_n-\vec{x}_m)\}\right\}\right\}\right\}.
\end{aligned} \tag{2.119}$$

As in the discussion of Eq. (2.101), the double summation in the inner expectation has two types of terms: 1) K terms for which $n = m$; and 2) $K^2 - K$ terms for which $n \neq m$. Terms for which $n = m$ are evaluated trivially and contribute a total of K to the double summation. A single term in the inner expectation for which $n \neq m$ is

$$
\begin{aligned}
&\mathrm{E}_{x|K,\mathcal{H}}\left\{\exp\{-j2\pi\vec{f}\cdot(\vec{x}_n - \vec{x}_m)\}\right\} = \\
&= \frac{\int \mathrm{d}\vec{x}_n\, \lambda(\vec{x}_n)\exp\{-j2\pi\vec{f}\cdot\vec{x}_n\}}{\int \mathrm{d}\vec{x}_n\, \lambda(\vec{x}_n)} \times \frac{\int \mathrm{d}\vec{x}_m\, \lambda(\vec{x}_m)\exp\{j2\pi\vec{f}\cdot\vec{x}_m\}}{\int \mathrm{d}\vec{x}_m\, \lambda(\vec{x}_m)} \\
&= |\Lambda_n(\vec{f})|^2 \\
&= |\mathcal{H}(\vec{f})O_n(\vec{f})|^2,
\end{aligned}
\tag{2.120}
$$

where the statistical independence of non-overlapping photoevents given in Eq. (2.102) has been used. Since there are $K^2 - K$ terms for which $n \neq m$, the inner expectation yields

$$
\mathrm{E}_{x|K,\mathcal{H}}\left\{|D(\vec{f})|^2\right\} = K + (K^2 - K)|\mathcal{H}(\vec{f})O_n(\vec{f})|^2\,. \tag{2.121}
$$

The middle expectation is evaluated as

$$
\begin{aligned}
\mathrm{E}_{K|\mathcal{H}}\left\{\mathrm{E}_{x|K,\mathcal{H}}\{|D(\vec{f})|^2\}\right\} &= \mathrm{E}_{K|\mathcal{H}}\left\{K + (K^2 - K)|\mathcal{H}(\vec{f})O_n(\vec{f})|^2\right\} \\
&= \overline{K} + (\overline{K^2} - \overline{K})|\mathcal{H}(\vec{f})O_n(\vec{f})|^2 \\
&= \overline{K} + (\overline{K})^2|\mathcal{H}(\vec{f})O_n(\vec{f})|^2\,.
\end{aligned}
\tag{2.122}
$$

Finally, the outer expectation is given by

$$
\begin{aligned}
\mathrm{E}_{\mathcal{H}}\left\{\mathrm{E}_{K|\mathcal{H}}\left\{\mathrm{E}_{x|K,\mathcal{H}}\{|D(\vec{f})|^2\}\right\}\right\} &= \mathrm{E}_{\mathcal{H}}\left\{\overline{K} + (\overline{K})^2|\mathcal{H}(\vec{f})O_n(\vec{f})|^2\right\} \\
&= \overline{K} + (\overline{K})^2\mathrm{E}_{\mathcal{H}}\left\{|\mathcal{H}(\vec{f})|^2\right\}|O_n(\vec{f})|^2 \\
&= \overline{K} + (\overline{K})^2\overline{|\mathcal{H}(\vec{f})|^2}|O_n(\vec{f})|^2\,.
\end{aligned}
\tag{2.123}
$$

The variance of $D(\vec{f})$ is determined from the first and second moments of $D(\vec{f})$:

$$
\begin{aligned}
\mathrm{var}\{D(\vec{f})\} &= \mathrm{E}\{|D(\vec{f})\}|^2\} - |\mathrm{E}\{D(\vec{f})\}|^2 \\
&= \overline{K} + (\overline{K})^2|O_n(\vec{f})|^2\mathrm{var}\{\mathcal{H}(\vec{f})\}\,.
\end{aligned}
\tag{2.124}
$$

The first term in Eq. (2.124) is due to Poisson noise, and the second term in Eq. (2.124) is due to randomness in the OTF. These effects combine to increase the total variance of $D(\vec{f})$ compared to the fixed OTF case, as expected. In the limit of a deterministic OTF, $\mathrm{var}\{\mathcal{H}(\vec{f})\} = 0$, and the fixed OTF result of Eq. (2.108) is obtained.

The image spectrum SNR for an image measured under conditions of a doubly stochastic Poisson random process is derived from the definition of the image spectrum SNR in Eq. (2.107), and Eqs. (2.118) and (2.124)

$$
\mathrm{SNR}_D(\vec{f}) = \frac{\overline{K}|\overline{\mathcal{H}(\vec{f})}||O_n(\vec{f})|}{\sqrt{\overline{K} + (\overline{K})^2|O_n(\vec{f})|^2\mathrm{var}\{\mathcal{H}(\vec{f})\}}} \tag{2.125}
$$

If additive noise is present in the image measurement, as defined in Eq. (2.109) , then the variance of $D(\vec{f})$ is increased by the amount $P\sigma_n^2$, where P is the number of pixels in the image, and σ_n^2 is the additive noise variance in photoelectrons per pixel per readout. The resulting expression for the image spectrum SNR for random OTF and additive noise is

$$\mathrm{SNR}_D(\vec{f}) = \frac{\overline{K}|\overline{\mathcal{H}}(\vec{f})||O_n(\vec{f})|}{\sqrt{\overline{K} + (\overline{K})^2|O_n(\vec{f})|^2\mathrm{var}\{\mathcal{H}(\vec{f})\} + P\sigma_n^2}} . \qquad (2.126)$$

Two observations are in order regarding Eq. (2.126):

1. The effect of random fluctuations in the OTF lowers the SNR of the image spectrum.
2. Increasing the average number of photoevents in the image cannot overcome the effects of randomness in the OTF. In fact, the $\overline{K} \to \infty$ limit of Eq. (2.126) is:

$$\lim_{\overline{K}\to\infty} \{\mathrm{SNR}_D(\vec{f})\} = \frac{|\overline{\mathcal{H}}(\vec{f})|}{\sqrt{\mathrm{var}\{\mathcal{H}(\vec{f})\}}} = \mathrm{SNR}_{\mathcal{H}}(\vec{f}) , \qquad (2.127)$$

   where $\mathrm{SNR}_{\mathcal{H}}(\vec{f})$ is referred to as the SNR of the random OTF.

The analysis of the image spectrum SNR provided here has yielded useful results and served to establish the analysis techniques which will be applied in subsequent chapters. However, our discussion of SNRs is by no means complete. Further analyses of SNRs for the various imaging techniques are deferred to later chapters.

Bibliography

[1] J. W. Goodman, *Introduction to Fourier Optics*. New York: McGraw-Hill Book Co., 1968.

[2] J. D. Gaskill, *Linear Systems, Fourier Transforms, and Optics*. New York: John Wiley & Sons, 1978.

[3] J. W. Goodman, *Statistical Optics*. New York: John Wiley & Sons, 1985.

[4] M. Born and E. Wolf, *Principles of Optics*. New York: Pergamon Press, 1964.

[5] B. E. A. Saleh and M. C. Teich, *Fundamentals of Photonics*. New York: John Wiley & Sons, 1991.

[6] B. R. Frieden, *Probability, Statistical Optics, and Data Testing*. New York: Springer-Verlag, 1991.

[7] A. C. S. Readhead, T. S. Nakajima, T. J. Pearson, G. N. amd J. B. Oke, and W. L. W. Sargent, "Diffraction-limited imaging with ground-based optical telescopes," *Astron. J.*, vol. 95, pp. 1278–1296, 1988.

[8] L. Mandel, "The case for and against semiclassical radiation theory," in *Progress in Optics* (E. Wolf, ed.), vol. XIII, Amsterdam: North-Holland, 1976.

[9] R. H. Kingston, *Detection of Optical and Infrared Radiation*. Berlin: Springer-Verlag, 1978.

[10] L. Mandel, "Fluctuations of photon beams: the distribution of the photo-electrons," *Proc. Phys. Soc., London*, vol. 74, pp. 233–243, 1958.

[11] A. Papoulis, *Probability, Random Variables, and Stochastic Processes*. New York: McGraw-Hill, third ed., 1991.

[12] P. S. Idell and A. Webster, "Resolution limits for coherent optical imaging: signal-to-noise analysis in the spatial-frequency domain," *J. Opt. Soc. Am. A*, vol. 9, pp. 43–56, 1992.

3

Turbulence Effects on Imaging Systems

3.1 Introduction

Atmospheric turbulence affects, to some degree, all optical systems which use light that has propagated through long atmospheric paths. For example, the smallest angle which can be resolved by ground-based astronomical telescopes operating at visible wavelengths is much larger than predicted by diffraction alone. In the long time average, as shown in Fig. 1.1, turbulence causes the point spread function of an imaging system to be very broad and smooth. In very short exposures turbulence causes the instantaneous point spread function to be broad, but the irradiance pattern is also modulated, giving the image a speckled appearance. These effects limit our ability to measure high angular resolution information about space objects, providing motivation for efforts to overcome these effects.

Atmospheric turbulence affects imaging systems by causing spatially and temporally random fluctuations in the index of refraction of the atmosphere. Index of refraction fluctuations result in random spatial and temporal variations in the optical path length (OPL) of the atmosphere through the continuum of altitudes, h. These OPL variations are denoted $n_1(\vec{x}, h)\Delta h$, where $n_1(\vec{x}, h)$ represents the fluctuation of the index of refraction about the mean value at some spatial location; Δh is the thickness of a thin layer of the atmosphere; and $\vec{x}$ is a two dimensional vector location in the plane of the turbulence layer. Propagation of an optical wavefront through a thin layer of the atmosphere changes the phase of the wave by an amount $\psi(\vec{x}, h) = kn_1(\vec{x}, h)\Delta h$, where $k = 2\pi/\lambda$ is the wavenumber of light with wavelength, λ. Though all atmospheric turbulence-induced wavefront aberrations begin as phase aberrations, propagation of the aberrated wavefront can, over long distances, cause amplitude variations in the wavefront. The field perturbations in the pupil of the telescope are the result of the phase aberration contributed by each layer of the atmosphere and the propagation from layer-to-layer. This field is typically modeled using [1]

$$u(\vec{x}) = W(\vec{x}) \exp\{\chi(\vec{x}) + j\psi(\vec{x})\}, \tag{3.1}$$

where $W(\vec{x})$ is the telescope aperture weighting function, $\psi(\vec{x})$ accounts for the phase fluctuations of the wave, and $\chi(\vec{x})$ accounts for the logarithm of the amplitude fluctuations.

Turbulence-induced wavefront variations across the aperture of a telescope distort the point spread function of an imaging system from its ideal diffraction-limited shape. This distortion corrupts images and reduces the resolution of imaging systems. Unlike the fixed aberrations discussed in Section 2.2.6, turbulence-induced wavefront aberrations fluctuate randomly from point-to-point and from instant-to-instant, causing the point spread function of an imaging system to be random. Statistical models and methods are required to fully understand these effects. In this chapter we study the statistical models for turbulence and wavefront propagation through the turbulent atmosphere, derive the first-order effects of turbulence on imaging systems, and present two methods for representing turbulence effects on a wavefront using modal expansion techniques.

Because of the pervasive nature of the atmospheric turbulence problem, modeling turbulence effects has received a great deal of attention. The body of literature dealing with optical propagation through turbulence is vast. However, the majority of the published literature has its roots in the seminal efforts of Kolmogorov [2], Tatarskii [3], and Fried [1, 4]. Kolmogorov developed a statistical model for the spatial structure of turbulent air flows [2]. Tatarskii applied Kolmogorov's results to the problems of wave propagation and imaging through random index of refraction distributions [3]. Fried has extended Tatarskii's results to a multitude of atmospheric optics problems and made the theory more accessible to the optics community [1, 4]. An extensive body of experimental evidence supporting the validity of the Kolmogorov theory exists and is reviewed in Ref. [5].

This chapter is organized as follows. In Section 3.2 the physical origin of the index of refraction fluctuations is presented, and spectral models used to describe turbulence effects are discussed. Models for characterizing the strength of the index of refraction fluctuations are also presented. In Section 3.3 the statistics of homogeneous, isotropic turbulence are presented. The effects of random index of refraction fluctuations on a propagating wave are presented in Section 3.4. First-order effects of turbulence on imaging systems are discussed in Section 3.5. In Section 3.6 modal expansions for turbulence-induced phase aberrations are presented. In the final section, Section 3.7, we present techniques for generating phase screens to simulate the effects of atmospheric turbulence on imaging systems.

3.2 Index of refraction fluctuations in the atmosphere

Index of refraction fluctuations in the atmosphere originate with turbulent air motion. The source of energy for this air motion is the differential heating and cooling of the surface of the Earth caused by sunlight and the diurnal cycle, resulting in large scale inhomogeneities in the temperature of the air. These large scale inhomogeneities are broken into smaller scale inhomogeneities by air movement we experience as wind. As shown below, these air motions nearly always have complex structure and are unpredictable: the characteristics of turbulent air motion. Turbulent air motion, in turn, gives rise to randomly distributed pockets of air, each having a characteristic temperature. The index of refraction of air is quite sensitive to temperature, and thus the index of refraction of the atmosphere is random.

The statistics of the spatial structure of turbulence are of critical importance to modeling the optical effects of atmospheric turbulence. The present understanding of the spatial structure of turbulence arises from the study of fluid motion. If the average velocity, v_{avg}, of a viscous fluid moving with characteristic size, l, is gradually increased from very slow to very fast, two distinct states of fluid motion are observed. At very low v_{avg} the fluid motion is found to be laminar, that is, smooth and regular. However, as v_{avg} is increased past some critical value, the flow becomes unstable and random. This random fluid motion is called turbulence [3, 6]. The critical average velocity bounding the region between laminar and turbulent flow is a function of v_{avg}, l, and the kinematic viscosity of the fluid, k_v. These quantities are related through the non-dimensional Reynolds number, Re:

$$\mathrm{Re} = \frac{v_{avg} l}{k_v}. \tag{3.2}$$

When Re exceeds some critical value, which depends on the geometry of the flow, the fluid motion becomes unstable. The kinematic viscosity of air is $k_v = 1.5 \times 10^{-5} \mathrm{m}^2/\mathrm{s}$, and if we assume a scale size of $l = 10$ m and a velocity of $v_{avg} = 1$ m/s, a Reynolds number of $\mathrm{Re} = 6.7 \times 10^5$ is obtained. This Reynolds number is sufficiently high to ensure that atmospheric air flow is nearly always turbulent [6].

Kolmogorov suggested that the kinetic energy in large spatial scale motions of a turbulent flow is transferred to smaller scale motions [2]. He also assumed that the small scale turbulent motions are both homogeneous and isotropic. The term homogeneous is analogous to stationarity and implies the statistical characteristics of the turbulent flow are independent of position within the flow field.

The term isotropic is a stronger assumption that implies homogeneity in addition to requiring that the second and higher order statistical moments only have a dependence on the radial distance between any two points in the field. As the scale size, l, of the air flow becomes smaller, the Reynolds number associated with that region of air flow is reduced. When the Reynolds number falls below some critical value, turbulent motion stops, and the remaining kinetic energy is dissipated as heat. Kolmogorov's theory provides a mathematical form for the spatial frequency domain statistics of the index of refraction variations in the atmosphere [7].

The physical source of the index of refraction variations in the atmosphere is temperature inhomogeneities arising from turbulent air motion. The index of refraction of the atmosphere, $n(\vec{r},t)$, is modeled as the sum of a mean index of refraction, n_0, and a randomly fluctuating term, $n_1(\vec{r},t)$:

$$n(\vec{r},t) = n_0 + n_1(\vec{r},t) \, , \tag{3.3}$$

where $\vec{r}$ is a three dimensional vector position, and t is time. The mean index of refraction of air is $n_0 = 1$. At optical wavelengths, the dependence of the index of refraction of air upon pressure and temperature is given by [6]

$$n_1 = n - 1 = \frac{77.6P}{T} \times 10^{-6}, \tag{3.4}$$

where T is the temperature of the air in Kelvins, and P is the pressure of the air in millibars. Temperature fluctuations are the dominant source of index of refraction variations, as can be seen by differentiating Eq. (3.4) with respect to temperature:

$$\frac{dn_1}{dT} = -\frac{77.6P}{T^2} \times 10^{-6} \, . \tag{3.5}$$

Pockets of air which have uniform index of refraction are generally referred to as *turbulent eddies*. The statistical distribution of the size and number of these turbulent eddies is characterized by the spatial power spectral density (PSD) of $n_1(\vec{r})$, denoted $\Phi_n(\vec{\kappa})$. The independent variable $\vec{\kappa}$ is the spatial wavenumber vector with orthogonal components $(\kappa_x, \kappa_y, \kappa_z)$. $\Phi_n(\vec{\kappa})$ may be considered to be a measure of the relative abundances of turbulent eddies with scale sizes $l_x = 2\pi/\kappa_x$, $l_y = 2\pi/\kappa_y$, and $l_z = 2\pi/\kappa_z$. Under conditions where homogeneous, isotropic turbulence may be assumed, the power spectral density of the index of refraction fluctuations is a function of the scalar wavenumber,

$$\kappa = \sqrt{\kappa_x^2 + \kappa_y^2 + \kappa_z^2}, \tag{3.6}$$

which is related to the isotropic scale size by $l = 2\pi/\kappa$.

In the Kolmogorov theory the mathematical form of $\Phi_n(\vec{\kappa})$ has three regimes. For very small κ, $\kappa < 2\pi/L_0$, corresponding to large scale sizes, Kolmogorov's theory does not predict a mathematical form for $\Phi_n(\vec{\kappa})$. The quantity L_0 is referred to as the *outer scale* of the turbulence and represents the characteristic dimension of the largest turbulent eddies which break up following the Kolmogorov theory. In this regime, $\Phi_n(\vec{\kappa})$ depends upon local geographical and meteorological conditions [8]. Inhomogeneities with scale sizes greater than L_0 are not believed to be homogeneous. However, the smaller scale index of refraction inhomogeneities affecting optical systems originate with these large scale inhomogeneities. Near the ground L_0 is on the order of the height above the ground, and in the free atmosphere L_0 is believed to be between several meters and a few tens of meters [8].

For large κ, $\kappa > 2\pi/l_0$, $\Phi_n(\vec{\kappa})$ is also not predicted by the Kolmogorov theory. The quantity l_0 is referred to as the *inner scale* of the turbulence and represents the characteristic dimension of the smallest turbulent eddies. In this regime of scale sizes the turbulent eddies disappear, and the remaining energy in the fluid motion is dissipated as heat. $\Phi_n(\vec{\kappa})$ is known to be quite small in this regime [8]. The inner scale, l_0, is on the order of a few millimeters.

When κ is in the range, $2\pi/L_0 \leq \kappa \leq 2\pi/l_0$, the form of $\Phi_n(\vec{\kappa})$ is predicted by the Kolmogorov theory [6]. Within this range of wavenumbers the Kolmogorov theory predicts that $\Phi_n(\vec{\kappa})$ takes the form:

$$\Phi_n^K(\vec{\kappa}) = 0.033 C_n^2 \kappa^{-11/3} \, , \tag{3.7}$$

where the superscript K indicates the Kolmogorov spectrum. The quantity C_n^2 is called the *structure constant* of the index of refraction fluctuations with units of $\mathrm{m}^{-2/3}$. C_n^2 characterizes the strength of the index of refraction fluctuations. Noting that Eq. (3.7) is a function of the scalar magnitude of $\vec{\kappa}$ implies that the turbulent eddies are isotropic and homogeneous. However, as already noted, eddies with scale size greater than L_0 are not believed to be homogeneous. Thus, care must be exercised in interpreting the behavior of $\Phi_n^K(\vec{\kappa})$ in the $\kappa < 2\pi/L_0$ regime. Henceforth, we will only discuss homogeneous, isotropic turbulence statistics, and as such, the scalar wavenumber, κ, will replace the vector wavenumber, $\vec{\kappa}$, in the argument of Φ_n.

Mathematical problems arise when Eq. (3.7) is used to model the spectrum of the index of refraction fluctuations in the $\kappa \to 0$ regime because of the non-integrable pole at $\kappa = 0$. To overcome this difficulty an alternate form for $\Phi_n(\kappa)$, referred to as the von Karman spectrum, has been used [6]. The von Karman spectrum is given by

$$\Phi_n^V(\kappa) = \frac{0.033C_n^2}{(\kappa^2 + \kappa_0^2)^{11/6}} \exp\left\{-\frac{\kappa^2}{\kappa_m^2}\right\}, \tag{3.8}$$

where $\kappa_0 = 2\pi/L_0$, $\kappa_m = 5.92/l_0$, and the superscript V indicates the von Karman spectrum. A non-zero value of κ_0 results in a finite value of $\Phi_n^V(0)$ while a non-zero κ_m has the effect of rapidly rolling off the spectrum for $\kappa > \kappa_m$. It should be noted that even though Eq. (3.8) is finite for all κ, its form for wavenumbers in the range $\kappa < 2\pi/L_0$ should be considered approximate. Eq. (3.8) also contains the explicit assumption of homogeneous index of refraction fluctuations in the range of scale sizes greater than L_0. However, as noted earlier, turbulent eddies in the range of scale sizes greater than L_0 are not believed to be homogeneous.

In many systems of practical interest the strength of the turbulence is a function of distance from the aperture, z. This is particularly true for astronomical observing conditions where it is necessary to view space objects through a vertical column of air. The notation $C_n^2(z)$ is used to indicate the dependence of the turbulence strength upon distance, and the general notation $\Phi_n(\kappa, z)$ is used to indicate the distance-dependent PSD. Under these conditions the Kolmogorov spectrum may be written

$$\Phi_n^K(\kappa, z) = 0.033C_n^2(z)\kappa^{-11/3}, \tag{3.9}$$

and the von Karman spectrum may be written

$$\Phi_n^V(\kappa, z) = \frac{0.033C_n^2(z)}{(\kappa^2 + \kappa_0^2)^{11/6}} \exp\left\{-\frac{\kappa^2}{\kappa_m^2}\right\}. \tag{3.10}$$

Both the Kolmogorov and the von Karman spectra may be written as the product of a term depending only on distance, $f(z)$, and a normalized term depending only on wavenumber, $\Phi_0^{K,V}(\kappa)$:

$$\Phi_n^{K,V}(\kappa, z) = f(z)\,\Phi_0^{K,V}(\kappa). \tag{3.11}$$

The terms in Eq. (3.11) are given by

$$f(z) = 0.033C_n^2(z), \tag{3.12}$$

$$\Phi_0^K(\kappa) = \kappa^{-11/3}, \tag{3.13}$$

for the Kolmogorov spectrum, and

$$\Phi_0^V(\kappa) = \frac{1}{(\kappa^2 + \kappa_0^2)^{11/6}} \exp\left\{-\frac{\kappa^2}{\kappa_m^2}\right\}, \tag{3.14}$$

for the von Karman spectrum.

The shapes of the normalized Kolmogorov and von Karman spectra are shown in Fig. 3.1. In Fig. 3.1 $\Phi_0^K(\kappa)$ and $\Phi_0^V(\kappa)$ are plotted on the same scale. For the von Karman spectra curves, the inner scale is $l_0 = 5$ mm, and the outer scale is varied with $L_0 = 2, 5,$ and 10 m. The roles of the inner and

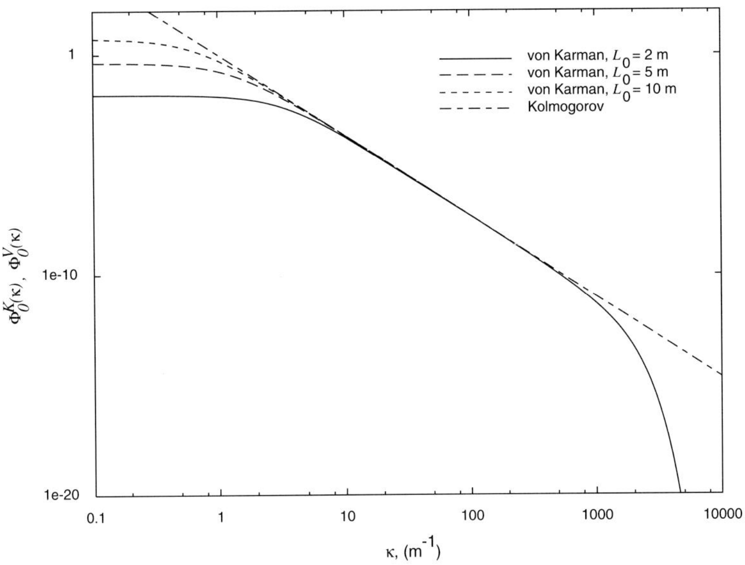

FIGURE 3.1
Normalized Kolmogorov and von Karman turbulence spectra, $\Phi_0^K(\kappa)$ and $\Phi_0^V(\kappa)$. For the von Karman spectra the inner scale is $l_0 = 5$ mm, and the outer scale is varied, with $L_0 = 2\text{m},\ 5\text{m},\ 10\text{m}$.

outer scales in the von Karman spectrum are also shown in Fig. 3.1. The effect of the inner scale is to reduce the value of $\Phi_0^V(\kappa)$ compared to $\Phi_0^K(\kappa)$ for wavenumbers in the $\kappa > 2\pi/l_0$ regime. This is consistent with the physical observation that turbulent air motion ceases once scale sizes on the order of l_0 have been attained. The effect of the outer scale is to reduce the the value of $\Phi_0^V(\kappa)$ compared to $\Phi_0^K(\kappa)$ for wavenumbers in the $\kappa < 2\pi/L_0$ regime. As L_0 increases, the amount of power in the lower wavenumber components of $\Phi_0^V(\kappa)$ increases, and at very low wavenumbers $\Phi_0^V(\kappa) \to \Phi_0^K(\kappa)$ as $L_0 \to \infty$. However, care must be used in choosing L_0 for use in the von Karman spectrum. Experimental evidence indicates that L_0 is at least several meters in the free atmosphere [5, 9, 10, 11]. The exact value of L_0 depends on local conditions and geographical location.

The important role of the turbulence strength, as characterized by C_n^2, in predicting the optical effects of turbulence has motivated the extensive study of C_n^2. C_n^2 has been found to vary as functions of altitude, location, and time of day. Several mathematical models have been developed from experimental measurements of C_n^2. Some of the commonly used models are presented here. These profiles are plotted as a function of altitude, h, in Fig. 3.2. If the telescope is looking straight up, or in the so-called zenith direction, the distance from the telescope aperture, z, and the altitude, h, are identical. However, if the direction of view makes an angle, θ_z, with the zenith direction, then z and h are related by

$$z = \frac{h}{\cos\theta_z}. \tag{3.15}$$

Daytime C_n^2 distributions as a function of altitude have been modeled using the Hufnagel-Valley and the Submarine Laser Communication-Day (SLC-Day) profiles. The Hufnagel-Valley turbulence profile is given by

$$\begin{aligned} C_n^2(h) = \ & 5.94 \times 10^{-53}\,(v/27)^2\, h^{10} \exp\{-h/1000\} \\ & + 2.7 \times 10^{-16} \exp\{-h/1500\} + A \exp\{-h/100\}\,, \end{aligned} \tag{3.16}$$

where A and v are free parameters [12]. The parameter A sets the strength of the turbulence near ground

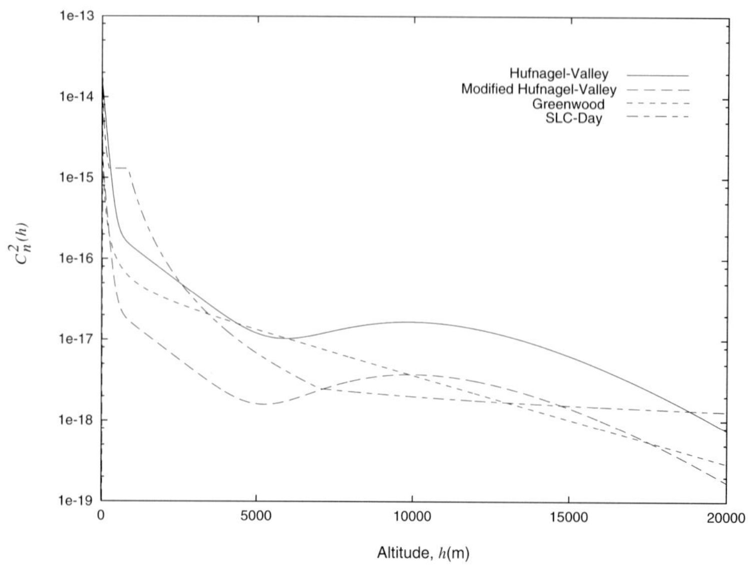

FIGURE 3.2
C_n^2 profiles.

level. A commonly used value for the A parameter is $A = 1.7 \times 10^{-14}\ \mathrm{m}^{-2/3}$. The parameter v (m/s) represents the high altitude wind speed and can be varied to adjust the high altitude turbulence strength. A commonly used value for v is $v = 21$ m/s. The SLC–Day model is given by [12]

$$C_n^2(h) = \begin{cases} 0 & 0\ \mathrm{m} < h < 19\ \mathrm{m} \\ 4.008 \times 10^{-13}\, h^{-1.054} & 19\ \mathrm{m} < h < 230\ \mathrm{m} \\ 1.300 \times 10^{-15} & 230\ \mathrm{m} < h < 850\ \mathrm{m} \\ 6.352 \times 10^{-7}\, h^{-2.966} & 850\ \mathrm{m} < h < 7000\ \mathrm{m} \\ 6.209 \times 10^{-16}\, h^{-0.6229} & 7000\ \mathrm{m} < h < 20,000\ \mathrm{m} \end{cases} \tag{3.17}$$

Nighttime C_n^2 distributions have been modeled using a modified version of the Hufnagel-Valley profile and the Greenwood profile. The modified Hufnagel-Valley profile is given by [12]

$$\begin{aligned} C_n^2(h) = {} & 8.16 \times 10^{-54}\, h^{10} \exp\{-h/1000\} + 3.02 \times 10^{-17} \exp\{-h/1500\} \\ & +1.90 \times 10^{-15} \exp\{-h/100\}. \end{aligned} \tag{3.18}$$

The Greenwood model is given by [13]

$$C_n^2(h) = [2.2 \times 10^{-13}\, (h + 10)^{-1.3} + 4.3 \times 10^{-17}] \times \exp\{-h/4000\}. \tag{3.19}$$

3.3 Statistics of index of refraction fluctuations

Thus far we have presented only the spectral statistics of the turbulence-induced index of refraction fluctuations. These spectral representations have associated spatial representations in the form of spatial correlation functions and spatial structure functions. The spectral and spatial representations are related through the Fourier transform. These relationships are presented in this section and used in the succeeding sections to investigate wave propagation and imaging through turbulence. We will consider the second order statistics of the random three dimensional index of refraction fluctuations, $n_1(\vec{r})$.

Two definitions relating to special cases of three dimensional random processes will be required. Let a three dimensional random process be given by $X(\vec{r})$, where $\vec{r}$ is a three dimensional position vector. The random process $X(\vec{r})$ is said to be *homogeneous* if it is stationary. Mathematically, homogeneity implies that the autocorrelation of $X(\vec{r})$,

$$\Gamma_X(\vec{r}_1, \vec{r}_2) = E\{X(\vec{r}_1)X(\vec{r}_2)\}\,, \tag{3.20}$$

is a function only of the separation, $\vec{r}_1 - \vec{r}_2$, and as a result takes the simpler form:

$$\Gamma_X(\vec{r}) = E\{X(\vec{r}_1)X(\vec{r}_1 - \vec{r})\}\,. \tag{3.21}$$

If the autocorrelation of the random process is spherically symmetric, then the random process is said to be *isotropic*. For a homogeneous, isotropic random process the autocorrelation function, $\Gamma_X(\vec{r})$, is a function only of the scalar distance, $r = |\vec{r}|$. The index of refraction fluctuations in the atmosphere are assumed to be homogeneous and isotropic in all cases.

Under the assumption that the random process $n_1(\vec{r})$ is homogeneous, the autocorrelation of $n_1(\vec{r})$ is given by

$$\Gamma_n(\vec{r}) = E\{n_1(\vec{r}_1)n_1(\vec{r}_1 - \vec{r})\}. \tag{3.22}$$

The autocorrelation, $\Gamma_n(\vec{r})$, and the power spectral density, $\Phi_n(\vec{\kappa})$, of $n_1(\vec{r})$ are related by the Wiener-Khinchin theorem [8]:

$$\Gamma_n(\vec{r}) = \int \mathrm{d}\vec{\kappa}\, \Phi_n(\vec{\kappa})\, \exp\{-j\vec{\kappa}\cdot\vec{r}\}. \tag{3.23}$$

The power spectral density, $\Phi_n(\vec{\kappa})$, is obtained from $\Gamma_n(\vec{r})$ by:

$$\Phi_n(\vec{\kappa}) = \frac{1}{(2\pi)^3}\int \mathrm{d}\vec{r}\, \Gamma_n(\vec{r})\, \exp\{j\vec{\kappa}\cdot\vec{r}\}. \tag{3.24}$$

In words, Eqs. (3.23) and (3.24) state that the autocorrelation of $n_1(\vec{r})$ and the power spectral density of $n_1(\vec{r})$ form a three dimensional Fourier transform pair.

The definition of the Fourier transform used in this chapter is not consistent with the definition given in Eq. (2.5). This difference is in the choice of the kernel for the forward Fourier transform. In Chapter 2 the kernel used for the forward transform was $\exp\{-j\vec{\kappa}\cdot\vec{r}\}$, while in this chapter the forward transform kernel $\exp\{j\vec{\kappa}\cdot\vec{r}\}$ is used. The forward transform kernel used here is consistent with that used by Tatarskii [3] and has been adopted by others [8]. Since there is little practical consequence of the choice of sign in the Fourier transform kernel, we adopt Tatarskii's convention in this chapter to allow easy access to his work.

It should be noted that both the autocorrelation function and the power spectral density for homogeneous random processes are guaranteed to be even functions [3, 14]. This fact allows us to rewrite Eq. (3.23) as

$$\Gamma_n(\vec{r}) = \int \mathrm{d}\vec{\kappa}\, \Phi_n(\vec{\kappa})\, \cos(\vec{\kappa}\cdot\vec{r}). \tag{3.25}$$

Equation (3.24) may also be rewritten as

$$\Phi_n(\vec{\kappa}) = \frac{1}{(2\pi)^3}\int \mathrm{d}\vec{r}\, \Gamma_n(\vec{r})\, \cos(\vec{\kappa}\cdot\vec{r}). \tag{3.26}$$

The additional assumption that $n_1(\vec{r})$ is an isotropic random process allows the autocorrelation and power spectral density of $n_1(\vec{r})$ to be written in terms of the scalar variables $r = |\vec{r}|$ and $\kappa = |\vec{\kappa}|$ [3]:

$$\Gamma_n(r) = \frac{4\pi}{r}\int \kappa \mathrm{d}\kappa\, \Phi_n(\kappa)\, \sin(\kappa r), \tag{3.27}$$

$$\Phi_n(\kappa) = \frac{1}{2\pi^2\kappa}\int r\mathrm{d}r\,\Gamma_n(r)\sin(\kappa r). \tag{3.28}$$

The results presented in Eqs. (3.27) and (3.28) are easily obtained by writing the integrals in Eqs. (3.25) and (3.26) in spherical coordinates and integrating over the angular dimensions [3].

We now introduce the structure function of the random index of refraction distribution. The structure function has a special role in the analysis of systems which form images through atmospheric turbulence since, as shown in Section 3.5, the structure function of the turbulence-corrupted index of refraction arises naturally in the analysis of the average optical transfer function of a telescope looking through turbulence. The structure function of the homogeneous index of refraction fluctuation is defined as

$$D_n(\vec{r}) = E\{[n_1(\vec{r}_1) - n_1(\vec{r}_1 + \vec{r})]^2\}. \tag{3.29}$$

Using the definitions in Eqs. (3.22) and (3.29), it is easy to see that $D_n(\vec{r})$ is related to $\Gamma_n(\vec{r})$ by

$$D_n(\vec{r}) = 2[\Gamma_n(0) - \Gamma_n(\vec{r})]. \tag{3.30}$$

Equation (3.23) can be used to express $D_n(\vec{r})$ as a function of the power spectral density, $\Phi_n(\vec{\kappa})$ by

$$\begin{aligned} D_n(\vec{r}) &= 2\left[\int \mathrm{d}\vec{\kappa}\,\Phi_n(\vec{\kappa}) - \int \mathrm{d}\vec{\kappa}\,\Phi_n(\vec{\kappa})\,e^{-j\vec{\kappa}\cdot\vec{r}}\right], \\ &= 2\int \mathrm{d}\vec{\kappa}\,\left[1 - e^{-j\vec{\kappa}\cdot\vec{r}}\right]\,\Phi_n(\vec{\kappa}), \\ &= 2\int \mathrm{d}\vec{\kappa}\,[1 - \cos(\vec{\kappa}\cdot\vec{r})]\,\Phi_n(\vec{\kappa}). \end{aligned} \tag{3.31}$$

In writing the last line of Eq. (3.31) we have exploited the fact that $\Phi_n(\vec{\kappa})$ is an even function. If $n_1(\vec{r})$ is also isotropic, the structure function depends only on the scalar variable $r = |\vec{r}|$ and is given by [3]

$$D_n(r) = 8\pi\int \mathrm{d}\kappa\,\Phi_n(\kappa)\left[1 - \frac{\sin(\kappa r)}{\kappa r}\right]. \tag{3.32}$$

Finally, if we use the Kolmogorov power spectrum, $\Phi_n^K(\vec{\kappa})$, in Eq. (3.32) in a region where C_n^2 is assumed constant, we obtain the result:

$$D_n(r) = C_n^2 r^{2/3}. \tag{3.33}$$

It should be noted in evaluating Eq. (3.32) that

$$\lim_{r\to 0}\frac{\sin(\kappa r)}{\kappa r} = 1. \tag{3.34}$$

Thus, the structure function provides a degree of insensitivity to the small wavenumber behavior of $\Phi_n(\kappa)$. This insensitivity is not achieved in the use of $\Gamma_n(\vec{r})$. To illustrate this point, consider the Wiener-Khinchin theorem stated in Eq. (3.23). The variance of the random process at any point in space is obtained from Eq. (3.23) by setting $r = 0$, which yields

$$\Gamma_n(0) = \mathrm{var}\{n_1\} = \int \mathrm{d}\kappa\,\Phi_n(\kappa), \tag{3.35}$$

where the notation $\mathrm{var}\{n_1\}$ has been used to represent the variance of the index of refraction fluctuation, n_1. Substitution of the Kolmogorov spectrum, given in Eq. (3.7), into Eq. (3.35) shows that the variance of n_1 is infinite. However, the structure function of n_1 at $\vec{r} = \vec{0}$ is finite and identically zero. This well behaved characteristic of the structure function is useful for analysis of telescopes having finite apertures.

Thus far we have not addressed the temporal evolution of turbulence. However, the temporal behavior of turbulence can be quite important for accurately modeling certain aspects of imaging through turbulence. Examples of problems where the temporal behavior of turbulence is important include computing the optimal exposure time for speckle imaging [15, 16, 17, 18, 19, 20], and choosing the optimal closed loop control bandwidth for an adaptive optics system [21, 22, 23].

The temporal effects of turbulence are generally modeled using Taylor's frozen flow hypothesis [8, 6]. In words, Taylor's frozen flow hypothesis states that over short time intervals the index of refraction fluctuation remains fixed except for translation with uniform transverse velocity $\vec{v}$. In this case, the term transverse velocity refers to the component of the wind velocity normal to the line-of-sight of the imaging system. For a point in space, $\vec{r}$, the frozen flow hypothesis means that the index of refraction fluctuation at $t_2 > t_1$, $n_1(\vec{r}, t_2)$ is related to the index of refraction fluctuation at t_1, $n_1(\vec{r}, t_1)$ by

$$n_1(\vec{r}, t_2) = n_1(\vec{r} - \vec{v}(t_2 - t_1), t_1). \tag{3.36}$$

The generalized space-time covariance of the index fluctuation, $\Gamma_n^S(\vec{r}_1, t_1; \vec{r}_2, t_2)$, is defined by

$$\Gamma_n^S(\vec{r}_1, t_1; \vec{r}_2, t_2) = E\left\{n_1(\vec{r}_1, t_1) n_1(\vec{r}_2, t_2)\right\}, \tag{3.37}$$

where the times, t_1 and t_2, are included in the argument of Γ_n^S to indicate the time dependence, and $\vec{r}_1$ and $\vec{r}_2$ indicate different points in space. Using the frozen flow hypothesis of Eq. (3.36) in Eq. (3.37) yields

$$\begin{aligned}\Gamma_n^S(\vec{r}_1, t_1; \vec{r}_2, t_2) &= E\left\{n_1(\vec{r}_1, t_1) n_1(\vec{r}_2, t_2)\right\}, \\ &= E\left\{n_1(\vec{r}_1, t_1) n_1(\vec{r}_2 - \vec{v}(t_2 - t_1), t_1)\right\}.\end{aligned} \tag{3.38}$$

Eq. (3.38) shows that the frozen flow hypothesis allows time differences to be represented as spatial shifts. This is an extremely important result which allows analytic simplifications to many problems. Applying the assumption of homogeneous turbulence to Eq. (3.38) yields

$$\Gamma_n^S(\vec{r}_1, t_1; \vec{r}_2, t_2) = \Gamma_n(\vec{r}_1 - \vec{r}_2 + \vec{v}(t_2 - t_1)). \tag{3.39}$$

Thus, the space-time covariance is given by the spatial covariance of Eq. (3.22) with an extra distance term, $\vec{v}(t_2 - t_1)$ which accounts for temporal effects.

3.4 Wave propagation through random media

In this section we shift our focus from describing the statistical characteristics of the index of refraction variations to describing the statistical effects of turbulence induced perturbations on a propagating optical field. As discussed in Section 3.2, atmospheric turbulence is characterized by random variations in the index of refraction. As an optical field propagates through a region of turbulence, these variations cause phase perturbations in the wave front. Propagation causes the phase perturbations to evolve into both phase and amplitude perturbations. In the case of astronomical imaging, we are interested in the situation in which the optical field propagates from a distant source through the turbulence to the pupil plane of a telescope. Figure 3.3 illustrates the geometry of interest. In this section we use the second order statistical description of the index of refraction variations presented in the previous section to develop a statistical description of the optical field in the pupil of the telescope.

In the presentation of the "interferometric view" of imaging of Section 2.3.1, we stressed how the spatial correlation properties of the pupil field control the outcome of an imaging experiment. Propagation through turbulence alters the correlation properties of the pupil field and as a result, motivates us to investigate the correlation properties of the turbulence induced field perturbations. To conduct the investigation we consider the problem of a plane wave propagating through a turbulent region as shown in Fig. 3.3. The goal of this section is to derive the pupil field correlation function describing the turbulence induced perturbations of the plane wave. Later in the section the results are

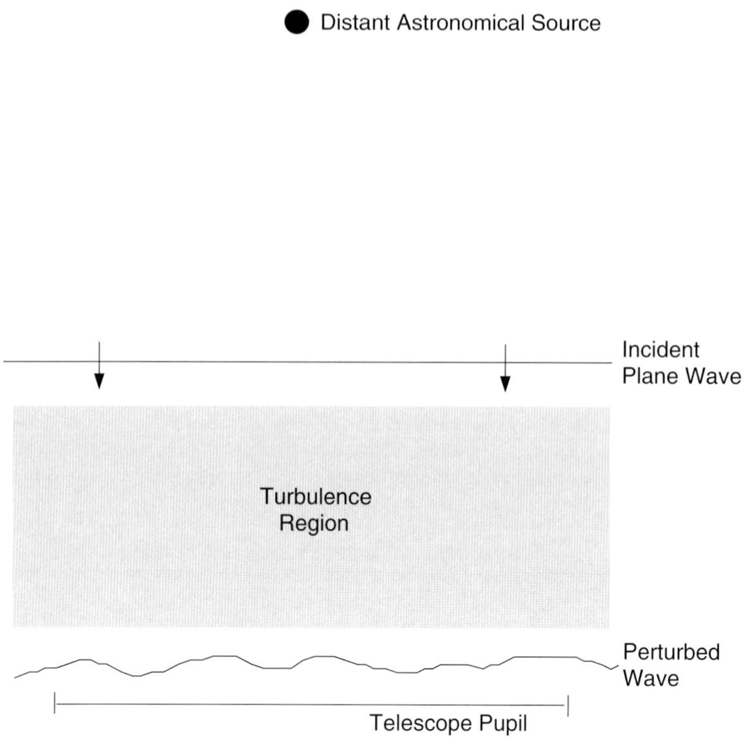

FIGURE 3.3
Astronomical imaging geometry. A distant source gives rise to an incident plane wave. The problem at hand is to find the correlation function describing the atmospheric induced field perturbations in the pupil.

also generalized to account for spherical wave propagation. These two cases are of interest for studying imaging of celestial sources, modeled using plane wave propagation, and imaging of finite altitude sources, modeled as spherical wave propagation.

In Subsection 3.4.1 a layered model of the turbulence region is introduced. The layered model is used in Subsection 3.4.2 to derive the spatial correlation properties of the pupil field. This derivation is based on a geometrical optics approach. The geometrical optics approach is equivalent to assuming that refraction due to propagation through turbulence can be ignored which is also referred to as the "near field" assumption. Following the geometrical optics derivation, we comment on the implications of the near field assumption and also introduce results for non-near or "far" field conditions. Finally, the section concludes with an investigation of temporal and anisoplanatic effects on the pupil field correlation properties. These two effects are related in that they both are related to the correlation between the wave front phase perturbations for two distinct wave fronts. In the case of temporal effects, we are referring to wave front perturbations at two different times and in the case of anisoplanatic effects, we consider perturbations for two different propagation paths. Understanding temporal and anisoplanatic effects is important for characterizing the adaptive optical system performance discussed in Chapter 5.

3.4.1 Layered model for turbulence

In Section 3.2 we discussed the distribution of the turbulence strength as a function of altitude. The structure constant of the index of refraction fluctuations, C_n^2, is used to characterize the strength of the turbulence. A representative set of C_n^2 profiles are given in Eqs. (3.16) through (3.19) and shown in

TABLE 3.1
Weights for 4-layered atmospheric models. Layer 1 is placed at 200 meters, layer 2 at 2 km, layer 3 at 10 km, and layer 4 at 18 km. HV-21 and HV-54 represent the Hufnagel-Valley model [12] calculated with a 21 and 54 mph upper atmospheric wind, respectively. Model SLC-N represents a fit to the AMOS night data, and SLC-D represents a fit to the AMOS night data with the addition of an altered boundary layer to simulate daytime conditions [12]. Model Greenwood is Greenwood's 'good seeing' model [13]. The layer altitudes are referenced to the altitude of the site.

C_n^2 Model	Layer 1 W_1	Layer 2 W_2	Layer 3 W_3	Layer 4 W_4	I_C 10^{-12} $\mathrm{m}^{1/3}$	Total r_0 cm
HV-21	0.8902	0.0443	0.0591	0.0064	2.233	5.0
HV-54	0.6877	0.0204	0.2554	0.0365	2.953	4.2
SLC-N	0.4965	0.4623	0.0299	0.0113	0.6879	10.3
SLC-D	0.7397	0.2513	0.0048	0.0042	2.220	5.06
Greenwood	0.8615	0.0980	0.0394	0.0011	1.078	7.7

Fig. 3.2. For the derivation given in the next subsection we break the C_n^2 profile into a finite number of slabs or layers. Each layer is characterized by a turbulence strength that is approximately constant within the layer. The notation $C_{n_i}^2$ is used to designate the structure constant for the ith layer. The notations z_i and Δz_i are used to designate the altitude and thickness of the ith layer. The motivation for breaking the turbulence region into layers is the desire to investigate turbulence subregions that have approximately homogeneous statistics. The effect of field propagation through these homogeneous regions can be calculated separately and then combined to characterize propagation through the entire turbulence region (assuming the index of refraction fluctuations for each layer are statistically independent). This layering of the turbulence greatly simplifies the calculations to follow.

For some of the results presented in later chapters we adopt a four-layer model of the C_n^2 profiles given in Eqs. (3.16) through (3.19). It has been shown [24] that a four layer model is adequate for a wide range of imaging performance calculations. The values of the weight of the ith layer, $C_{n_i}^2$, and the altitude, z_i, are chosen in such a way that the zeroth through seventh-order moments, m, of the continuous model match the layered model:

$$\int_0^L \mathrm{d}z\, z^m C_n^2(z) = \sum_{i=1}^{4} z_i^m C_{n_i}^2 \Delta z_i \,, \tag{3.40}$$

where $0 \leq m \leq 7$ and L is the length of the propagation path through the turbulence. The fraction, W_i, of the total integrated turbulence strength in each layer can be written:

$$W_i = \frac{C_{n_i}^2 \Delta z_i}{I_C} \,, \tag{3.41}$$

where I_C is the total integrated value of C_n^2 (the zeroth order moment of Eq. (3.40)). Troxel *et. al.* [24] have calculated the weights and heights of the four layers used to model the continuous profiles given by Eqs. (3.16) through (3.19). Table 3.1 gives the weights and heights of the four layers for each of these continuous C_n^2 profiles.

3.4.2 Derivation of the pupil field spatial correlation function

The problem of calculating the spatial correlation properties of the turbulence induced field perturbations can be reduced to calculating the correlation properties for propagation through a single turbulence layer and then extending the results to account for all layers. The basic assumption required to use this approach concerns the statistical correlation of the index of refraction fluctuations between individual layers. Tatarskii [3] and others [8, 24] have investigated this assumption and have argued that individual

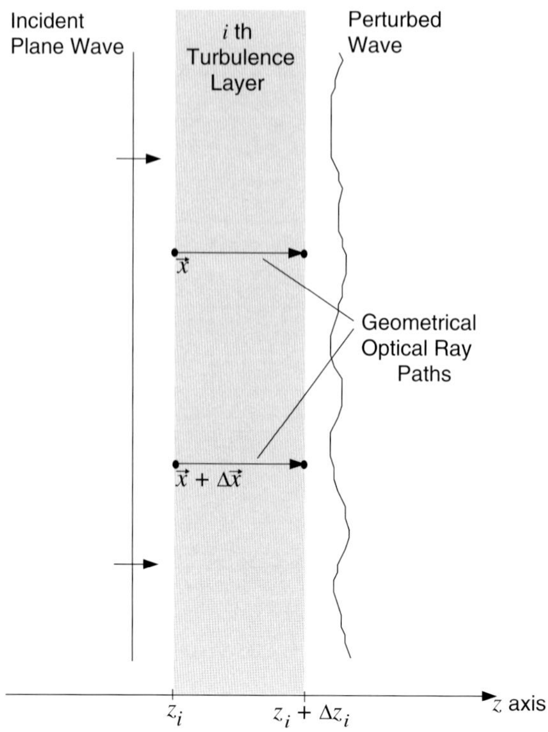

FIGURE 3.4
Plane wave propagation through a single turbulence layer.

layers can be treated as being approximately independent if the separation of layer centers is greater than the largest distance between field points in the pupil. The independence condition allows us to characterize the spatial correlation properties of the field perturbations for each layer separately and multiply the results to account for propagation through the entire turbulence region. Mathematically we can write

$$\Gamma_p(\Delta\vec{x}) = \prod_{i=1}^{N} \Gamma_{p_i}(\Delta\vec{x}), \tag{3.42}$$

where $\Gamma_p(\Delta\vec{x})$ is the spatial correlation function of optical field in the pupil, $\Gamma_{p_i}(\Delta\vec{x})$ is the spatial correlation function for the field perturbations associated with the ith layer, $\Delta\vec{x}$ is the vector separation of two points in a plane parallel to the pupil plane, and N is the number of layers comprising the turbulence region. The problem now reduces to computing the spatial correlation function for the turbulence induced perturbations of a single turbulence layer.

Plane wave propagation

Consider the geometry of the single turbulent layer shown in Fig. 3.4. A plane wave is incident from the left. Using geometrical optics, the optical path length (OPL) variations caused by propagation along a ray path extending from the left edge to the right edge of the layer is given by

$$\psi_i(\vec{x}) = k \int_{z_i}^{z_i+\Delta z_i} \mathrm{d}z\, n_1(\vec{x}, z), \tag{3.43}$$

where $\psi_i(\vec{x})$ is the phase perturbation at the transverse vector location $\vec{x}$, k is the optical wave number $(2\pi/\lambda)$, $n_1(\vec{x}, z)$ is the index of refraction variation as a function position, $(\vec{x}, z)$. In writing the 3-dimensional argument of n_1 we have chosen to show the transverse vector component, $\vec{x}$, and the z component separately. For a unit amplitude plane wave incident on the left, the transmitted optical field on the right can be expressed as

$$u_t(\vec{x}) = \exp\left[j\psi_i(\vec{x})\right]. \tag{3.44}$$

The spatial correlation function for $u_t(\vec{x})$ is given by $\Gamma_{p_i}(\Delta\vec{x}) = \langle u_t(\vec{x}) u_t^*(\vec{x} - \Delta\vec{x})\rangle$. In expressing the spatial correlation function in this way, we have implicitly assumed that the turbulence induced perturbations are at least wide sense stationary (WSS). Substituting the expression for the field found in Eq. (3.44) into $\Gamma_{p_i}(\Delta\vec{x})$ gives

$$\begin{aligned}\Gamma_{p_i}(\Delta\vec{x}) &= \langle \exp\left[j\psi_i(\vec{x})\right] \exp\left[-j\psi_i(\vec{x} - \Delta\vec{x})\right]\rangle, \\ &= \left\langle \exp\left[jk\left(\int_{z_i}^{z_i+\Delta z_i} \mathrm{d}z \left[n_1(\vec{x}, z) - n_1(\vec{x} - \Delta\vec{x}, z)\right]\right)\right]\right\rangle.\end{aligned} \tag{3.45}$$

Without an assumption regarding the statistics of the index of refraction fluctuation, $n_1(\vec{x}, z)$, further progress is impossible. The standard approach is to assume that $n_1(\vec{x}, z)$ is a Gaussian random process [1, 3, 8]. This assumption allows us to use well known properties of characteristic functions for Gaussian random processes [14]. Using these properties it is easy to show the following general result:

$$\mathrm{E}\{e^{a+jb}\} = \exp\{\overline{a} + j\overline{b}\} \exp\left\{\frac{1}{2}\left(\sigma_a^2 - j2\rho\sigma_a\sigma_b - \sigma_b^2\right)\right\} \tag{3.46}$$

where a and b are jointly Gaussian random variables. The mean and variance of a and b are $\overline{a}$ and σ_a^2, and $\overline{b}$ and σ_b^2, respectively. The correlation between a and b is described by the correlation coefficient ρ:

$$\rho = \frac{\langle (a - \overline{a})(b - \overline{b})\rangle}{\sqrt{\sigma_a^2 \sigma_b^2}}. \tag{3.47}$$

Using Eq. (3.46) we can write Eq. (3.45) as

$$\begin{aligned}\Gamma_{p_i}(\Delta\vec{x}) &= \exp\left[-\frac{1}{2}k^2\left\langle\left(\int_{z_i}^{z_i+\Delta z_i} \mathrm{d}z \left[n_1(\vec{x}, z) - n_1(\vec{x} - \Delta\vec{x}, z)\right]\right)^2\right\rangle\right], \\ &= \exp\left[-\frac{1}{2}D_{\psi_i}(\Delta\vec{x})\right],\end{aligned} \tag{3.48}$$

where $D_{\psi_i}(\Delta\vec{x})$ is called the phase structure function and is defined from Eq. (3.48) as

$$\begin{aligned}D_{\psi_i}(\Delta\vec{x}) &= \langle (\psi_i(\vec{x}) - \psi_i(\vec{x} - \Delta\vec{x}))^2\rangle, \\ &= k^2\left\langle\left(\int_{z_i}^{z_i+\Delta z_i} \mathrm{d}z \left[n_1(\vec{x}, z) - n_1(\vec{x} - \Delta\vec{x}, z)\right]\right)^2\right\rangle.\end{aligned} \tag{3.49}$$

The problem of characterizing $\Gamma_{p_i}(\Delta\vec{x})$ now becomes one of calculating the phase structure function $D_{\psi_i}(\Delta\vec{x})$. Expanding the expression for $D_{\psi_i}(\Delta\vec{x})$ results in

$$D_{\psi_i}(\Delta\vec{x}) = k^2 \left\langle \int_{z_i}^{z_i+\Delta z_i} \int_{z_i}^{z_i+\Delta z_i} \mathrm{d}z\mathrm{d}z' \left[n_1(\vec{x}, z) - n_1(\vec{x} - \Delta\vec{x}, z)\right] \times \left[n_1(\vec{x}, z') - n_1(\vec{x} - \Delta\vec{x}, z')\right] \right\rangle. \tag{3.50}$$

Distributing the ensemble average angle brackets over the terms in the integrand of Eq. (3.50) gives

$$D_{\psi_i}(\Delta\vec{x}) = k^2 \int_{z_i}^{z_i+\Delta z_i} \int_{z_i}^{z_i+\Delta z_i} \mathrm{d}z\mathrm{d}z' \left[2\Gamma_n(0, z - z') - \Gamma_n(\Delta\vec{x}, z - z') - \Gamma_n(-\Delta\vec{x}, z - z')\right], \tag{3.51}$$

where $\Gamma_n(\Delta\vec{x}, z)$ is the correlation function for the index of refraction fluctuations defined by Eq. (3.22). The index of refraction correlation function $\Gamma_n(\Delta\vec{x}, z)$ is related to the index structure function $D_n(\Delta\vec{x}, z)$ by Eq. (3.30). Solving for $\Gamma_n(\vec{x}, z)$ from Eq. (3.30) and substituting this result into Eq. (3.51) gives

$$D_{\psi_i}(\Delta\vec{x}) = -k^2 \int_{z_i}^{z_i+\Delta z_i} \int_{z_i}^{z_i+\Delta z_i} \mathrm{d}z\mathrm{d}z' \left[D_n(0, z - z') - \frac{1}{2}D_n(\Delta\vec{x}, z - z') - \frac{1}{2}D_n(-\Delta\vec{x}, z - z')\right]. \tag{3.52}$$

The change of variables $\tilde{z} = z - z'$ gives

$$D_{\psi_i}(\Delta\vec{x}) = -k^2 \Delta z_i \int_{-\Delta z_i}^{\Delta z_i} \mathrm{d}\tilde{z} \left(1 - \frac{|\tilde{z}|}{\Delta z_i}\right) \left[D_n(0, \tilde{z}) - D_n(\Delta\vec{x}, \tilde{z})\right]. \tag{3.53}$$

where we have recognized, for isotropic statistics, that $D_n(\Delta\vec{x}, z) = D_n(-\Delta\vec{x}, z)$. If the Kolmogorov power spectral density characterizes the index of refraction fluctuations, the index structure function, $D_n(\Delta\vec{x}, z)$, is given by Eq. (3.33). Substituting Eq. (3.33) into Eq. (3.53) gives

$$D_{\psi_i}(\Delta\vec{x}) = -k^2 \Delta z_i C_{n_i}^2 \int_{-\Delta z_i}^{\Delta z_i} \mathrm{d}\tilde{z} \left(1 - \frac{|\tilde{z}|}{\Delta z_i}\right) \left[\tilde{z}^{2/3} - (|\Delta\vec{x}|^2 + \tilde{z}^2)^{1/3}\right]. \tag{3.54}$$

The change variables $u = \tilde{z}/|\Delta\vec{x}|$ finally gives

$$D_{\psi_i}(\Delta\vec{x}) = -k^2 \Delta z_i C_{n_i}^2 |\Delta\vec{x}|^{5/3} \int_{-\Delta z_i/|\Delta\vec{x}|}^{\Delta z_i/|\Delta\vec{x}|} \mathrm{d}u \left(1 - \frac{|u\Delta\vec{x}|}{\Delta z_i}\right) \left[u^{2/3} - (1 + u^2)^{1/3}\right]. \tag{3.55}$$

The integral term in Eq. (3.55) is numerically evaluated and shown in Fig. 3.5 as a function of $\Delta z_i/|\Delta\vec{x}|$. As the thickness of the turbulent layer becomes significantly larger than the separation of the two field points, $|\Delta\vec{x}|$, the integral converges to a value of -2.91. Using this value gives the result first reported in the literature by Fried [1]:

$$D_{\psi_i}(\Delta\vec{x}) = 2.91 k^2 \Delta z_i C_{n_i}^2 |\Delta\vec{x}|^{5/3}. \tag{3.56}$$

Fried goes on to express the phase structure function as a function of a quantity called the seeing cell size. The seeing cell size is also commonly called the Fried parameter. The Fried parameter for the ith layer is designated with the notation r_{o_i} and is defined as [8]

$$r_{o_i} = 0.185 \left[\frac{4\pi^2}{k^2 C_{n_i}^2 \Delta z_i}\right]^{3/5}. \tag{3.57}$$

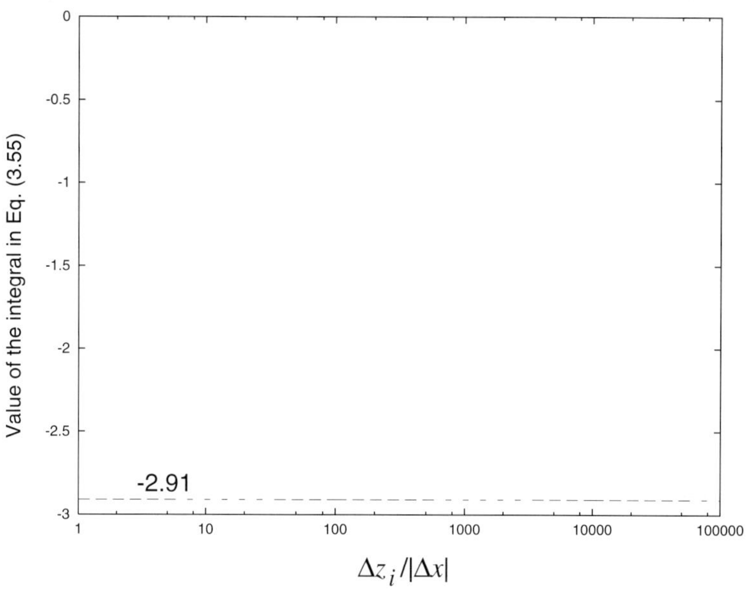

FIGURE 3.5
Numerical evaluation of the integral term of Eq. (3.55) as a function of $\Delta z_i/|\Delta\vec{x}|$.

The Fried parameter can be interpreted as the aperture size beyond which further increases in diameter result in no further increase in resolution. Using r_{o_i} to rewrite Eq. (3.56) gives the form of the phase structure function found most often in the literature:

$$D_{\psi_i}(\Delta\vec{x}) = 6.88\left(\frac{|\Delta\vec{x}|}{r_{o_i}}\right)^{5/3}. \tag{3.58}$$

Later in Section 3.5.3 we find that the Fried parameter plays the key role in determining the frequency domain performance of an incoherent imaging system. The Fried parameter is found to define the roll-off of the optical transfer function (OTF) associated with imaging through atmospheric turbulence.

With the phase structure function for the ith layer established, we can now compute the field correlation function $\Gamma_{p_i}(\Delta\vec{x})$. Substituting Eq. (3.56) into Eq. (3.48) gives

$$\Gamma_{p_i}(\Delta\vec{x}) = \exp\left[-\frac{1}{2}2.91k^2\Delta z_i C_{n_i}^2|\Delta\vec{x}|^{5/3}\right]. \tag{3.59}$$

Substituting this result into the correlation expression for propagation through the entire turbulence region given by Eq. (3.42) gives

$$\begin{aligned}\Gamma_p(\Delta\vec{x}) &= \prod_{i=1}^{N}\exp\left[-\frac{1}{2}2.91k^2\Delta z_i C_{n_i}^2|\Delta\vec{x}|^{5/3}\right],\\ &= \exp\left[-\frac{1}{2}2.91k^2|\Delta\vec{x}|^{5/3}\sum_{i=1}^{N}\Delta z_i C_{n_i}^2\right].\end{aligned} \tag{3.60}$$

From the last result it is possible to define the Fried parameter for propagation through the entire turbulence region:

$$r_o = 0.185 \left[\frac{4\pi^2}{k^2 \sum_{i=1}^{N} C_{n_i}^2 \Delta z_i} \right]^{3/5},$$

$$= 0.185 \left[\frac{4\pi^2}{k^2 \int_0^L dz \, C_n^2(z)} \right]^{3/5}, \tag{3.61}$$

where L is the propagation distance through the turbulence region and $C_n^2(z)$ is the continuous profile of the structure constant. Substituting Eq. (3.61) into Eq. (3.60) gives

$$\Gamma_p(\Delta\vec{x}) = \exp\left[-\frac{1}{2} 6.88 \left(\frac{|\Delta\vec{x}|}{r_0} \right)^{5/3} \right],$$

$$= \exp\left[-\frac{1}{2} D_\psi(\Delta\vec{x}) \right]. \tag{3.62}$$

The term $D_\psi(\Delta\vec{x})$ is the plane wave phase structure function for the entire turbulence region and is given by

$$D_\psi(\Delta\vec{x}) = 6.88 \left(\frac{|\Delta\vec{x}|}{r_o} \right)^{5/3}. \tag{3.63}$$

We finally note that the Fried parameter for the individual turbulence layers, r_{o_i}, is related to the overall Fried parameter r_0 by

$$r_0^{-5/3} = \sum_{i=1}^{N} r_{o_i}^{-5/3}. \tag{3.64}$$

Equations (3.62) and (3.63) represent the two main results for plane wave propagation. These two results show that the structure function $D_\psi(\Delta\vec{x})$ is all that is required to characterize the pupil field correlation function $\Gamma_p(\Delta\vec{x})$. In Section 3.5 our investigation of imaging performance reveals that $\Gamma_p(\Delta\vec{x})$ is the key statistical quantity that quantifies the performance for imaging through atmospheric turbulence. The correlation function $\Gamma_p(\Delta\vec{x})$ will be shown to be essentially an "optical transfer function".

The results in this section have been derived using geometrical optics and, as noted earlier, the use of geometrical optics is equivalent to assuming that near field conditions exist. The phrase "near field" refers to the proximity of the turbulence relative to the pupil. Near field conditions are said to exist if the total thickness of the turbulence region satisfies [25]

$$\sum_{n=1}^{N} \Delta z_i \leq \frac{|\Delta\vec{x}_{max}|^2}{\lambda\pi}, \tag{3.65}$$

where $|\Delta\vec{x}_{max}|$ is the maximum transverse separation of interest in the pupil. In the case of imaging, $|\Delta\vec{x}_{max}|$ corresponds to the telescope diameter. In near field conditions the turbulence region is close enough to the pupil plane that refraction caused by the index of refraction variations can be ignored. Ignoring refraction allows us to derive $\Gamma_p(\Delta\vec{x})$ based on straight ray path calculations that lie along the direction of propagation as shown in Fig. 3.4. As the turbulence region extends away from the pupil, the propagation path becomes long enough so that refraction causes the ray paths to diverge from straight lines. As a result, refraction cannot be ignored and the propagation paths cross and intersect each other in the pupil plane and cause interference. The interference in turn causes field amplitude perturbations. We address the implication of far field turbulence later in this section.

Spherical wave propagation

In the above derivation we assumed an incident plane wave as shown in Fig. 3.4. In many instances the incident optical field will arise from a source that is close enough to the turbulence that a spherical wave more accurately models the incident field. In this case the geometrical optics derivation given above can be repeated for ray paths corresponding to spherical wave propagation. References [26] and [27] give derivations of the spherical phase structure functions. Rather than repeating a derivation similar to the development given above, we simply present the phase structure function for spherical wave propagation [8]:

$$D_{\psi}(\Delta\vec{x}) = 6.88 \left(\frac{|\Delta\vec{x}|}{r_0}\right)^{5/3}. \tag{3.66}$$

Note that this result, on the surface, is the same as found for the plane wave phase structure function. The difference is contained in the expression for r_0. The Fried parameter for spherical wave propagation is found to be

$$r_o = 0.185 \left[\frac{4\pi^2}{k^2 \int_0^L \mathrm{d}z \left(\frac{L-z}{L}\right)^{5/3} C_n^2(z)}\right]^{3/5}, \tag{3.67}$$

where the integration is from the pupil plane at $z = 0$ to $z = L$. For completeness we also give the Fried parameter for the ith layer:

$$r_{o_i} = 0.185 \left[\frac{4\pi^2}{k^2 C_{n_i}^2 \left(\frac{L-z_i}{L}\right) \Delta z_i}\right]^{3/5}. \tag{3.68}$$

It is interesting to note that if C_n^2 is constant throughout the turbulence region the spherical and plane wave definitions of r_0 are related by

$$r_0 \text{ (spherical wave)} = \left(\frac{8}{3}\right)^{3/5} r_0 \text{ (plane wave)}. \tag{3.69}$$

Far field turbulence

The implication of the near field assumption in the developments of the two previous subsections is that the perturbations in the pupil plane are entirely phase perturbations. This fact is reflected by the fact that the structure function $D_{\psi}(\Delta\vec{x})$ is a phase structure function. If near field conditions do not exist, the perturbations in the pupil plane in general consist of both phase and amplitude perturbations. The bending of the optical rays caused by refraction along with the subsequent propagation to the the pupil plane cause interference effects. These interference effects in turn give rise to amplitude variations in the pupil field. In the next section of the chapter (Section 3.5), we show that the correlation function describing the pupil field *amplitude* and *phase* perturbations can be expressed using a quantity known as the *wave* structure function. We do not derive this more general result here since the development in this section is focused on the geometrical optics approach to computing the pupil field correlation properties.

In the development of the next section (starting with Eq. (3.115)), amplitude and phase perturbations are both accounted for in the calculation of the pupil field correlation function $\Gamma_p(\Delta\vec{x})$. The wave structure function is the main result from this derivation. Just as we have seen in the geometrical optics derivations of this section, the correlation function $\Gamma_p(\Delta\vec{x})$ is shown to be given by [1, 8]

$$\Gamma_p(\Delta\vec{x}) = \exp\left[-\frac{1}{2}D(\Delta\vec{x})\right], \tag{3.70}$$

where $D(\Delta\vec{x}, t)$ is now interpreted as the wave structure function. The wave structure function is a combination of the amplitude and phase structure functions. In order to discuss the wave structure function, we must have a notation for representing the field amplitude and phase perturbations in the pupil. We have already extensively discussed the phase perturbations and will again use the designation $\psi(\vec{x})$ for this perturbation. The turbulence induced field amplitude perturbation is designated with a "log-amplitude" perturbation, $\chi(\vec{x})$. In other words the perturbed pupil field, $u(\vec{x})$, can be written

$$u(\vec{x}) = \exp(\chi(\vec{x}))\exp(j\psi(\vec{x})), \tag{3.71}$$

where $\vec{x}$ is a point in the pupil. In Section 3.5.3 we show that the wave structure function can be expressed as

$$\begin{aligned} D(\Delta\vec{x}) &= \left\langle(\chi(\vec{x}) - \chi(\vec{x}-\Delta\vec{x}))^2\right\rangle + \left\langle(\psi(\vec{x}) - \psi(\vec{x}-\Delta\vec{x}))^2\right\rangle, \\ &= D_\chi(\Delta\vec{x}) + D_\psi(\Delta\vec{x}), \end{aligned} \tag{3.72}$$

where $D_\chi(\Delta\vec{x})$ is the log amplitude structure function. For the purposes of this section the result shown in Eq. (3.72) is presented without proof. The development of Eq. (3.70) is left to Section 3.5.

In order to use Eq. (3.72) we need an expression for the amplitude and phase structure functions, $D_\chi(\Delta\vec{x})$ and $D_\psi(\Delta\vec{x})$. Tatarskii [3] has studied the problem of obtaining the wave structure function by starting with the inhomogeneous wave equation. Use of the Rytov approximation to solve the wave equation in an inhomogeneous medium results in a Fresnel-like propagation equation that expresses both the log amplitude and phase perturbations induced by the turbulence. This propagation equation is then used in conjunction with the statistics of the index of refraction fluctuations given in Section 3.3 to derive the wave structure function. A more intuitive approach is presented by Lee and Harp [28]. They present a wave optics analysis of wave propagation through the atmospheric layers. The wave optics analysis is performed in the frequency domain and is based on a decomposition of the turbulent layers into Fourier components. The diffraction results for an incident plane wave for each of the layer's Fourier components are calculated. The diffracted fields are propagated to the pupil plane via the Fresnel diffraction formula. In the pupil plane the diffraction effects from all of the Fourier components of each slab are summed. Amplitude and phase perturbations of the optical field are computed from the summed field contributions. In either Tatarskii's or Lee and Harp's approach the final results are the same.

Rather than presenting the lengthy theoretical development of Tatarskii's or Lee and Harp's, we refer the reader to the references given above. The final results of these developments can be summarized with the phase and log amplitude structure functions for plane wave propagation [24, 28]:

$$D_\psi(\Delta\vec{x}) = 8\pi^2 k^2 \int_0^L dz \int_0^\infty d\kappa\, \kappa\, (1 - J_o(\kappa|\Delta\vec{x}|))\, \Phi_n(\kappa, z)\cos^2\left(\frac{\kappa^2 z}{2k}\right), \tag{3.73}$$

$$D_\chi(\Delta\vec{x}) = 8\pi^2 k^2 \int_0^L dz \int_0^\infty d\kappa\, \kappa\, (1 - J_o(\kappa|\Delta\vec{x}|))\, \Phi_n(\kappa, z)\sin^2\left(\frac{\kappa^2 z}{2k}\right), \tag{3.74}$$

where $J_o(x)$ is a Bessel function of the first kind of zero order, $\Phi_n(\kappa)$ is the power spectrum of the index of refraction fluctuations introduced in Section 3.2, and L is the length of the optical path through the turbulence. Combining the phase and amplitude structure functions gives the wave structure function:

$$D(\Delta\vec{x}) = D_\psi(\Delta\vec{x}) + D_\chi(\Delta\vec{x})$$

$$= 8\pi^2 k^2 \int_0^L dz \int_0^\infty d\kappa\, \kappa\, (1 - J_o(\kappa|\Delta\vec{x}|))\, \Phi_n(\kappa, z). \quad (3.75)$$

Substituting the Kolmogorov index of refraction power spectrum given by Eq. (3.9) into Eq. (3.75) gives

$$D(\Delta\vec{x}) = 8(0.033)\pi^2 k^2 \int_0^L dz C_n^2(z) \int_0^\infty d\kappa\, \kappa^{-8/3}\, (1 - J_o(\kappa|\Delta\vec{x}|)). \quad (3.76)$$

The second integral in Eq. (3.76) can be analytically evaluated to obtain [3, 8]

$$D(\Delta\vec{x}) = 2.91 k^2 |\Delta\vec{x}|^{5/3} \int_0^L dz C_n^2(z). \quad (3.77)$$

The remarkable fact about this result is that it is identical to the geometrical optics result derived earlier. Using the definition of the plane wave Fried parameter, r_0, given by Eq. (3.61) in Eq. (3.77) transforms it into the exact form given by the geometrical optics result in Eq. (3.63)

The wave structure function for spherical wave propagation is also of interest and is given by

$$D(\Delta\vec{x}) = 8\pi^2 k^2 \int_0^L dz \int_0^\infty d\kappa\, \kappa \left(1 - J_o\left(\frac{\kappa|\Delta\vec{x}|(L-z)}{L}\right)\right) \Phi_n(\kappa, z), \quad (3.78)$$

where the integration is from the pupil plane at $z = 0$ to $z = L$. Again substituting in the Kolmogorov power spectrum yields

$$D(\Delta\vec{x}) = 8(0.033)\pi^2 k^2 \int_0^L dz C_n^2(z) \int_0^\infty d\kappa\, \kappa^{-8/3} \left(1 - J_o\left(\frac{\kappa|\Delta\vec{x}|(L-z)}{L}\right)\right). \quad (3.79)$$

Just as above for plane wave propagation, the second integral in Eq. (3.79) can be analytically evaluated to obtain

$$D(\Delta\vec{x}) = 2.91 k^2 |\Delta\vec{x}|^{5/3} \int_0^L dz \left(\frac{L-z}{L}\right)^{5/3} C_n^2(z). \quad (3.80)$$

The spherical wave Fried parameter r_0 defined by Eq. (3.67) can be used to transform Eq. (3.80) to a form that matches the geometrical optics result in Eq. (3.66).

Even though the wave structure functions of Eqs. (3.77) and (3.80) are identical to the geometrical optics results of Eqs. (3.63) and (3.66), it should be noted that the wave structure functions represent a second order statistical characterization of both the field amplitude and phase perturbations. Inherent in the geometrical optics results of Eqs. (3.63) and (3.66) is the implication that the field perturbations are entirely phase distortions. The phase structure function derived earlier can be interpreted as a limiting case of the wave structure function in which the amplitude function, $D_\chi(\Delta\vec{x})$, is negligible compared to the phase function, $D_\psi(\Delta\vec{x})$ [8, 28]. In the most general sense, the wave structure function must be interpreted as containing both amplitude and phase contributions.

3.4.3 Temporal and anisoplanatic effects

In Section 3.3 we discussed the temporal correlation properties of the index of refraction fluctuations. Understanding how these temporal properties impact the temporal properties of the optical field is important for investigating a number of problems in speckle imaging and adaptive optics. For speckle imaging, the temporal evolution of the speckled images is a direct result of the temporal properties of the index of refraction fluctuations. Understanding the nature of the temporally changing speckled images is useful for determining optimal speckle exposure times [15, 16, 17, 18, 19, 20]. In the case of adaptive optics, the temporally changing pupil field plays a key role in determining the overall performance of an adaptive optical imaging (AOI) system [21, 22, 23]. An AOI system senses the pupil field phase perturbations and, at some later time, applies a compensation. During the time delay the pupil field evolves and the compensation applied is not exactly the required compensation. As a result, the AOI system suffers a performance degradation.

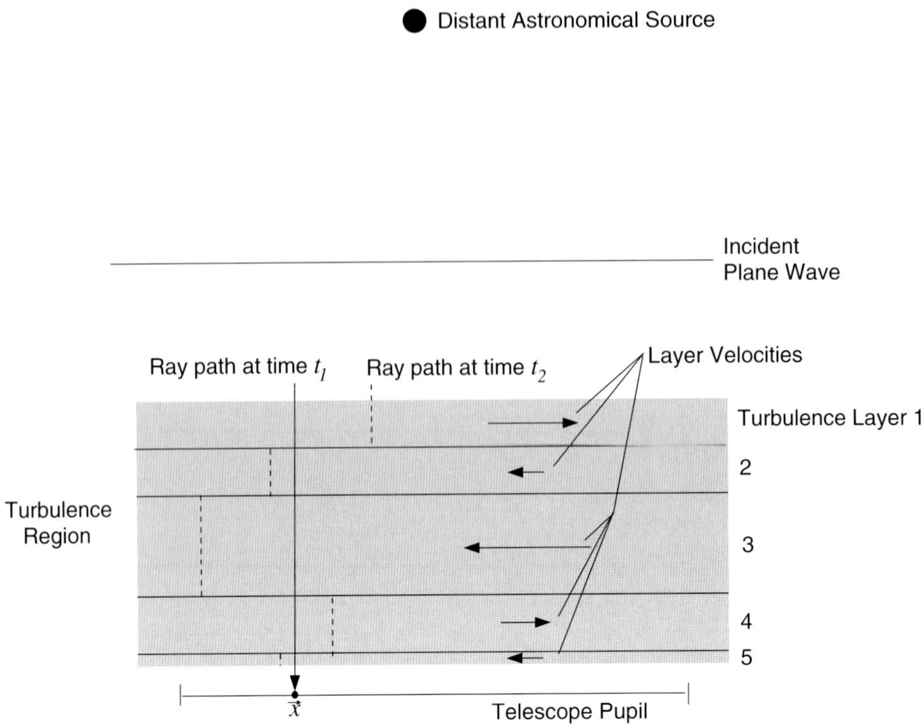

FIGURE 3.6
Geometry for understanding the temporal evolution of the wave front phase in the pupil of the imaging system. Each turbulence layer has an associated transverse wind velocity, $\vec{v}_i$.

The correlation properties of the optical fields arising from separate propagation directions are also of interest. In the operation of an AOI system the applied wave front compensation is based on sensing the wave front phase perturbations arising from a beacon or guide star. The sensed perturbations are used to compensate the wave front perturbations associated with a separate observation direction. The observation and beacon directions are generally distinct from each other and, as a result, the phase perturbations associated with the two directions are different. This difference results in a performance degradation which is referred to as an anisoplanatic effect. To fully understand the extent of anisoplanatic effects, the correlation between the pupil fields arising from the two directions must be known.

In the following subsection we develop expressions that characterize how the temporal correlation properties of the index of refraction fluctuations impact the temporal correlation properties of the optical field in the pupil of an imaging system. The last subsection presents a development of the effects of anisoplanatism on the pupil field correlation properties.

Pupil field temporal correlation function

To develop the field correlation function describing the effects of temporal changes in the index of refraction, we use Taylor's frozen flow hypothesis [8, 6] in combination with a geometrical optics analysis. Taylor's frozen flow hypothesis is discussed in Section 3.3. Figure 3.6 shows the geometry used to model the temporal wave front phase evolution. The ith turbulence layer shown in the figure has an associated transverse wind velocity, $\vec{v}_i$. The term "transverse" refers to the component of the wind velocity perpendicular to the optical axis. The straight ray path shown in Fig. 3.6 represents the path over which the optical field propagates at time t_1. Using Eq. (3.43) we can write the pupil phase perturbation associated with the ith layer at time t_1 as:

$$\psi_i(\vec{x}, t_1) = k \int_{z_i}^{z_i+\Delta z_i} \mathrm{d}z\, n_1(\vec{x}, z, t_1), \tag{3.81}$$

where z_i is the bottom edge of the ith layer, Δz_i is the thickness of the ith layer, and $n_1(\vec{x}, z, t_1)$ is the index of refraction fluctuation at time t_1. To calculate the phase perturbation at time t_2 ($t_2 > t_1$), the translation of the layer must be factored into the computation. Taylor's frozen flow hypothesis calls for translating the ith layer by $\vec{v}_i(t_2 - t_1)$ (see Section 3.3). Mathematically the phase perturbation for the ith layer can be written as

$$\begin{aligned} \psi_i(\vec{x}, t_2) &= k \int_{z_i}^{z_i+\Delta z_i} \mathrm{d}z\, n_1(\vec{x}, z, t_2), \\ &= k \int_{z_i}^{z_i+\Delta z_i} \mathrm{d}z\, n_1(\vec{x} - \vec{v}_i(t_2 - t_1), z, t_1). \end{aligned} \tag{3.82}$$

By accounting for the translations of each of the layers, the piecewise ray path shown in Fig. 3.6 is the resulting ray path at time t_2. As before the goal is to compute the field correlation function, $\Gamma_{p_i}(\Delta\vec{x})$. In contrast to earlier results, we are interested in the space-time correlation function:

$$\Gamma_{p_i}(\Delta\vec{x}, t_2 - t_1) = \langle \exp[j\psi_i(\vec{x}, t_1)] \exp[-j\psi_i(\vec{x} - \Delta\vec{x}, t_2)] \rangle, \tag{3.83}$$

where $\Delta\vec{x}$ is the spatial separation of two points in the pupil. Substituting Eqs. (3.81) and (3.82) into Eq. (3.83) and making the Gaussian assumptions discussed previously gives an expression similar to the earlier space-time result in Eq. (3.48):

$$\Gamma_{p_i}(\Delta\vec{x}, t_2 - t_1) = \exp\left[-\frac{1}{2}k^2 \left\langle \left(\int_{z_i}^{z_i+\Delta z_i} \mathrm{d}z \left[n_1(\vec{x}, z, t_1) - n_1(\vec{x} - \Delta\vec{x} - \vec{v}_i(t_2 - t_1), z, t_1) \right] \right)^2 \right\rangle \right]. \tag{3.84}$$

Starting with Eq. (3.84) and following a development identical to that used to obtain the plane wave correlation function shown in Eq. (3.60) (the steps from Eq. (3.48) to Eq. (3.60)), it is straightforward to show that the pupil plane space-time correlation function, $\Gamma_p(\Delta\vec{x}, t_2 - t_1)$, has the following familiar form:

$$\Gamma_p(\Delta\vec{x}, t_2 - t_1) = \exp\left[-\frac{1}{2} D_\psi(\Delta\vec{x}, t_2 - t_1) \right], \tag{3.85}$$

where $D_\psi(\Delta\vec{x}, t_2 - t_1)$ is now interpreted as the space-time phase structure function given by

$$\begin{aligned} D_\psi(\Delta\vec{x}, t_2 - t_1) &= 2.91k^2 \sum_{i=1}^{N} |\Delta\vec{x} + \vec{v}_i(t_2 - t_1)|^{5/3} \Delta z_i C_{n_i}^2, \\ &= 2.91k^2 \int_0^L \mathrm{d}z |\Delta\vec{x} + \vec{v}(z)(t_2 - t_1)|^{5/3} C_n^2(z). \end{aligned} \tag{3.86}$$

The second line of Eq. (3.86) results from generalizing the layered result for continuous $C_n^2(z)$ and $\vec{v}(z)$ profiles. Using the definition of r_{o_i} and r_0 given by Eqs. (3.57) and (3.61), Eq. (3.86) can be rewritten as

$$\begin{aligned} D_\psi(\Delta\vec{x}, t_2 - t_1) &= 6.88 \sum_{i=1}^{N} \left(\frac{|\Delta\vec{x} + \vec{v}_i(t_2 - t_1)|}{r_{o_i}} \right)^{5/3}, \\ &= 6.88 \int_0^L \mathrm{d}z \left(\frac{|\Delta\vec{x} + \vec{v}(z)(t_2 - t_1)|}{r_0} \right)^{5/3} \tilde{C}_n^2(z) \text{ (plane wave result)}, \end{aligned} \tag{3.87}$$

where $\tilde{C}_n^2(z)$ is a normalized version of $C_n^2(z)$:

$$\tilde{C}_n^2(z) = \frac{C_n^2(z)}{\int_0^L \mathrm{d}z\, C_n^2(z)} = \frac{C_n^2(z)}{I_C}. \tag{3.88}$$

The spherical wave phase structure function for a continuous $C_n^2(z)$ and $v(z)$ profile is a generalization of the second line of Eq. (3.87):

$$\begin{aligned} &D_\psi(\Delta\vec{x}, t_2 - t_1) = \\ &6.88 \int_0^L \mathrm{d}z \left(\frac{|\Delta\vec{x}\frac{L-z}{L} + \vec{v}(z)(t_2 - t_1)|}{r_0} \right)^{5/3} \tilde{C}_n^2(z) \text{ (spherical wave result)}, \end{aligned} \tag{3.89}$$

where again r_0 is given by Eq. (3.61).

In the case of plane wave propagation the result given in Eq. (3.87) reduces to a simple expression for a constant wind velocity. In this case we obtain

$$D_\psi(\Delta\vec{x}, t_2 - t_1) = 6.88 \left(\frac{|\Delta\vec{x} + \vec{v}(t_2 - t_1)|}{r_0} \right)^{5/3} \text{ (plane wave result)}. \tag{3.90}$$

The Fried parameter, r_0, required for substitution in Eq. (3.90) is given in Eqs. (3.61).

Anisoplanatic effects on the pupil field correlation function

In our investigation of anisoplanatic effects we again use a geometrical optics approach to account for the phase perturbations associated with separate propagation directions. A Fourier optics treatment of anisoplanatic effects that accounts for refraction is found in reference [24]. The method of calculating the pupil field correlation function for the temporal effects can be extended directly to the anisoplanatic case. Just as in the computation of the temporal correlation function the quantity that must be obtained is the phase structure function. Here, the phase structure function characterizes the statistical correlation properties of the wave front phase for two optical fields propagating from two distinct directions. In other words, we are interested in $D_\psi(\vec{x}_o, \vec{x}_b) = \langle(\psi_o(\vec{x}_o) - \psi_b(\vec{x}_b))^2\rangle$, where $\psi_o(\vec{x})$ and $\psi_b(\vec{x})$ are the wave front phases of the two distinct optical fields. Figure 3.7 shows the geometry under consideration. This figure shows the imaging system pupil and two optical source points. Ray paths are shown from the source points to two points in the pupil. In the case of adaptive optical imaging, one of the source points models the beacon. The other source point represents the object being imaged. The generalized geometry shown in Fig. 3.7 can be used to model any beacon-object-pupil geometry that might be encountered in an adaptive optical imaging scenario. In the example of an AOI system using an on axis laser guide star, one of the sources models a finite altitude beacon. The other source is used to model the celestial object and is effectively placed at infinity.

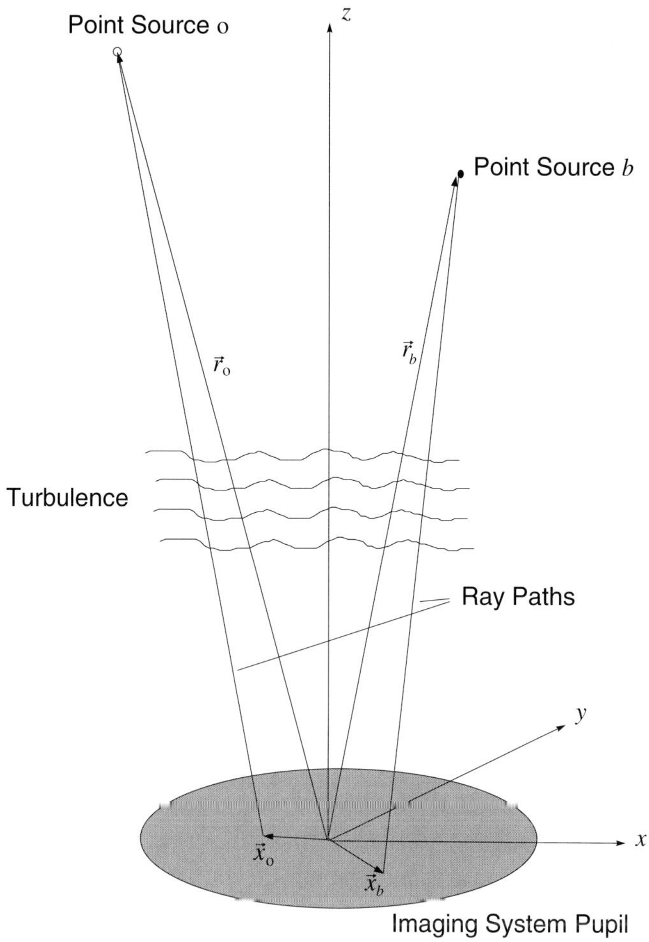

FIGURE 3.7
Geometry for deriving the anisoplanatic effects on the wave front phase in the pupil of the imaging system.

As seen from Fig. 3.7 the two optical sources are located at the vector positions $\vec{r}_o$ and $\vec{r}_b$. The z components of these two vectors are designated z_o and z_b, respectively. The vector positions $\vec{x}_o$ and $\vec{x}_b$ designate two field points in the pupil plane. The subscripts o and b are arbitrary but are, in later chapters, related to the object and the beacon wave front sources for an AOI system. To calculate the phase structure function, the transverse distance between the two ray paths must be characterized as a function of z. The transverse separation of the two ray paths has been shown to be [29]

$$\Delta\vec{p}_{ob}(z) = \vec{x}_o - \vec{x}_b + \frac{z}{z_o}(\vec{r}_o - \vec{x}_o) - \frac{z}{z_b}(\vec{r}_b - \vec{x}_b). \tag{3.91}$$

Using a geometrical optics derivation identical to that used to obtain the plane and spherical wave phase structure functions of the previous section, Lutomirski and Buser [30] have shown that the phase structure function for the geometry shown in Fig. 3.7 is given by

$$\begin{aligned} D_\psi(\vec{x}_o, \vec{x}_b) &= 2.91k^2 \sum_{i=1}^{N} |\Delta\vec{p}_{ob}(z_i)|^{5/3} \Delta z_i C_{n_i}^2, \\ &= 2.91k^2 \int_0^L \mathrm{d}z |\Delta\vec{p}_{ob}(z)|^{5/3} C_n^2(z), \end{aligned} \tag{3.92}$$

where L is the length of the vertical propagation path to the source at the minimum altitude. The second line of Eq. (3.92) results from generalizing the layered result for continuous $C_n^2(z)$ and $\vec{v}(z)$ profiles. Incorporating the plane wave definition of r_{o_i} and r_0 given by Eqs. (3.57) and (3.61) into Eq. (3.92) gives [29]

$$\begin{aligned} D_\psi(\vec{x}_o, \vec{x}_b) &= 6.88 \sum_{i=1}^{N} \left(\frac{|\Delta\vec{p}_{ob}(z)|}{r_{o_i}} \right)^{5/3}, \\ &= 6.88 \int_0^L \mathrm{d}z \left(\frac{|\Delta\vec{p}_{ob}(z)|}{r_0} \right)^{5/3} \tilde{C}_n^2(z). \end{aligned} \tag{3.93}$$

Equation (3.93) is the main result of this subsection. It can be used, along with the geometry in Fig. 3.7, to characterize the phase structure function for almost any beacon-object-pupil geometry. This relationship will be used in Chapter 5 to study the anisoplanatic effects on AOI system performance. The limitations of Eq. (3.93) are related to the approximations invoked by Lutomirski and Buser. In particular, the pupil points, $\vec{x}_o$ and $\vec{x}_b$, and the source points must be close to the z axis relative to the distance to the two sources, z_o and z_b. It should also be noted that in the development of Eq. (3.93) the turbulence height is assumed to be less than the height of the lower of the two source points, L. Finally, it is interesting to consider Eq. (3.93) for some special geometries. Consider first the geometry in which the two sources are placed at infinity at the same zenith angle. This case is identical to the plane wave discussion of Section 3.4.2. In this case, the ray path separation $\Delta\vec{p}_{ob}$ given by Eq. (3.91) reduces $\vec{x}_o - \vec{x}_b$ and Eq. (3.93) reduces to the plane wave result

$$D_\psi(\vec{x}_o, \vec{x}_b) = 6.88 \left(\frac{|\vec{x}_o - \vec{x}_b|}{r_0} \right)^{5/3}, \tag{3.94}$$

where r_0 is the plane Fried parameter given by Eq. (3.61). In the second geometry we consider the case in which the two sources are located at the same point at some finite altitude on the telescope axis. This geometry is the same as that used to derive the spherical wave structure function. In this case the ray path separation is given by

$$\Delta\vec{p}_{ob}(z) = (\vec{x}_o - \vec{x}_b) \left(\frac{L - z}{L} \right), \tag{3.95}$$

and Eq. (3.93) reduces to the result given by Eq. (3.94) with r_0 now given by the spherical wave Fried parameter of Eq. (3.67).

3.5 First–order turbulence effects on incoherent imaging

The interferometric view of imaging introduced in Section 2.3.1 demonstrates the key role of the optical field's spatial coherence in determining imaging system performance. It was shown that the spatial coherence of the pupil field establishes the expected outcome of an imaging experiment. For incoherent imaging, Eq. (2.74) establishes the following relationship between the measured image spectrum, I, and the pupil field spatial correlation function, $\Gamma_p(\Delta\vec{x})$:

$$I(\vec{f}) = \Gamma_p(\overline{\lambda} d_i \vec{f}) \int d\vec{x}\, W(\vec{x}) W(\vec{x} - \overline{\lambda} d_i \vec{f}), \tag{3.96}$$

where $W(\vec{x})$ is the pupil function. The unique relationship between $\Gamma_p(\Delta\vec{x})$ and the object irradiance distribution is given by the Van Cittert-Zernike theorem presented in Eq. (2.71) of Section 2.3.1. The interferometric view of imaging described by Eq. (3.96) establishes the key concept that an imaging system forms an image via the interference of pairs of field points in the pupil. These interference effects are manifested as fringes that are characterized by both an amplitude (visibility) and phase (spatial shift). Optical aberrations that alter $\Gamma_p(\Delta\vec{x})$ have the effect of perturbing the amplitude and phase of the fringes. In addition, the aberration nullifies the relationship between the object irradiance distribution and $\Gamma_p(\Delta\vec{x})$ that exists for free space propagation (i.e., no atmospheric turbulence). As a result, the imaging system produces a distorted image. The fringes from each pair of field points in the pupil no longer add together in the way required for aberration free imaging.

Optical wave propagation through the atmosphere has the effect of altering the spatial coherence of the field in the pupil. The statistical nature of this alteration is presented in Section 3.4 via the development of the pupil field correlation function $\Gamma_p(\Delta\vec{x})$. In this section we present an analysis of the degradations suffered in an imaging system for an altered pupil field correlation function. We focus on incoherent imaging and quasimonochromatic conditions are assumed unless stated otherwise. In Subsection 3.5.1, a random screen model for the turbulence is presented. The random screen is placed in the pupil plane of the imaging system and effectively models the atmospheric induced field perturbations. In Subsection 3.5.2 the statistical characteristics of the screen are used to characterize imaging performance via the optical transfer function (OTF). The statistical characteristics specific to atmospheric induced perturbations developed in Section 3.4 are incorporated into the screen model. The result of incorporating these statistics is a "transfer function" applicable for propagation through the Earth's atmosphere. The imaging performance results presented in Subsection 3.5.3 are derived under the assumption of "long exposure" imaging. In long exposure imaging the image is captured over an exposure time that is long compared to the coherence time of the atmosphere. In Subsection 3.5.4 we present an analysis for characterizing "short exposure" imaging performance.

3.5.1 Thin screen turbulence model

In Section 3.4 we developed expressions describing the space and time correlation properties of a perturbed field in the pupil plane of an optical system. In this earlier development it was assumed that a plane or spherical wave enters the atmospheric turbulence region and propagates to the pupil plane of the system. In this section we consider the more general situation in which the incident optical field may be something other than a plane or spherical wave. In fact, in most cases of interest the incident field will not be one of these two fundamental waves. Any imaged object, that is other than a point source, gives rise to an optical field that departs from the simple plane and spherical wave models. In order to use the plane and spherical wave results of the previous section, it is convenient to cast the turbulence propagation problem in terms of propagation through an equivalent thin, random screen. In other words, we will use the screen to effectively replace the turbulence region. The screen takes on the statistical correlation properties found in Section 3.4 and is used to characterize the amplitude and phase perturbations caused by propagation through the turbulence. The screen model is useful since it has the effect of breaking the problem of analyzing imaging performance into two fundamental problems: 1) characterizing the turbulence perturbations in the pupil and 2) using this characterization to investigate imaging performance. The first problem is the primary subject of Section 3.4. The second problem is the subject of this section.

The goal in developing the random screen model is to capture the statistical characteristics of the turbulence induced perturbations. Figure 3.8 shows the propagation geometry of interest. The figure

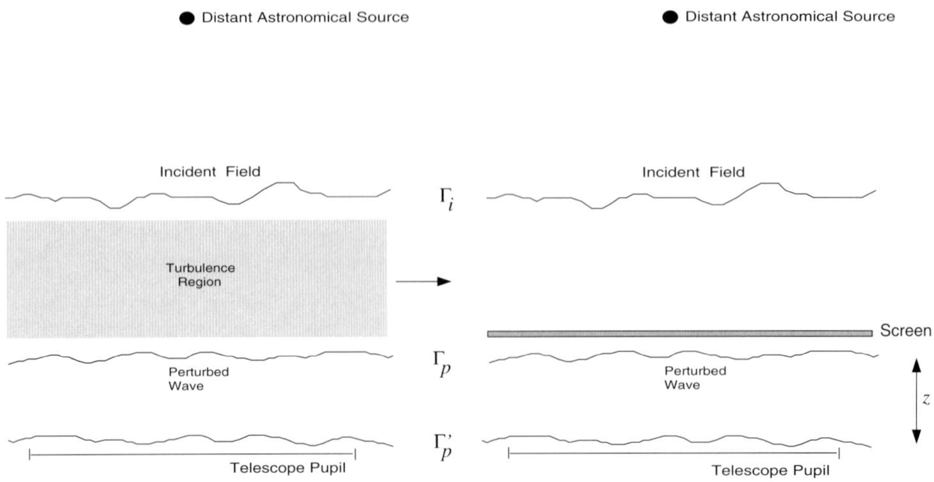

FIGURE 3.8
Geometry for characterizing the relationship between the input and output spatial coherence properties of an optical field for propagation through a homogeneous, isotropic turbulence region.

shows that the screen model is used to effectively replace the turbulence region. The screen perturbs the amplitude and phase of the incident optical field in the same way that propagation through turbulence does. In mathematical terms the screen relates the incident field, $u_i(\vec{x})$, and the pupil field, $u_p(\vec{x})$, by

$$u_p(\vec{x}) = u_i(\vec{x})t_s(\vec{x}), \tag{3.97}$$

where $t_s(\vec{x})$ is the screen transparency function describing the random field perturbation. It is important to note that $t_s(\vec{x})$, in its most general form, is complex and, as such, can describe both amplitude and phase perturbations. For the analysis to follow, the random process describing the screen is assumed spatially wide sense stationary (WSS). The incident optical field is also assumed spatially WSS and independent of the screen. Using straightforward random process concepts [14], it is easy to show the relationship between the incident and transmitted spatial correlation functions illustrated in Fig. 3.8 is given by

$$\Gamma_p(\Delta\vec{x}) = \Gamma_s(\Delta\vec{x})\Gamma_i(\Delta\vec{x}), \tag{3.98}$$

where $\Gamma_i(\Delta\vec{x})$ and $\Gamma_p(\Delta\vec{x})$ are the incident and transmitted optical field correlation functions, respectively, and $\Gamma_s(\Delta\vec{x})$ is the spatial correlation function of the random screen. The imaging performance analysis to follow is accomplished in a manner that is completely independent of the form of $\Gamma_s(\Delta\vec{x})$. Only at the end do we substitute the expressions for the spatial correlation function $\Gamma_s(\Delta\vec{x})$ as developed and discussed in Section 3.4. Note that in this section the subscript p is being used to designate the pupil field properties arising from both the atmospheric induced perturbations and the incident field. In the previous section the subscript p was used to only describe the atmospheric induced field perturbations since the incident fields were assumed to be either a plane or spherical wave. In the context of this section the subscript s will be reserved for describing the random screen, and as such, it is only associated with the turbulence induced field perturbations. Equation (3.98) is the main result of this section and is used in the next subsection to account for the effects of the turbulence induced field perturbations in the pupil of an imaging system.

Before proceeding to the next subsection we present an interesting result. In some instances it may be of interest to know how the field correlation function $\Gamma_p(\Delta\vec{x})$ changes with propagation through a region of non-turbulence. In other words, we may be interested in how $\Gamma_p(\Delta\vec{x})$ changes with free space propagation. Understanding this question is particularly useful for the case in which the perturbed optical field propagates through non-turbulent regions before arriving at the pupil plane of the imaging system. Fortunately the question of how the field correlation properties evolve with free space propagation is easily answered using a relationship well established in the field of linear, shift-invariant systems. In Section 2.2.1 we showed that free space propagation can be viewed within a shift-invariant, linear systems framework. The free space transfer function, $H_d(\vec{f})$, for Fresnel propagation is given by Eq. (2.11). The corresponding impulse response, $h_d(\vec{x})$, is given by Eq. (2.10). By viewing free space propagation within this framework, we can relate the input correlation function and the output correlation function some propagation distance z away by [14]

$$\Gamma'_p(\Delta\vec{x}) = \Gamma_p(\Delta\vec{x}) * h_d(\Delta\vec{x}) * h^*_d(-\Delta\vec{x}), \tag{3.99}$$

where the output and input field correlation functions $\Gamma'_p(\Delta\vec{x})$ and $\Gamma_p(\Delta\vec{x})$ are shown in Fig. 3.8. Fourier transforming Eq. (3.99) gives

$$\tilde{\Gamma}'_p(\vec{f}) = \tilde{\Gamma}_p(\vec{f})|H_d(\vec{f})|^2, \tag{3.100}$$

where $\tilde{\Gamma}'_p(\vec{f})$ and $\tilde{\Gamma}_p(\vec{f})$ are the Fourier transforms of $\Gamma'_p(\Delta\vec{x})$ and $\Gamma_p(\Delta\vec{x})$, respectively. Noting from Eq. (2.11) that the modulus of $H_d(\vec{f})$ is unity for $|\vec{f}| < \lambda^{-1}$, it is easy to see that $\tilde{\Gamma}'_p(\vec{f}) = \tilde{\Gamma}_p(\vec{f})$ and thus, $\Gamma'_p(\Delta\vec{x}) = \Gamma_p(\Delta\vec{x})$. In other words, the spatial coherence of the transmitted field is invariant with free space propagation.

Using the screen model and relationship between the incident and transmitted spatial correlation functions given in Eq. (3.98) allows us to now consider the effects of turbulence on an incoherent imaging system.

3.5.2 Long exposure imaging through a random medium

In discussing long exposure imaging we first need to identify the specific sense of the word “long” used here. The phrase “long exposure imaging” is used here to mean an imaging experiment in which the captured image is the result of a large number of pupil plane perturbation realizations. As discussed in Section 3.4, atmospheric turbulence effects evolve with time. This temporal evolution is characterized by a correlation time, τ_c. Qualitatively we can think of the atmosphere as generating a new field perturbation realization every τ_c seconds. If an imaging system exposes the image plane detector for a large number of atmospheric correlation times, the image is called a long exposure image. In this long exposure case, it is assumed the atmosphere evolves through a large number of independent realizations and as a result, the imaging system experiences a large number of turbulence induced realizations. In this case the statistical result describing the pupil field in Eq. (3.98) is valid since the imaging system, over the exposure time, experiences the correlations described by $\Gamma_s(\Delta\vec{x})$. On the other hand, Eq. (3.98) is not valid for a single, short exposure image since the turbulence induced perturbations are “frozen” by the short exposure time. In a subsequent section we treat a special case of short exposure imaging that allows for a statistical modeling of the turbulence effects.

Two approaches can be used to analyze the performance degradations caused by atmospheric turbulence. In the first, we capitalize on the intuitive understanding of imaging afforded by the “interferometric view of imaging” described in the last subsection of Section 2.3.1. In the second approach the derivation starts with the OTF expression given by Eq. (2.44) in Section 2.2.6.

Approach 1: We start by considering Eq. (3.96). The correlation function $\Gamma_p(\Delta\vec{x})$ in Eq. (3.96) describes the spatial coherence of the pupil plane field. The correlation function is a combination of both the coherence properties arising from the object (as predicted by the Van Cittert-Zernike theorem) and the field perturbations caused by the turbulence. The effects of field propagation through the turbulence are modeled with the random screen introduced in the previous subsection. As such, the optical field in the pupil is characterized using Eq. (3.98)

$$\Gamma_p(\Delta\vec{x}) = \Gamma_s(\Delta\vec{x})\Gamma_i(\Delta\vec{x}), \tag{3.101}$$

where $\Gamma_i(\Delta\vec{x})$ describes the spatial coherence of the optical field in the absence of turbulence and $\Gamma_s(\Delta\vec{x})$ is the spatial coherence of the optical field perturbations arising from propagation through the turbulence. Substituting Eq. (3.101) into Eq. (3.96) results in the following expression for the measured image spectrum, $I(\vec{f})$:

$$I(\vec{f}) = \Gamma_i(\overline{\lambda}d_i\vec{f})\Gamma_s(\overline{\lambda}d_i\vec{f}) \int \mathrm{d}\vec{x}\, W(\vec{x})W(\vec{x} - \overline{\lambda}d_i\vec{f}). \tag{3.102}$$

The incident spatial coherence described by $\Gamma_i(\Delta\vec{x})$ is the Fourier transform of the object irradiance distribution, $o(\vec{x})$, (by the Van Cittert-Zernike theorem). Using this fact and an assumption of unity image magnification gives

$$I(\vec{f}) = O(\vec{f})\Gamma_s(\overline{\lambda}d_i\vec{f}) \int \mathrm{d}\vec{x}\, W(\vec{x})W(\vec{x} - \overline{\lambda}d_i\vec{f}), \tag{3.103}$$

where $O(\vec{f})$ designates the Fourier transform of the object irradiance distribution. In using the Van Cittert-Zernike theorem (Eq. (2.71)) to obtain Eq. (3.103), we have ignored the quadratic phase term ψ. As discussed in Section 2.3.1, this term can be ignored in the case of imaging since the lens in the system cancels the quadrature phase. The integral in Eq. (3.103) is a scaled version of the system's diffraction limited OTF (see the OTF expression given in Eq. (2.44)). Substituting the designation $\mathcal{H}_o(\vec{f})$ for the diffraction limited OTF into Eq. (3.103) finally gives

$$I(\vec{f}) = O(\vec{f})\Gamma_s(\overline{\lambda}d_i\vec{f})\mathcal{H}_o(\vec{f}). \tag{3.104}$$

The frequency response expression given in Eq. (3.104) is the main result of this subsection. It completely describes the frequency response of the turbulence degraded imaging system using long exposure times. The interpretation of the turbulence effects afforded by Eq. (3.104) is appealing. The measured image spectrum, $I(\vec{f})$, is a filtered version of the object spectrum, $O(\vec{f})$. The filtering consists of two components: 1) the filter due to the aberration free optics, $\mathcal{H}_o(\vec{f})$ and 2) the filter due to the turbulence induced field perturbations, $\Gamma_s(\Delta\vec{x})$. A normalized version of $\Gamma_s(\Delta\vec{x})$ is referred to as the long exposure atmospheric transfer function and designated here as

$$\mathcal{H}_{LE}(\vec{f}) = \frac{\Gamma_s(\overline{\lambda}d_i\vec{f})}{\Gamma_s(0)}. \tag{3.105}$$

Rewriting Eq. (3.104) using $\mathcal{H}_{LE}(\vec{f})$ gives the result

$$I(\vec{f}) = O(\vec{f})\mathcal{H}_{LE}(\vec{f})\mathcal{H}_o(\vec{f}), \tag{3.106}$$

where we have, as throughout, ignored any constant factors. Eqs. (3.104) and (3.106) show explicitly how the spatial correlation properties of the field perturbations affect the imaging system. The atmospheric transfer function $\mathcal{H}_{LE}(\vec{f})$ will have values near unity for highly correlated field perturbations and nearly diffraction limited imaging will result. On the other hand, if the optical field perturbations are poorly correlated over the pupil, $\mathcal{H}_{LE}(\vec{f})$ will attenuate spatial frequencies for $|\vec{f}| > 0$. This attenuation arises from the amplitude and phase perturbations of the sinusoidal fringes associated with each pair of points in the pupil. The decorrelation of the field at a pair of points in the pupil causes a random amplitude and phase shift of the associated fringe in the image plane. Every pair of points in the aperture suffers this decorrelation and as a result, the fringes no longer sum in an "in phase" manner. The effect of the random amplitude and phase on the fringe summation is a significant reduction of the amplitude of the associated image spatial frequency component.

Approach 2: The effects of fixed pupil plane aberrations are discussed in the OTF discussion of Section 2.2.6. This previous discussion points out a straightforward mathematical approach to characterizing turbulence effects. In this approach the pupil function, $W(\vec{x})$, is generalized to account for both the pupil aperture and the pupil plane aberration. Using the screen model introduced above we can write the generalized pupil function as

$$W(\vec{x}) = W_p(\vec{x})t_s(\vec{x}). \tag{3.107}$$

where $W_p(\vec{x})$ is a real valued function describing the pupil aperture and $t_s(\vec{x})$ is a complex valued screen function that describes the field aberration in the pupil plane. Substituting Eq. (3.107) into the expression for the OTF given in Eq. (2.44) gives the OTF for one realization of the turbulence induced aberration:

$$\mathcal{H}(\vec{f}) = \frac{(W_p(\vec{f}\lambda d_i)t_s(\vec{f}\lambda d_i)) \star (W_p(\vec{f}\lambda d_i)t_s(\vec{f}\lambda d_i))}{(W_p(0)t_s(0)) \star W_p(0)t_s(0))}. \tag{3.108}$$

Given the long exposure assumption, $t_s(\vec{x})$ will evolve through many independent realizations over an exposure time. As a result it is appropriate to consider the ensemble average of Eq. (3.108). Taking the ensemble average of both sides of Eq. (3.108) while at the same time expanding the correlation operator (defined in Eq. (2.43)) gives

$$\langle \mathcal{H}(\vec{f}) \rangle = \left\langle \frac{\int \mathrm{d}\vec{x}\, W_p(\vec{x})W_p(\vec{x} - \vec{f}\lambda d_i)t_s(\vec{x})t_s^*(\vec{x} - \vec{f}\lambda d_i)}{\int \mathrm{d}\vec{x}\, W_p^2(\vec{x})|t_s(\vec{x})|^2} \right\rangle. \tag{3.109}$$

In order to continue with the evaluation of the right hand side of Eq. (3.109), it is necessary to make the approximation that the ensemble average is equal to the average of the numerator divided by the average of the denominator. The approximation is justified in cases in which the statistical fluctuation of $|t_s(\vec{x})|^2$ is much less than $t_s(\vec{x})t_s^*(\vec{x} - \vec{f}\lambda d_i)$. In the special case that the perturbations are only phase perturbations (i.e., $t_s(\vec{x}) = \exp(j\psi(\vec{x}))$), then the approximation is exact. Rewriting Eq. (3.109) using this approximation gives

$$\langle \mathcal{H}(\vec{f}) \rangle = \frac{\int \mathrm{d}\vec{x}\, W_p(\vec{x})W_p(\vec{x} - \vec{f}\lambda d_i)\langle t_s(\vec{x})t_s^*(\vec{x} - \vec{f}\lambda d_i)\rangle}{\int \mathrm{d}\vec{x}\, W_p^2(\vec{x})\langle |t_s(\vec{x})|^2\rangle}. \tag{3.110}$$

Noting the existence of the correlation function $\Gamma_s(\Delta\vec{x})$ in Eq. (3.110) and using the WSS property assumed for $t_s(\vec{x})$ allows us to write

$$\langle \mathcal{H}(\vec{f}) \rangle = \frac{\Gamma_s(\vec{f}\lambda d_i)}{\Gamma_s(0)} \frac{\int \mathrm{d}\vec{x}\, W_p(\vec{x})W_p(\vec{x} - \vec{f}\lambda d_i)}{\int \mathrm{d}\vec{x}\, W_p^2(\vec{x})}. \tag{3.111}$$

The first term in Eq. (3.111) is the atmospheric transfer function defined previously in Eq. (3.105) and the second term is the definition of the OTF for an unaberrated imaging system. Using the notation $\mathcal{H}_{LE}(\vec{f})$ for the atmospheric transfer function and $\mathcal{H}_o(\vec{f})$ for the diffraction limited OTF finally gives

$$\langle \mathcal{H}(\vec{f}) \rangle = \mathcal{H}_{LE}(\vec{f})\mathcal{H}_o(\vec{f}). \tag{3.112}$$

This result is exactly the same as the result given by Eq. (3.106).

To complete the characterization of long exposure imaging the statistics of the atmospheric induced field perturbations must be incorporated into $\mathcal{H}_{LE}(\vec{f})$. In the next subsection the turbulence induced statistics presented in Section 3.4 are used to develop an expression for the atmospheric transfer function, $\mathcal{H}_{LE}(\vec{f})$, applicable for imaging in the Earth's atmosphere.

3.5.3 Long exposure OTF, $\mathcal{H}_{LE}(\vec{f})$

In the previous subsection we found the transfer function associated with long exposure imaging through a random medium to be

$$\mathcal{H}_{LE}(\vec{f}) = \frac{\Gamma_s(\overline{\lambda} d_i \vec{f})}{\Gamma_s(0)}, \tag{3.113}$$

where $\Gamma_s(\Delta\vec{x})$ is the spatial correlation function for the turbulence induced field perturbations in the pupil of the imaging system. The field correlation expressions derived in Section 3.4 can now be substituted into Eq. (3.113) to obtain the long exposure transfer function associated with imaging through atmospheric turbulence. Before making those substitutions, we need to first derive a result that is only stated in Section 3.4. The field correlation expressions developed in Section 3.4 all have the form

$$\Gamma_s(\Delta\vec{x}) = \exp\left[-\frac{1}{2}D(\Delta\vec{x})\right], \tag{3.114}$$

where $D(\Delta\vec{x})$ is either the phase or the wave structure function. If the turbulence conditions are such that near field conditions exist (as discussed in Section 3.4.2), then the structure function $D(\Delta\vec{x})$ is a phase structure function and the geometrical optics results derived in Section 3.4 are applicable. If on the other hand, non-near field conditions exist, both amplitude and phase perturbations will exist in the pupil plane. In this case we asserted, without proof, in Section 3.4.2 that $D(\Delta\vec{x})$ becomes a wave structure function as defined by Eq. (3.72). In the following pages we prove this assertion using an approach similar to that first presented by Fried [1].

In order to proceed, we must go back to the representation of the field perturbations in the pupil. The turbulence induced field perturbation can be characterized with a log-amplitude perturbation, $\chi(\vec{x})$, and phase perturbation, $\psi(\vec{x})$. In other words, the random screen $t_s(\vec{x})$ can be written

$$t_s(\vec{x}) = \exp(\chi(\vec{x}))\exp(j\psi(\vec{x})), \tag{3.115}$$

where $\vec{x}$ is a point in the pupil. Using Eq. (3.115) to express the coherence function $\Gamma_s(\Delta\vec{x})$ gives

$$\begin{aligned}\Gamma_s(\Delta x) &= \langle t_s(\vec{x})t_s^*(\vec{x}-\Delta\vec{x})\rangle, \\ &= \langle \exp\big(\chi(\vec{x})+\chi(\vec{x}-\Delta\vec{x})\big)\exp\big(j(\psi(\vec{x})-\psi(\vec{x}-\Delta\vec{x}))\big)\rangle.\end{aligned} \tag{3.116}$$

Without assuming anything about the statistics of $\chi(\vec{x})$ and $\psi(\vec{x})$ further progress is impossible. The standard approach is to assume that $\chi(\vec{x})$ and $\psi(\vec{x})$ are jointly WSS Gaussian random processes [1, 8]. The Gaussian assumption can be justified in most situations of interest using a central limit theorem argument [3]. Equation (3.46) is a well known result established from characteristic functions of Gaussian random variables [8, 14] and can be used to show that $\Gamma_s(\Delta\vec{x})$ reduces to

$$\begin{aligned}\Gamma_s(\Delta x) = {} & \exp\{\overline{\chi(\vec{x})}+\overline{\chi(\vec{x}-\Delta\vec{x})}\}\exp\{j(\overline{\psi(\vec{x})}-\overline{\psi(\vec{x}-\Delta\vec{x})}\} \\ & \times\exp\left\{+\frac{1}{2}\left(\overline{(\chi(\vec{x})+\chi(\vec{x}-\Delta\vec{x}))^2}-\left(\overline{\chi(\vec{x})}+\overline{\chi(\vec{x}-\Delta\vec{x})}\right)^2\right)\right\} \\ & \times\exp\left\{-j\rho'\left(\overline{(\psi(\vec{x})-\psi(\vec{x}-\Delta\vec{x}))^2}-\left(\overline{\psi(\vec{x})}-\overline{\psi(\vec{x}-\Delta\vec{x})}\right)^2\right)^{1/2}\right. \\ & \qquad\left.\times\left(\overline{(\chi(\vec{x})+\chi(\vec{x}-\Delta\vec{x}))^2}-\left(\overline{\chi(\vec{x})}+\overline{\chi(\vec{x}-\Delta\vec{x})}\right)^2\right)^{1/2}\right\} \\ & \times\exp\left\{-\frac{1}{2}\left(\overline{(\psi(\vec{x})-\psi(\vec{x}-\Delta\vec{x}))^2}-\left(\overline{\psi(\vec{x})}-\overline{\psi(\vec{x}-\Delta\vec{x})}\right)^2\right)\right\}.\end{aligned} \tag{3.117}$$

where ρ' is the correlation coefficient describing the correlation between $\chi(\vec{x}) + \chi(\vec{x} - \Delta\vec{x})$ and $\psi(\vec{x}) - \psi(\vec{x} - \Delta\vec{x})$. Fried [1] has shown that to be consistent with the joint WSS assumption the correlation coefficient ρ' must be zero. The WSS assumption also implies $\overline{\psi(\vec{x})} = \overline{\psi(\vec{x} - \Delta\vec{x})}$ and $\overline{\chi(\vec{x})} = \overline{\chi(\vec{x} - \Delta\vec{x})}$. Note that the overbar notation used above is equivalent to the ensemble average angle brackets used previously. The switch proves to be less cumbersome than the angle brackets in this development. Equation (3.117) can now be reduced to

$$\Gamma_s(\Delta x) = \exp\{\overline{\chi(\vec{x})} + \overline{\chi(\vec{x} - \Delta\vec{x})}\} \times \exp\left\{+\frac{1}{2}\left(\overline{(\chi(\vec{x}) + \chi(\vec{x} - \Delta\vec{x}))^2} - 4\overline{\chi(\vec{x})}^2\right)\right\} \exp\left\{-\frac{1}{2}D_\psi(\Delta\vec{x})\right\}, \quad (3.118)$$

where we have incorporated the phase structure function $D_\psi(\Delta\vec{x})$ defined in Eq. (3.49) as

$$D_\psi(\Delta\vec{x}) = \overline{(\psi(\vec{x}) - \psi(\vec{x} + \Delta\vec{x}))^2}. \quad (3.119)$$

Equation (3.118) can be further simplified by invoking a "conservation of energy" argument [1]. Since the turbulent atmosphere is assumed to be a non absorbing medium, the average value of the irradiance of a unit amplitude optical field transmitted through the turbulence must be unity. Mathematically this argument can be stated as

$$\left\langle |t_s(\vec{x})|^2 \right\rangle = \langle \exp[2\chi(\vec{x})] \rangle = 1. \quad (3.120)$$

Making use of the fact that $\chi(\vec{x})$ is Gaussian we can use Eq. (3.46) to show that Eq. (3.120) can be written

$$\langle \exp[2\chi(\vec{x})] \rangle = \exp\{2\langle \chi(\vec{x}) \rangle\} \exp\{2\sigma_\chi^2\} = 1, \quad (3.121)$$

where σ_χ^2 is the variance of $\chi(\vec{x})$. Equation (3.121) indicates that $\langle \chi(\vec{x}) \rangle = -\sigma_\chi^2$. Substituting this result back into Eq. (3.118) reduces $\Gamma_s(\Delta\vec{x})$ to

$$\Gamma_s(\Delta\vec{x}) = \exp\left\{-\frac{1}{2}(D_\chi(\Delta\vec{x}) + D_\psi(\Delta\vec{x}))\right\}, \quad (3.122)$$

where $D_\chi(\Delta\vec{x})$ is the log amplitude structure function defined by

$$D_\chi(\Delta\vec{x}) = \overline{(\chi(\vec{x}) - \chi(\vec{x} + \Delta\vec{x}))^2}. \quad (3.123)$$

Note that the expression in Eq. (3.122) is the same result introduced without proof in Eq. (3.70) of Section 3.4. The sum of amplitude and phase structure functions is designated the wave structure function, $D(\Delta\vec{x}) = D_\chi(\Delta\vec{x}) + D_\psi(\Delta\vec{x})$.

Substituting the expression for $\Gamma_s(\Delta\vec{x})$ given by Eq. (3.122) into Eq. (3.113) finally results in the long exposure atmospheric transfer function $\mathcal{H}_{LE}(\vec{f})$:

$$\begin{aligned} \mathcal{H}_{LE}(\vec{f}) &= \exp\left\{-\frac{1}{2}(D_\chi(\overline{\lambda} d_i \vec{f}) + D_\psi(\overline{\lambda} d_i \vec{f}))\right\}, \\ &= \exp\left\{-\frac{1}{2}D(\overline{\lambda} d_i \vec{f})\right\}. \end{aligned} \quad (3.124)$$

To complete this investigation of long exposure imaging we now make use of the wave structure function developed for wave propagation through atmospheric turbulence. Equations (3.63) and (3.66) give the plane and spherical wave structure functions for wave propagation through a locally homogeneous, isotropic medium described with Kolmogorov index of refraction power spectrum. Substituting these two results in Eq. (3.124) gives the well known long exposure OTF developed by Fried [1]:

$$\mathcal{H}_{LE}(\vec{f}) = \exp\left\{-\frac{1}{2}6.88\left(\frac{\overline{\lambda} d_i |\vec{f}|}{r_o}\right)^{5/3}\right\}. \quad (3.125)$$

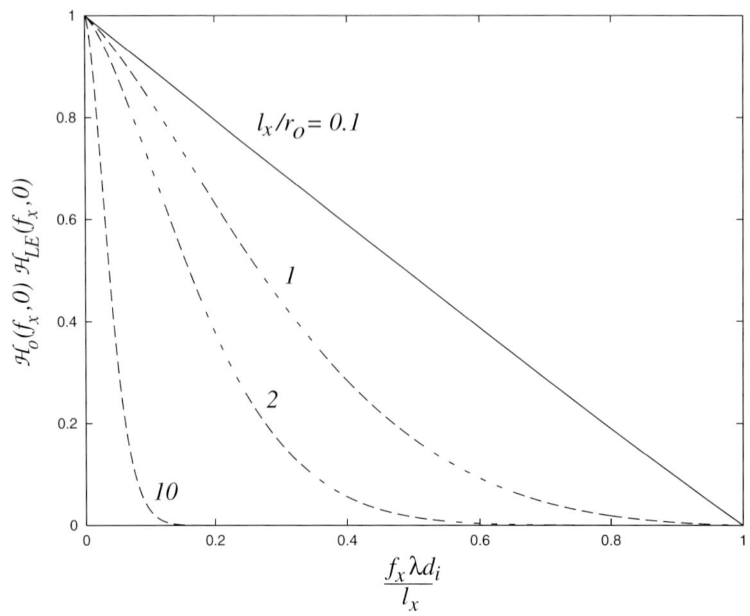

FIGURE 3.9
Family of long exposure OTFs, $\mathcal{H}_o(\vec{f})\mathcal{H}_{LE}(\vec{f})$, plotted versus the normalized spatial frequency $f_x\lambda d_i/l_x$. The ratio l_x/r_o ranges from 0.1 to 10.

where r_o is given either by the plane or spherical wave definitions of Eqs. (3.61) or (3.67), respectively. Fig. 3.9 shows a family of long exposure OTFs for the rectangular aperture described in the incoherent imaging example of Section 2.2.6. In this figure the product $\mathcal{H}_o(\vec{f})\mathcal{H}_{LE}(\vec{f})$ is plotted along the f_x axis. The normalized aperture dimension l_x/r_o ranges from 0.1 to 10. Notice the significant attenuation of spatial frequencies for cases in which $l_x/r_o > 2$. Fig. 3.10 illustrates cross-sections of the corresponding PSFs for the OTFs shown in Fig. 3.9.

In subsequent sections of this chapter and subsequent chapters of the book we will generally assume that the turbulence induced perturbations in the pupil of the imaging system are dominated by phase perturbations. In other words, the amplitude perturbation $\chi(\vec{x})$ is assumed small compared to $\psi(\vec{x})$. This assumption is equivalent to the near field assumptions discussed in Section 3.4. The near field assumption is generally applicable for imaging vertically through the Earth's atmosphere from good observatory sites. In terms of the wave structure function, the near field assumption is equivalent to assuming $D(\Delta\vec{x})$ is dominated by the phase structure function, $D_\psi(\Delta\vec{x})$.

3.5.4 Short exposure imaging through atmospheric turbulence

It was stated earlier that the statistical techniques presented in the previous subsection are not applicable for analyzing short exposure imaging. The phrase "short exposure" imaging refers to the situation in which the exposure time is short enough to freeze the effects of the atmosphere. In the long exposure discussion of the previous subsection the approach for analyzing imaging performance was based on the assumption that the imaging system experienced a large number of independent realizations of the atmospheric induced perturbations. In the case of short exposure imaging, the imaging system experiences one realization of the atmospheric induced perturbations. It appears that statistical techniques of the previous subsection will not aid us in analyzing the performance for short exposure imaging. Rather than considering short exposure imaging in the strict sense described above we consider another

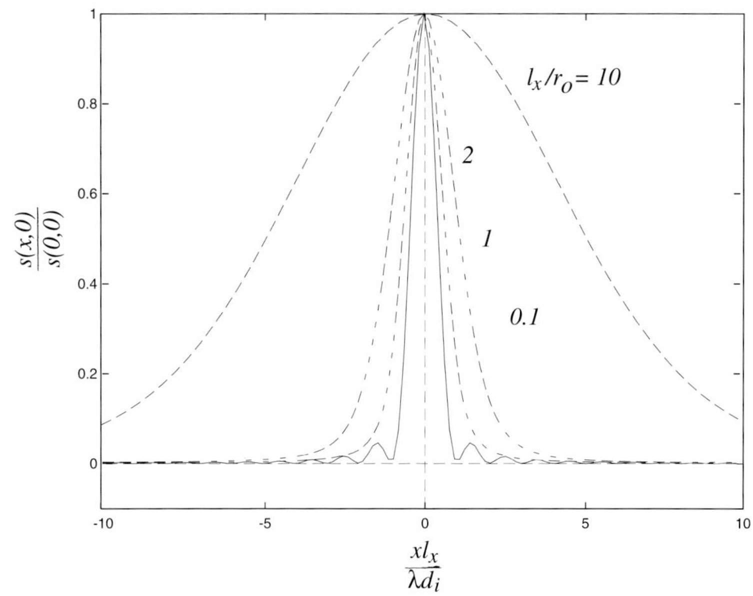

FIGURE 3.10
Family of long exposure PSFs plotted versus the normalized image plane vector $xl_x/\lambda d_i$. The ratio l_x/r_o ranges from 0.1 to 10.

scenario in which the statistical techniques can be used.

Very early it was recognized that the one of the dominant effects of the atmosphere is random wave front tilt in the pupil. The tilt is described mathematically as the vector slope of the best fit plane to the wave front phase $\psi(\vec{x})$. This component of the phase perturbation has the effect of shifting the location of the image in the image plane. It is this random shifting of the image that accounts for most of the degradations incurred for long exposure imaging. However, the shift does not distort the short exposure image. The short exposure image is distorted because the wave front tilt component is accompanied by higher spatial frequency perturbations. Fried [31] has shown that the tilt accounts for 90% of the power in the phase perturbations. With this fact in mind, it is natural to consider what performance can be achieved for an imaging system that compensates for the random shift of the image.

The compensation can take two forms. In the first, the imaging system measures the wave front tilt and adjusts, in real time, a flat tilt correction mirror. The tilt mirror is controlled in such a way that image motion is removed. A tilt compensation system of this type is a simple adaptive optical system. It is the performance of this type of system in conjunction with long exposure images that is of interest. In the second approach, the imaging system takes short exposure images. The short exposures freeze the effects of the turbulence induced perturbations for each image. Each short exposure image is subsequently re-centroided, thereby effectively removing the image shift caused by the random tilt. The ensemble of images are then summed. The summing of the centroided images is equivalent to the long exposure imaging of the real time tilt compensation system described above. The imaging performance of either one of these systems can be described with a transfer function just as in the long exposure case. The transfer function for this type of imaging is referred to in the literature with the misnomer "short exposure OTF". The "short exposure" label derives from the short exposures used to capture the images described above. It is important to remember that a reference to a "short exposure" OTF is, in most cases, a reference to a long exposure OTF for the case of tilt compensation.

In order to perform the imaging analysis of a tilt compensating system, we first need a screen model that accounts for tilt removal. Assuming that the wave front perturbation is a phase perturbation, we can express the tilt-removed phase screen in the following way:

$$t_s(\vec{x}) = \exp\left\{j(\psi(\vec{x}) - \vec{a}\cdot\vec{x})\right\}, \tag{3.126}$$

where the vector $\vec{a}$ describes the tilt of the wave front phase over the pupil. One method of relating $\vec{a}$ to the phase perturbation $\psi(\vec{x})$ is by a least squares fit of a tilted plane to $\psi(\vec{x})$ over the pupil $W(\vec{x})$. The least squares fit gives

$$\vec{a} = \hat{x}K_x \int \mathrm{d}\vec{x}\, W(\vec{x})\left[\vec{x}\cdot\hat{x}\right]\psi(\vec{x}) + \hat{y}K_y \int \mathrm{d}\vec{x}\, W(\vec{x})\left[\vec{x}\cdot\hat{y}\right]\psi(\vec{x}), \tag{3.127}$$

where $\hat{x}$ and $\hat{y}$ are x- and y-directed unit vectors in the pupil plane, $W(\vec{x})$ is the pupil aperture function, and K_x and K_y are normalization factors required to satisfy the equality in Eq. (3.127) if $\psi(\vec{x})$ is replaced by $\vec{a}\cdot\vec{x}$. These normalization factors are given by

$$\begin{aligned} K_x &= \left[\int \mathrm{d}\vec{x}\, W(\vec{x})(\vec{x}\cdot\hat{x})^2\right]^{-1}, \\ K_y &= \left[\int \mathrm{d}\vec{x}\, W(\vec{x})(\vec{x}\cdot\hat{y})^2\right]^{-1}. \end{aligned} \tag{3.128}$$

With the screen model of Eq. (3.126) we are now able to perform an analysis similar to the long exposure OTF analysis of the previous subsection. Just as in the long exposure case, the problem reduces to the calculation of $\Gamma_s(\Delta\vec{x})$. Substituting the expression for $t_s(\vec{x})$ given above into the definition of $\Gamma_s(\Delta\vec{x})$ found in Eq. (3.116) results in

$$\Gamma_s(\Delta x) = \left\langle \exp\{j(\psi(\vec{x}) - \psi(\vec{x}-\Delta\vec{x}) - \vec{a}\cdot\Delta\vec{x})\}\right\rangle. \tag{3.129}$$

Since $\psi(\vec{x})$ is assumed to be Gaussian random process, $\vec{a}$ will be a Gaussian random variable. Using the Gaussian statistical result of Eq. (3.46) and the assumption of stationarity for $\psi(\vec{x})$ allows us to write

$$\begin{aligned} \Gamma_s(\Delta\vec{x}) &= \exp\left\{-\frac{1}{2}\left(\overline{(\psi(\vec{x}) - \psi(\vec{x}-\Delta\vec{x}))^2} + \overline{(\vec{a}\cdot\Delta\vec{x})^2}\right.\right. \\ &\qquad \left.\left. - \overline{2(\psi(\vec{x}) - \psi(\vec{x}-\Delta\vec{x}))(\vec{a}\cdot\Delta\vec{x})}\right)\right\}, \\ &= \exp\left\{-\frac{1}{2}\left(D_\psi(\Delta\vec{x}) + \frac{1}{2}\overline{|\vec{a}|^2}|\Delta\vec{x}|^2\right.\right. \\ &\qquad \left.\left. - \overline{2(\psi(\vec{x}) - \psi(\vec{x}-\Delta\vec{x}))(\vec{a}\cdot\Delta\vec{x})}\right)\right\}. \end{aligned} \tag{3.130}$$

Fried [1] was the first to work with the expression given above. In order to simplify the analysis he assumed the residual phase after tilt removal,

$$\psi_r(\vec{x}) = \psi(\vec{x}) - \vec{a}\cdot\vec{x}, \tag{3.131}$$

to be uncorrelated with the tilt, $\vec{a}$. Since both $\psi_r(\vec{x})$ and $\vec{a}$ are Gaussian, and assumed uncorrelated, they are also statistically independent. This assumption considerably simplifies the evaluation of Eq. (3.130). From a mathematical viewpoint the assumption is obviously not valid since $\vec{a}$ depends on $\psi_r(\vec{x})$. Fried argued on an qualitative level that the excursions of $\psi(\vec{x})$ about the best fit plane are uncorrelated with the plane, and in reality this correlation is small for pupil sizes large with respect to r_o. Heidbreder [32] and others [7, 8, 25] have discussed this assumption. In particular, Heidbreder has laid out the

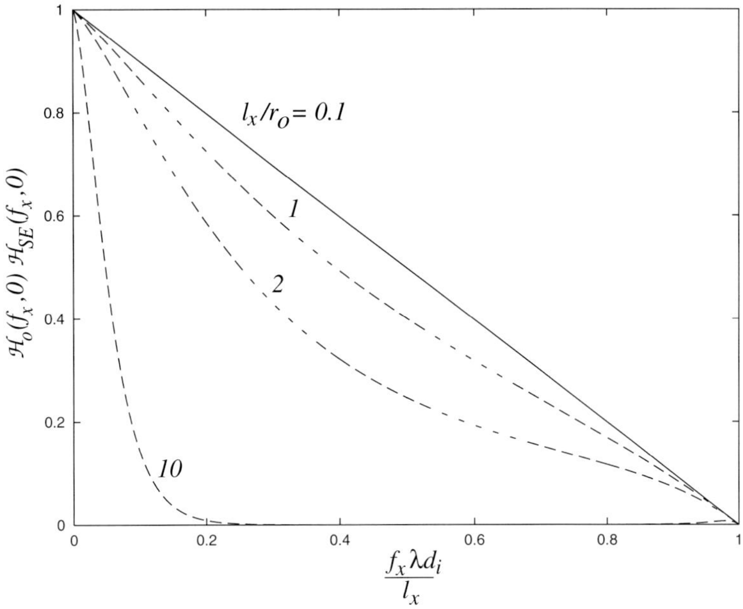

FIGURE 3.11
Family of short exposure OTFs, $\mathcal{H}_o(\vec{f})\mathcal{H}_{SE}(\vec{f})$, plotted versus the normalized spatial frequency $f_x\lambda d_i/l_x$. The ratio l_x/r_o ranges from 0.1 to 10.

mathematics for computing Eq. (3.130) without the assumption. Our analysis follows Fried [1], while recognizing that it is an approximation that improves as the aperture size increases.

Writing $\psi(\vec{x})$ as a function of $\psi_r(\vec{x})$ and $\vec{a}$ gives

$$\psi(\vec{x}) = \psi_r(\vec{x}) + \vec{a}\cdot\vec{x}. \tag{3.132}$$

Substituting this expression into Eq. (3.130) and making use of the assumption that $\psi_r(\vec{x})$ and $\vec{a}$ are uncorrelated gives

$$\Gamma_s(\Delta\vec{x}) = \exp\left\{-\frac{1}{2}\left(D_\psi(\Delta\vec{x}) - \frac{1}{2}\overline{|\vec{a}|^2}|\Delta\vec{x}|^2\right)\right\}. \tag{3.133}$$

Fried has computed $\overline{|\vec{a}|^2}/2$ to be [1]

$$\frac{\overline{|\vec{a}|^2}}{2} = 6.88 r_o^{-5/3} D^{-1/3}. \tag{3.134}$$

Substituting the expression for $D_\psi(\Delta\vec{x})$ found in Eq. (3.58) and substituting Eq. (3.134) into Eq. (3.133) gives the "short exposure" OTF [1]

$$\mathcal{H}_{SE}(\vec{f}) = \Gamma_s(\overline{\lambda d_i \vec{f}}) = \exp\left\{-\frac{1}{2}6.88\left(\frac{|\overline{\lambda d_i \vec{f}}|}{r_o}\right)^{5/3}\left[1-\left(\frac{|\overline{\lambda d_i \vec{f}}|}{D}\right)^{1/3}\right]\right\}, \tag{3.135}$$

where D is the pupil diameter. Fig. 3.11 shows a family of short exposure OTFs for the rectangular aperture described in the incoherent imaging example of Section 2.2.6. In this figure the product $\mathcal{H}_o(\vec{f})\mathcal{H}_{SE}(\vec{f})$ is plotted along the f_x axis. The normalized aperture dimension l_x/r_o ranges from

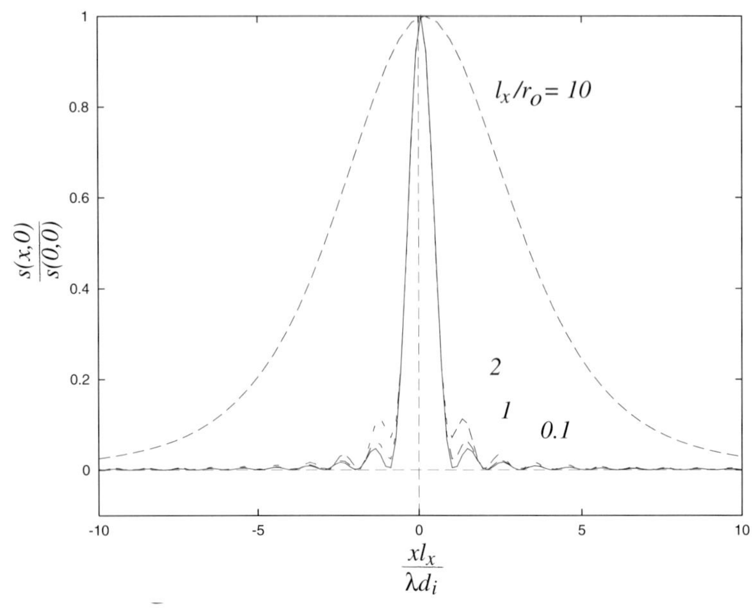

FIGURE 3.12
Family of short exposure PSFs plotted versus the normalized image plane vector $xl_x/\lambda d_i$. The ratio l_x/r_o ranges from 0.1 to 10.

0.1 to 10. Compare these OTF curves to the long exposure results of Figure 3.9. Fig. 3.12 illustrates cross-sections of the corresponding PSFs for the OTFs shown in Fig. 3.11.

3.6 Modal expansions of the phase perturbation

In analyzing the effects of turbulence, it is often useful to express the phase perturbation of an optical wave front, $\psi(\vec{x})$, as a linear combination of orthonormal basis functions. Expanding the phase of the wave in this way can provide deeper understanding of the structure and effects of the turbulence-corrupted wave front. This is particularly true if the chosen set of basis functions is widely used and understood. This approach has been taken in the description of fixed aberrations, where the phase of the generalized pupil function given in Eq. (2.45) has been expanded using the Zernike polynomials [33]. The Zernike polynomials have been widely studied and used to represent fixed aberrations in imaging systems, and the effect of the low order Zernike polynomials on imagery is well understood.

The general mathematical form of an expansion of $\psi(\vec{x})$ is

$$\psi(\vec{x}) \approx \sum_{i=1}^{N} a_i f_i(\vec{x}), \tag{3.136}$$

where $f_i(\vec{x})$ is the ith member of a set of orthonormal basis functions, $\{f_i(\vec{x})\}$, N is the number of basis functions used in the expansion, and a_i is the weight associated with $f_i(\vec{x})$. The functions in the set $\{f_i(\vec{x})\}$ are often referred to as modes. The subscript, i, refers to the ith mode, or the ith order of the expansion. The approximation in Eq. (3.136) becomes exact as $N \to \infty$. Decompositions of $\psi(\vec{x})$ of the type shown in Eq. (3.136) are referred to as modal expansions of $\psi(\vec{x})$.

Mathematically, the condition that the functions in the set $\{f_i(\vec{x})\}$ be orthonormal requires that

$$\int d\vec{x}\, W(\vec{x}) f_i(\vec{x}) f_j(\vec{x}) = \delta_{ij}, \tag{3.137}$$

where δ_{ij} is the Kronecker delta function defined by

$$\delta_{ij} = \begin{cases} 1 & i = j \\ 0 & i \neq j \end{cases}. \tag{3.138}$$

The function $W(\vec{x})$ in Eq. (3.137) is the aperture weighting function defined so that $W(\vec{x}) = 0$ outside the telescope pupil, and inside the pupil $W(\vec{x})$ is normalized so that

$$\int d\vec{x}\, W(\vec{x}) = 1. \tag{3.139}$$

The integration is taken to be over all space in every case. The coefficients, a_i, are given by the projection of $\psi(\vec{x})$ onto the various $f_i(\vec{x})$'s [34]:

$$a_i = \int d\vec{x}\, W(\vec{x})\psi(\vec{x}) f_i(\vec{x}). \tag{3.140}$$

If the input phase, $\psi(\vec{x})$, is zero-mean, then the mean value of all of the a_i's, $\overline{a}_i$, is

$$\begin{aligned} \overline{a}_i &= \mathrm{E}\left\{\int d\vec{x}\, W(\vec{x})\psi(\vec{x}) f_i(\vec{x})\right\}, \\ &= \int d\vec{x}\, W(\vec{x}) \mathrm{E}\{\psi(\vec{x})\} f_i(\vec{x}), \\ &= 0, \end{aligned} \tag{3.141}$$

where the overline notation, as in $\overline{a}_i$, is used to represent the expected value operator. Hence, all of the a_i's are also zero-mean. The second order statistics of the a_i's, that is, the variances and covariances of the coefficients, are of primary interest. The covariance of a_i and a_j is defined as

$$\begin{aligned} \overline{a_i a_j} &= \mathrm{E}\left\{\int d\vec{x}\, W(\vec{x})\psi(\vec{x}) f_i(\vec{x}) \int d\vec{x}'\, W(\vec{x}')\psi(\vec{x}') f_j(\vec{x}')\right\} \\ &= \int d\vec{x} \int d\vec{x}'\, W(\vec{x}) W(\vec{x}') f_i(\vec{x}) f_j(\vec{x}') \mathrm{E}\{\psi(\vec{x})\psi(\vec{x}')\}. \end{aligned} \tag{3.142}$$

If $i = j$, then Eq. (3.142) gives the variance of a_i.

With these mathematical preliminaries, we can now show another motivation for modal decompositions of the turbulence-corrupted wave front. Consider the aperture averaged mean square phase value, ϵ^2, given by

$$\epsilon^2 = \int d\vec{x}\, W(\vec{x}) \overline{\psi^2(\vec{x})}, \tag{3.143}$$

where $\overline{\psi^2(\vec{x})}$ is the mean square phase within the aperture. The aperture averaged mean square phase value ϵ^2 is a useful performance measure of imaging system performance. Small ϵ^2 provides superior image quality compared to a large value of ϵ^2, and $\epsilon^2 < 1$ rad^2 has been used as a criterion for attaining a nearly diffraction-limited image [1]. In view of this fact, adaptive optics systems have been developed which seek to minimize ϵ^2 [35, 36].

Substituting for $\psi(\vec{x})$ using Eq. (3.136) yields

$$\epsilon^2 = \int d\vec{x}\, W(\vec{x}) \left(\sum_{i=1}^{\infty}\sum_{j=1}^{\infty} \overline{a_i a_j} f_i(\vec{x}) f_j(\vec{x})\right). \tag{3.144}$$

Using the fact that the $f_i(\vec{x})$'s are orthonormal on $W(\vec{x})$ yields the result

$$\epsilon^2 = \sum_{i=1}^{\infty} \overline{(a_i^2)}. \tag{3.145}$$

In words, Eq. (3.145) means that the aperture averaged mean square phase error is simply the sum of the mean square values of the coefficients, a_i. Further, if it were somehow possible to perfectly correct or remove some subset of the modes present in $\psi(\vec{x})$, then the residual phase, $\psi_R(\vec{x})$ would be given by

$$\psi_R(\vec{x}) = \psi(\vec{x}) - \sum_{\{i_c\}} a_i f_i(\vec{x}), \tag{3.146}$$

where the notation $\{i_c\}$ represents summation over the corrected modes. In this case, the aperture averaged mean square phase error for the residual phase, ϵ_R^2, is given by

$$\begin{aligned} \epsilon_R^2 &= \int \mathrm{d}\vec{x}\, W(\vec{x})\overline{\psi_R^2(\vec{x})}, \\ &= \int \mathrm{d}\vec{x}\, W(\vec{x})\overline{\psi^2(\vec{x})} - \sum_{\{i_c\}} \overline{(a_i^2)}. \end{aligned} \tag{3.147}$$

Thus, if some modes can be removed using mechanical techniques, such as adaptive optics, simple expressions for ϵ_R^2 result. The Strehl ratio for an imaging system can also be related to ϵ_R^2 for small values of ϵ_R^2 [33].

In this section we examine two methods for expanding the phase of the turbulence-corrupted wave front. The first is a direct expansion of $\psi(\vec{x})$ as a linear combination of Zernike polynomials. We will find that the coefficients of the Zernike polynomial expansion are slightly correlated, suggesting that the Zernike polynomials provide a suboptimal basis set for expanding the phase of a turbulence-corrupted wave front. The second is an extension of the Zernike polynomial expansion using the method of Karhunen and Loève [14]. The Karhunen-Loève expansion provides basis functions which are optimal in the sense that the coefficients of the expansion are uncorrelated. The Karhunen-Loève expansion will also be shown to be slightly more efficient than the Zernike polynomial expansion because ϵ_R^2 is slightly smaller when the first N Karhunen-Loève modes are removed than when the first N Zernike modes are removed. We also show that optimal basis functions provided by the Karhunen-Loève expansion can be obtained from linear combinations of Zernike polynomials.

3.6.1 Zernike polynomials

Zernike polynomials have been widely used to represent fixed aberrations in optical systems [33]. They can also be used to represent random aberrations arising from atmospheric turbulence. The lowest order Zernike polynomials are generally referred to by the familiar names of tilt, defocus, astigmatism, coma, and spherical aberration. Due to the widespread usage of Zernike polynomials, optical researchers have developed deep insight into how aberrations described by the low order Zernike polynomials affect imaging systems. Further, the Zernike polynomials have convenient mathematical properties for the analysis which follows. This familiarity, and the mathematical properties of Zernike polynomials make them an attractive basis set for expanding the phase of the turbulence-corrupted wave front. In this section we present the Zernike polynomials, and obtain the covariance of the Zernike coefficients when used for atmospheric turbulence decomposition.

The Zernike polynomials are two-dimensional polynomials defined to be orthonormal on a circle of unit radius. Zernike polynomials are generally written in polar coordinates, (r, θ), where r is the radius, $0 \le r \le 1$, and θ is the angle some point makes with respect to the x-axis. The polar coordinates, (r, θ), are related to the Cartesian coordinates, $\vec{x} = (x, y)$, by

$$x = r\cos\theta,$$

$$y = r \sin\theta,$$
$$r = \sqrt{x^2 + y^2},$$
$$\theta = \tan^{-1}\frac{y}{x}. \tag{3.148}$$

In polar coordinates the Zernike polynomials are the product of a radial term, which is solely a function of r, and an azimuthal term, which is only a function of θ. In the absence of an ordering scheme, two terms are required to define the order of a particular Zernike polynomial. These terms are the azimuthal order, m, and the radial order, n.

Two normalization schemes for the Zernike polynomials have been proposed [33, 34]. Both of these normalizations satisfy the condition of providing orthonormal functions on the unit circle. We adopt the normalization scheme of Noll [34] here to be consistent with the bulk of the literature on this subject [22, 37, 38]. We will also adopt Noll's technique for numbering the Zernike polynomials using a single index rather than the ordered pair, (m, n). This ordering scheme allows the Zernike polynomials to be uniquely identified without the cumbersome use of two numbers.

The Zernike polynomials, $Z_i(r,\theta)$, are defined under Noll's ordering scheme as:

$$\left.\begin{array}{l} Z_{i=\text{even}}(r,\theta) = \sqrt{n+1}R_n^m(r)\cos(m\theta) \\ Z_{i=\text{odd}}(r,\theta) = \sqrt{n+1}R_n^m(r)\sin(m\theta) \end{array}\right\} m \neq 0, \tag{3.149}$$

and

$$Z_i(r,\theta) = R_n^0(r) \qquad m = 0. \tag{3.150}$$

The functions $R_n^m(r)$ are referred to as the radial functions, which are defined by

$$R_n^m(r) = \sum_{s=0}^{(n-m)/2} \frac{(-1)^s (n-s)!}{s![(n+m)/2 - s]!\,[(n-m)/2 - s]!} r^{n-2s}. \tag{3.151}$$

The azimuthal and radial orders, m and n, are non-negative integers which satisfy $m \leq n$ and $(n - m) = \text{even}$.

Since the Zernike polynomials are orthonormal only on the unit circle, the radius of the aperture must be normalized to apply Zernike polynomials to non-unit radius apertures. The appropriate transformation is [34]

$$\rho = \frac{r}{R}, \tag{3.152}$$

where R is the radius of the aperture. The expansion of the turbulence-induced phase, $\psi(\vec{x})$, is written in polar coordinates as

$$\psi(R\rho,\theta) = \sum_{i=1}^{N} a_i Z_i(\rho,\theta). \tag{3.153}$$

In polar coordinates the a_i's are written

$$a_i = \int d\vec{\rho}\, W(\rho)\psi(R\rho,\theta)Z_i(\rho,\theta). \tag{3.154}$$

Note that because of the underlying Gaussian distributed nature of $\psi(R\rho,\theta)$ at every point in space, the random variables, a_i, are also Gaussian distributed.

In Noll's ordering scheme the Zernike polynomials are numbered according to the rule that for a given n the modes with smaller m are counted first. Note that for cases where $m > 0$ there are *two* Zernike polynomials for each (m, n) pair, as stated in Eq. (3.149). The first 36 Zernike polynomials arranged by (m, n) pair and indexed using Noll's ordering are shown in Table 3.2. Table 3.2 contains all of the Zernike polynomials through $(m, n) = (7, 7)$.

TABLE 3.2
First 36 Zernike polynomials, through $(m, n) = (7, 7)$.

n	m	i	Zernike polynomial
0	0	1	1
1	1	2	$2r\cos\theta$
1	1	3	$2r\sin\theta$
2	0	4	$3.464r^2 - 1.732$
2	2	5	$2.449r^2 \sin 2\theta$
2	2	6	$2.449r^2 \cos 2\theta$
3	1	7	$(8.485r^3 - 5.657r)\sin\theta$
3	1	8	$(8.485r^3 - 5.657r)\cos\theta$
3	3	9	$2.828r^3 \sin 3\theta$
3	3	10	$2.828r^3 \cos 3\theta$
4	0	11	$3.416r^4 - 13.416r^2 + 2.236$
4	2	12	$(12.649r^4 - 9.487r^2)\cos 2\theta$
4	2	13	$(12.649r^4 - 9.487r^2)\sin 2\theta$
4	4	14	$3.162r^4 \cos 4\theta$
4	4	15	$3.162r^4 \sin 4\theta$
5	1	16	$(34.641r^5 - 41.569r^3 + 10.392r)\cos\theta$
5	1	17	$(34.641r^5 - 41.569r^3 + 10.392r)\sin\theta$
5	3	18	$(17.321r^5 - 13.856r^3)\cos 3\theta$
5	3	19	$(17.321r^5 - 13.856r^3)\sin 3\theta$
5	5	20	$3.464r^5 \cos 5\theta$
5	5	21	$3.464r^5 \sin 5\theta$
6	0	22	$52.915r^6 - 79.373r^4 + 31.749r^2 - 2.646$
6	2	23	$(56.125r^6 - 74.833r^4 + 22.450r^2)\sin 2\theta$
6	2	24	$(56.125r^6 - 74.833r^4 + 22.450r^2)\cos 2\theta$
6	4	25	$(22.450r^6 - 18.708r^4)\sin 4\theta$
6	4	26	$(22.450r^6 - 18.708r^4)\cos 4\theta$
6	6	27	$3.742r^6 \sin 6\theta$
6	6	28	$3.742r^6 \cos 6\theta$
7	1	29	$(140.0r^7 - 240.0r^5 + 120.0r^3 - 16.0r)\sin\theta$
7	1	30	$(140.0r^7 - 240.0r^5 + 120.0r^3 - 16.0r)\cos\theta$
7	3	31	$(84.0r^7 - 120.0r^5 + 40.0r^3)\sin 3\theta$
7	3	32	$(84.0r^7 - 120.0r^5 + 40.0r^3)\cos 3\theta$
7	5	33	$(28.0r^7 - 24.0r^5)\sin 5\theta$
7	5	34	$(28.0r^7 - 24.0r^5)\cos 5\theta$
7	7	35	$4.0r^7 \sin 7\theta$
7	7	36	$4.0r^7 \cos 7\theta$

The first Zernike mode, $Z_1(\rho, \theta)$, is referred to as piston, with coefficient given by

$$a_1 = \int d\vec{\rho}\, W(\rho)\psi(R\rho, \theta). \tag{3.155}$$

Physically, piston corresponds to the aperture averaged wave front phase. Single aperture imaging systems are insensitive to piston [33]. The coefficient of the piston term, a_1, has infinite variance for the case of Kolmogorov turbulence [34]. Hence, the piston term is generally explicitly subtracted in studies of turbulence effects on imaging systems. This step is taken in some of the analysis which follows. We adopt the notation $\phi(R\rho, \theta)$ to indicate the piston removed phase, where

$$\phi(R\rho, \theta) = \psi(R\rho, \theta) - \int d\vec{\rho}\, W(\rho)\psi(R\rho, \theta). \tag{3.156}$$

The Zernike modes $Z_2(\rho, \theta)$ and $Z_3(\rho, \theta)$ are referred to as tilt. These two modes correspond to the orthogonal components of the least-squares sense best fit plane to $\psi(\vec{x})$ in the telescope aperture plane. As presented in the short exposure imaging discussion of Section 3.5.4, tilt does not affect image quality [33], but does cause the image to be displaced from the location corresponding to the $a_2 = a_3 = 0$ condition. Other Zernike modes with well known names include $Z_4(\rho, \theta)$, which is called defocus; $Z_5(\rho, \theta)$ and $Z_6(\rho, \theta)$, which are the orthogonal components of astigmatism; $Z_7(\rho, \theta)$ and $Z_8(\rho, \theta)$, which are the orthogonal components of coma; and $Z_{11}(\rho, \theta)$, which is referred to as spherical aberration.

The covariance matrix of the Zernike polynomial coefficients, Γ_a, is given by

$$\Gamma_a = \overline{\mathbf{a}\mathbf{a}^T}, \tag{3.157}$$

where $\mathbf{a}$ is a column vector of expansion coefficients, with the first entry of $\mathbf{a}$ corresponding to $i = 1$, and the last entry of $\mathbf{a}$ corresponding to $i = N$. In Eq. (3.157) the matrix Γ_a is formed by letting i be the row index and j be the column index, so that the entry in the ith row and j^{th} column of Γ_a is given by $\overline{a_i a_j}$. The covariance matrix Γ_a provides information about the strength of the various modes and about correlation between modes. The elements of the Γ_a matrix have been calculated for index of refraction fluctuations characterized by the Kolmogorov power spectral density and are given by [34, 37]

$$\begin{aligned} \overline{a_i a_j} = {} & 0.0072 \left(\frac{D}{r_0}\right)^{5/3} (-1)^{(n_i + n_j - 2m_i)/2} [(n_i + 1)(n_j + 1)]^{1/2} \pi^{8/3} \delta_{m_i m_j} \\ & \times \frac{\Gamma(14/3)\Gamma[(n_i + n_j - 5/3)/2]}{\Gamma[(n_i - n_j + 17/3)/2]\Gamma[(n_j - n_i + 17/3)/2]\Gamma[(n_i + n_j + 23/3)/2]}, \end{aligned} \tag{3.158}$$

for $i - j =$ even, and for $i - j =$ odd

$$\overline{a_i a_j} = 0. \tag{3.159}$$

In the above equations m_i and n_i refer to the azimuthal and radial orders associated with the ith Zernike polynomial, respectively, m_j and n_j refer to the azimuthal and radial orders associated with the jth Zernike polynomial, and $\delta_{m_i m_j}$ is the Kronecker delta function. The covariances of the first nine Zernike modes, excluding piston is shown in Table 3.3.

Equation (3.158) can be used to obtain expressions for the aperture averaged mean square phase error after removal of the first N Zernike modes, as shown in Eq. (3.147). The aperture averaged mean square residual phase error for this case will be represented by $\epsilon_N^2(Z)$, where the (Z) notation is used to indicate the association with Zernike polynomials. The results of this analysis are presented in Table 3.4. In Table 3.4 expressions for $\epsilon_N^2(Z)$ are shown for the first 21 Zernike modes. For $N > 21$, $\epsilon_N^2(Z)$ is given by [34]

$$\epsilon_N^2(Z) \approx 0.2944 N^{-\sqrt{3}/2} \left(\frac{D}{r_0}\right). \tag{3.160}$$

TABLE 3.3
$\overline{a_i a_j}(D/r_0)^{-5/3}$ **for the first 10 Zernike polynomials, excluding the piston term,** Z_1**. Rows correspond to** i **and columns correspond to** j**.**

	2	3	4	5	6	7	8	9	10
2	0.448	0	0	0	0	0	-0.0141	0	0
3	0	0.448	0	0	0	-0.0141	0	0	0
4	0	0	0.0232	0	0	0	0	0	0
5	0	0	0	0.0232	0	0	0	0	0
6	0	0	0	0	0.0232	0	0	0	0
7	0	-0.0141	0	0	0	0.00618	0	0	0
8	-0.0141	0	0	0	0	0	0.00618	0	0
9	0	0	0	0	0	0	0	0.00618	0
10	0	0	0	0	0	0	0	0	0.00618

In Table 3.4 $D = 2R$ is the diameter of the circular aperture, and r_0 is the atmospheric coherence length. The information presented in Table 3.4 is plotted in Fig. 3.13. In Fig. 3.13 the pupil averaged mean square phase after removal of the first N Zernike modes is plotted as a function of the number of modes. Fig. 3.13 illustrates that the main contributions to ϵ^2 arise from the lowest order terms in the Zernike polynomial expansion of the turbulence-induced phase. Note that over 86% of the total ϵ^2 in the piston-removed phase is due to Z_2 and Z_3, the tilt modes of the turbulence. For $N > 15$ the marginal reduction in ϵ^2 which results from removing additional Zernike modes is small, being on the order of $10^{-3}(D/r_0)^{-5/3}$, or less.

As an example of how to use Table 3.4 , consider $\epsilon_3^2(Z)$ for the $D/r_0 = 10$ case. Physically, this case corresponds to removal of the piston mode, Z_1, and two orthogonal tilt modes, Z_2 and Z_3. The aperture averaged mean square phase error for this case is given by

$$\epsilon_3^2(Z) = 0.134 \left(\frac{D}{r_0}\right)^{5/3} = 6.22 \text{ rad}^2, \tag{3.161}$$

which is a large mean square wave front error. Note that if the first 21 Zernike modes were removed then the residual mean square phase error would be $\epsilon_{21}^2(Z) = 0.965$ rad^2, which is still a sizable wave front error. From this example we can conclude that under many seeing conditions of interest, such as $D/r_0 > 10$, the high order Zernike modes make a considerable contribution to the residual mean square phase error.

3.6.2 Karhunen-Loève expansion

Table 3.3 shows the covariance matrix of the Zernike coefficients, Γ_a, contains non zero terms off the main diagonal. This implies that some of the coefficients of the Zernike polynomial expansion of $\psi(\vec{x})$ are correlated, which in turn suggests that there may exist a more efficient basis set for expanding $\psi(\vec{x})$ using uncorrelated coefficients. The term efficiency is used here in the sense that a more efficient modal expansion of the turbulence would require fewer modes to represent $\psi(\vec{x})$ with a specified degree of fidelity. In this subsection we show that it is possible to obtain an expansion based on Zernike polynomials which has uncorrelated coefficients. The method of Karhunen and Loève is applied here to obtain the expansion [8, 14].

The Karhunen-Loève expansion provides a means of using the properties of the eigenspace of covariance matrices to derive expansion coefficients which are uncorrelated, and to obtain the associated basis functions. The expansion which follows is based on Zernike polynomials, so we will use polar coordinates to represent the turbulence-corrupted phase. The analysis which follows is explicitly applied to the piston-removed phase, $\phi(R\rho, \theta)$. Mathematically, the problem is formulated by seeking

TABLE 3.4
$\epsilon_N^2(Z)$ for removal of the first N Zernike modes from the turbulence-corrupted wave front phase.

Mode index, i	$\epsilon_N^2(Z)$, rad^2
1	1.0299 $(D/r_0)^{5/3}$
2	0.582 $(D/r_0)^{5/3}$
3	0.134 $(D/r_0)^{5/3}$
4	0.111 $(D/r_0)^{5/3}$
5	0.0880 $(D/r_0)^{5/3}$
6	0.0648 $(D/r_0)^{5/3}$
7	0.0587 $(D/r_0)^{5/3}$
8	0.0525 $(D/r_0)^{5/3}$
9	0.0463 $(D/r_0)^{5/3}$
10	0.0401 $(D/r_0)^{5/3}$
11	0.0377 $(D/r_0)^{5/3}$
12	0.0352 $(D/r_0)^{5/3}$
13	0.0328 $(D/r_0)^{5/3}$
14	0.0304 $(D/r_0)^{5/3}$
15	0.0279 $(D/r_0)^{5/3}$
16	0.0267 $(D/r_0)^{5/3}$
17	0.0255 $(D/r_0)^{5/3}$
18	0.0243 $(D/r_0)^{5/3}$
19	0.0232 $(D/r_0)^{5/3}$
20	0.0220 $(D/r_0)^{5/3}$
21	0.0208 $(D/r_0)^{5/3}$

an expansion of the form

$$\phi(R\rho, \theta) \approx \sum_{i=1}^{M} b_i e_i(\rho, \theta), \tag{3.162}$$

where $e_i(\rho, \theta)$ are the new basis functions, and b_i are the associated expansion coefficients. The condition that the b_i's be uncorrelated requires that

$$\Gamma_b = \overline{\mathbf{b}\mathbf{b}^T} = D, \tag{3.163}$$

where the notation $\mathbf{b}$ is used to indicate a vector containing the coefficients b_i, and D is a diagonal matrix. The essence of the Karhunen-Loève expansion in continuous space is that uncorrelated expansion coefficients can be obtained by using the eigenfunctions of the autocorrelation of the underlying random process as the basis set for the expansion. This concept extends directly to vector spaces, as shown below.

As suggested in Ref. [38], we begin with the covariance matrix for the Zernike polynomial coefficients, Γ_a. Note that the covariance and the correlation are identical quantities for zero-mean random processes. Γ_a is a Hermitian matrix, that is,

$$\Gamma_a = \Gamma_a^T. \tag{3.164}$$

For all Hermitian matrices there exists a unitary diagonalizing matrix, U, such that

$$U^T \Gamma_a U = \Lambda, \tag{3.165}$$

where Λ is a diagonal matrix [39]. The mathematical condition that U be unitary requires that the

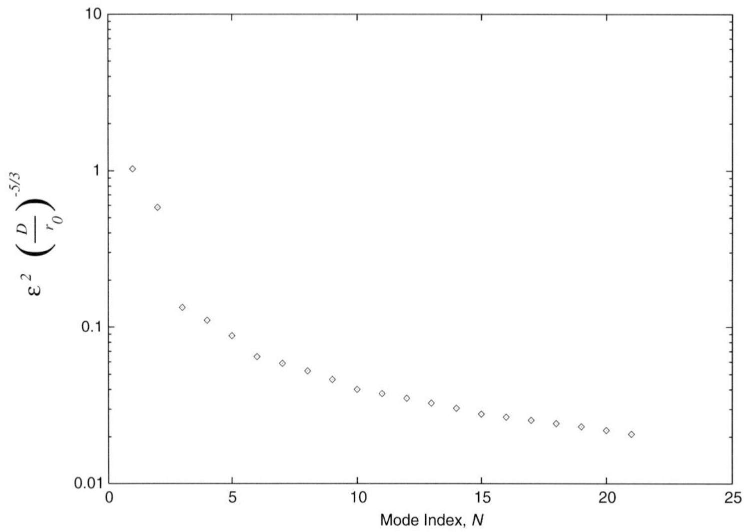

FIGURE 3.13
Pupil averaged mean square phase after removal of the first N Zernike modes as a function of the number of modes removed.

columns of U be orthonormal, and further that

$$U^{-1} = U^T, \tag{3.166}$$

where U^{-1} is the matrix inverse of U. Note that Eq. (3.165) implies that

$$\Gamma_a = U \Lambda U^T. \tag{3.167}$$

If the eigenvectors of Γ_a are orthonormal, then a suitable way to construct U is to use the eigenvectors of Γ_a as the columns of U [39]. Since Γ_a has repeated eigenvalues [38] there is no mathematical guarantee that the eigenvectors of Γ_a will be orthonormal. However, an exhaustive search of the eigenvectors of Γ_a through $N = 300$ has shown that the eigenvectors of Γ_a are mutually orthogonal. Using the eigenvectors of Γ_a as the columns of U results in Λ being ordered so that the diagonal element of any column of Λ is the eigenvalue of the eigenvector in the equivalent column of U.

To proceed, we now define a column vector, $\mathbf{a}$, of length $N-1$ containing the Zernike polynomial coefficients

$$\mathbf{a} = \begin{pmatrix} a_2 \\ a_3 \\ \cdot \\ \cdot \\ \cdot \\ a_N \end{pmatrix}, \tag{3.168}$$

where the piston coefficient, a_1, has been excluded. In this notation the covariance of $\mathbf{a}$ is defined by

$$\overline{\mathbf{a}\mathbf{a}^T} = \Gamma_a. \tag{3.169}$$

Now consider the following transformation on $\mathbf{a}$:

$$\mathbf{b} = U^T \mathbf{a}. \tag{3.170}$$

The covariance of $\mathbf{b}$ is given by

$$\begin{aligned}\overline{\mathbf{b}\mathbf{b}^T} &= \overline{(U^T\mathbf{a})(U^T\mathbf{a})^T},\\ &= U^T\overline{\mathbf{a}\mathbf{a}^T}U,\\ &= U^T\Gamma_a U,\\ &= \Lambda.\end{aligned} \tag{3.171}$$

Since Λ is a diagonal matrix, the coefficients, $\mathbf{b}$, are uncorrelated. By virtue of the underlying Gaussian distributed nature of the elements of $\mathbf{b}$, the elements of $\mathbf{b}$ are also statistically independent.

The mapping $\mathbf{b} = U^T\mathbf{a}$ accomplishes the goal of providing an expansion with uncorrelated coefficients which can be derived from the Zernike polynomials. We refer to this expansion as the Karhunen-Loève expansion, and the orthonormal basis functions for the expansion are referred to as the Karhunen-Loève modes. The Karhunen-Loève modes are linear combinations of the Zernike polynomials. This is easily shown by noting that Eq. (3.170) implies

$$\mathbf{a} = U\mathbf{b}. \tag{3.172}$$

Recalling the form of the Zernike polynomial expansion from Eq. (3.153), and using a summation to represent vector-matrix product in Eq. (3.172) yields

$$\begin{aligned}\phi(R\rho,\theta) &= \sum_{i=2}^{N} a_i Z_i(\rho,\theta),\\ &= \sum_{i=2}^{N}\left(\sum_{j=2}^{N} U_{ij}b_j\right) Z_i(\rho,\theta),\\ &= \sum_{j=2}^{N} b_j\left(\sum_{i=2}^{N} U_{ij}Z_i(\rho,\theta)\right),\end{aligned} \tag{3.173}$$

where the notation U_{ij} denotes the entry in the ith row and j^{th} column of the U matrix, and piston removal is indicated by starting the summation at $i = 2$. Comparing Eqs. (3.162) and (3.173) leads to the conclusion that the Karhunen-Loève modes are mathematically described by

$$e_j(\rho,\theta) = \sum_{i=2}^{N} U_{ij}Z_i(\rho,\theta). \tag{3.174}$$

Thus, the Karhunen-Loève modes are linear combinations of the Zernike polynomials.

The exact amount of each Zernike polynomial in a Karhunen-Loève mode can be obtained by computing the U matrix. Note that the exact form of the U matrix is a function of the number of Zernike polynomials used to obtain the Karhunen-Loève expansion. As an example, the form of the first two Karhunen-Loève modes for the case of using 300 Zernike polynomials is given by

$$\begin{aligned}e_1(\rho,\theta) &= 0.99949Z_2(\rho,\theta) - 0.03195Z_8(\rho,\theta) + \text{HOT},\\ e_2(\rho,\theta) &= 0.99949Z_3(\rho,\theta) - 0.03195Z_7(\rho,\theta) + \text{HOT},\end{aligned} \tag{3.175}$$

where HOT represents higher order terms with much weaker contributions to e_1 and e_2. Thus, the first two Karhunen-Loève modes are composed primarily of tilt, $Z_2(\rho,\theta)$ and $Z_3(\rho,\theta)$, and coma, $Z_7(\rho,\theta)$ and $Z_8(\rho,\theta)$.

Finally, we examine the efficiency of the Karhunen-Loève expansion in representing $\phi(R\rho,\theta)$. To address this issue we calculate the aperture averaged mean square residual phase remaining after

TABLE 3.5
$\epsilon_N^2(KL)$ for removal of the first N Karhunen-Loève modes from the turbulence-corrupted wave front phase.

Mode index	$\epsilon_N^2(KL)$, rad^2
1	$1.0299\,(D/r_0)^{5/3}$
2	$0.581\,(D/r_0)^{5/3}$
3	$0.133\,(D/r_0)^{5/3}$
4	$0.109\,(D/r_0)^{5/3}$
5	$0.0849\,(D/r_0)^{5/3}$
6	$0.0617\,(D/r_0)^{5/3}$
7	$0.0551\,(D/r_0)^{5/3}$
8	$0.0485\,(D/r_0)^{5/3}$
9	$0.0422\,(D/r_0)^{5/3}$
10	$0.0360\,(D/r_0)^{5/3}$
11	$0.0333\,(D/r_0)^{5/3}$
12	$0.0306\,(D/r_0)^{5/3}$
13	$0.0281\,(D/r_0)^{5/3}$
14	$0.0260\,(D/r_0)^{5/3}$
15	$0.0239\,(D/r_0)^{5/3}$
16	$0.0225\,(D/r_0)^{5/3}$
17	$0.0211\,(D/r_0)^{5/3}$
18	$0.0202\,(D/r_0)^{5/3}$
19	$0.0192\,(D/r_0)^{5/3}$
20	$0.0182\,(D/r_0)^{5/3}$
21	$0.0172\,(D/r_0)^{5/3}$

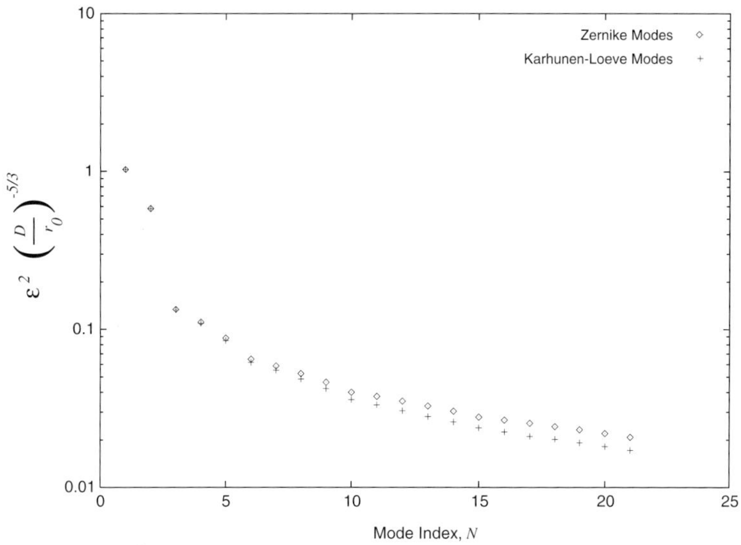

FIGURE 3.14
Pupil averaged mean square phase after removal of the first N Karhunen-Loève modes as a function of the number of modes removed. The similar result for removing the first N Zernike modes is also shown for comparison.

removal of the first N Karhunen-Loève modes, $\epsilon_N^2(KL)$. $\epsilon_N^2(KL)$ was defined for a general expansion in Eq. (3.147), and is given for this case by

$$\epsilon_N^2(KL) = 1.0299\left(\frac{D}{r_0}\right) - \sum_{i=1}^{N}\overline{(b_i^2)}, \tag{3.176}$$

where we have used the piston-removed phase error as the starting point of our evaluation, similar to the results shown in Table 3.4. If the Karhunen-Loève expansion provides lower aperture averaged mean square residual phase error than the Zernike polynomial expansion for equivalent numbers of modes removed, then the Karhunen-Loève expansion is more efficient than the Zernike polynomial expansion.

The results of the analysis of $\epsilon_N^2(KL)$ are presented in Table 3.5 for removal of the first 21 Karhunen-Loève modes. In this case the Karhunen-Loève expansion was computed using 300 Zernike polynomials. Comparison of Tables 3.4 and 3.5 shows that the Karhunen-Loève expansion is slightly more efficient than the Zernike polynomials by this measure. For example, after removal of the first 21 Zernike modes there is an aperture averaged mean square residual phase error of $0.0208(D/r_0)^{5/3}$ rad^2, while for the case of removing the first 21 Karhunen-Loève modes the aperture averaged mean square residual phase error is $0.0172(D/r_0)^{5/3}$ rad^2. The information presented in Table 3.5 is plotted in Fig. 3.14. In Fig. 3.14 the pupil averaged mean square phase after removal of the first N Karhunen-Loève modes is plotted as a function of the number of modes. Similar to the Zernike polynomial expansion, Fig. 3.13 illustrates that the main contribution to ϵ^2 arises from the lowest order terms in the Karhunen-Loève expansion of the turbulence-induced phase.

3.7 Phase screen generation

In Section 3.5 we derived analytical results describing atmospheric turbulence effects on incoherent imaging systems. In particular, results for the average OTF for long and short exposure imaging are presented in Eqs. (3.125) and (3.135). Analytical results like these can be derived because of the assumption that the wave front phase and log-amplitude have Gaussian statistics and the phase and amplitude correlations can be expressed with simple closed form expressions. As we start to consider more complicated situations involving image post processing and adaptive optics, we will find that the statistics of the wave front phase and amplitude perturbations cannot be expressed in closed form. When this condition arises it becomes advantageous to consider other methods for performing the type of first order performance calculations presented in Section 3.5. In fact as we consider higher order performance calculations, such as computing the variance of the OTF, we find that the analytical calculations become extremely difficult due to the number of integral expressions that cannot be solved in closed form. An example of the situation in which the pupil phase correlation function cannot be expressed with a simple expression is the residual phase in the pupil of an adaptive optical imaging system. Adaptive optical imaging systems were introduced in Chapter 1 and are discussed in detail in Chapter 5. The purpose of the adaptive optics is to compensate for the wave front phase perturbations in the pupil. The degree to which the adaptive optical system is successful in achieving this goal is a complicated function of the components of the system, atmospheric conditions and light levels. To estimate the average OTF, or for that matter any other performance measure, the phase structure function for the residual phase in the system pupil is required. Unfortunately, the structure function cannot be expressed in closed form due to the complicated dependence on the individual components of the adaptive optical system. Numerical evaluation of the complicated expressions is possible in some cases as demonstrated in references [36, 40], but in many other cases the computations are simply beyond the capacity of today's high speed computer workstations. In particular, we have found that analytical performance computations for adaptive optical imaging systems having pupil diameters greater than 1 meter can take days to complete on widely available computer workstations.

Another reason for considering other approaches for performance calculations is that some types of calculations simply do not lend themselves to tractable solutions even if we have simple expressions for the phase structure function. Another motivation for considering an alternate approach is that the analytical computations discussed above yield statistical results. For example, OTF calculations based on the phase structure function yield statistical quantities such as the average and variance of the OTF. It may also be of interest to compute how an adaptive optical system responds on an instant-by-instant basis. Being able to study the performance of the adaptive optics system for individual realizations of turbulence provides valuable insight into how the system will perform.

Monte Carlo simulations have found widespread use for investigating the performance of optical systems operating in the presence of atmospheric turbulence. In these simulations the components of an optical system are modeled in with high fidelity [40]. The simulations are executed by generating random realizations of the wave front phase in the pupil of the system. Using the component models of the optical system it is possible to employ the Fourier optics concepts presented in Chapter 2 to propagate each realization of the pupil field through the system. Statistical performance results can be obtained by calculating the required performance values for each realization of the pupil phase and accumulating the results. Simulations are widely used to predict the performance of adaptive optics systems and various post processing schemes for imaging through turbulence. Many of the practical problems associated with the analytical computations discussed above are solved with the use of simulations. In the case of adaptive optical systems, solving the integrals that result for many of the analytical computations is sufficiently difficult and time consuming that analytic calculations have only been performed for a relatively small number of points. In the case of computing the average OTF, previous analytical calculations have only been performed for cross-sections of the 2-dimensional OTF [36, 40]. Simulations allow the average OTF (and higher order quantities) to be evaluated everywhere on a two dimensional grid of sample points simultaneously. The spacing of the samples in a simulation is generally much smaller than the sample spacing used in analytic calculations. In the case of an adaptive optical imaging system having a large pupil (greater than 1 m), a simulation offers the only practical means of predicting performance due to the fact simulations often run faster than analytic calculations. Finally, an *image* is often desired, rather than the statistical quantities that result from analytical computations. Simulations can be used to provide images which show the effects of all noises and system limitations present in the simulated optical system.

The purpose of this section is to present the key component of a Monte Carlo simulation, the phase screen generator. The phase screen generator is a computer program which creates random arrays of phase values on a grid of sample points which have the same statistics as the turbulence-induced atmospheric phase, $\psi(\vec{x}, t)$. The sample grid points can be spaced in both space, $\vec{x}$, and time, t. We discuss methods for creating random phase screens in this section. The statistics of $\psi(\vec{x}, t)$ required to create the desired phase screens have been presented in Sections 3.3 and 3.4.

There are a number of approaches used to generate random phase screens with the proper point statistics and spatial and temporal correlation properties. Goldring and Carlson [41] review a number of the more common approaches. Of these approaches the one that offers the best performance in terms of giving the spatial and temporal statistics that match the desired statistics is an approach based on relatively simple linear algebra concepts. We only present this approach since it is simple and readily implemented on widely available high speed computer workstations. The technique is computationally intense in terms of the number of calculations which must be performed prior to creating random phase screens. However, when the required matrices are available the technique is quite fast since phase screens are created by a vector-matrix multiplication. Also, this method is very powerful in that it can be used to create "stacks" of random phase screens delayed in time relative to each other, and having the correct temporal correlation statistics. It can also be used to create properly correlated multiple phase screens due to wave fronts arriving from different directions, as required for investigating anisoplanatic effects.

3.7.1 Phase Screen Generation Approach

A random realization of the wave front phase can be generated over an aperture using a linear combination of orthonormal basis functions. A set of Q realizations of the random phase can be written

$$\tilde{\psi}(\vec{x}, t_q) = \sum_{p=1}^{P} a_p(t_q) f_p(\vec{x}), \qquad \text{for } q = 1, \ldots, Q \tag{3.177}$$

where $\tilde{\psi}(\vec{x}, t)$ is a single realization of the phase as a function of pupil position $\vec{x}$ at time t, $f_p(\vec{x})$ is the pth member of a set of orthonormal basis functions, $a_p(t_q)$ is the pth weight for the qth time instant, P is the number of basis functions used in the phase screen generation, and Q is the number of time instants for which random phase screens are being generated. A detailed discussion of the mathematical properties of orthonormal basis functions is presented in Section 3.6. The phase screen generation problem requires an algorithm that makes random draws of the weights $a_p(t_q)$, for $p = 1, \ldots, P$ and $q = 1, \ldots, Q$, such that the spatial and temporal correlation properties, as well as the point statistics of the generated phase, have approximately the desired characteristics. For the correlation properties we want

$$\Gamma_{\tilde{\psi}}(\vec{x}_1, t_q; \vec{x}_2, t_{q'}) = \mathrm{E}\{\tilde{\psi}(\vec{x}_1, t_q)\tilde{\psi}(\vec{x}_2, t_{q'})\} \approx \mathrm{E}\{\psi(\vec{x}_1, t_q)\psi(\vec{x}_2, t_{q'})\} = \Gamma_{\psi}(\vec{x}_1, t_q; \vec{x}_2, t_{q'}), \tag{3.178}$$

where $\psi(\vec{x}, t)$ denotes the random phase process that is being simulated, the vectors $\vec{x}_1$ and $\vec{x}_2$ are positions within the aperture, and t_q and $t_{q'}$ designate any two time instants contained in the set of time instants $t_q, q = 1, \ldots, Q$.

The most common basis used in phase screen generators is a set of non-overlapping 2-dimensional rect functions spaced on a rectangular grid of points. In the limit as the dimensions of the rect functions approach zero this basis reduces to a set of Dirac-delta functions (i.e., a "bed of nails") and corresponds to generating the random phase over a rectangular grid of points. The weight $a_p(t_q)$ in this case corresponds to the actual value of the generated phase at the center of the pth basis function. (Note that this interpretation is slightly inconsistent with the discussion of orthonormal basis functions presented in Section 3.6. If the aperture function $W(\vec{x})$ is unity within the aperture instead of satisfying Eq. (3.139), then the orthonormal discussion is applicable for this case.) It turns out that this particular basis function gives equality in Eq. (3.178) for $\vec{x}_1$ and $\vec{x}_2$'s that take on values corresponding to the centers of the basis functions. Another commonly used basis is the Zernike polynomials described in Section 3.6.1. In this case, the phase is generated over a circular aperture and Eq. (3.178) is satisfied exactly only if the number of basis functions goes to infinity.

In order to develop an algorithm to generate random draws of the weights $a_p(t_q)$ for $p = 1, \cdots, P$ and $q = 1, \cdots, Q$ let us first consider the covariance matrix for the weights. Let this matrix be designated Γ_a. The matrix Γ_a will have $(P \times Q)^2$ elements with each element given by

$$\mathrm{E}\{a_p(t_q)a_{p'}(t_{q'})\} \qquad \text{for } p, p' = 1, \ldots, P \text{ and } q, q' = 1, \ldots, Q \tag{3.179}$$

The ordering of the elements of the matrix is arbitrary. In the case of the delta function basis set, the algorithm simultaneously creates Q phase screens: one for each time, t_q. These phase screens have the correct spatial statistics within each phase screen, and the correct space-time statistics across the stack of phase screens. Unfortunately, computing the temporal statistics of the Zernike polynomial coefficients is computationally quite complicated. Also, as noted in Section 3.6, some of the Zernike polynomial coefficients are correlated, further complicating the computation of the required space-time covariance matrix, Γ_a, required to create stacks of phase screens using the Zernike polynomial basis set. Hence, our discussion of phase screen generation using Zernike polynomials will be limited to single phase screens.

The covariance matrix, Γ_a, is guaranteed to be real, symmetric, and positive definite [14]. As a consequence, the Cholesky factorization can be used to factor Γ_a into the product of two square matrices [39]

$$\Gamma_a = RR^T, \tag{3.180}$$

where the superscript T represents the matrix transpose operator. The Cholesky factorization of a matrix can be readily computed using widely available software [42]. Consider now a random draw of a vector of uncorrelated, zero mean, unit variance, Gaussian random variables, $\mathbf{b}$. The vector $\mathbf{b}$ is defined as a column vector of length $P \times Q$. This random vector can be generated in a straightforward manner using standard random number generators [42]. The covariance of the random vector $\mathbf{b}$ is given by:

$$\mathrm{E}\{\mathbf{b}\mathbf{b}^T\} = I_M, \tag{3.181}$$

where I_M is the identity matrix.

Now consider forming a random draw of the weights $a_p(t_q)$ by using the vector $\mathbf{b}$ in the following way:

$$\mathbf{a} = R\mathbf{b}, \tag{3.182}$$

where $\mathbf{a}$ is a column vector formed by the $P \times Q$ weights $a_p(t_q)$. The motivation for this particular transformation is the hope that the covariance of $\mathbf{a}$ matches the desired covariance given by Γ_a. The covariance of $\mathbf{a}$ can be derived using Eq. (3.182):

$$\mathrm{E}\{\mathbf{a}\mathbf{a}^T\} = \mathrm{E}\{R\mathbf{b}(R\mathbf{b})^T\} = \mathrm{E}\{R\mathbf{b}\mathbf{b}^T R^T\}. \tag{3.183}$$

Eq. (3.183) can be simplified by noting that R is a constant matrix, so that

$$\mathrm{E}\{\mathbf{a}\mathbf{a}^T\} = R\mathrm{E}\{\mathbf{b}\mathbf{b}^T\}R^T. \tag{3.184}$$

Substituting Eq. (3.181) into Eq. (3.184) gives

$$\mathrm{E}\{\mathbf{a}\mathbf{a}^T\} = RI_M R^T = RR^T = \Gamma_a, \tag{3.185}$$

demonstrating that this technique provides random draws of $\mathbf{a}$ with the desired covariance, Γ_a. The final step in generating a phase screen realization is substituting the random weights $a_p(t_q)$ generated by the process above into Eq. (3.177) to compute the phase.

In the following two subsections we present the theoretical phase covariance expressions that can be used to generate the matrix Γ_a. We first present the covariance expression required when generating the phase over a grid of points. This case corresponds to using basis functions consisting of a collection of delta functions located at the grid points. The second subsection presents the covariance expressions required for using a basis of Zernike polynomials.

3.7.2 Covariance expression required for generating the phase over a grid of points

To compute the elements of Γ_a, we let $f_p(\vec{x}) = \delta(\vec{x} - \vec{x}_p)$ in the expression for $\mathrm{E}[a_p(t_q)a_{p'}(t_{q'})]$. This expression is found in Eq. (3.142) and the substitution gives

$$\begin{aligned}\mathrm{E}[a_p(t_q)a_{p'}(t_{q'})] &= \mathrm{E}[\psi(\vec{x}_p, t_q)\psi(\vec{x}_{p'}, t_{q'})], \\ &= \Gamma_\psi(\vec{x}_p, t_q; \vec{x}_{p'}, t_{q'})\end{aligned} \tag{3.186}$$

where $\vec{x}_p$ and $\vec{x}_{p'}$ are spatial grid points at which the random phase is being generated. Recall that for this basis the actual values of the weights, $a_p(t_q)$, correspond to the phase. Equation (3.186) indicates the problem at hand is simply to develop a covariance expression for the phase $\psi(\vec{x}, t)$.

In the development to follow the covariance expression is derived for both the von Karman and the Kolmogorov turbulence power spectra, $\Phi_n(\kappa, z)$. These two power spectra are given in Eqs. (3.9) and (3.10), respectively. Recall that z is the distance from the aperture and κ is the spatial wave number of the index-of-refraction fluctuations. The strength of the turbulence as a function of z is characterized by the structure constant of the index of refraction fluctuations, $C_n^2(z)$. The transverse velocity of the turbulence caused by wind is characterized by velocity profile, $\vec{v}(z)$. The derivations of the phase covariance differ for the von Karman and Kolmogorov spectra because the Kolmogorov spectrum contains a non-integrable pole at $\vec{\kappa} = 0$ [8]. Explicitly removing the piston term of the phase over the aperture in which the phase is being generated removes the problem associated with integrating the Kolmogorov spectrum [34], but requires a different derivation involving the structure function of the phase fluctuations, D_ψ. Early in the development we find it useful to use the layered turbulence model introduced in Section 3.4. We conclude this section by showing how to extend the covariance expression to account for anisoplanatic effects.

We begin with the turbulence-induced wavefront phase fluctuation given in Eq. (3.82) where we are considering the entire optical path from the aperture at $z = 0$ to the most distant turbulence at $z = h$:

$$\psi(\vec{x}, t) = k \int_0^h \mathrm{d}z\, n_1(\vec{x} - \vec{v}(z)t, z; 0), \tag{3.187}$$

where n_1 is the index-of-refraction fluctuation, $k = 2\pi/\lambda$ and λ is the mean wavelength.

As noted above, it is necessary to explicitly remove the piston component of the incident phase when the Kolmogorov power spectrum is used. However, piston removal is not required for the von Karman power spectrum. With these two cases in mind, the possibly piston-removed phase is represented by

$$\phi(\vec{x}, t) = \psi(\vec{x}, t) - p \int \mathrm{d}\vec{x}'\, W(\vec{x}')\psi(\vec{x}', t), \tag{3.188}$$

where $W(\vec{x})$ is a weighting function defining the physical extent of the phase screen, $p = 1$ signifies piston removal, while $p = 0$ signifies piston is not removed, and the integration is over all space. The weighting function $W(\vec{x})$ is a non zero constant for points on the phase screen and zero outside and is normalized so that

$$\int \mathrm{d}\vec{x}'\, W(\vec{x}') = 1. \tag{3.189}$$

Using Eq. (3.188) we can write the covariance of $\phi(\vec{x}, t)$:

$$\begin{aligned}
&\Gamma_\phi(\vec{x}_p, t_q; \vec{x}_{p'}, t_{q'}) = \\
&k^2 \mathrm{E}\left\{\left[\int_0^h \mathrm{d}z_1\, n_1(\vec{x}_p - \vec{v}(z_1)t_q, z_1; 0) - p\int \mathrm{d}\vec{x}' \int_0^h \mathrm{d}z_1\, W(\vec{x}')n_1(\vec{x}' - \vec{v}(z_1)t_q, z_1; 0)\right]\right. \\
&\times \left.\left[\int_0^h \mathrm{d}z_2\, n_1(\vec{x}_{p'} - \vec{v}(z_2)t_{q'}, z_2; 0) - p\int \mathrm{d}\vec{x}'' \int_0^h \mathrm{d}z_2\, W(\vec{x}'')n_1(\vec{x}'' - \vec{v}(z_2)t_{q'}, z_2; 0)\right]\right\}
\end{aligned} \tag{3.190}$$

Eq. (3.190) provides the most general form for $\Gamma_\phi(\vec{x}_p, t_q; \vec{x}_{p'}, t_{q'})$. Just as we did in Section 3.4.1 we find it convenient to break the atmosphere into a set of slabs or layers. This step allows us to later assert that the individual layers are statistically independent, greatly simplifying the analysis. Tatarskii [3] and others [8, 24] have investigated this assumption, and have argued that individual layers can be treated as being approximately independent if the separation of layer centers are greater than the inverse of the largest distance between field points in the pupil. A layered turbulence model converts the integration over z in Eq. (3.190) to a summation. The phase fluctuation due to the ith layer, $\psi_i(\vec{x}, t)$, is given by Eq. (3.82):

$$\psi_i(\vec{x}; t) = k\int_{z_i}^{z_i+\Delta z_i} \mathrm{d}z\, n_1(\vec{x}, z; t), \tag{3.191}$$

where z_i is the distance of the ith layer, and Δz_i is the thickness of the ith layer. Substituting Eq. (3.191) into Eq. (3.190) and expanding yields

$$\begin{aligned}
&\Gamma_\phi(\vec{x}_p, t_q; \vec{x}_{p'}, t_{q'}) = \\
&\sum_{i=1}^{N}\sum_{j=1}^{N} \mathrm{E}\{\psi_i(\vec{x}_p - \vec{v}(z_i)t_q; 0)\psi_j(\vec{x}_{p'} - \vec{v}(z_j)t_{q'}; 0)\} \\
&-p\int \mathrm{d}\vec{x}' W(\vec{x}') \sum_{i=1}^{N}\sum_{j=1}^{N} \mathrm{E}\{\psi_i(\vec{x}' - \vec{v}(z_i)t_q; 0)\psi_j(\vec{x}_{p'} - \vec{v}(z_j)t_{q'}; 0)\} \\
&-p\int \mathrm{d}\vec{x}'' W(\vec{x}'') \sum_{i=1}^{N}\sum_{j=1}^{N} \mathrm{E}\{\psi_i(\vec{x}_p - \vec{v}(z_i)t_q; 0)\psi_j(\vec{x}'' - \vec{v}(z_j)t_{q'}; 0)\} \\
&+p^2\int \mathrm{d}\vec{x}' \int \mathrm{d}\vec{x}''\, W(\vec{x}')W(\vec{x}'') \sum_{i=1}^{N}\sum_{j=1}^{N} \mathrm{E}\{\psi_i(\vec{x}' - \vec{v}(z_i)t_q; 0)\psi_j(\vec{x}'' - \vec{v}(z_j)t_{q'}; 0)\},
\end{aligned} \tag{3.192}$$

where $\vec{v}(z_i)$ is the velocity of the ith layer and N is the number of layers. Invoking the assumptions that the phase fluctuations are zero mean, and Gaussian distributed, and that the layers are sufficiently far apart to be uncorrelated [3, 8, 24] implies

$$\mathrm{E}\{\psi_i(\vec{x}; t)\psi_j(\vec{x}; t)\} = 0 \quad \text{if } i \neq j, \tag{3.193}$$

which provides the following result:

$$
\begin{aligned}
\Gamma_{\phi}(\vec{x}_p, t_q; \vec{x}_{p'}, t_{q'}) &= \sum_{i=1}^{N} \Gamma_{\psi_i}(|\vec{x}_p - \vec{x}_{p'} - \vec{v}(z_i)(t_q - t_{q'})|) \\
&-p \int d\vec{x}'\, W(\vec{x}') \left\{ \sum_{i=1}^{N} \Gamma_{\psi_i}(|\vec{x}' - \vec{x}_{p'} - \vec{v}(z_i)(t_q - t_{q'})|) \right\} \\
&-p \int d\vec{x}''\, W(\vec{x}'') \left\{ \sum_{i=1}^{N} \Gamma_{\psi_i}(|\vec{x}_p - \vec{x}'' - \vec{v}(z_i)(t_q - t_{q'})|) \right\} \\
&+p \int d\vec{x}' \int d\vec{x}''\, W(\vec{x}')W(\vec{x}'') \left\{ \sum_{i=1}^{N} \Gamma_{\psi_i}(|\vec{x}' - \vec{x}'' - \vec{v}(z_i)(t_q - t_{q'})|) \right\},
\end{aligned} \tag{3.194}
$$

where Γ_{ψ_i} is the covariance of the phase fluctuations due to the ith layer:

$$
\Gamma_{\psi_i}(|\vec{x}_p - \vec{x}_{p'} - \vec{v}(z_i)(t_q - t_{q'})|) = \mathrm{E}\{\psi_i(\vec{x}_p - \vec{v}(z_i)t_q; 0)\psi_i(\vec{x}_{p'} - \vec{v}(z_i)t_{q'}; 0)\}. \tag{3.195}
$$

Equation (3.194) is the main result of this section.

Phase covariance for the von Karman spectrum

To compute the phase covariance given in Eq. (3.194) requires an expression for the covariance of the phase fluctuations due to the ith layer, Γ_{ψ_i}. To obtain this expression we start with the power spectrum of the index-of-refraction. The von Karman power spectrum is given by Eq. (3.10). A simplified form of the von Karman spectrum that ignores the inner scale effect is given by [43]

$$
\Phi_n^V(\kappa, z) = \frac{0.033 C_n^2(z)}{(\kappa^2 + 4\pi^2/L_0^2)^{11/6}}, \tag{3.196}
$$

where L_0 is the outer scale of the turbulence. This form differs from the form given in Eq. (3.10) by the absence of the $\exp\{-\kappa^2/\kappa_m^2\}$ term. Both expressions have been widely used in the literature. Note that the $\exp\{-\kappa^2/\kappa_m^2\}$ term only affects the high wavenumber portion of spectrum, where the value is quite small. Since we will only be using the von Karman spectrum inside integrals over κ, the effects of ignoring this term are negligible. Also, the form given in Eq. (3.196) permits analytic evaluation of a key integral in the derivation which follows.

If the outer scale, L_0, is independent of z, Eq. (3.196) can be factored into the product of a term, $f(z)$, depending only on z, and a normalized spectral density, $\Phi_0^V(\kappa)$, which depends only upon the wavenumber, κ:

$$
\Phi_n^V(\kappa, z) = f(z)\Phi_0^V(\kappa), \tag{3.197}
$$

where:

$$
f(z) = 0.033 C_n^2(z), \tag{3.198}
$$

and:

$$
\Phi_0^V(\kappa) = \frac{1}{(\kappa^2 + 4\pi^2/L_0^2)^{11/6}}. \tag{3.199}
$$

Since $\Phi_n^V(\kappa, z)$ is finite at $\kappa = 0$, there is no need to explicitly remove the piston component of the phase, $\psi(\vec{x}; t)$. Thus, p may be set to zero in Eq. (3.194).

The phase covariance for propagation through the entire atmosphere is characterized by the phase structure function given in Eq. (3.73). For near field conditions the $\cos^2(\kappa^2 z/2k)$ can be ignored [25]. Using this simplified structure function and it's relationship to the correlation function

$$D_\psi(\vec{\rho}) = 2[\Gamma_\psi(0) - \Gamma_\psi(\vec{\rho})], \tag{3.200}$$

we can write [6]:

$$\begin{aligned}\Gamma_\psi(\rho) &= (2\pi)^2 k^2 \int_0^h \mathrm{d}z \int_0^\infty \kappa \mathrm{d}\kappa\, J_0(\kappa\rho)\Phi_n^V(\kappa, z) \\ &= (2\pi)^2 k^2 \int_0^h \mathrm{d}z \int_0^\infty \kappa \mathrm{d}\kappa\, J_0(\kappa\rho)\Phi_0^V(\kappa) f(z),\end{aligned} \tag{3.201}$$

where $J_0(x)$ is the zeroth order Bessel function of the first kind and $\rho = |\vec{\rho}|$ is the radial separation between two field points. Note that the integration over z cannot be pulled out of the integral in Eq. (3.201), and evaluated independently of the integration over κ since ρ will in general depend on z as seen in Eq. (3.195). Applying the layered atmosphere model of Section 3.4.1 to Eq. (3.201), and using the assumption of independent layers yields

$$\Gamma_\psi(\rho) = 0.033(2\pi)^2 k^2 \left(\sum_{i=1}^{N} C_{n_i}^2 \Delta z_i \int_0^\infty \kappa \mathrm{d}\kappa\, J_0(\kappa\rho)\Phi_0^V(\kappa)\right) \tag{3.202}$$

A single element of the summation in Eq. (3.202) is interpreted as the phase covariance due to the ith layer:

$$\Gamma_{\psi_i}(\rho) = 0.033(2\pi)^2 k^2 C_{n_i}^2 \Delta z_i \int_0^\infty \kappa \mathrm{d}\kappa\, J_0(\kappa\rho)\Phi_0^V(\kappa). \tag{3.203}$$

For an N-layer atmosphere, the covariance is found by substituting Eq. (3.203) into Eq. (3.194) and setting $p = 0$ to obtain

$$\begin{aligned}&\Gamma_\phi(\vec{x}_p, t_q; \vec{x}_{p'}, t_{q'}) = \\ &0.033(2\pi)^2 k^2 \left(\sum_{i=1}^{N} C_{n_i}^2 \Delta z_i \int_0^\infty \kappa \mathrm{d}\kappa\, J_0(\kappa|\vec{x}_p - \vec{x}_{p'} - \vec{v}(z_i)(t_q - t_{q'})|)\Phi_0^V(\kappa)\right).\end{aligned} \tag{3.204}$$

Equation (3.204) can also be written in terms of the Fried parameter for the ith layer, r_{0_i}:

$$\begin{aligned}&\Gamma_\phi(\vec{x}_p, t_q; \vec{x}_{p'}, t_{q'}) = \\ &3.089 \left(\sum_{i=1}^{N} r_{0_i}^{-5/3} \int_0^\infty \kappa \mathrm{d}\kappa\, J_0(\kappa|\vec{x}_p - \vec{x}_{p'} - \vec{v}(z_i)(t_q - t_{q'})|)\Phi_0^V(\kappa)\right).\end{aligned} \tag{3.205}$$

where r_{0_i} is defined by Eq. (3.57). The integral in Eqs. (3.204) and (3.205) has a closed form solution given by [44, 45]:

$$\int_0^\infty \kappa \mathrm{d}\kappa\, J_0(\kappa\rho)(\kappa^2 + 4\pi^2/L_0^2)^{-11/6} = \frac{(L_0/2\pi)^{5/6} K_{5/6}\left[2\pi\rho/L_0\right]\rho^{5/6}}{2^{5/6}\Gamma\left[11/6\right]}, \tag{3.206}$$

where $\rho = |\vec{x}_p - \vec{x}_{p'} - \vec{v}(z_i)(t_q - t_{q'})|$, $K_{5/6}[\cdot]$ is the modified Bessel function of the second kind of order $5/6$, and $\Gamma[\cdot]$ is the gamma function. The modified Bessel function and the gamma function can be evaluated using widely available mathematical computation software packages [46].

There is evidence that the outer scale, L_0, is not independent of altitude, z, as assumed above [7, 8]. If L_0 varies with altitude, then $\Phi_n^V(\kappa, z)$ cannot be factored as shown in Eq. (3.197). This effect can be accounted for by using the appropriate L_0 for each layer in the calculation of the theoretical covariance in Eqs. (3.204) and (3.205).

Equations (3.204) and (3.205) provide the final expressions for the covariance of the phase fluctuations for the von Karman spectrum. Note that each layer can have a unique vector velocity whose effects will be represented in the covariance matrix calculation. Also note the covariance expression can be scaled for different seeing conditions by adjusting the structure constant for each layer, $C_{n_i}^2$ or equivalently adjusting the Fried parameter for each layer, r_{0_i}.

Phase covariance for the Kolmogorov turbulence spectrum.

The Kolmogorov spectrum $\Phi_n^K(\kappa, z)$ is given by Eq. (3.9):

$$\Phi_n^K(\kappa, z) = 0.033 C_n^2(z) \kappa^{-11/3} . \tag{3.207}$$

As previously noted, this spectrum can also be factored into the product of a term, $f(z)$, depending only on distance, z, and a normalized spectral density, $\Phi_0^K(\kappa)$, which depends only upon the wavenumber, κ:

$$\Phi_n^K(\kappa, z) = f(z) \Phi_0^K(\kappa) \tag{3.208}$$

where:

$$\Phi_0^K(\kappa) = \kappa^{-11/3} . \tag{3.209}$$

The function $\Phi_n^K(\kappa, z)$ has a non-integrable pole at $\kappa = 0$. Hence, the piston component of the incident phase, $\psi(\vec{x}, t)$ must be explicitly removed by setting $p = 1$ in Eq. (3.194).

To evaluate the resulting expression for the phase covariance Eq. (3.194) is rearranged:

$$\begin{aligned}\Gamma_\phi(\vec{x}_p, t_q; \vec{x}_{p'}, t_{q'}) &= \int \mathrm{d}\vec{x}' \int \mathrm{d}\vec{x}'' \, W(\vec{x}') W(\vec{x}'') \\ &\times \sum_{i=1}^{N} \Big\{ \Gamma_{\psi_i}(|\vec{x}_p - \vec{x}_{p'} - \vec{v}(z_i)(t_q - t_{q'})|) - \Gamma_{\psi_i}(|\vec{x}' - \vec{x}_{p'} - \vec{v}(z_i)(t_q - t_{q'})|) \\ &- \Gamma_{\psi_i}(|\vec{x}_p - \vec{x}'' - \vec{v}(z_i)(t_q - t_{q'})|) + \Gamma_{\psi_i}(|\vec{x}' - \vec{x}'' - \vec{v}(z_i)(t_q - t_{q'})|) \Big\} .\end{aligned} \tag{3.210}$$

Equation (3.210) can be written in terms of phase structure functions using Eq. (3.200) with the result:

$$\begin{aligned}\Gamma_\phi(\vec{x}_p, t_q; \vec{x}_{p'}, t_{q'}) &= -\frac{1}{2} \int \mathrm{d}\vec{x}' \int \mathrm{d}\vec{x}'' \, W(\vec{x}') W(\vec{x}'') \\ &\times \sum_{i=1}^{N} \Big\{ D_{\psi_i}(|\vec{x}_p - \vec{x}_{p'} - \vec{v}(z_i)(t_q - t_{q'})|) - D_{\psi_i}(|\vec{x}' - \vec{x}_{p'} - \vec{v}(z_i)(t_q - t_{q'})|) \\ &- D_{\psi_i}(|\vec{x}_p - \vec{x}'' - \vec{v}(z_i)(t_q - t_{q'})|) + D_{\psi_i}(|\vec{x}' - \vec{x}'' - \vec{v}(z_i)(t_q - t_{q'})|) \Big\} ,\end{aligned} \tag{3.211}$$

where $D_{\psi_i}(\rho)$ is the structure function of the phase fluctuations for the ith layer given by Eq. (3.58):

$$D_{\psi_i}(\rho) = \mathrm{E}[(\psi_i(\vec{x}) - \psi_i(\vec{x} + \vec{\rho}))^2] = 6.88 \left(\frac{\rho}{r_{0_i}} \right)^{5/3} , \tag{3.212}$$

and r_{0_i} is the Fried parameter for the ith layer. Distributing the integration over the individual terms of Eq. (3.211) and substituting in Eq. (3.212) gives:

$$\Gamma_\phi(\vec{x}_p, t_q; \vec{x}_{p'}, t_{q'}) = 6.88 \sum_{i=1}^{N} r_{0_i}^{-5/3} \left\{ -\frac{1}{2} |\vec{x}_p - \vec{x}_{p'} - \vec{v}(z_i)(t_q - t_{q'})|^{5/3} \right.$$
$$+ \frac{1}{2} \int \mathrm{d}\vec{x}' \, W(\vec{x}') |\vec{x}' - \vec{x}_{p'} - \vec{v}(z_i)(t_q - t_{q'})|^{5/3}$$
$$+ \frac{1}{2} \int \mathrm{d}\vec{x}'' \, W(\vec{x}'') |\vec{x}_p - \vec{x}'' - \vec{v}(z_i)(t_q - t_{q'})|^{5/3}$$
$$\left. - \frac{1}{2} \int \mathrm{d}\vec{x}' \int \mathrm{d}\vec{x}'' \, W(\vec{x}') W(\vec{x}'') |\vec{x}' - \vec{x}'' - \vec{v}(z_i)(t_q - t_{q'})|^{5/3} \right\}, \qquad (3.213)$$

which is the final form of the phase covariance for the Kolmogorov power spectrum. For a circular aperture the integrals in Eq. (3.213) can be evaluated using the Gegenbauer polynomial technique described in Ref. [47], and summarized below.

Applying Gegenbauer polynomials to evaluate Eq. (3.213) requires that the aperture described by $W(\vec{x})$ be circular [47]. Let the aperture radius be R. Using the appropriate Gegenbauer polynomials, as shown in Ref. [47], to evaluate Eq. (3.213) yields

$$\Gamma_\phi(\vec{x}_p, t_q; \vec{x}_{p'}, t_{q'}) = 6.88 \sum_{i=1}^{N} r_{0_i}^{-5/3} \left\{ -\frac{1}{2} |\vec{x}_p - \vec{x}_{p'} - \vec{v}(z_i)(t_q - t_{q'})|^{5/3} \right.$$
$$+ \frac{1}{2} R^{5/3} F_1\left(\frac{\vec{x}_{p'}}{R} + \frac{\vec{v}(z_i)}{R}(t_q - t_{q'}) \right)$$
$$+ \frac{1}{2} R^{5/3} F_1\left(-\frac{\vec{x}_p}{R} + \frac{\vec{v}(z_i)}{R}(t_q - t_{q'}) \right)$$
$$\left. - \frac{1}{2} R^{5/3} F_2\left(-\frac{\vec{v}(z_i)}{R}(t_q - t_{q'}) \right) \right\}, \qquad (3.214)$$

where

$$F_1(\vec{\Omega}) = \begin{cases} \frac{6}{11} \, {}_2F_1(\frac{-11}{6}, \frac{-5}{6}; 1; |\vec{\Omega}|^2) & |\vec{\Omega}| \le 1 \\ |\vec{\Omega}|^{5/3} \, {}_2F_1(\frac{-5}{6}, \frac{-5}{6}; 2; |\vec{\Omega}|^{-2}) & |\vec{\Omega}| \ge 1 \end{cases}, \qquad (3.215)$$

$${}_2F_1(a, b; c; z) = \frac{\Gamma(c)}{\Gamma(a)\Gamma(b)} \sum_{n=0}^{\infty} \frac{\Gamma(a+n)\Gamma(b+n)}{\Gamma(c+n)} \frac{z^n}{n!}, \qquad (3.216)$$

and

$$F_2(\vec{\Omega}) = \frac{2}{\pi} \int_0^1 K_1(\vec{\rho}) H(\vec{\rho}, \vec{\Omega}) \mathrm{d}\vec{\rho}, \qquad (3.217)$$

$$K_1(\vec{\rho}) = \begin{cases} \pi & |\vec{\rho}| \le 0 \\ 2\cos^{-1}\left(\frac{|\vec{\rho}|}{2}\right) - |\vec{\rho}| \sqrt{1 - \left(\frac{|\vec{\rho}|}{2}\right)^2} & 0 < |\vec{\rho}| < 2 \\ 0 & \text{otherwise} \end{cases}, \qquad (3.218)$$

$$H(\vec{\rho}, \vec{\Omega}) = \begin{cases} |\vec{\rho}|^{8/3} \, {}_2F_1\left(\frac{-5}{6}, \frac{-5}{6}; 1; \frac{|\vec{\Omega}|^2}{|\vec{\rho}|^2}\right) & |\vec{\Omega}| \le |\vec{\rho}| \\ |\vec{\rho}| |\vec{\Omega}|^{5/3} \, {}_2F_1\left(\frac{-5}{6}, \frac{-5}{6}; 1; \frac{|\vec{\rho}|^2}{|\vec{\Omega}|^2}\right) & |\vec{\Omega}| \ge |\vec{\rho}| \end{cases}, \qquad (3.219)$$

Phase covariance calculation extended to the anisoplanatic case

In some situations it is useful to generate random phase screens that arise from distinct propagation directions. These screens could be used to simulate the performance of an adaptive optical system observing in one direction while sensing the light from a beacon in another direction. The degradation suffered by the system due to the difference in direction of the beacon and object wave fronts is referred to as an anisoplanatic effect. Anisoplanatic effects are discussed in Section 3.4.3. To simulate anisoplanatism it is necessary to make random draws of a phase screen for each direction of interest. From the development in Section 3.4.3, it is apparent that to incorporate anisoplanatic effects in the general covariance expression of Eq. (3.194) it is a simple matter of substituting $\Delta p_{qq'}(z)$ for $v(z)(t_q - t_{q'})$, where $\Delta p_{qq'}(z)$ is the transverse separation of the qth and q'^{th} ray paths associated with the qth and q'^{th} wave front sources (see Fig. 3.7 for geometry). In Section 3.4.3 the subscripts q and q' were related to the object (subscript o) and beacon (subscript b) sources that are required for operation of an adaptive optical imaging system.

The final forms of the phase covariance for the von Karman and Kolmogorov turbulence spectra for the anisoplanatic case can be obtained by substituting $\Delta p_{qq'}(z)$ for $\vec{v}(z_i)(t_q - t_{q'})$ in Eqs. (3.205) and (3.213).

Scaling covariance matrices for changes in r_0

Once the covariance matrix Γ_a has been computed from the covariance expressions given in Eqs. (3.205) and (3.213), it is straightforward to scale the matrix for different values of the Fried parameter r_0. If the value of the Fried parameter used to compute Γ_a is designated $r_{0_{old}}$, then the ratio

$$\left(\frac{r_{0_{new}}}{r_{0_{old}}}\right)^{-5/3} \tag{3.220}$$

can be used to the scale Γ_a to the new Fried parameter designated $r_{0_{new}}$:

$$\Gamma_{a_{new}} = \left(\frac{r_{0_{new}}}{r_{0_{old}}}\right)^{-5/3} \Gamma_{a_{old}}, \tag{3.221}$$

Phase screen generation examples

In this section we present some examples of phase screen generation using the delta function basis set. The results will show random draws of phase screens which are separated in time. For the results presented here the von Karman spectrum and the Kolmogorov spectra were used with the SLC-Night C_n^2 profile shown in Fig. 3.2. The layered model of this profile is characterized by Table 3.1. The value of the Fried parameter is assumed $r_0 = 0.103$ m. The outer scale for the von Karman spectrum is set at $L_0 = 100$ m. For the von Karman spectrum results, the length of one side of the square aperture is set equal to 1 m, and there are 21 phase sample points across one side of the aperture (i.e., $P = 21 \times 21$), which provided a sample spacing of $\Delta x = 5$ cm. The Kolmogorov spectrum results were calculated using the Gegenbauer polynomial technique for a 1 m radius aperture. The Kolmogorov spectrum results presented are for the 1 m by 1 m square centered within the 1 m radius circle. Results for three temporal realizations of the phase (i.e., $Q = 3$) are shown, which required the computation and factorization of 1323×1323 element covariance arrays.

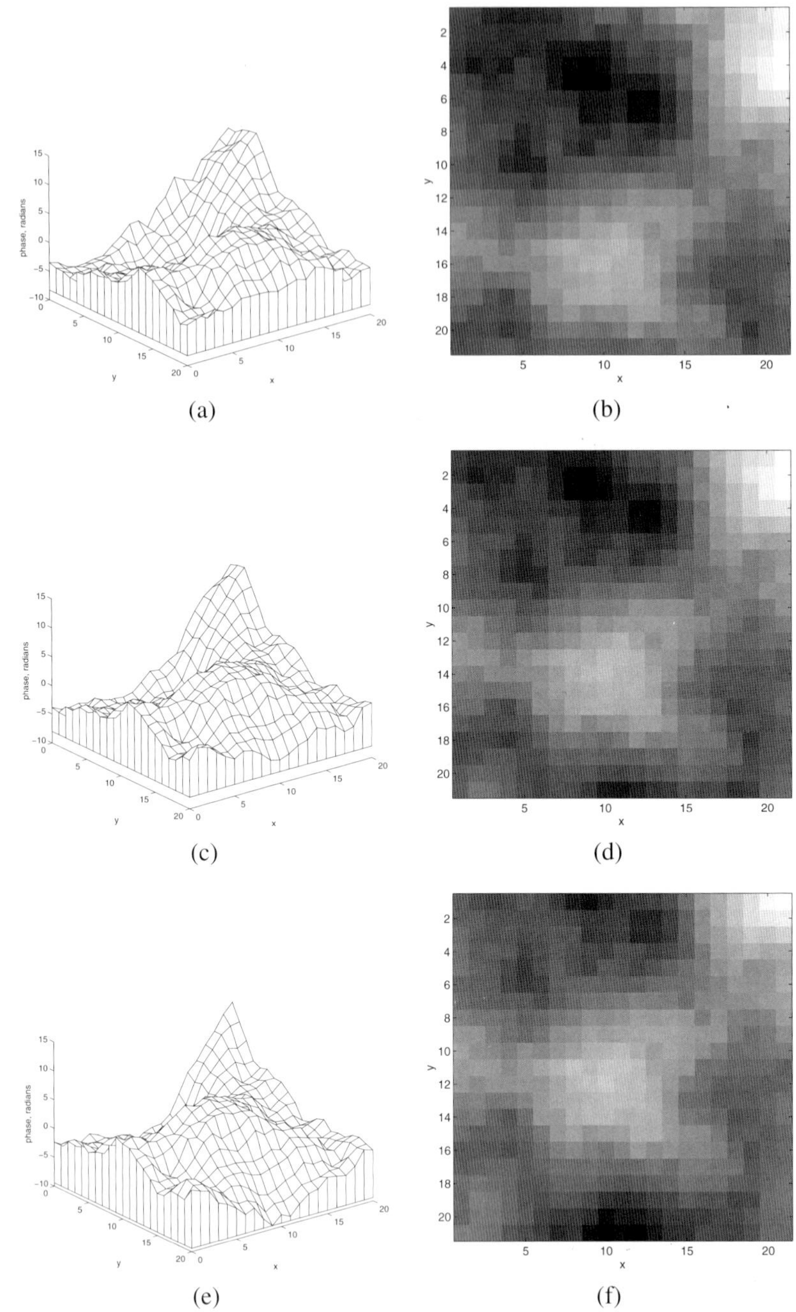

FIGURE 3.15
Three random, time sequential images for the case of atmospheric layers moving in the same direction and speed, and a von Karman spectrum: (a) surface plot of phase screen, $t_1 = 0$ s; (b) grey scale image of (a); (c) surface plot of phase screen, $t_2 = 0.01$ s; (d) grey scale image of (c); (e) surface plot of phase screen, $t_3 = 0.02$ s; (f) grey scale image of (e).

Figure 3.15 shows three random time-sequential phase screens created using the von Karman spectrum with the following inputs. The wind velocities for each of the four layers are identical, with $v_x = 0$ and $v_y = 10$ m/s, where v_x and v_y are the x- and y-directed component of $\vec{v}(z)$. Thus, Taylor's frozen field hypothesis has been applied to the whole atmosphere in this case, and the atmosphere is presumed to be moving as a monolithic block in the y direction (i.e., upward). Fixed structures in the phase screen should thus move in the y direction. The time between phase screens was set to be 0.01 s, so that each successive phase screen is shifted by exactly 10 cm in the y direction relative to the previous phase screen. This shift distance corresponds to a shift of exactly two sample spacings in the simulated phase screen arrays. The random piston component of each phase screen was removed prior to displaying it in Fig. 3.15. In Fig. 3.15 both surface plots and grey scale images of the phase screens are shown. The grey scale plots were created by linearly stretching the phase screen values between 0 and 255 gray levels. Since each phase screen has a slightly different dynamic range, portions of the phase screens which have identical phase fluctuations do not necessarily appear identical in the grey scale image. However, it is easy to note the movement of major structures in the phase screen in the upward direction (corresponding to the positive y direction) exactly two sample spacings per time step, in correspondence with the input parameters. Figure 3.16 shows the same case for the Kolmogorov spectrum obtained using the same random number seed used to generate Fig. 3.15.

One important performance issue is the accuracy with which the calculated input covariance can be simulated. To evaluate the accuracy of the simulation, we computed the covariance matrix for the simulated phase, $\Gamma_{\tilde{\phi}}$, given by

$$\Gamma_{\tilde{\phi}} = \mathrm{E}[\tilde{\phi}\tilde{\phi}^T], \tag{3.222}$$

where $\tilde{\phi}$ is a vector of the random generated phases. The computation given above is computed across a data set of 1000 random draws. The mean square error per covariance array element, ϵ^2, was computed, where ϵ^2 is given by

$$\epsilon^2 = \frac{\sum(\Gamma_\phi - \Gamma_{\tilde{\phi}})^2}{(P \times Q)^2}, \tag{3.223}$$

where the summation is taken over all points in the covariance matrices. For the calculation corresponding to the results above, the root mean square error per covariance matrix element was $\epsilon = 5.315 \times 10^{-3}$, which is $6.361 \times 10^{-5}\%$ of the smallest element in the theoretical covariance matrix, Γ_ϕ.

As final examples of the performance of the technique, we provide a case where the wind velocities of the four layers are different. The velocities of the layers are given in Table 3.6. Three phase screens with time separations of 0.01 s are shown in Fig. 3.17 for the von Karman spectrum. A similar set of phase screens is shown in Fig. 3.18 for the Kolmogorov spectrum. Note that the temporal evolution of the phase screens in Figs. 3.17 and 3.18 is more complicated than the temporal evolution of the phase screen in Figs. 3.15 and 3.16. This is consistent with the use of a more complicated wind model. The same random number seed was used to create all of the random phase screens shown in Figs. 3.15, 3.16, (3.17), and (3.18). Hence, for each power spectrum, the first phase screen in each case should be nearly identical, as can be seen by inspection of the first time steps in Figs. 3.15 and 3.17, and in Figs. 3.16 and 3.18.

3.7.3 Phase screen generation using a Zernike polynomial basis

Zernike polynomials have been used frequently in phase screen simulations [38]. The basic phase screen generation approach is the same as presented in Section 3.7.1. To implement the approach outlined in this section, the covariance matrix Γ_a must be computed. In this case, the elements of Γ_a are the correlations of the amplitude of each Zernike polynomial mode. An expression for the mode correlations at a single instant of time has already been presented in Section 3.6.1 and is given in Eq. (3.158). This result is derived directly from Eq. (3.142) using the Kolmogorov power spectrum for the index-of-refraction fluctuations and the definition of the Zernike polynomials given in Section 3.6.1.

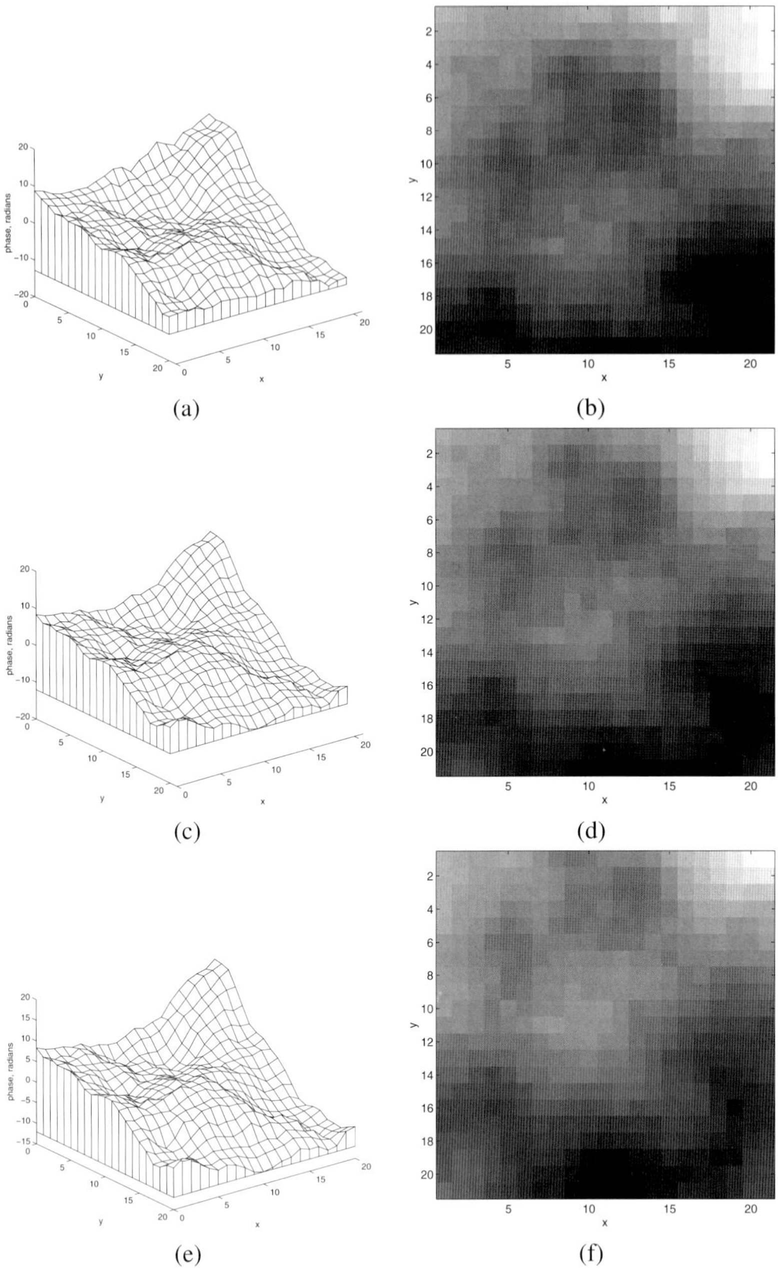

FIGURE 3.16
Three random, time sequential images for the case of atmospheric layers moving in the same direction and speed, and a Kolmogorov spectrum: (a) surface plot of phase screen, $t_1 = 0$ s; (b) grey scale image of (a); (c) surface plot of phase screen, $t_2 = 0.01$ s; (d) grey scale image of (c); (e) surface plot of phase screen, $t_3 = 0.02$ s; (f) grey scale image of (e).

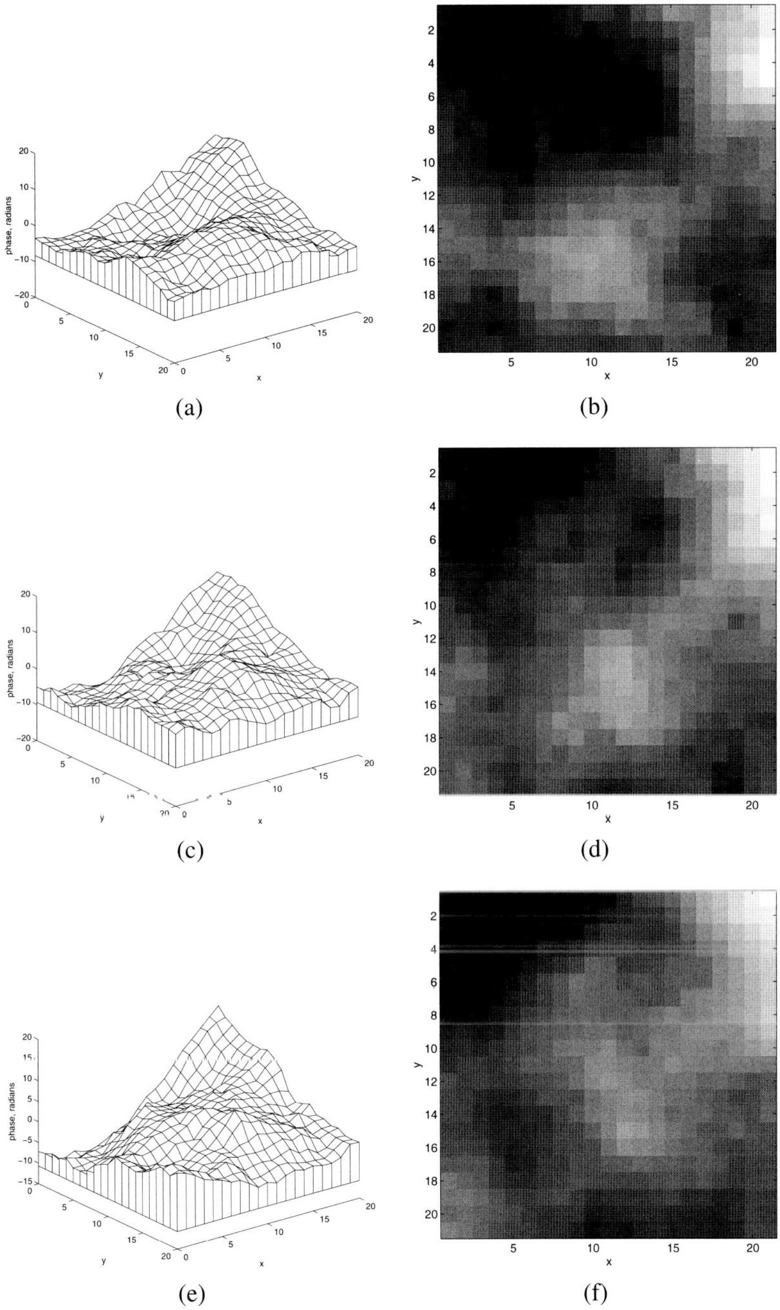

FIGURE 3.17
Three random, time sequential images for the case of atmospheric layers moving in different directions and speeds, and for a von Karman power spectrum: (a) surface plot of phase screen, $t_1 = 0$ s; (b) grey scale image of (a); (c) surface plot of phase screen, $t_2 = 0.01$ s; (d) grey scale image of (c); (e) surface plot of phase screen, $t_3 = 0.02$ s; (f) grey scale image of (e).

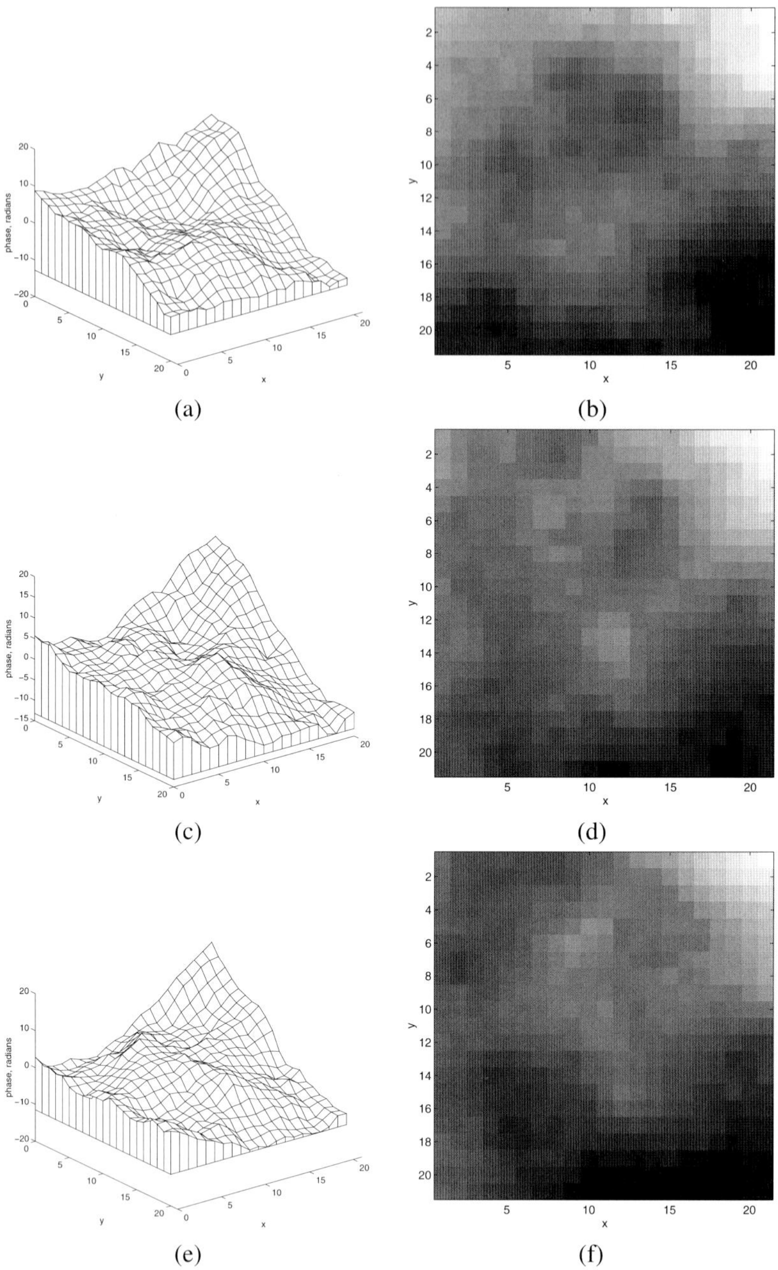

FIGURE 3.18
Three random, time sequential images for the case of atmospheric layers moving in different directions and speeds, and for a Kolmogorov power spectrum: (a) surface plot of phase screen, $t_1 = 0\,\mathrm{s}$; (b) grey scale image of (a); (c) surface plot of phase screen, $t_2 = 0.01\,\mathrm{s}$; (d) grey scale image of (c); (e) surface plot of phase screen, $t_3 = 0.02\,\mathrm{s}$; (f) grey scale image of (e).

TABLE 3.6
Wind velocities of the four atmospheric layers for the phase screens shown in Figs. 3.17 and 3.18.

Layer	v_x (m/s)	v_y (m/s)
1	0	10
2	0	-10
3	$\sqrt{10}$	$\sqrt{10}$
4	$-\sqrt{10}$	$-\sqrt{10}$

Unfortunately, this result is not applicable for computing the correlations required to simulate temporal correlated phase screens (i.e., computing $\mathrm{E}[a_p(t_q)a_{p'}(t_{q'})]$ for $q \neq q'$). These temporal correlations can be found in published work by Roddier [22] and Hogge and Butts [48]. The approach used by these authors is a straightforward application of frozen flow concepts presented earlier to the weight correlation expression given in Eq. (3.142).

Bibliography

[1] D. L. Fried, "Optical resolution through a randomly inhomogeneous medium for very long and very short exposures," *J. Opt. Soc. Am.*, vol. 56, pp. 1372–1379, 1966.

[2] A. N. Kolmogorov, "The local structure of turbulence in incompressible viscous fluids for very large reynolds' numbers," in *Turbulence, Classic Papers on Statistical Theory* (S. K. Friedlander and L. Topper, eds.), pp. 151–155, New York: Wiley-Interscience, 1961.

[3] V. I. Tatarskii, *Wave Propagation in a Turbulent Medium.* New York: Dover Publications, 1967.

[4] D. L. Fried, "Anisoplanatism in adaptive optics," *J. Opt. Soc. Am.*, vol. 72, pp. 52–61, 1982.

[5] V. I. Tatarskii and V. U. Zavorotny, "Atmospheric turbulence and the resolution limits of large ground-based telescopes: comment," *J. Opt. Soc. Am. Communications*, vol. 10, pp. 2410–2414, 1993.

[6] A. Ishimaru, *Wave Propagation and Scattering in Random Media*, vol. 2. New York: Academic Press, 1978.

[7] F. Roddier, "The effects of atmospheric turbulence in optical astronomy," in *Progress in Optics* (E. Wolf, ed.), vol. XIX, New York: North-Holland, 1981.

[8] J. W. Goodman, *Statistical Optics.* New York: John Wiley & Sons, 1985.

[9] C. A. Haniff, J. E. Baldwin, P. J. Warner, and T. R. Scott, "Atmospheric phase fluctuation measurement: interferometric results from the wht and coast telescopes," in *Proc. SPIE*, vol. 2200, pp. 407–417, 1994.

[10] D. Dayton, B. Pierson, B. Spielbusch, and J. Gonglewski, "Atmospheric structure function measurements with a shack-hartmann wave-front sensor," *Opt. Lett.*, vol. 17, pp. 1737–1739, 1992.

[11] M. M. Colavita, "Design considerations for very long baseline fringe-tracking interferometers," in *Proc. SPIE on Amplitude and Intensity Interferometry*, vol. 1237, pp. 80–86, 1990.

[12] R. R. Parenti and R. J. Sasiella, "Laser-guide-star systems for astronomical applications," *J. Opt. Soc. Am. A*, vol. 11, pp. 288–309, 1994.

[13] D. P. Greenwood, "Bandwidth specifications for adaptive optics systems," *J. Opt. Soc. Am.*, vol. 67, pp. 390–392, 1977.

[14] A. Papoulis, *Probability, Random Variables, and Stochastic Processes.* New York: McGraw-Hill, third ed., 1991.

[15] D. W. Tyler and C. L. Matson, "Speckle imaging detector optimization and comparison," *Opt. Eng.*, vol. 32, pp. 864–869, 1993.

[16] G. M. Cochran, B. L. Ellerbroek, and D. L. Fried, "Spectral and temporal windows in knox-thompson imaging," Tech. Rep. TR-819R, The Optical Sciences Company, 1988.

[17] T. W. Lawrence, D. M. Goodman, E. M. Johansson, and J. P. Fitch, "Speckle imaging of satellites at the u.s. air force maui optical station," *Appl. Opt.*, vol. 31, pp. 6307–6321, 1992.

[18] D. L. Fried and D. L. Hench, "Spectral bandwidth and adaptive optics considerations for white light speckle imagery," Tech. Rep. TR-1033, the Optical Sciences Company, 1990.

[19] K. A. O'Donnell and J. C. Dainty, "A space-time analysis of photon-limited stellar speckle interferometry," *J. Opt. Soc. Am.*, vol. 70, pp. 1354–1361, 1980.

[20] J. G. Walker, "Optimum exposure time and filter bandwidth in speckle interferometry," in *High Angular Resolution Stellar Interferometry*, vol. 5, pp. 25–1 to 25–24, 1978. Proceedings of Colloquium #50, International Astronomical Union, Univ. of Maryland, College Park, MD.

[21] B. L. Ellerbroek, "First-order performance evaluation of adaptive optics systems for atmospheric turbulence compensation in extended field-of-view astronomical telescopes," *J. Opt. Soc. Am. A*,

vol. 11, pp. 783–805, 1994.

[22] F. Roddier, M. J. Northcott, J. E. Graves, and D. L. McKenna, "One-dimensional spectra of turbulence-induced zernike aberrations: time-delay and isoplanicity error in partial adaptive compensation," *J. Opt. Soc. Am. A*, vol. 10, pp. 957–965, 1993.

[23] P. M. Harrington and B. M. Welsh, "A frequency domain analysis of an adaptive optical system's temporal response," *Opt. Eng.*, vol. 33, pp. 2336–2342, 1994.

[24] S. E. Troxel, B. M. Welsh, and M. C. Roggemann, "Off-axis optical transfer function calculations in an adaptive-optics system by means of a diffraction calculation for weak index fluctuations," *J. Opt. Soc. Am. A*, vol. 11, pp. 2100–2111, 1994.

[25] A. T. Young, "Seeing: Its cause and cure," *Astrophys. J.*, vol. 189, pp. 587–604, 1974.

[26] D. L. Fried, "Limiting resolution looking down through the atmosphere," *J. Opt. Soc. Am.*, vol. 56, pp. 1380–1384, 1966.

[27] H. T. Yura, "An elementary derivation of phase fluctuations of an optical wave in the atmosphere," *SPIE Proc.*, vol. 75, pp. 9–15, 1976.

[28] R. W. Lee and J. C. Harp, "Weak scattering in random media, with applications to remote probing," *Proc. IEEE*, vol. 57, pp. 375–406, 1969.

[29] B. M. Welsh and C. S. Gardner, "Effects of turbulence induced anisoplanatism on the imaging performance of adaptive astronomical telescopes using laser guide stars," *J. Opt. Soc. Am. A*, vol. 8, pp. 69–80, 1991.

[30] R. F. Lutomirski and R. G. Buser, "Mutual coherence function of a finite optical beam and application to coherent detection," *Appl. Opt.*, vol. 12, pp. 2153–2160, 1973.

[31] D. L. Fried, "Statistics of a geometric representation of wavefront distortion," *J. Opt. Soc. Am.*, vol. 55, pp. 1427–1435, 1965.

[32] G. R. Heidbreder, "Image degradation with random wavefront tilt compensation," *IEEE Trans. Ant. Prop.*, vol. AP-15, pp. 90–98, 1967.

[33] M. Born and E. Wolf, *Principles of Optics*. New York: Pergamon Press, 1964.

[34] R. J. Noll, "Zernike polynomials and atmospheric turbulence," *J. Opt. Soc. Am.*, vol. 66, pp. 207–211, 1976.

[35] E. P. Wallner, "Optimal wave front correction using slope measurements," *J. Opt. Soc. Am.*, vol. 73, pp. 1771–1776, 1983.

[36] B. M. Welsh and C. S. Gardner, "Performance analysis of adaptive optics systems using slope sensors," *J. Opt. Soc. Am. A*, vol. 6, pp. 1913–1923, 1989.

[37] J. Y. Wang and J. K. Markey, "Modal compensation of atmospheric turbulence phase distortion," *J. Opt. Soc. Am.*, vol. 68, pp. 78–87, 1978.

[38] N. Roddier, "Atmospheric wavefront simulation using zernike polynomials," *Opt. Eng.*, vol. 29, pp. 1174–1180, 1990.

[39] G. Strang, *Linear Algebra and its Applications*. New York: Academic Press, second ed., 1980.

[40] M. C. Roggemann, "Optical performance of fully and partially compensated adaptive optics systems using least-squares and minimum variance phase reconstruction," *Comp. & Elec. Eng.*, vol. 18, pp. 451–466, 1992.

[41] T. Goldring and L. Carlson, "Analysis and implementation of non-kolmogorov phase screens appropriate to structured environments," in *SPIE proceedings on Nonlinear Optical Beam Manipulation and High Energy Beam Propagation Through the Atmosphere*, vol. 1060, pp. 244–264, 1989.

[42] W. Press, B. Flannery, S. Teukolsky, and W. Vetterling, *Numerical Recipes - The Art of Scientific Computing*. Cambridge, UK: Cambridge University Press, 1986.

[43] R. J. Sasiela and J. D. Shelton, "Transverse spectral filtering and mellin transform techniques applied to the effect of outer scale on tilt and tilt anisoplanatism," *J. Opt. Soc. Am. A*, vol. 10, pp. 646–660, 1993.

[44] R. J. Sasiela and J. D. Shelton, "Mellin transform methods applied to integral evaluation: Taylor series and asymptotic approximations," *J. Math Phys.*, vol. 34, pp. 2572–2619, 1993.

[45] R. J. Sasiela, "A unified approach to electromagnetic wave propagation in turbulence and the evaluation of multiparameter integrals," Tech. Rep. TR-807, Lincoln Laboratory, Massachusetts Institute of Technology, 1988.

[46] "Math library/special functions," in *IMSL FORTRAN Subroutines for Mathematical Applications*, 1991.

[47] G. A. Tyler, "Merging: a new method for tomography through random media," *J. Opt. Soc. Am. A*, vol. 10, pp. 409–425, 1993.

[48] C. B. Hogge and R. R. Butts, "Frequency spectra for the geometric representation of wavefront distortions due to atmospheric turbulence," *IEEE Tran. Ant. Prop.*, vol. AP-24, pp. 144–154, 1976.

4

Speckle Imaging Techniques

4.1 Introduction

Imaging of space objects has been hindered by atmospheric turbulence since the invention of the telescope. As mentioned in the introduction to Chapter 1, Issac Newton (1730) was aware that the atmosphere was the cause of reduced resolution in astronomical imaging systems [1]. Following the invention of photographic film in the early 1800's, and continuing through until the 1950's, the general mode of operation for astronomical observations was to measure images using very long exposures - on the order of minutes to hours. Use of such long exposure times was driven, in part, by the desire to see the dimmest possible objects in space. Astronomers were aware that the images of stars were much broader than the point spread functions which would arise from diffraction due to the telescope optics alone [2]. The images of stars clearly showed the effective point spread function of the atmosphere-telescope system, with most of the broadening in the star images attributed to atmospheric turbulence effects. However, prior to the early 1960's there was no theory of wave propagation through the turbulent atmosphere, and hence there was no quantitative theory predicting the optical effects of turbulence.

By the late 1950's the first truly short exposure images of space objects had been measured. The term *short exposure image* refers to images measured using exposure times short enough to effectively freeze the turbulence, typically on the order of a few milliseconds. Short exposure images were found to have a significantly different character than long exposure images. In particular, the short exposure images were broader than the point spread function of the optics alone, but were "speckled" in appearance, possessing high spatial frequency modulation not present in long exposure images. Further, the centroid of the intensity of the short exposure images was observed to move randomly about the image plane.

During the early and mid 1960's great advances were made in the theoretical understanding of turbulence effects on optical systems. The work of Tatarskii [3] stands as the seminal contribution, uniting the work of Kolmogorov in the statistics of turbulent fluid motion [4], and the theory of electro magnetic wave propagation. Tatarskii's theory correctly predicted the broadening of the point spread function of an imaging system due to turbulence. Subsequent work by Fried [5, 6] provided the basis for understanding the speckled nature and random centroid motion of short exposure images. The combined efforts of Tatarskii, Fried, and others (see for example, the excellent review paper by Roddier in Ref. [2]) have provided an elegant mathematical framework for understanding both the effects of turbulence on imaging systems, and the performance of various techniques used to overcome these effects.

The first efforts to overcome the effects of atmospheric turbulence relied purely on post detection processing of short exposure images measured through turbulence. Pure post processing techniques are now referred to as *speckle imaging* methods, in reference to the speckled appearance of the images processed. In speckle imaging a high resolution image of the object is not measured directly. Rather, an estimate of the Fourier transform of the object is computed, and an image is obtained by the inverse Fourier transform operation. Speckle imaging is not a single technique, but rather a collection of post processing techniques for recovering an estimate of both the modulus and the phase of the Fourier transform of the object. The Fourier transform of an object or image will be referred to as the spectrum of the appropriate quantity. An estimate of the object is then obtained by applying the inverse Fourier transform to the spectral data.

All of the techniques used in speckle imaging arise from Labeyrie's observation that the speckles in short exposure images contain more high spatial frequency information about the object than long exposure images [7]. In his seminal paper Labeyrie proposed a method for estimating the modulus of the Fourier transform, or modulus spectrum of an object. Labeyrie's technique is today referred to as speckle interferometry. Speckle interferometry is widely used for obtaining information about the separation and relative brightnesses of binary stars. As we shall see, estimating the separation and relative brightnesses of a binary star is an application which does not require the phase of the object spectrum.

However, the phase of the Fourier transform of an object, often referred to as the phase spectrum of the object, is needed for applications where an image is required. The two most widely used methods for estimating the phase spectrum of an object are referred to as the Knox-Thompson or cross-spectrum method, [8] and the bispectrum method [9]. Both the Knox-Thompson and the bispectrum method cannot provide a direct estimate of the phase spectrum of the object. Rather they provide quantities which contain linear combinations of the elements of the phase spectrum of the object. A distinct operation is required to reconstruct the object phase spectrum from the cross-spectrum or bispectrum.

In this chapter we present speckle imaging techniques, and provide examples of the performance of speckle imaging. The remainder of this chapter is organized as follows. In the next section we present a brief overview of speckle imaging. In Section 4.3 we discuss speckle interferometry, which is used to obtain an estimate of the modulus of the Fourier transform of the object. Results are presented showing the signal-to-noise ratio performance of the speckle interferometry method across a range of seeing and light level conditions. In Section 4.4 we present the Knox-Thompson and bispectrum methods for estimating the phase spectrum of an object. Section 4.5 provides the details of reconstructing an image from speckle imaging measurements and computations, and presents examples of speckle imaging performance.

4.2 Overview of speckle imaging.

In this section we qualitatively describe speckle imaging and provide the foundation for the algorithms and performance analysis presented in the succeeding sections. The speckle imaging technique is shown in block diagram form in Fig. 4.1. Two sets of image data are required to perform speckle imaging: (1) a set of N short exposure images of the object of interest, with the nth image of the object denoted by $i_n(\vec{x})$; and (2) a set of M short exposure images of a bright nearby reference star, with the mth image of the reference star denoted by $r_m(\vec{x})$. The vector $\vec{x}$ is used to represent a two dimensional spatial position in the image plane. The size of the data sets is generally driven by signal-to-noise ratio considerations, with typical values of N and M on the order of a several tens to a few thousand. Exposure times on the order of a few milliseconds to a few tens of milliseconds are commonly used. All correction for turbulence effects is accomplished by the speckle imaging estimators. Reference star data is used to boost strongly attenuated spectral components at high spatial frequencies in the estimated object spectrum to their proper values. This boosting step makes the reconstructed images appear much sharper than they would without this step. In practice, the brightest possible reference star is used to insure that the signal-to-noise ratio of the reference data is much higher than the signal-to-noise ratio of the object data. It is also good practice to measure the reference star data soon before or after the object data is measured to maximize the likelihood that the atmospheric statistics for the object and the reference star data are the same. The image data is generally recorded and processed at a later time.

The first step in post processing is to compute the Fourier transform of the images. The Fourier

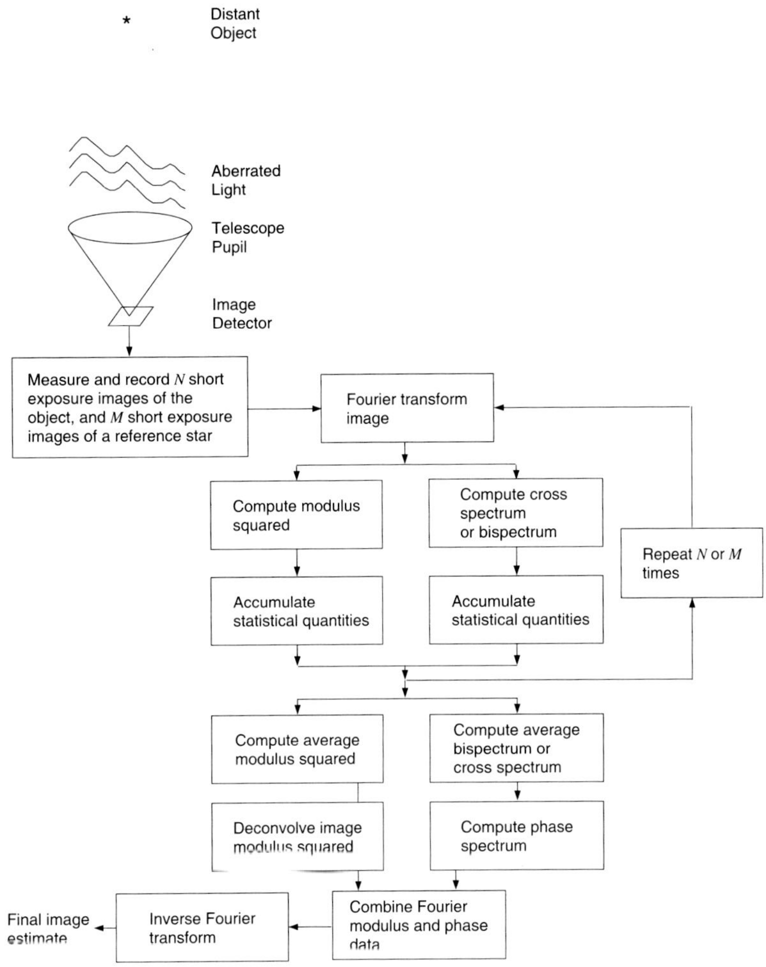

FIGURE 4.1
Block diagram for speckle imaging.

transform, $I(\vec{f})$, of an image, $i(\vec{x})$, is given by

$$I(\vec{f}) = \int \mathrm{d}\vec{x}\, i(\vec{x}) \exp\{-j2\pi \vec{f} \cdot \vec{x}\}, \tag{4.1}$$

where $\vec{f}$ is a coordinate in frequency space, and capital letters are used to represent Fourier transformed quantities. The Fourier transform is, in general, a complex quantity which has both modulus $|I(\vec{f})|$, and phase $\phi_i(\vec{f})$, given by

$$I(\vec{f}) = |I(\vec{f})| \exp\{j\phi_i(\vec{f})\}. \tag{4.2}$$

The phase of the Fourier transform of an image is referred to as the *phase spectrum* of the image. The phase spectrum of the object is defined similarly.

Two processing steps are performed on the Fourier transformed images. One step is to estimate the average modulus squared, or second moment of the Fourier transform of the object data and the reference star data. These quantities are denoted by $\mathrm{E}\{|I(\vec{f})|^2\}$ for the object, and $\mathrm{E}\{|R(\vec{f})|^2\}$ for the reference, where the $\mathrm{E}\{\cdot\}$ notation is used to represent the statistical expectation operator.

One difference between speckle imaging and conventional imaging through turbulence can be understood by recalling the relationship between the image spectrum $I(\vec{f})$, the object spectrum $O(\vec{f})$, and the optical transfer function (OTF) $\mathcal{H}(\vec{f})$, given in Eq. (2.41), and repeated here:

$$I(\vec{f}) = O(\vec{f})\mathcal{H}(\vec{f}). \tag{4.3}$$

As discussed in Section 3.5 for $D \gg r_0$, the mean OTF, $\text{E}\{\mathcal{H}(\vec{f})\}$, is very nearly zero for frequencies in the range $|\vec{f}| > r_0/\lambda d_i$, where r_0 is the Fried seeing parameter defined in Section 3.5, λ is the mean wavelength, d_i is the distance between the exit pupil and the image plane, and D is the telescope diameter. Thus, high spatial frequency information is not passed by the atmosphere-telescope system to the image, severely limiting the resolution of conventional imaging through turbulence.

However, the second moment of the image spectrum is of interest in speckle imaging. In Eq. (4.3) the spectrum of the object irradiance distribution, $O(\vec{f})$, is constant, while $I(\vec{f})$ is random due to turbulence-induced random fluctuations in the OTF, $\mathcal{H}(\vec{f})$. Computing the average modulus squared of $I(\vec{f})$ yields the following relationship between the second moments of the image spectrum and the OTF, and the modulus squared of the object spectrum:

$$\begin{aligned} \text{E}\left\{|I(\vec{f})|^2\right\} &= \text{E}\left\{|O(\vec{f})\mathcal{H}(\vec{f})|^2\right\}, \\ &= |O(\vec{f})|^2\text{E}\left\{|\mathcal{H}(\vec{f})|^2\right\}. \end{aligned} \tag{4.4}$$

As shown later, the second moment of the OTF is finite out to the diffraction limited cutoff frequency of the telescope, $D/\lambda d_i$ [10]. Hence, if $|O(\vec{f})|^2$ is greater than zero for frequencies approaching the cutoff frequency of the telescope, then the second moment of the image spectrum will contain diffraction-limited information about the object. Note that the phase spectrum of the object is lost in computing the second moment of the image. Hence, an image cannot be obtained directly from $\text{E}\{|I(\vec{f})|^2\}$.

We now consider the requirement for imagery of a reference star. Note that in general $|O(\vec{f})|^2$ is not a constant function of spatial frequency. Rather, $|O(\vec{f})|^2$ takes its maximum value at $|\vec{f}| = 0$, and generally diminishes with increasing $|\vec{f}|$. The spectral components of $\text{E}\{|I(\vec{f})|^2\}$ are further attenuated by multiplication with $\text{E}\{|\mathcal{H}(\vec{f})|^2\}$, which takes its maximum value of unity at $|\vec{f}| = 0$, and monotonically decreases with increasing $|\vec{f}|$. An image obtained by inverse Fourier transforming such strongly attenuated spectral data (assuming for a moment that phase spectrum information is available) would appear badly blurred, even if the spectral data was *perfectly known*. Now consider the case of an unresolved source, such as a distant star, which has an irradiance distribution, $r(\vec{x})$, that is proportional to a delta function. The Fourier transform of the reference star irradiance distribution is, in this case, a constant function of $\vec{f}$, say C_R. Hence, the average modulus squared of the reference star image spectrum, $\text{E}\{|R(\vec{f})|^2\}$, is given by

$$\text{E}\left\{|R(\vec{f})|^2\right\} = C_R^2\text{E}\left\{|\mathcal{H}(\vec{f})|^2\right\}. \tag{4.5}$$

$\text{E}\{|R(\vec{f})|^2\}$ can easily be normalized to take the value of unity at $|\vec{f}| = 0$ to obtain the second moment of the OTF, $\text{E}\{|\mathcal{H}(\vec{f})|^2\}$.

If the ratio of $\text{E}\{|I(\vec{f})|^2\}$ and the normalized version of $\text{E}\{|R(\vec{f})|^2\}$ is now formed we obtain

$$\frac{\text{E}\left\{|I(\vec{f})|^2\right\}}{C_R^{-2}\text{E}\left\{|R(\vec{f})|^2\right\}} = \frac{|O(\vec{f})|^2\text{E}\left\{|\mathcal{H}(\vec{f})|^2\right\}}{\text{E}\left\{|\mathcal{H}(\vec{f})|^2\right\}} = |O(\vec{f})|^2, \tag{4.6}$$

which is the modulus squared of the object spectrum, the desired quantity. Thus, the purpose of collecting and processing reference star data is to provide a means of calibrating the average modulus squared of the object image spectrum for the effects of the transfer function of the combined atmosphere-telescope system. The effect of this calibration is to boost the mid and high spatial frequency components of $\text{E}\{|I(\vec{f})|^2\}$ to take the value of $|O(\vec{f})|^2$. Equation (4.6) forms the essence of Labeyrie's speckle interferometry technique. Note that the ratio in Eq. (4.6) can only be taken in regions of frequency space where $E\{|\mathcal{H}(\vec{f})|^2\}$ is non-zero and has signal-to-noise ratio of greater than unity.

The expectations above are strictly statistical expectations over all possible realizations of the relevant random processes. Two sources of randomness are present in the measured images: (1) randomness arising from turbulence, which causes the imaging system point spread function to change randomly from instant-to-instant, and is governed by turbulence statistics; and (2) randomness due to the random arrival times and locations of photoevents in the image plane, which are governed by Poisson statistics. Both sources of randomness must be accounted for to calculate the required expectations and predict the performance of speckle interferometry.

In practice, the expectations required to implement speckle imaging must be estimated using the sample-based average. The sample based estimate of $E\{|I(\vec{f})|^2\}$ is defined as

$$E\left\{|I(\vec{f})|^2\right\} \approx \frac{1}{N}\sum_{n=1}^{N}|I_n(\vec{f})|^2 = \langle|I(\vec{f})|^2\rangle. \tag{4.7}$$

Sample-based estimates for the mean of other random processes are defined similarly, and are represented with the angle bracket notation, $\langle\cdot\rangle$. Note that the sample-based average is equivalent to a time average since the N frames are collected sequentially in time. Throughout this chapter it is assumed that all of the random processes are ergodic, allowing the statistical expectation and time average operations to be interchanged.

Care must be used in computing the second moment of the Fourier transform of processes governed by Poisson statistics. Images measured with devices sensitive to individual photo-detection events, which includes most conventional electro-optical image detectors, are governed by Poisson statistics. Under this condition $E\{|I(\vec{f})|^2\}$ is not directly proportional to $|O(\vec{f})|^2$ [11]. Rather, $E\{|I(\vec{f})|^2\}$ also contains an additive term, referred to as the photon noise bias, arising from the Poisson statistics governing the measured image. Higher order moments of the spectrum of processes governed by Poisson statistics, such as the bispectrum, are also biased by photon noise effects. Specific steps must be taken in processing the data to remove these biases.

If an image is desired, steps must be taken to compute the phase spectrum of the object. The first step in the phase spectrum recovery process is to compute either the average cross spectrum or the average bispectrum of the object data. As mentioned previously, the cross spectrum and the bispectrum are specialized moments of the image spectra which contain information about the phase of the object spectrum needed to create an image. The cross spectrum of an image, $C(\vec{f}, \Delta\vec{f})$, is defined as [8]

$$C(\vec{f}, \Delta\vec{f}) = I(\vec{f})I^*(\vec{f} + \Delta\vec{f}), \tag{4.8}$$

where the notation, $I^*(\vec{f})$, indicates the complex conjugate of $I(\vec{f})$. The vector, $\Delta\vec{f}$, is a constant shift vector defined in frequency space. The purpose of forming the specialized product shown in Eq. (4.8) is to encode the phase of the object spectrum as *phase differences* in the phase of $C(\vec{f}, \Delta\vec{f})$. Specifically, the phase of $C(\vec{f}, \Delta\vec{f})$ is the difference of the phase of the image spectrum at $\vec{f}$ and the phase of the image spectrum at $(\vec{f} + \Delta\vec{f})$. As shown later, the phase spectrum of the object can be recovered from the phase differences encoded in the cross spectrum. If $|\Delta\vec{f}|$ is small compared to $r_0/\lambda d_i$ then the phase of the average cross spectrum will contain diffraction limited information [8].

The bispectrum of an image is defined as [9]

$$B(\vec{f_1}, \vec{f_2}) = I(\vec{f_1})I(\vec{f_2})I^*(\vec{f_1} + \vec{f_2}). \tag{4.9}$$

Similar to the cross spectrum, the phase of the bispectrum encodes the phase of the object spectrum as phase differences. However, in the case of the bispectrum the phase differences are between *three* points in frequency space. If $\vec{f_1}$ is allowed to sweep a wide range of frequency space and $\vec{f_2}$ is a small offset vector such that $|\vec{f_2}| < r_0/\lambda d_i$, then the bispectrum contains diffraction limited information about the object of interest [12, 13].

No calibration of the cross spectrum and bispectrum analogous to Eq. (4.6) is required for a diffraction-limited telescope. Hence, the cross spectrum and bispectrum are generally computed only for the images of the object. We denote the mean cross spectrum by $\mathrm{E}\{C(\vec{f}, \Delta\vec{f})\}$ and the mean bispectrum by $\mathrm{E}\{B(\vec{f}_1, \vec{f}_2)\}$. Sample-based estimates of the cross spectrum and bispectrum must be used in practice.

Additional processing steps are required to create an estimate of the object, as shown in Fig. 4.1. The object phase spectrum, $\phi_o(\vec{f})$, must be obtained from $\mathrm{E}\{C(\vec{f}, \Delta\vec{f})\}$ or $\mathrm{E}\{B(\vec{f}_1, \vec{f}_2)\}$. The Fourier transform of the object, $O(\vec{f})$, is then formed by combining the Fourier modulus spectrum data obtained using Eq. (4.6), and the phase spectrum data obtained from the cross spectrum or the bispectrum using

$$O(\vec{f}) = |O(\vec{f})| \exp\{j\phi_o(\vec{f})\}. \tag{4.10}$$

An estimate of the object irradiance distribution, $o(\vec{x})$, is then obtained by applying the inverse Fourier transform operation to $O(\vec{f})$:

$$o(\vec{x}) = \int \mathrm{d}\vec{f}\, O(\vec{f}) \exp\{j2\pi \vec{f} \cdot \vec{x}\}. \tag{4.11}$$

This concludes the overview of the speckle imaging technique. Succeeding sections will explain and justify the steps of speckle imaging in more detail. In the next section the topic of speckle interferometry is discussed in detail.

4.3 Speckle interferometry

As discussed in Section 4.2, speckle interferometry is a technique for estimating the modulus of the Fourier transform of an object from a set of speckled images of the object. Speckle interferometry was treated qualitatively in the previous section. In this section we establish two key points required to understand speckle interferometry: (1) the atmosphere-telescope system allows diffraction-limited information to be present in the second moment of the image spectrum; and (2) the signal-to-noise ratio imposed limits to speckle interferometry permit practical application of the method to imaging problems. An example of speckle interferometry is also provided.

The remainder of this section is organized as follows. In the next subsection we analyze the second moment of the OTF $\mathrm{E}\{|\mathcal{H}(\vec{f})|^2\}$, referred to as the speckle transfer function. The speckle transfer function relates the modulus squared of the object spectrum to the second moment of the image spectrum, as shown in Eq. (4.4). We will show that the speckle transfer function is finite out to spatial frequencies approaching the diffraction-limited cutoff frequency, and hence allows diffraction limited information to be measured. In the subsection that follows, the signal-to-noise ratio performance of the speckle interferometry technique is derived, and representative signal-to-noise ratio results are provided. An example of speckle interferometry performance is provided in the final subsection.

4.3.1 Speckle transfer function

As indicated in Eq. (4.4), the average performance of speckle interferometry depends critically upon the speckle transfer function, $\mathrm{E}\{|\mathcal{H}(\vec{f})|^2\}$. If $\mathrm{E}\{|\mathcal{H}(\vec{f})|^2\}$ is non-zero at spatial frequencies approaching the diffraction-limited cutoff frequency, then the second moment of the image spectrum will contain nearly diffraction-limited information. In this subsection $\mathrm{E}\{|\mathcal{H}(\vec{f})|^2\}$ is evaluated and shown to have finite value at high spatial frequencies.

The analysis presented here is based on the geometrical optics model for wave propagation through the atmosphere. This model for wave front propagation was discussed in Section 3.4. The phase fluctuation in the aperture plane arising from propagation through the atmosphere, $\psi(\vec{x})$, is given by

$$\psi(\vec{x}) = k \int_0^L \mathrm{d}z\, n_1(\vec{x}, z), \tag{4.12}$$

where k is the wavenumber of the light, $k = 2\pi/\lambda$, L is the greatest distance of turbulence affecting the system, and $n_1(\vec{x}, z)$ is the turbulence-induced index of refraction fluctuation discussed in Section 3.3. The aperture function of the telescope is represented by $W(\vec{x})$, where $W(\vec{x})$ is zero outside the aperture, and is unity inside the aperture. As in Section 3.4, $\psi(\vec{x})$ is modeled as a zero mean Gaussian random process. The correlation properties of $\psi(\vec{x})$ are described by its structure function, $D_\psi(\Delta\vec{x})$, defined in Eq. (3.49) as

$$D_\psi(\Delta\vec{x}) = \mathrm{E}\left\{[\psi(\vec{x}) - \psi(\vec{x} - \Delta\vec{x})]^2\right\}. \tag{4.13}$$

For the Kolmogorov spectrum the structure function is given in Eq. (3.58) by

$$D_\psi(\Delta\vec{x}) = 6.88\left(\frac{|\Delta\vec{x}|}{r_0}\right)^{5/3}. \tag{4.14}$$

Recall from Eq. (2.42) that the OTF, $\mathcal{H}(\vec{f})$, is given by the normalized autocorrelation of the generalized pupil function. The mathematical expression for the OTF is

$$\begin{aligned}\mathcal{H}(\vec{f}) &= \frac{\int \mathrm{d}\vec{x}\, W(\vec{x})W(\vec{x} - \vec{f}\lambda d_i)\exp\{j[\psi(\vec{x}) - \psi(\vec{x} - \vec{f}\lambda d_i)]\}}{\int \mathrm{d}\vec{x}|W(\vec{x})\exp\{j\psi(\vec{x})\}|^2} \\ &= N_F^{-1}\int \mathrm{d}\vec{x}\, W(\vec{x})W(\vec{x} - \vec{f}\lambda d_i)\exp\{j[\psi(\vec{x}) - \psi(\vec{x} - \vec{f}\lambda d_i)]\},\end{aligned} \tag{4.15}$$

where d_i is the distance from the pupil plane to the image plane. N_F is a normalizing constant given by

$$N_F = \int \mathrm{d}\vec{x}\, |W(\vec{x})|^2, \tag{4.16}$$

which guarantees that $\mathcal{H}(\vec{f})$ takes the value of unity at $\vec{f} = 0$. Note that N_F is a deterministic quantity, and hence may be taken outside the expectation operators in the analysis which follows. With these models for turbulence effects we may now derive the second moment of the OTF.

The second moment of the OTF may be written as

$$\begin{aligned}\mathrm{E}\{|\mathcal{H}(\vec{f})|^2\} = N_F^{-2}\mathrm{E}\bigg\{\int \mathrm{d}\vec{x}\, W(\vec{x})W(\vec{x} - \vec{f}\lambda d_i)\exp\{j[\psi(\vec{x}) - \psi(\vec{x} - \vec{f}\lambda d_i)]\} \\ \times \int \mathrm{d}\vec{x}'\, W(\vec{x}')W(\vec{x}' - \vec{f}\lambda d_i)\exp\{-j[\psi(\vec{x}') - \psi(\vec{x}' - \vec{f}\lambda d_i)]\}\bigg\}.\end{aligned} \tag{4.17}$$

Since integration and expectation are linear operators their orders may be exchanged in Eq. (4.17). Further, the only random quantities in Eq. (4.17) are the turbulence corrupted phases represented by ψ. Applying these observations to Eq. (4.17) yields

$$\begin{aligned}\mathrm{E}\{|\mathcal{H}(\vec{f})|^2\} = N_F^{-2}\int \mathrm{d}\vec{x}\int \mathrm{d}\vec{x}'\, W(\vec{x})W(\vec{x} - \vec{f}\lambda d_i)W(\vec{x}')W(\vec{x}' - \vec{f}\lambda d_i) \\ \times\ \mathrm{E}\Big\{\exp\{j[\psi(\vec{x}) - \psi(\vec{x} - \vec{f}\lambda d_i) - \psi(\vec{x}') + \psi(\vec{x}' - \vec{f}\lambda d_i)]\}\Big\}.\end{aligned} \tag{4.18}$$

The problem of computing $\mathrm{E}\{|\mathcal{H}(\vec{f})|^2\}$ requires evaluation of the expectation inside the integral of Eq. (4.18). Recall that the turbulence-induced phase $\psi(\vec{x})$ is assumed to be Gaussian distributed. Hence, the expectation in Eq. (4.18) can be evaluated using Eq. (3.46):

$$\begin{aligned}\mathrm{E}\left\{\exp\{j[\psi(\vec{x}) - \psi(\vec{x}) - \vec{f}\lambda d_i) - \psi(\vec{x}' + \psi(\vec{x}' - \vec{f}\lambda d_i)]\}\right\} = \\ \exp\left\{-\frac{1}{2}\mathrm{E}\{[\psi(\vec{x}) - \psi(\vec{x} - \vec{f}\lambda d_i) - \psi(\vec{x}') + \psi(\vec{x}' - \vec{f}\lambda d_i)]^2\}\right\}.\end{aligned} \tag{4.19}$$

The expectation on the right hand side of Eq. (4.19) can be expressed in terms of the structure function of $\psi(\vec{x})$ by expanding the square and recombining terms. With the following substitutions:

$$\begin{aligned}\psi_1 &= \psi(\vec{x}),\\ \psi_2 &= \psi(\vec{x} - \vec{f}\lambda d_i),\\ \psi_3 &= \psi(\vec{x}'),\\ \psi_4 &= \psi(\vec{x}' - \vec{f}\lambda d_i),\end{aligned} \tag{4.20}$$

Equation (4.19) becomes

$$\begin{aligned}&\mathrm{E}\{(\psi_1 - \psi_2 - \psi_3 + \psi_4)^2\}\\ &= \mathrm{E}\{\psi_1^2 + \psi_2^2 + \psi_3^2 + \psi_4^2 - 2\psi_1\psi_2 - 2\psi_1\psi_3 + 2\psi_1\psi_4 + 2\psi_2\psi_3 - 2\psi_2\psi_4 - 2\psi_3\psi_4\},\\ &= \mathrm{E}\{(\psi_1 - \psi_2)^2 + (\psi_1 - \psi_3)^2 - (\psi_1 - \psi_4)^2\\ &\qquad -(\psi_2 - \psi_3)^2 + (\psi_2 - \psi_4)^2 + (\psi_3 - \psi_4)^2\},\\ &= 2D_\psi(|\vec{f}\lambda d_i|) + 2D_\psi(|\vec{x} - \vec{x}'|) - D_\psi(|\vec{x} - \vec{x}' + \vec{f}\lambda d_i|) - D_\psi(|\vec{x} - \vec{x}' - \vec{f}\lambda d_i|).\end{aligned} \tag{4.21}$$

Thus, the final expression for the speckle transfer function is [10, 11]

$$\begin{aligned}\mathrm{E}\{|\mathcal{H}(\vec{f})|^2\} = N_F^{-2} \int \mathrm{d}\vec{x} \int \mathrm{d}\vec{x}'\, & W(\vec{x})W(\vec{x} - \vec{f}\lambda d_i)W(\vec{x}')W(\vec{x}' - \vec{f}\lambda d_i)\\ \times \exp\Big\{ & -D_\psi(|\vec{f}\lambda d_i|) - D_\psi(|\vec{x} - \vec{x}'|)\\ & +\frac{1}{2}D_\psi(|\vec{x} - \vec{x}' + \vec{f}\lambda d_i|)\\ & +\frac{1}{2}D_\psi(|\vec{x} - \vec{x}' - \vec{f}\lambda d_i|)\Big\}.\end{aligned} \tag{4.22}$$

Interestingly, a more detailed calculation based on the solution of the wave equation for propagation through a randomly inhomogeneous medium yields exactly the same result for $\mathrm{E}\{|\mathcal{H}(\vec{f})|^2\}$ [10].

The expression for $\mathrm{E}\{|\mathcal{H}(\vec{f})|^2\}$ given in Eq. (4.22) can be evaluated numerically for the case of a circular aperture. The results of such a calculation are presented in Fig. 4.2 for $D/r_0 = 2, 5, 10$, and 15. The horizontal axis of Fig. 4.2 is expressed in terms of the normalized spatial frequency, $(f_x\lambda d_i/D)$, where f_x is one orthogonal component of the spatial frequency vector $\vec{f}$. Note that $\mathrm{E}\{|\mathcal{H}(\vec{f})|^2\}$ is greater than 10^{-3} out to normalized spatial frequencies of greater than 0.7 for $D/r_0 = 10$. This result can be compared to the normalized spatial frequency at which the mean OTF, $\mathrm{E}\{\mathcal{H}(\vec{f})\}$ goes to essentially zero. Inspection of Fig. 3.9 shows that for $D/r_0 = 10$, $\mathrm{E}\{\mathcal{H}(\vec{f})\}$, goes to zero at a normalized spatial frequency of less than 0.15. Thus, we conclude that speckle interferometry measurements will provide higher spatial frequency information than conventional long exposure imaging through turbulence.

The discussion in this subsection demonstrates that the speckle interferometry method can, on average, provide high spatial frequency information about objects imaged through turbulence. However, the analysis presented thus far does not address the *fidelity* of the data. A detailed analysis of the effects of randomness arising from both turbulence and measurement noise, leading to an expression for the signal-to-noise ratio for speckle interferometry, is required. In the next subsection this analysis is presented.

4.3.2 Signal-to-noise ratio considerations

In this subsection we undertake the analysis of the signal-to-noise ratio for speckle interferometry. Before beginning the mathematical analysis, several introductory comments are in order. Since speckle

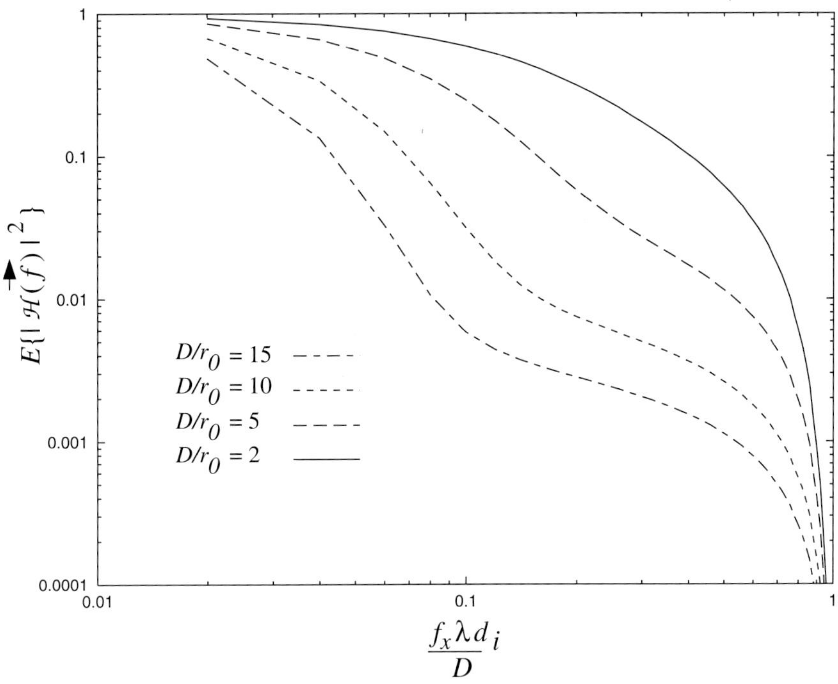

FIGURE 4.2
Second moment of OTF for various values of D/r_0. The horizontal axis is normalized to the diffraction limited cutoff frequency, $D/\lambda d_i$.

interferometry is a frequency domain technique, the analysis is conducted in the Fourier transform domain of the measured images. The resulting signal-to-noise ratio expression will be spatial frequency dependent. The basic methods for this analysis were covered in Section 2.3.2. We shall rely heavily on the techniques and results established in Section 2.3.2 to conduct this analysis.

Two sources of randomness are present in the measured images: (1) turbulence-induced randomness in the OTF, or, equivalently, in the point spread function, and (2) randomness due to the random locations and arrival times of photoevents in the image plane, referred to as photon noise. Photoevents are governed by Poisson statistics, as discussed in Section 2.3.2. However, the rate function, which governs the average number of photoevents per unit time at any location in the image, is also random due to the random fluctuations of the point spread function. Such processes are referred to as doubly stochastic Poisson random processes [11], which were also discussed in Section 2.3.2. The second moment of the measured image spectrum is not directly proportional to the modulus squared of the object spectrum, $|O(\vec{f})|^2$. Rather, a photon noise bias is present which must be removed in the data processing. With these considerations in mind, we may proceed with the analysis.

The analysis will follow that presented in Ref. [14]. It will be shown that an appropriate limiting case of this analysis yields the simpler result obtained in Ref. [15]. Attention is initially restricted to photon-limited image detection. This assumption is then relaxed to include the effects of additive noise in the image measurement. The photon-limited detection model for the measured image, $d(\vec{x})$, is given in Eq. (2.83) as

$$d(\vec{x}) = \sum_{n=1}^{K} \delta(\vec{x} - \vec{x}_n), \tag{4.23}$$

where $\vec{x}_n$ is the spatial location of the nth photoevent in the image plane, and K is the number of photoevents contributing to the image. The Fourier transform of the detected image is given in Eq. (2.85) as

$$D(\vec{f}) = \sum_{n=1}^{K} \exp\{-j2\pi\vec{f}\cdot\vec{x}_n\}. \tag{4.24}$$

The signal-to-noise ratio of the modulus squared of the detected image spectrum, $\text{SNR}_{|D|^2}(\vec{f})$, is defined as [11]

$$\text{SNR}_{|D|^2}(\vec{f}) = \frac{\text{E}\{|D(\vec{f})|^2\}}{\text{var}\{|D(\vec{f})|^2\}^{1/2}}, \tag{4.25}$$

where the notation var$\{\cdot\}$ refers to the variance of a random quantity. The square root of the variance is referred to as the standard deviation, and is a measure of how strongly a random quantity fluctuates about its mean. This definition of the signal-to-noise ratio has been widely used in the literature [13, 14, 15, 16, 17]. Equation (4.25) is intuitively appealing because it provides a comparison of the strength of the mean value of the quantity of interest to a measure of the strength of its fluctuations. In the analysis that follows, the numerator and denominator of Eq. (4.25) are evaluated separately, and results are combined at the end.

The second moment of the detected image spectrum is required for the numerator of Eq. (4.25). This quantity was evaluated in Section 2.3.2 for a doubly stochastic Poisson random process, and it is stated in Eq. (2.123) as

$$\text{E}\{|D(\vec{f})|^2\} = \overline{K} + (\overline{K})^2\text{E}\{|\mathcal{H}(\vec{f})|^2\}|O_n(\vec{f})|^2, \tag{4.26}$$

where $\overline{K}$ is the average number of photoevents per image, $\text{E}\{|\mathcal{H}(\vec{f})|^2\}$ is the second moment of the OTF, and $O_n(\vec{f})$ is the object spectrum normalized to have the value of unity at $\vec{f} = 0$. The expected value notation, E$\{\cdot\}$, and the overline notation, as in $\overline{K}$, are used interchangeably to represent the expected value of a random quantity. Note that the second moment of the detected image spectrum contains the desired term, which is proportional to $|O_n(\vec{f})|^2$, and an additive term given by $\overline{K}$. The $\overline{K}$ term is the photon noise bias in the estimate of $(\text{E}\{|\mathcal{H}(\vec{f})|^2\}|O_n(\vec{f})|^2)$ which arises from the fact that photon detection processes are governed by Poisson statistics. The $\overline{K}$ term must be removed to obtain an *unbiased* estimate of $(\text{E}\{|\mathcal{H}(\vec{f})|^2\}|O_n(\vec{f})|^2)$.

Two bias removal approaches have been used. The approach suggested by Goodman and Belsher requires forming a new estimator for $(\text{E}\{|\mathcal{H}(\vec{f})|^2\}|O_n(\vec{f})|^2)$, $Q_1(\vec{f})$, given by [11, 18]

$$Q_1(\vec{f}) = |D(\vec{f})|^2 - \overline{K}. \tag{4.27}$$

Clearly, the mean of $Q_1(\vec{f})$ is

$$\text{E}\{Q_1(\vec{f})\} = (\overline{K})^2\text{E}\{|\mathcal{H}(\vec{f})|^2\}|O_n(\vec{f})|^2, \tag{4.28}$$

which is the desired result. Later, Dainty and Greenaway used an alternative unbiased estimator, $Q(\vec{f})$, given by [15]

$$Q(\vec{f}) = |D(\vec{f})|^2 - K, \tag{4.29}$$

where K is the actual number of photoevents detected per image. The mean of $Q(\vec{f})$ is also given by Eq. (4.28). Hence, these two estimators differ from each other only by whether the *average* or the *actual* number of photoevents per image is subtracted from $|D(\vec{f})|^2$ in each frame.

Two comments are in order regarding Eqs. (4.27) and (4.29). First, the physical origin of the bias term is when photoevent locations coincide when Eq. (4.24) is used to compute $|D(\vec{f})|^2$. This same effect will later be shown to cause biases in the estimates of the cross spectrum and the bispectrum. Second, if the actual number of photoevents per image is known, then this bias is *deterministic* [19]. The actual number of photoevents per image can be obtained very accurately from the outputs of widely available CCD camera systems. Because the true value of K can be determined in practice, we could intuitively conclude that the estimator of Dainty and Greenaway, given in Eq. (4.29), would provide better signal-to-noise ratio performance.

Detailed comparative analysis of $Q_1(\vec{f})$ and $Q(\vec{f})$ has supported this intuitive conclusion, and it has been rigorously demonstrated that $Q(\vec{f})$ provides the higher signal-to-noise ratio [15]. Hence, the Dainty and Greenaway estimator, $Q(\vec{f})$, is used for further analysis. The quantity $Q(\vec{f})$ will be referred to as the unbiased speckle interferometry estimator. The relevant signal-to-noise ratio for speckle interferometry is thus the signal-to-noise ratio of $Q(\vec{f})$, given by

$$\mathrm{SNR}_Q(\vec{f}) = \frac{\mathrm{E}\{Q(\vec{f})\}}{\mathrm{var}\{Q(\vec{f})\}^{1/2}}. \tag{4.30}$$

The expected value of $Q(\vec{f})$ is easily obtained from Eq. (2.123) as

$$\mathrm{E}\{Q(\vec{f})\} = (\overline{K})^2 \mathrm{E}\{|\mathcal{H}(\vec{f})|^2\}|O_n(\vec{f})|^2. \tag{4.31}$$

The variance of $Q(\vec{f})$, required to compute $\mathrm{SNR}_Q(\vec{f})$, is more difficult to evaluate. The variance of $Q(\vec{f})$ is defined as

$$\mathrm{var}\{Q(\vec{f})\} = \mathrm{E}\{[Q(\vec{f})]^2\} - (\mathrm{E}\{Q(\vec{f})\})^2. \tag{4.32}$$

Hence, to evaluate $\mathrm{var}\{Q(\vec{f})\}$ the second moment of $Q(\vec{f})$ must be computed. The second moment of $Q(\vec{f})$ is given by

$$\begin{aligned}\mathrm{E}\{[Q(\vec{f})]^2\} &= \mathrm{E}\{(|D(\vec{f})|^2 - K)^2\} \\ &= \mathrm{E}\{|D(\vec{f})|^4 - 2K|D(\vec{f})|^2 + K^2\}.\end{aligned} \tag{4.33}$$

The terms of Eq. (4.33) are now evaluated individually.

The second and third terms of Eq. (4.33) can be evaluated directly. The third term is given by Eq. (2.81) as

$$\mathrm{E}\{K^2\} = \overline{K^2} = \overline{K} + (\overline{K})^2. \tag{4.34}$$

The second term in Eq. (2.81) can be obtained from direct application of the method in Eqs. (2.120) to (2.123), and the moment theorem for Poisson random processes [11]:

$$\mathrm{E}\{K(K-1)(K-2)\cdots(K-k+1)\} = (\overline{K})^k. \tag{4.35}$$

The result of this analysis is [14, 15]

$$\mathrm{E}\{K|D(\vec{f})|^2\} = \overline{K} + (\overline{K})^2 + (\overline{K})^2(\overline{K}+2)\mathrm{E}\{|\mathcal{H}(\vec{f})|^2\}|O_n(\vec{f})|^2. \tag{4.36}$$

Evaluation of the first term of Eq. (4.33) requires special attention. A model for the statistics of $\mathcal{H}(\vec{f})$ is required since the fourth moment of $D(\vec{f})$ will contain the fourth moment of $\mathcal{H}(\vec{f})$. The general model for the statistics of $\mathcal{H}(\vec{f})$ used in Ref. [14] is adopted, since the result obtained with this model is valid both for conventional imaging through turbulence and imaging through turbulence using adaptive optics compensation. It is shown that an appropriate limiting case of the general model yields a simpler result for speckle imaging obtained in earlier references [18]. The statistical model for the OTF of the telescope-atmosphere system is

$$\mathcal{H}(\vec{f}) = s(\vec{f}) + [n_r(\vec{f}) + jn_i(\vec{f})], \tag{4.37}$$

where $s(\vec{f})$ is the entirely real mean OTF, and $n_r(\vec{f})$ and $n_i(\vec{f})$ are the fluctuations in the real and imaginary parts of the OTF, respectively. The random variables $n_r(\vec{f})$ and $n_i(\vec{f})$ are assumed to be zero mean Gaussian random variables with variances given by $\sigma_r^2(\vec{f})$ and $\sigma_i^2(\vec{f})$, respectively. It is further assumed that $n_r(\vec{f})$ and $n_i(\vec{f})$ are uncorrelated with each other. The overall variance of the OTF, $\sigma^2_{\mathcal{H}}(\vec{f})$, is given by

$$\sigma^2_{\mathcal{H}}(\vec{f}) = \mathrm{E}\{|\mathcal{H}(\vec{f})|^2\} - |\mathrm{E}\{\mathcal{H}(\vec{f})\}|^2 = \sigma_r^2(\vec{f}) + \sigma_i^2(\vec{f}). \tag{4.38}$$

The first term in Eq. (4.33), $\mathrm{E}\{|D(\vec{f})|^4\}$, can be expanded using Eq. (16) of Ref. [15] and Eq. (C-5) of Ref. [11] as [14]:

$$\begin{aligned}\mathrm{E}\{|D(\vec{f})|^4\} = {} & \overline{K} + 2(\overline{K})^2 \\ & +4(1+\overline{K})(\overline{K})^2\mathrm{E}\{|\mathcal{H}(\vec{f})|^2\}|O_n(\vec{f})|^2 \\ & +(\overline{K})^2\mathrm{E}\{|\mathcal{H}(2\vec{f})|^2\}|O_n(2\vec{f})|^2 \\ & +(\overline{K})^4|O_n(\vec{f})|^4\mathrm{E}\{|\mathcal{H}(\vec{f})|^4\},\end{aligned} \tag{4.39}$$

where in the final term, $\mathrm{E}\{|\mathcal{H}(\vec{f})|^4\}$ is the fourth moment of the OTF. It remains to exhibit the fourth moment of the OTF. This is accomplished by writing:

$$\mathrm{E}\{|\mathcal{H}(\vec{f})|^4\} = \mathrm{E}\left\{[(s(\vec{f}) + n_r(\vec{f}))^2 + (n_i(\vec{f}))^2]^2\right\}. \tag{4.40}$$

Expanding the right side of the above equation yields an expression containing moments of $n_r(\vec{f})$ and $n_i(\vec{f})$:

$$\mathrm{E}\{|\mathcal{H}(\vec{f})|^4\} = \mathrm{E}\{n_i^4 + 2n_i^2n_r^2 + n_r^4 + 4n_i^2n_rs + 4n_r^3s + 2n_i^2s^2 + 6n_r^2s^2 + 4n_rs^3 + s^4\}, \tag{4.41}$$

where the spatial frequency dependence has been suppressed. Odd order moments of zero-mean Gaussian random variables are zero [20], so that accomplishing the expectation eliminates the fourth, fifth, and eighth terms in the above equation. The fourth order moments are given by the Gaussian moment theorem [11]:

$$\mathrm{E}\{n_r^4\} = 3\sigma_r^4, \tag{4.42}$$

$$\mathrm{E}\{n_i^4\} = 3\sigma_i^4. \tag{4.43}$$

Incorporating these results, and combining terms, yields:

$$\mathrm{E}\{|\mathcal{H}(\vec{f})|^4\} = 3\sigma_r^4 + 6\sigma_r^2s^2 + s^4 + 2(\sigma_r^2 + s^2)\sigma_i^2 + 3\sigma_i^4. \tag{4.44}$$

Thus, the fourth moment of the OTF can be expressed entirely in terms of the first and second moments of the real and imaginary parts of the OTF.

The results of Eqs. (4.32) through (4.38) and Eq. (4.44) can be combined to obtain the following expression for $\mathrm{var}\{Q(\vec{f})\}$:

$$\begin{aligned}\mathrm{var}\{Q(\vec{f})\} = {} & (\overline{K})^2 + (\overline{K})^2|O_n(2\vec{f})|^2\mathrm{E}\{|\mathcal{H}(2\vec{f})|^2\} \\ & +2(\overline{K})^3|O_n(\vec{f})|^2\mathrm{E}\{|\mathcal{H}(\vec{f})|^2\} \\ & +(\overline{K})^4|O_n(\vec{f})|^4\left[\mathrm{E}\{|\mathcal{H}(\vec{f})|^4\} - (\mathrm{E}\{|\mathcal{H}(\vec{f})|^2\})^2\right].\end{aligned} \tag{4.45}$$

The most general form of $\mathrm{SNR}_Q(\vec{f})$ is thus given by

$$\begin{aligned}\mathrm{SNR}_Q(\vec{f}) = {} & (\overline{K})^2\mathrm{E}\{|\mathcal{H}(\vec{f})|^2\}|O_n(\vec{f})|^2 \\ & \times\Big((\overline{K})^2 + (\overline{K})^2|O_n(2\vec{f})|^2\mathrm{E}\{|\mathcal{H}(2\vec{f})|^2\} + 2(\overline{K})^3|O_n(\vec{f})|^2\mathrm{E}\{|\mathcal{H}(\vec{f})|^2\} \\ & + (\overline{K})^4|O_n(\vec{f})|^4\left[\mathrm{E}\{|\mathcal{H}(\vec{f})|^4\} - (\mathrm{E}\{|\mathcal{H}(\vec{f})|^2\})^2\right]\Big)^{-1/2}.\end{aligned} \tag{4.46}$$

Monte Carlo simulations of imaging through turbulence offer the only practical means of evaluating $\sigma_r^2(\vec{f})$ and $\sigma_i^2(\vec{f})$, since these statistics cannot be expressed in a way which permits analytic evaluation. Such simulations have been used to evaluate these statistics for both uncompensated (that is, no adaptive optics) imaging, and compensated imaging using adaptive optics [14].

In the case of uncompensated imaging through turbulence a simpler expression can be obtained for $\mathrm{E}\{|\mathcal{H}(\vec{f})|^4\}$ than is given in Eq. (4.44). For spatial frequencies, $|\vec{f}| > r_0/\lambda d_i$, the mean OTF is essentially zero, or mathematically

$$s(\vec{f}) \approx 0, \qquad \text{for } |\vec{f}| > \frac{r_0}{\lambda d_i}. \tag{4.47}$$

Further, the variances of the real and imaginary parts of the OTF are known to be equal for $|\vec{f}| > r_0/\lambda d_i$, thus

$$\sigma_r^2(\vec{f}) = \sigma_i^2(\vec{f}) = \frac{1}{2}\sigma_{\mathcal{H}}^2. \tag{4.48}$$

Applying Eqs. (4.47) and (4.48) to Eq. (4.44) yields

$$\mathrm{E}\{|\mathcal{H}(\vec{f})|^4\} = 2\sigma_{\mathcal{H}}^4 = 2[\mathrm{E}\{|\mathcal{H}(\vec{f})|^2\}]^2, \quad |\vec{f}| > \frac{r_0}{\lambda d_i}. \tag{4.49}$$

It is shown in Refs. [10, 15] that a simple expression for $\mathrm{E}\{|\mathcal{H}(\vec{f})|^2\}$ can be obtained in terms of the telescope diameter, D, and the Fried seeing parameter, r_0, for frequencies in the range $|\vec{f}| > r_0/\lambda d_i$. This expression is given by

$$\mathrm{E}\{|\mathcal{H}(\vec{f})|^2\} = 0.435\mathcal{H}_0(\vec{f})\left(\frac{r_0}{D}\right)^2, \tag{4.50}$$

where $\mathcal{H}_0(\vec{f})$ is the diffraction limited OTF of the telescope.

Several simplifications can now be applied to Eq. (4.45). If attention is limited to frequencies greater than $0.5D/\lambda d_i$ the second term of Eq. (4.45) is eliminated, since the speckle transfer function is zero at frequencies greater than the diffraction limited cutoff. This yields the result

$$\begin{aligned}\mathrm{var}\{Q(\vec{f})\} &= (\overline{K})^2 + 2(\overline{K})^3|O_n(\vec{f})|^2\mathrm{E}\{|\mathcal{H}(\vec{f})|^2\} \\ &\quad + (\overline{K})^4|O_n(\vec{f})|^4[\mathrm{E}\{|\mathcal{H}(\vec{f})|^2\}]^2.\end{aligned} \tag{4.51}$$

The square root of Eq. (4.51) can now be easily taken and included in the expression for $\mathrm{SNR}_Q(\vec{f})$ to obtain [11, 15]

$$\mathrm{SNR}_Q(\vec{f}) = \frac{\overline{K}|O_n(\vec{f})|^2\mathrm{E}\{|\mathcal{H}(\vec{f})|^2\}}{1 + \overline{K}|O_n(\vec{f})|^2\mathrm{E}\{|\mathcal{H}(\vec{f})|^2\}}, \tag{4.52}$$

where $\mathrm{E}\{|\mathcal{H}(\vec{f})|^2\}$ is obtained from Eq. (4.50). Eq. (4.52) is the key result of this section.

An examination of Eq. (4.52) is in order. Note that $\mathrm{SNR}_Q(\vec{f})$ is bounded from above by unity. Hence, even for infinitely bright objects, characterized by $\overline{K} \to \infty$, the maximum possible $\mathrm{SNR}_Q(\vec{f})$ is unity. For objects with finite $\overline{K}$, and for extended objects, which have $|O_n(\vec{f})| << 1$ at mid and high spatial frequencies, $\mathrm{SNR}_Q(\vec{f})$ is substantially less than one. This is a considerable practical impediment to performing speckle interferometry. To boost $\mathrm{SNR}_Q(\vec{f})$ many independent realizations of $Q(\vec{f})$ must be averaged. Averaging boosts $\mathrm{SNR}_Q(\vec{f})$ according to

$$\mathrm{SNR}_Q^N(\vec{f}) = \sqrt{N}\mathrm{SNR}_Q(\vec{f}), \tag{4.53}$$

where N is the number of independent realizations averaged, and the notation $\mathrm{SNR}_Q^N(\vec{f})$ is used to represent the signal-to-noise ratio of the averaged $Q(\vec{f})$. Practical values of N range from a few hundred to several thousand.

The spatial frequency domain signal-to-noise ratio behavior of speckle interferometry is important because it shows the highest spatial frequency component of $Q(\vec{f})$ which is reliably estimated. The signal-to-noise ratio limited cutoff frequency, f_{SNR}, is defined as the spatial frequency at which $\text{SNR}_Q^N(\vec{f})$ goes below some threshold, often chosen to be unity [17]. The signal-to-noise ratio limited cutoff frequency is related to the noise effective minimum resolvable spatial separation, Δx_{SNR} by

$$\Delta x_{\text{SNR}} \approx \frac{1}{f_{\text{SNR}}}. \tag{4.54}$$

Thus, higher f_{SNR} implies better effective spatial resolution, showing the importance of steps taken to improve the signal-to-noise ratio of speckle interferometry.

Effects of additive noise

Thus far in the analysis of speckle interferometry only photon-limited detection has been considered. The photon-limited detection model is quite useful as it provides an upper bound on the performance of speckle interferometry. However, the photon-limited detection model is unrealistic for most widely available light detection devices. The most widely used light detectors for speckle interferometry are based on charge-coupled device (CCD) readouts. CCD-based cameras provide a signal which is proportional to the number of photoevents collected during the measurement time, summed with an additive random noise. The physical origin of this noise lies in the detector readout and amplification electronics, and is generally referred to as read noise. Read noise affects speckle interferometry in two ways: (1) there is an additional bias present in the modulus squared of the measured image spectrum which must be removed; and (2) the variance of the unbiased speckle interferometry estimator, $Q(\vec{f})$, is increased with the result that the signal-to-noise ratio of $Q(\vec{f})$ is reduced. We analyze both of these effects in this subsection.

The detected image model for the case of additive noisewas given in Eq. (2.109), and is repeated here for completeness. The model for a detected image corrupted with additive noise is given by

$$d(\vec{x}) = \sum_{n=1}^{K} \delta(\vec{x} - \vec{x}_n) + \sum_{p=1}^{P} n_p \delta(\vec{x} - \vec{x}_p), \tag{4.55}$$

where $\vec{x}_p$ is the location of the pth pixel in the image, n_p is a random variable representing the amount of additive noise at the pth pixel in units of photoevents, and P is the total number of pixels in the image. The first term in Eq. (4.55) describes the photon-limited image, and the second term describes the additive noise. The random variable, n_p, is assumed to have the following properties:

1. $\text{E}\{n_p\} = 0$: n_p is assumed to be zero mean,
2. $\text{E}\{n_p n_{p'}\} = 0$ if $p \neq p'$: additive noise is assumed to be spatially uncorrelated,
3. $\text{E}\{n_p n_{p'}\} = \sigma_n^2$ if $p = p'$: a single variance describes the additive noise at each pixel, with units of photoelectrons squared per pixel, and
4. n_p is statistically independent of K and $\vec{x}_n$.

Read noise in CCD-based cameras is usually given by specifying σ_n in units of electrons per pixel per readout.

Additive noise affects the second moment of the detected image spectrum, $\mathrm{E}\{|D(\vec{f})|^2\}$, in the following way:

$$\begin{aligned}\mathrm{E}\{|D(\vec{f})|^2\} &= \mathrm{E}\left\{\left|\sum_{n=1}^{K}\exp\{-j2\pi\vec{f}\cdot\vec{x}_n\} + \sum_{p=1}^{P} n_p\exp\{-j2\pi\vec{f}\cdot\vec{x}_p\}\right|^2\right\} \\ &= \mathrm{E}\left\{\left|\sum_{n=1}^{K}\exp\{-j2\pi\vec{f}\cdot\vec{x}_n\}\right|^2 + \left|\sum_{p=1}^{P} n_p\exp\{-j2\pi\vec{f}\cdot\vec{x}_p\}\right|^2\right\}, \qquad (4.56)\end{aligned}$$

where the independence of n_p, K, and x_n has been used to write the second line of the above equation. The first term in Eq. (4.56) has already been simplified using Eq. (2.123), which was repeated in Eq. (4.26). The second term in Eq. (4.56) was evaluated in Eq. (2.112) as

$$\mathrm{E}\left\{\left|\sum_{p=1}^{P} n_p\exp\{-j2\pi\vec{f}\cdot\vec{x}_p\}\right|^2\right\} = P\sigma_n^2. \qquad (4.57)$$

Hence, the second moment of the detected image spectrum when read noise is present is given by

$$\mathrm{E}\{|D(\vec{f})|^2\} = \overline{K} + (\overline{K})^2\mathrm{E}\{|\mathcal{H}(\vec{f})|^2\}|O_n(\vec{f})|^2 + P\sigma_n^2. \qquad (4.58)$$

The term $P\sigma_n^2$ thus provides an additional bias in the estimate of $(\mathrm{E}\{|\mathcal{H}(\vec{f})|^2\}|O_n(\vec{f})|^2)$ which must be removed. Hence, the proper unbiased estimator for $(\mathrm{E}\{|\mathcal{H}(\vec{f})|^2\}|O_n(\vec{f})|^2)$ when read noise is present, also referred to as $Q(\vec{f})$, is given by [14, 21]

$$Q(\vec{f}) = |D(\vec{f})|^2 - K - P\sigma_n^2. \qquad (4.59)$$

When the additive noise bias is properly removed there is no effect due to additive read noise on the mean value of $Q(\vec{f})$.

Additive noise also increases the variance of $Q(\vec{f})$. The statistical independence of the additive noise from K and $\vec{x}_n$ means that additive noise causes a purely additive increase in the variance of $Q(\vec{f})$. The increase in the variance on $Q(\vec{f})$ is obtained by calculating the variance of the additive noise contribution to the detected image spectrum. The specific form of the variance which must be calculated is given by

$$\begin{aligned}\mathrm{var}\left\{\left|\sum_{p=1}^{P} n_p\exp\{-j2\pi\vec{f}\cdot\vec{x}_p\}\right|^2\right\} &= \mathrm{E}\left\{\left|\sum_{p=1}^{P} n_p\exp\{-j2\pi\vec{f}\cdot\vec{x}_p\}\right|^4\right\} \\ &- \left(\mathrm{E}\left\{\left|\sum_{p=1}^{P} n_p\exp\{-j2\pi\vec{f}\cdot\vec{x}_p\}\right|^2\right\}\right)^2. \qquad (4.60)\end{aligned}$$

This variance term has been evaluated, and is given by [22]

$$\mathrm{var}\left\{\left|\sum_{p=1}^{P} n_p\exp\{-j2\pi\vec{f}\cdot\vec{x}_p\}\right|^2\right\} = \left(P\sigma_n^2\right)^2. \qquad (4.61)$$

When additive read noise is present the variance of $Q(\vec{f})$ is obtained by adding $(P\sigma_n^2)^2$ to the expressions for the variance expression given in Eqs. (4.45) and (4.52). Hence, the net effect of the additive noise is to reduce the signal-to-noise ratio of $Q(\vec{f})$.

Later in this chapter examples of the behavior of $\mathrm{SNR}_Q(\vec{f})$ are presented. A simulation was used to obtain the results presented. Before presenting the $\mathrm{SNR}_Q(\vec{f})$ results the problem of simulating speckle

FIGURE 4.3
Block diagram for simulating speckle interferometry.

interferometry is examined.

Simulating speckle interferometry

As discussed in Section 3.7, simulations are an important aspect of understanding the performance limits of techniques for imaging through turbulence. In this subsection we discuss a technique for simulating speckle interferometry. A block diagram of a simulation of speckle interferometry is shown in Fig. 4.3.

The first block of the simulation is devoted to one time computations required to setup the simulation. In this block the simulation parameters are read, such as r_0, $\overline{K}$, and the pupil diameter, D. The pupil model is established, the sample spacing in the pupil, Δx_{pup}, is determined, and the object is either created or read in from a disk file. Precomputations required for phase screen generation are performed, such as computing and factoring the covariance matrices as described in Section 3.7. Finally, the object is Fourier transformed so that the object spectrum can be available for forming the image spectrum later in the simulation.

Next, a series of steps are repeated for N independent phase screens. First, a random phase screen, $\phi(\vec{x})$, is generated. The pupil model is imposed on the phase screen, leaving the phase screen samples inside the pupil, and setting the phase screen samples outside the pupil to zero. This step defines the generalized pupil function, $W_p(\vec{x})e^{j\phi(\vec{x})}$, given in Eq. (2.45).

The OTF corresponding to $W_p(\vec{x})e^{j\phi(\vec{x})}$, $\mathcal{H}(\vec{f})$, is then computed. The relationship between the generalized pupil function and the OTF is given in Eq. (2.42) as the normalized autocorrelation of the generalized pupil function. In practice, Fourier methods are often used to compute this autocorrelation. The autocorrelation theorem of Fourier transforms states [23]

$$\mathcal{F}\{W_p(\vec{x})e^{j\phi(\vec{x})} \star W_p(\vec{x})e^{j\phi(\vec{x})}\} = |\mathcal{W}_p(\vec{f})|^2, \tag{4.62}$$

where $\mathcal{F}\{\cdot\}$ represents the Fourier transform operator, the $\star$ notation indicates the correlation operator, and $\mathcal{W}_p(\vec{f})$ is the Fourier transform of $W_p(\vec{x})e^{j\phi(\vec{x})}$. The OTF is then obtained by inverse Fourier transforming $|\mathcal{W}_p(\vec{f})|^2$.

The sample spacing in the frequency domain can be determined from the number of samples in the pupil, the pupil diameter, and the imaging wavelength. It is easiest to discuss sampling issues in terms of angles and angular frequencies, as this allows telescope specific parameters, such as focal length, to be removed from the analysis. Let the number of samples across the pupil diameter be odd, and represented by N_{pup}. This allows for a sample at the exact center of the pupil, and an even number of samples on both sides of the center. The sample spacing in the pupil, Δx_{pup}, is given by

$$\Delta x_{pup} = \frac{D}{(N_{pup} - 1)}. \tag{4.63}$$

The autocorrelation of $W_p(\vec{x})e^{j\phi(\vec{x})}$ has twice as many samples across the diameter as the function $W_p(\vec{x})e^{j\phi(\vec{x})}$. Hence, the number of samples across the diameter of the OTF, N_{OTF}, is given by

$$N_{OTF} = 2N_{pup} - 1. \tag{4.64}$$

Care must be exercised in choosing array sizes to prevent wrap-around error in the calculation of the OTF [24]. The angular cutoff frequency of the telescope is D/λ, so that the angular frequency sample spacing in the OTF array, Δf, is given by

$$\Delta f = \frac{D/\lambda}{(N_{OTF} - 1)/2}. \tag{4.65}$$

The angular sample spacing in object space can now be determined from the angular frequency sample spacing. Typically the fast Fourier transform (FFT) is used to perform the required Fourier transforms. Widely available FFT software [25] requires that the input and output arrays have dimensions which are given by integer powers of two. Let the number of samples across one side of the array holding the OTF be denoted by N_S, with

$$N_S = 2^n \text{ and } N_S \geq N_{OTF}, \tag{4.66}$$

where n is a positive integer. The relationship between the sample spacing in the angular frequency domain, Δf, and the sample spacing in object angle space, $\Delta\theta_O$, is governed by the general relationship for the FFT

$$\Delta\theta_O = \frac{1}{N_S \Delta f}, \tag{4.67}$$

where the notation θ_O is used to represent the angular sample spacing in object space.

To compute the noise-free image spectrum the product of the normalized object spectrum and the OTF is computed. The associated noise-free image, $i_{nf}(\vec{x})$, is given by the inverse Fourier transform

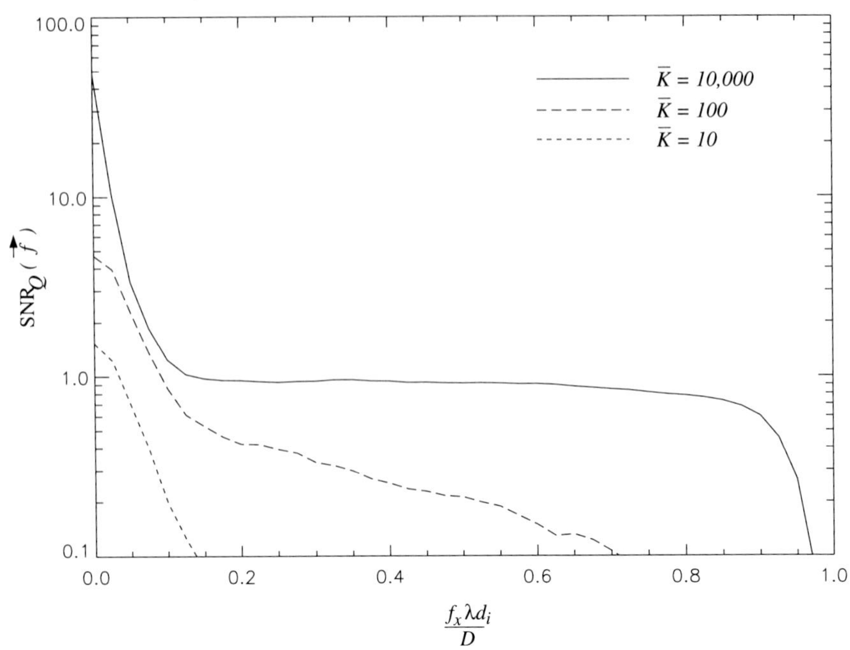

FIGURE 4.4
$SNR_Q(\vec{f})$ **for** $\overline{K} =$ **10,000, 100, and 10, for the fixed value of** $D/r_0 = 10$. **The object is an unresolved star, and the horizontal axis is normalized to the diffraction limited cutoff frequency.**

of the noise-free image spectrum. Photon noise effects can be included in the image by the following steps. A normalized version of the noise-free image, $i_{nf}^{norm}(\vec{x})$, is created by

$$i_{nf}^{norm}(\vec{x}) = \frac{i_{nf}(\vec{x})}{\sum i_{nf}(\vec{x})}, \tag{4.68}$$

where the summation is taken over all of the samples in the image. The average number of photoevents at each sample point in $i_{nf}^{norm}(\vec{x})$, $\overline{K}(\vec{x})$, is then computed by

$$\overline{K}(\vec{x}) = \overline{K} i_{nf}^{norm}(\vec{x}). \tag{4.69}$$

Finally, the number of photoevents at each sample of the image is computed by setting the value $\overline{K}(\vec{x})$ as the mean value which is input to a Poisson random number generator [25]. The result of this process is a simulated image, $i(\vec{x})$. The actual number of photoevents in this image, K, can be computed by

$$K = \sum i(\vec{x}), \tag{4.70}$$

where the summation is taken over all the samples in the image.

The next step in the speckle interferometry simulation is to compute the Fourier transform of $i(\vec{x})$, denoted $I(\vec{f})$. The modulus squared of the image spectrum, $|I(\vec{f})|^2$, is then computed, and a single realization of $Q(\vec{f})$ is obtained by

$$Q(\vec{f}) = |I(\vec{f})|^2 - K. \tag{4.71}$$

The instantaneous $Q(\vec{f})$ is accumulated so that the sample-based estimate of the mean of $Q(\vec{f})$, $\langle Q(\vec{f})\rangle$, can be computed. The modulus squared of $Q(\vec{f})$, $Q^2(\vec{f})$, is also generally computed to provide the

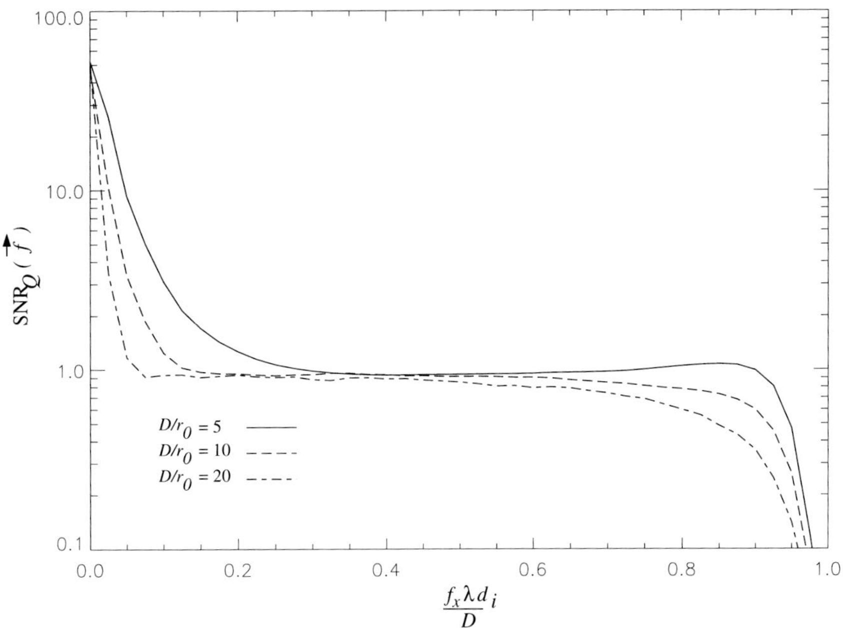

FIGURE 4.5
$SNR_Q(\vec{f})$ **for** $D/r_0 = $ **5, 10, and 20, with the fixed photon flux level,** $\overline{K} = 1000$. **The object is an unresolved star, and the horizontal axis is normalized to the diffraction limited cutoff frequency.**

sample-based estimate of var$\{Q(\vec{f})\}$ given by

$$\text{var}\{Q(\vec{f})\} = \frac{1}{N}\sum_{n=1}^{N} |Q_n(\vec{f})|^2 - \left[\frac{1}{N}\sum_{n=1}^{N} Q_n(\vec{f})\right]^2, \tag{4.72}$$

which can be used to compute the sample-based estimate of the signal-to-noise ratio of the data using Eqs. (4.30) and (4.53).

Examples of the behavior of $\text{SNR}_Q(\vec{f})$ obtained using the simulation are shown in Figs. 4.4 and 4.5. In both figures the object of interest is an unresolved star. In Fig. 4.4 $\text{SNR}_Q(\vec{f})$ is plotted for $\overline{K}$ = 10,000, 100, and 10, for the fixed value $D/r_0 = 10$. In Fig. 4.5 $\text{SNR}_Q(\vec{f})$ is plotted for D/r_0 = 5, 10, and 20 for the fixed value $\overline{K} = 1000$. Examining the $\overline{K} = 100$ curve in Fig. 4.4 shows that $\text{SNR}_Q(\vec{f})$ at a normalized spatial frequency of 0.5 is approximately 0.2. Thus, to obtain $\text{SNR}_Q(\vec{f})$ of unity at this frequency would require averaging approximately 25 independent frames. To obtain $\text{SNR}_Q(\vec{f}) = 10$ would require averaging 2500 independent frames. The lowest limit of detection is generally considered to be a signal-to-noise ratio of approximately unity. However, in practice, a signal-to-noise ratio of 10 or more is often desired. Thus, it can be seen that under conditions of low light level and poor seeing, several hundred to a several thousand frames of data may be required to obtain an acceptable signal-to-noise ratio at a given spatial frequency.

4.3.3 Example of speckle interferometry

As mentioned in the introduction to this chapter, speckle interferometry has been widely used to measure the separation and relative brightnesses of binary star systems. Binary stars are two stars whose separation is sufficiently small that there exists a strong mutual gravitational attraction, causing the stars to orbit their common center of mass. Binary stars are interesting, in part, because there is such a large number of them. It has been estimated that half of the stars in this galaxy are members of binary or higher order multiple star systems [26]. Thus, there is considerable scientific interest in detecting binary stars, and determining their parameters. The particular parameters of interest are the angular separation over time, and the relative brightnesses of the individual stars. As shown in this subsection, speckle interferometry can be used to measure these two parameters of binary star systems.

Speckle interferometry has been applied to the problem of detecting and measuring the parameters of binary stars because many binaries have angular separations, α_s, in the range $\lambda/D \leq \alpha_s \leq \lambda/r_0$. Recall from Section 3.5 that λ/r_0 is approximately the smallest angular separation of two points which can be detected by imaging through turbulence, while λ/D is approximately the smallest angular separation of two points which can be sensed by a diffraction limited telescope of diameter, D, in the absence of turbulence. Hence, many binary star systems cannot be detected and parameterized by conventional imaging through the atmosphere.

As established in Section 4.3.1, the speckle transfer function, $\mathrm{E}\{|\mathcal{H}(\vec{f})|^2\}$, passes a wider band of spatial frequencies than the average OTF. This observation provides the motivation for applying speckle interferometry to measuring binary stars. It must be noted that speckle interferometry does not provide an image. Rather, the function $Q(\vec{f})$ contains information which can be used to estimate the parameters of the binary star. We now show how to accomplish this task.

Consider the following object model, $o(\vec{x})$, for a binary star:

$$o(\vec{x}) = d_1\delta(x - x_1) + d_2\delta(x - x_2), \tag{4.73}$$

where x is one of the orthogonal components of the position vector, $\vec{x}$, x_1 and x_2 are the locations of the binary components along the x axis, and d_1 and d_2 are the intensities of the two binary star components. Modeling the spatial shape of the binary star components as δ functions is equivalent to modeling the components as point sources. This is suitable since the angular width of all single stars is much less than λ/D. Hence, the individual stars are unresolvable and the δ function model is appropriate.

The modulus squared of the normalized object spectrum $|O_n(\vec{f})|^2$ is in this case given by

$$\begin{aligned}
|O_n(\vec{f})|^2 &= \frac{|\int d\vec{x}[d_1\delta(x - x_1) + d_2\delta(x - x_2)]\exp\{-j2\pi\vec{f}\cdot\vec{x}\}|^2}{(d_1 + d_2)^2} \\
&= \frac{|d_1\exp\{-j2\pi f_x x_1\} + d_2\exp\{-j2\pi f_x x_2\}|^2}{(d_1 + d_2)^2} \\
&= \frac{d_1^2 + d_2^2 + 2d_1d_2\cos[2\pi f_x(x_1 - x_2)]}{(d_1 + d_2)^2},
\end{aligned} \tag{4.74}$$

where f_x is the orthogonal component of $\vec{f}$ corresponding to the Fourier transform along the x-axis. The quantity $Q(\vec{f})$ will be given by

$$\begin{aligned}
Q(\vec{f}) &= (\overline{K})^2|O_n(\vec{f})|^2\mathrm{E}\{|\mathcal{H}(\vec{f})|^2\} \\
&= \frac{(d_1^2 + d_2^2 + 2d_1d_2\cos[2\pi f_x(x_1 - x_2)])}{(d_1 + d_2)^2}(\overline{K})^2\mathrm{E}\{|\mathcal{H}(\vec{f})|^2\}.
\end{aligned} \tag{4.75}$$

Inspection of Eq. (4.74) shows that $|O_n(\vec{f})|^2$ has the appearance of an interference pattern. That is, if we could observe $|O_n(\vec{f})|^2$ directly it would appear as a sinusoidally modulated irradiance pattern. However, $|O_n(\vec{f})|^2$ cannot be observed directly - it can only be measured indirectly using $Q(\vec{f})$. As shown in Eq. (4.75), $Q(\vec{f})$ has sinusoidal oscillations, but $\mathrm{E}\{|\mathcal{H}(\vec{f})|^2\}$ attenuates these oscillations. To overcome this attenuation, the reference star data is used to calibrate the estimate of $Q(\vec{f})$ for effects due to the speckle transfer function, $\mathrm{E}\{|\mathcal{H}(\vec{f})|^2\}$, using

$$|\tilde{O}_n(\vec{f})|^2 = \frac{Q(\vec{f})}{\mathrm{E}\{|\mathcal{H}(\vec{f})|^2\}}, \tag{4.76}$$

where $|\tilde{O}_n(\vec{f})|^2$ is the estimated modulus squared of the object spectrum, and $\mathrm{E}\{|\mathcal{H}(\vec{f})|^2\}$ is obtained from the reference star data using

$$\mathrm{E}\{|\mathcal{H}(\vec{f})|^2\} = \frac{\mathrm{E}\{Q_R(\vec{f})\}}{\mathrm{E}\{Q_R(0)\}}. \tag{4.77}$$

Note that in Eq. (4.77) K is the number of photo events in each reference star image. For this calibration to be accurate it is required that the reference star data be collected under nearly identical atmospheric seeing conditions as the object data so that the estimate of $\mathrm{E}\{|\mathcal{H}(\vec{f})|^2\}$ obtained from the reference star data is the same as for the object data.

In Fig. 4.6 speckle interferometry results are shown for a simulated binary star with the following parameters: $D = 1$ m, $r_0 = 0.1$ m, $\lambda = 500$ nm, $\overline{K} = 10{,}000$, and the binary star components are separated by 1.5 μrad. Note that the diffraction-limited resolution for this case is approximately $\lambda/D = 0.5$ μrad. The relative brightnesses of the binary components are $d_1/d_2 = 1$, and 200 independent frames were averaged using the simulation. In Fig. 4.6(a) the negative of the long exposure image for this binary star system is shown. Note that the presence of a binary star cannot be determined from the long exposure image because the angular separation of the binary components is less than $\lambda/r_0 = 5\mu$rad. Fig. 4.6(b) shows an image with intensity proportional to the logarithm base 10 of $Q(\vec{f})$. Displaying $Q(\vec{f})$ using the logarithm of the intensity is required since the dynamic range of $Q(\vec{f})$ is approximately eight orders of magnitude due to the attenuating effects of the speckle transfer function. Note that the interference fringe-like character of $Q(\vec{f})$ is evident in Fig. 4.6(b) as can be seen by the large intensity near the center, $|\vec{f}| = 0$, position, and the decreasing intensity of the fringe pattern as $|\vec{f}|$ increases. Also, the attenuating effects of the speckle transfer function are evident in Fig. 4.6(b), as predicted by the analysis above. In Fig. 4.6(c) the calibrated quantity, $|\tilde{O}_n(\vec{f})|^2$, is shown. The image shown in Fig. 4.6(c) is proportional to $|\tilde{O}_n(\vec{f})|^2$. Taking the logarithm of the data is not required after calibration because the speckle transfer function effects have been removed from $Q(\vec{f})$ by the calibration process. Note that the calibration process has removed the attenuation at high spatial frequencies caused by the speckle transfer function, allowing the visibility and the modulation of the fringe pattern to be easily determined.

In Eq. (4.75) the separation of the binary star, $\Delta x = (x_1 - x_2)$, is encoded in the argument of the $\cos(\cdot)$ function. The period, P, of the $\cos(\cdot)$ function is (note that f_x is the independent variable)

$$P = \frac{1}{\Delta x}. \tag{4.78}$$

Thus, the separation of the binary components can be determined by measuring the period of the fringe in $Q(\vec{f})$. The relative brightness of the binary components are encoded in the modulation term, M, given by

$$M = \frac{2d_1 d_2}{(d_1 + d_2)^2}. \tag{4.79}$$

Equation (4.79) can be demonstrated by writing the relationship between the intensities as

$$d_1 = a d_2, \tag{4.80}$$

where a is a constant of proportionality relating the relative brightnesses of the components. Substituting Eq. (4.80) into Eq. (4.79) and performing the necessary algebraic manipulations yields the result that a is one of the solutions to the quadratic equation

$$a^2 + 2a + \left(1 - \frac{2a}{M}\right) = 0. \tag{4.81}$$

In the next section we present a technique for recovering the phase spectrum of objects from short exposure images of the object. The phase spectrum information can be combined with the modulus

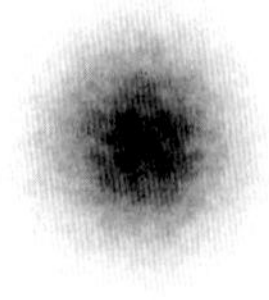

(a)

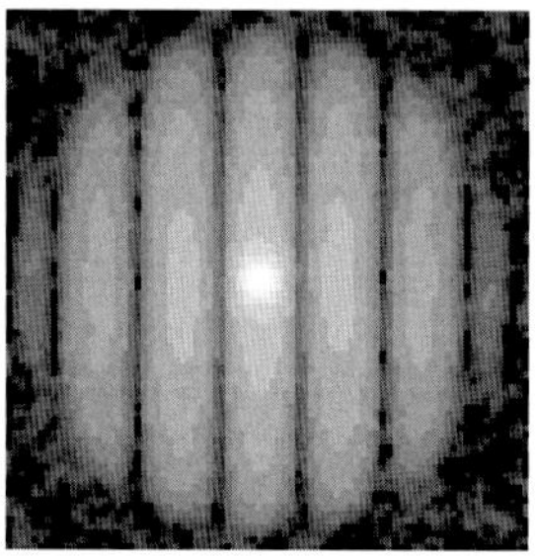

(b)

(c)

FIGURE 4.6
Speckle interferometry simulation results for a binary star with $D = 1$ m, $r_0 = $ 0.1 m, $\lambda = $ 500 nm, $\overline{K} = $ 10,000, and binary star components separated by 1.5 μrad. Note that the diffraction-limited resolution for this case is approximately $\lambda/D = 0.5$ μrad. The relative brightnesses of the binary components are $d_1/d_2 = $ 1, and 200 independent frames were averaged using the simulation to obtain this result. (a) negative of long exposure image; (b) logarithm base 10 of $Q(\vec{f})$, before calibration by reference star data; and (c) $|\tilde{O}_n(\vec{f})|^2$ obtained by calibration using reference star data.

spectrum information obtained from speckle interferometry to create estimates of objects.

4.4 Fourier phase estimation techniques

As discussed in the previous section, speckle interferometry provides only the modulus of the Fourier transform of an object. The phase of the Fourier transform is required to construct an image of the object. The Knox-Thompson, or cross spectrum method, and the bispectrum method are post detection image processing techniques for reconstructing the phase spectrum of an object viewed through turbulence. In this section an analysis of the cross spectrum and bispectrum phase reconstruction techniques is presented.

The cross spectrum and the bispectrum, defined in Eqs. (4.8) and (4.9), respectively, are specialized moments of the Fourier transform of the measured images. Both the cross spectrum and the bispectrum provide information about the phase spectrum of the object by encoding the phase spectrum as point-to-point phase differences. The cross spectrum and the bispectrum are known to contain high spatial frequency information [8, 9]. Hence, these methods have found wide use in space object imaging applications [21].

Two key points must be addressed to understand phase spectrum estimation using either the cross spectrum or the bispectrum. The first issue is to demonstrate that high spatial frequency information related to the object phase spectrum is present in these specialized moments of the image Fourier transform. Detailed proof of this point for the cross spectrum requires evaluation of a fourth order moment of the field in the telescope pupil. This proof is similar to the demonstration that the speckle transfer function contains high frequency information, and is provided here. To derive the bispectrum transfer function, evaluation of a sixth order moment of the pupil plane field is required. This analysis is not presented here. However, an adequate proof is available in the literature [9].

The second key issue is phase spectrum reconstruction. Neither the cross spectrum nor the bispectrum provide object phase spectrum information directly. A separate processing step is required to reconstruct the object phase spectrum. Phase spectrum reconstruction from both the cross spectrum and the bispectrum is treated in detail.

The remainder of this section is organized as follows. In the next subsection the cross spectrum method is presented. Calculation of the cross spectrum and practical issues associated with using the cross spectrum are discussed first. The transfer function associated with the cross spectrum is then derived and evaluated, showing that the cross spectrum preserves high spatial frequency information. Finally, phase spectrum reconstruction from the cross spectrum is discussed. In the following subsection the bispectrum method is presented. Details of calculating the bispectrum, and phase spectrum reconstruction from the bispectrum, are presented.

4.4.1 Knox-Thompson, or cross spectrum technique.

As presented in Eq. (4.8), the cross spectrum of a single image, $C(\vec{f}, \Delta\vec{f})$, is defined as [8]

$$C(\vec{f}, \Delta\vec{f}) = I(\vec{f})I^*(\vec{f} + \Delta\vec{f}), \tag{4.82}$$

where $\Delta\vec{f}$ is a small, constant offset spatial frequency. In the earlier discussion of the cross spectrum it was mentioned that the cross spectrum preserves object phase spectrum information in the form of phase differences. This is a critical aspect of speckle imaging which is now addressed.

Recall from Eq. (4.3) that the relationship between the image spectrum, the object spectrum, and the OTF is given by

$$I(\vec{f}) = O(\vec{f})\mathcal{H}(\vec{f}). \tag{4.83}$$

Both the OTF and the object spectrum are, in general, complex-valued quantities. Let the OTF be represented by the phasor notation

$$\mathcal{H}(\vec{f}) = |\mathcal{H}(\vec{f})| e^{j\phi_{\mathcal{H}}(\vec{f})}, \tag{4.84}$$

and let the object spectrum be represented by

$$O(\vec{f}) = |O(\vec{f})| e^{j\phi_o(\vec{f})}. \tag{4.85}$$

With these definitions, the cross spectrum of a single image realization is given by

$$\begin{aligned} C(\vec{f}, \Delta\vec{f}) = &|O(\vec{f})||O(\vec{f}+\Delta\vec{f})||\mathcal{H}(\vec{f})||\mathcal{H}(\vec{f}+\Delta\vec{f})| \\ &\times \exp\{j[\phi_o(\vec{f}) - \phi_o(\vec{f}+\Delta\vec{f}) + \phi_{\mathcal{H}}(\vec{f}) - \phi_{\mathcal{H}}(\vec{f}+\Delta\vec{f})]\}. \end{aligned} \tag{4.86}$$

Phase difference information, that is, the difference in phase between points in the object phase spectrum is encoded in the $\exp\{j[\phi_o(\vec{f}) - \phi_o(\vec{f}+\Delta\vec{f})]\}$ term in Eq. (4.86). However, in a single image realization this object phase difference information is corrupted by the random phase differences due to the atmosphere-telescope OTF contained in the $\exp\{j[\phi_{\mathcal{H}}(\vec{f}) - \phi_{\mathcal{H}}(\vec{f}+\Delta\vec{f})]\}$ term in Eq. (4.86).

Corruption of the object spectrum phase difference information by phase differences due to the OTF is overcome by averaging. To expand on this idea, consider the average cross spectrum defined by

$$\begin{aligned} \mathrm{E}\{C(\vec{f}, \Delta\vec{f})\} = \mathrm{E}\Big\{&|O(\vec{f})||O(\vec{f}+\Delta\vec{f})||\mathcal{H}(\vec{f})||\mathcal{H}(\vec{f}+\Delta\vec{f})| \\ &\times \exp\{j[\phi_o(\vec{f}) - \phi_o(\vec{f}+\Delta\vec{f}) + \phi_{\mathcal{H}}(\vec{f}) - \phi_{\mathcal{H}}(\vec{f}+\Delta\vec{f})]\}\Big\}. \end{aligned} \tag{4.87}$$

The object spectrum is deterministic, and thus may be taken outside the expectation in the above equation to obtain

$$\begin{aligned} \mathrm{E}\{C(\vec{f}, \Delta\vec{f})\} = &|O(\vec{f})||O(\vec{f}+\Delta\vec{f})| \exp\{j[\phi_o(\vec{f}) - \phi_o(\vec{f}+\Delta\vec{f})]\} \\ &\times \mathrm{E}\{\mathcal{H}(\vec{f})\mathcal{H}^*(\vec{f}+\Delta\vec{f})\}. \end{aligned} \tag{4.88}$$

The specialized moment of the OTF, $\mathrm{E}\{\mathcal{H}(\vec{f})\mathcal{H}^*(\vec{f}+\Delta\vec{f})\}$, is referred to as the cross spectrum transfer function since it relates the object spectrum to the cross spectrum. It is shown later that the cross spectrum transfer function is real-valued. Hence, the phase of the average cross spectrum, $\phi_C(\vec{f}, \Delta\vec{f})$, is given by

$$\phi_C(\vec{f}, \Delta\vec{f}) = \phi_o(\vec{f}) - \phi_o(\vec{f}+\Delta\vec{f}), \tag{4.89}$$

which clearly encodes the object phase spectrum.

Note that the cross spectrum has two-fold symmetry arising from the fact that the Fourier transform of a real image has Hermitian symmetry, thus

$$I(\vec{f}) = I^*(-\vec{f}). \tag{4.90}$$

The implication of this symmetry in the object spectrum for the cross spectrum is that

$$C(\vec{f}, \Delta\vec{f}) = C^*(-\vec{f}, -\Delta\vec{f}). \tag{4.91}$$

Thus, it is only necessary to compute half of the cross spectrum. For example, the first and second quadrants in the $\vec{f}$ plane contain all of the information in the cross spectrum.

Three issues relating to the cross spectrum are addressed in the following subsections. In the next subsection the unbiased estimators for the cross spectrum are obtained for the cases of photon-limited detection, and photon detection in the presence of additive noise. The cross spectrum transfer function is then derived and evaluated in the following subsection. Object phase spectrum reconstruction is addressed in the final subsection.

Unbiased estimator for the cross spectrum

When the cross spectrum is computed directly from photon-limited images a bias arises similar to the photon-noise bias present in speckle interferometry [13]. As a result of this bias the cross spectrum of the detected image is not directly proportional to the cross spectrum of the object. This bias term is referred to as the photon noise bias term. The origin and form of this bias can be understood by using the Fourier transform of a photon-limited image, given in Eq. (4.24), in the expression for the cross spectrum given in Eq. (4.8). Performing this operation yields the result

$$\begin{aligned} C(\vec{f},\Delta\vec{f}) &= D(\vec{f})D^*(\vec{f}+\Delta\vec{f}) \\ &= \sum_{n=1}^{K}\exp\{-j2\pi\vec{f}\cdot\vec{x}_n\}\times\sum_{l=1}^{K}\exp\{j2\pi(\vec{f}+\Delta\vec{f})\cdot\vec{x}_l\} \\ &= \sum_{n=1}^{K}\sum_{l=1}^{K}\exp\{-j2\pi[\vec{f}\cdot\vec{x}_n-\vec{f}\cdot\vec{x}_l-\Delta\vec{f}\cdot\vec{x}_l]\}. \end{aligned} \tag{4.92}$$

In Eq. (4.92) there are K terms for which $n = l$. Extracting these K terms from the double summation in Eq. (4.92) gives the term

$$\sum_{m=1}^{K}\exp\{j2\pi\Delta\vec{f}\cdot\vec{x}_m\}, \tag{4.93}$$

which is the conjugate of the spectrum of the photon-limited image evaluated at $\Delta\vec{f}$, or, mathematically,

$$D^*(\Delta\vec{f}) = \sum_{m=1}^{K}\exp\{j2\pi\Delta\vec{f}\cdot\vec{x}_m\}. \tag{4.94}$$

The term $D^*(\Delta\vec{f})$ is not desired, and must be removed. The remaining $(K^2 - K)$ terms in Eq. (4.92) for which $n \neq l$ provide the desired cross spectrum. Hence, the unbiased estimator for the cross spectrum for photon-limited images, $C_U(\vec{f},\Delta\vec{f})$, is [13, 19]

$$C_U(\vec{f},\Delta\vec{f}) = D(\vec{f})D^*(\vec{f}+\Delta\vec{f}) - D^*(\Delta\vec{f}). \tag{4.95}$$

The bias term $D^*(\Delta\vec{f})$ will vary from image-to-image, and should be removed from the cross spectrum of each image before averaging.

It is also natural to consider whether a bias arises in the cross spectrum due to additive noise in the image measurement. If additive noise is present in the image, then the detected image is given by Eq. (4.55). The detected image spectrum in this case is given by

$$D(\vec{f}) = \sum_{n=1}^{K}\exp\{-j2\pi\vec{f}\cdot\vec{x}_n\} + \sum_{p=1}^{P} n_p\exp\{-j2\pi\vec{f}\cdot\vec{x}_p\}, \tag{4.96}$$

where $\vec{x}_p$ is the location of the pth pixel, P is the number of pixels in the image, and the statistical properties of the random noise variable, n_p, are given in the discussion of Eq. (4.55). Since the additive noise term, n_p, is zero mean and statistically independent of both K and $\vec{x}_n$, we need only examine the effects of the second term in Eq. (4.96) in the average cross spectrum. The average cross spectrum of the noise term in Eq. (4.96), $\mathrm{E}\{C_N(\vec{f}, \Delta\vec{f})\}$, is given by

$$\begin{aligned}\mathrm{E}\{C_N(\vec{f}, \Delta\vec{f})\} &= \sum_{p=1}^{P}\sum_{q=1}^{P} \mathrm{E}\{n_p n_q\} \exp\{-j2\pi[\vec{f}\cdot\vec{x}_p - \vec{f}\cdot\vec{x}_q - \Delta\vec{f}\cdot\vec{x}_q]\} \\ &= \sum_{p=1}^{P} \sigma_n^2 \exp\{j2\pi\Delta\vec{f}\cdot\vec{x}_q\} \\ &= \sigma_n^2 \delta(\Delta\vec{f}),\end{aligned} \tag{4.97}$$

where σ_n^2 is the variance of the additive noise term. Since the offset vector is never chosen to be zero in the cross spectrum we see that no bias arises in the cross spectrum due to additive noise in the image measurement.

Cross spectrum transfer function

The expression for the average cross spectrum was given in Eq. (4.88) as

$$\begin{aligned}\mathrm{E}\{C(\vec{f}, \Delta\vec{f})\} = &|O(\vec{f})||O(\vec{f}+\Delta\vec{f})| \exp\{j[\phi_o(\vec{f}) - \phi_o(\vec{f}+\Delta\vec{f})]\} \\ &\times \mathrm{E}\{\mathcal{H}(\vec{f})\mathcal{H}^*(\vec{f}+\Delta\vec{f})\}.\end{aligned} \tag{4.98}$$

Eq. (4.88) is the product of two terms: one which is a function of only the object spectrum, and one which is a function of the atmosphere-telescope transfer function. This second term, $\mathrm{E}\{\mathcal{H}(\vec{f})\mathcal{H}^*(\vec{f}+\Delta\vec{f})\}$, is referred to as the cross spectrum transfer function, and is the topic of this subsection. If the object-dependent term,

$$|O(\vec{f})||O(\vec{f}+\Delta\vec{f})| \exp\{j[\phi_o(\vec{f}) - \phi_o(\vec{f}+\Delta\vec{f})]\}, \tag{4.99}$$

is finite out to spatial frequencies approaching the diffraction-limited cutoff frequency, then the high frequency content of the average cross spectrum is governed by the cross spectrum transfer function.

The key point to be demonstrated in this subsection is that the cross spectrum transfer function is non-zero out to spatial frequencies approaching the diffraction-limited cutoff frequency. Because the cross spectrum transfer function allows high spatial frequency information to be passed to the cross spectrum, the cross spectrum can be used to reconstruct high resolution images. The analysis technique used here is similar to the method used to compute the speckle transfer function given in Eq. (4.22), though a different result will be obtained.

The cross spectrum transfer function can be written in terms of the telescope pupil function, $P(\vec{x})$, and the turbulence-induced phase aberration, $\psi(\vec{x})$, as

$$\begin{aligned}\mathrm{E}\{\mathcal{H}(\vec{f})\mathcal{H}^*(\vec{f}+\Delta\vec{f})\} = N_F^{-2} &\int d\vec{x} \int d\vec{x}'\, P(\vec{x})P(\vec{x}-\vec{f}\lambda d_i)P(\vec{x}')P(\vec{x}'-\vec{f}\lambda d_i - \Delta\vec{f}\lambda d_i) \\ &\times \mathrm{E}\left\{\exp\{j[\psi(\vec{x}) - \psi(\vec{x}-\vec{f}\lambda d_i) - \psi(\vec{x}') + \psi(\vec{x}'-\vec{f}\lambda d_i - \Delta\vec{f}\lambda d_i)]\}\right\}\end{aligned} \tag{4.100}$$

where N_F is a normalizing constant given in Eq. (4.16). Note that Eq. (4.100) is a fourth order moment of the field in the pupil plane of the telescope. The expectation inside the integral in Eq. (4.100) can be simplified using Eq. (4.21) with the following substitutions

$$\begin{aligned}\psi_1 &= \psi(\vec{x}) \\ \psi_2 &= \psi(\vec{x}-\vec{f}\lambda d_i) \\ \psi_3 &= \psi(\vec{x}')\end{aligned}$$

$$\psi_4 = \psi(\vec{x}' - \vec{f}\lambda d_i - \Delta\vec{f}\lambda d_i). \tag{4.101}$$

The result of this operation is

$$\begin{aligned}
\mathrm{E}\{\exp\{j[\psi_1 - \psi_2 - \psi_3 + \psi_4]\}\} &= \exp\left\{-\frac{1}{2}\mathrm{E}\{[\psi_1 - \psi_2 - \psi_3 + \psi_4]^2\}\right\} \\
&= \exp\Bigg\{-\frac{1}{2}\mathrm{E}\{(\psi_1 - \psi_2)^2 + (\psi_1 - \psi_3)^2 - (\psi_1 - \psi_4)^2 \\
&\qquad -(\psi_2 - \psi_3)^2 + (\psi_2 - \psi_4)^2 + (\psi_3 - \psi_4)^2\}\Bigg\} \\
&= \exp\Bigg\{-\frac{1}{2}[D_\psi(|\vec{f}\lambda d_i|) + D_\psi(|\vec{x} - \vec{x}'|) \\
&\qquad -D_\psi(|\vec{x} - \vec{x}' + \vec{f}\lambda d_i + \Delta\vec{f}\lambda d_i|) \\
&\qquad -D_\psi(|\vec{x} - \vec{x}' - \vec{f}\lambda d_i|) + D_\psi(|\vec{x} - \vec{x}' + \Delta\vec{f}\lambda d_i|) \\
&\qquad +D_\psi(|\vec{f}\lambda d_i + \Delta\vec{f}\lambda d_i|)]\Bigg\}
\end{aligned} \tag{4.102}$$

The final expression for the cross spectrum transfer function may be obtained by substituting Eq. (4.102) into Eq. (4.100).

The cross spectrum transfer function can be evaluated numerically. Note that it is necessary to pick a fixed value for the offset frequency, $\Delta\vec{f}$, to obtain a two dimensional expression for the cross spectrum transfer function. A plot of such an evaluation is shown in Fig. 4.7. In Fig. 4.7 an f_x-axis slice of the cross spectrum transfer function is shown for the cases of $D/r_0 = 2, 5, 10$, and 15. For the $D/r_0 = 15$ case the offset vector was set at $|\Delta\vec{f}| = 0.033D/(\lambda d_i)$ in the f_x direction. For all other D/r_0 values the offset vector was set at $|\Delta\vec{f}| = 0.05D/(\lambda d_i)$ in the f_x direction. The horizontal axis in Fig. 4.7 has been normalized to the diffraction limited cutoff frequency, $D/(\lambda d_i)$. Observe the similarity between the cross spectrum transfer function and the speckle transfer function shown in Fig. 4.2, and that the average cross spectrum is non-zero at spatial frequencies approaching the diffraction-limited cutoff frequency. In fact, it is easy to show that in the $\Delta f = 0$ limit, the cross spectrum transfer function and the speckle transfer function are identical.

Note that in all cases in Fig. 4.7 the offset spatial frequency, $|\Delta\vec{f}|$, was chosen to be small compared to the frequency, $r_0/(\lambda d_i)$. It is possible to analytically show that the cross spectrum transfer function becomes small as $|\Delta\vec{f}|$ increases, though this exercise provides little insight into why this should be. Knox and Thompson [8] analyzed the autocorrelation of the atmosphere-telescope OTF, $\mathcal{H}(\vec{f})$, and found the width of this autocorrelation to be approximately $r_0/(\lambda d_i)$. For offset spatial frequencies, $|\Delta\vec{f}|$, less than $r_0/(\lambda d_i)$ the specialized moment of the OTF, $\mathrm{E}\{\mathcal{H}(\vec{f})\mathcal{H}^*(\vec{f} + \Delta\vec{f})\}$, is non-zero, as illustrated in Fig. 4.7. However, if the offset frequency, $|\Delta\vec{f}|$, is greater than $r_0/(\lambda d_i)$ the OTF components at $\vec{f}$ and $(\vec{f} + \Delta\vec{f})$ are approximately *uncorrelated*, so that

$$\mathrm{E}\{\mathcal{H}(\vec{f})\mathcal{H}^*(\vec{f} + \Delta\vec{f})\} \approx \mathrm{E}\{\mathcal{H}(\vec{f})\}\mathrm{E}\{\mathcal{H}^*(\vec{f} + \Delta\vec{f})\} \approx 0, \quad |\Delta\vec{f}| > r_0/(\lambda d_i), \tag{4.103}$$

since at spatial frequencies, $|\vec{f}| > r_0/(\lambda d_i)$, the average OTF is approximately zero. A similar conclusion was obtained by Ayers *et al.* [13] using a simulation of the cross spectrum technique. Thus, spatial frequency offsets in the range $|\Delta\vec{f}| < r_0/(\lambda d_i)$ are always used in the cross spectrum method.

As a final comment, we note that the cross spectrum transfer function is not shift indifferent. That is, the cross spectrum is a function of the tilt component of the turbulence-induced aberration. Turbulence-induced random tilt causes the image to move randomly about the image plane, but does not affect the image in any other way. The lack of shift indifference in the cross spectrum can be seen by defining a

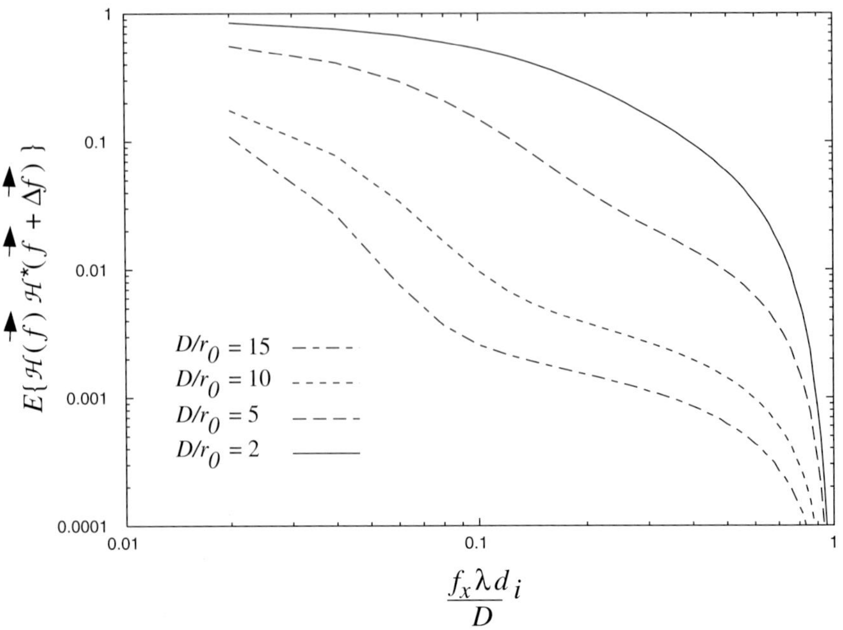

FIGURE 4.7
Cross spectrum transfer function for $D/r_0 = 2$, 5, 10, and 15. For the $D/r_0 = 15$ case the offset vector was set at $\Delta\vec{f} = 0.033D/(\lambda d_i)$ in the f_x direction. For all other D/r_0 values the offset vector was set at $\Delta\vec{f} = 0.05D/(\lambda d_i)$ in the f_x direction. The horizontal axis is normalized to the diffraction limited cutoff frequency, $D/(\lambda d_i)$.

shifted version, $i_s(\vec{x})$, of the unshifted image, $i(\vec{x})$, and computing the cross spectrum of $i_s(\vec{x})$. Let the shifted image be given by

$$i_s(\vec{x}) = i(\vec{x} + \Delta\vec{x}), \tag{4.104}$$

where $\Delta\vec{x}$ is a spatial shift. The Fourier transform of $i_s(\vec{x})$ is obtained using the shift theorem of Fourier transforms [23] as

$$\mathcal{F}\{i_s(\vec{x})\} = \exp\{-j2\pi\vec{f}\cdot\Delta\vec{x}\}I(\vec{f}), \tag{4.105}$$

where $I(\vec{f})$ is the Fourier transform of the unshifted image, $i(\vec{x})$. The cross spectrum of the shifted image, $C_s(\vec{f},\Delta\vec{f})$, is given by

$$C_s(\vec{f},\Delta\vec{f}) = \exp\{j2\pi\Delta\vec{f}\cdot\Delta\vec{x}\}I(\vec{f})I^*(\vec{f}+\Delta\vec{f}). \tag{4.106}$$

Randomness in the shift term, $\Delta\vec{x}$, in Eq. (4.106) results in attenuation of the average cross spectrum due to the $\exp\{j2\pi\Delta\vec{f}\cdot\Delta\vec{x}\}$ factor. To avoid this attenuation the images in a data set are generally shifted to have a common center of mass before the cross spectrum is computed.

The fact that the cross spectrum transfer function is found to be non-zero at spatial frequencies approaching the diffraction-limited cutoff frequency is the enabling factor in the usefulness of the cross spectrum technique. Through this analysis we have shown that the cross spectrum on average contains high frequency information. A suitable phase reconstruction algorithm is required to obtain the phase spectrum from the cross spectrum. Phase reconstruction is the topic of the next subsection.

Phase spectrum reconstruction from the cross spectrum

In this subsection the problem of reconstructing the object phase spectrum from the phase of the cross spectrum is addressed. To reconstruct a two dimensional phase map two offset vectors in orthogonal directions are required. The need for orthogonal offset vectors can be shown by the following analysis. Let one of the offset vectors be parallel to the f_x axis and denoted by Δf_x, and let the other offset vector be parallel to the f_y axis and denoted by Δf_y. The phase differences generated by these offset vectors are

$$\begin{aligned}\Delta\phi_x(f_x, f_y) &= \phi_o(f_x, f_y) - \phi_o(f_x + \Delta f_x, f_y) \approx \frac{\partial\phi_o(\vec{f})}{\partial f_x}\Delta f_x \\ \Delta\phi_y(f_x, f_y) &= \phi_o(f_x, f_y) - \phi_o(f_x, f_y + \Delta f_y) \approx \frac{\partial\phi_o(\vec{f})}{\partial f_y}\Delta f_y,\end{aligned} \tag{4.107}$$

where the notation $\Delta\phi_x(f_x, f_y)$ and $\Delta\phi_y(f_x, f_y)$ denotes phase differences in the f_x and f_y directions, respectively. The partial derivatives in Eq. (4.107) form the orthogonal components of the gradient of the object phase spectrum, $\nabla\phi_o(\vec{f})$.

In practical applications of the cross spectrum method all of the operations leading to Eq. (4.107) are performed in a sampled space. As a consequence, f_x, f_y, Δf_x, and Δf_y may only take on a discrete set of values. In the analyses which follow it is assumed that the frequency domain is sampled on a uniform grid with scalar sample spacing given by Δf. Hence, f_x, f_y, Δf_x, and Δf_y may only take on values given by integer multiples of Δf. Note that the phase differences in Eq. (4.107) exist at every point in the sampled space.

The phase differences given in Eq. (4.107) can be used to recursively reconstruct the phase of the object spectrum in the following way. As a starting point, note that the zero frequency component of the object spectrum is entirely real, and hence the phase of the object spectrum at that point is necessarily zero, or, mathematically, $\phi_o(0, 0) = 0$. This observation allows the phase at the two points along the f_x and f_y axes nearest to $\vec{f} = (0, 0)$ shown below to be determined using

$$\begin{aligned}\phi_o(\Delta f_x, 0) &= \phi_o(0, 0) - \Delta\phi_x(0, 0) \\ \phi_o(0, \Delta f_y) &= \phi_o(0, 0) - \Delta\phi_y(0, 0).\end{aligned} \tag{4.108}$$

The phase of a more distant point along the f_x and f_y axes can be computed by similar means. For example, consider the phase at $\vec{f} = (2\Delta f_x, 0)$. One way to write this phase is

$$\begin{aligned}\phi_o(2\Delta f_x, 0) &= \phi_o(\Delta f_x, 0) - \Delta\phi_x(\Delta f_x, 0) \\ &= \phi_o(0, 0) - \Delta\phi_x(0, 0) - \Delta\phi_x(\Delta f_x, 0) \\ &= -\Delta\phi_x(0, 0) - \Delta\phi_x(\Delta f_x, 0),\end{aligned} \tag{4.109}$$

where Eq. (4.108) was used to write the second line of Eq. (4.109). Eq. (4.109) shows that the phase at $\vec{f} = (2\Delta f_x, 0)$ can be expressed as the negative of the sum of the phase differences along a path from $\vec{f} = (0, 0)$ up to, but not including, the point $\vec{f} = (2\Delta f_x, 0)$. This approach generalizes to all points in the sampled space. One possible path to obtain the phase at any point in the sampled frequency space is

$$\phi_o(N_x\Delta f_x, N_y\Delta f_y) = -\sum_{i=0}^{N_x-1} \Delta\phi_x(i\Delta f_x, 0) - \sum_{j=0}^{N_y-1} \Delta\phi_y(0, j\Delta f_y), \tag{4.110}$$

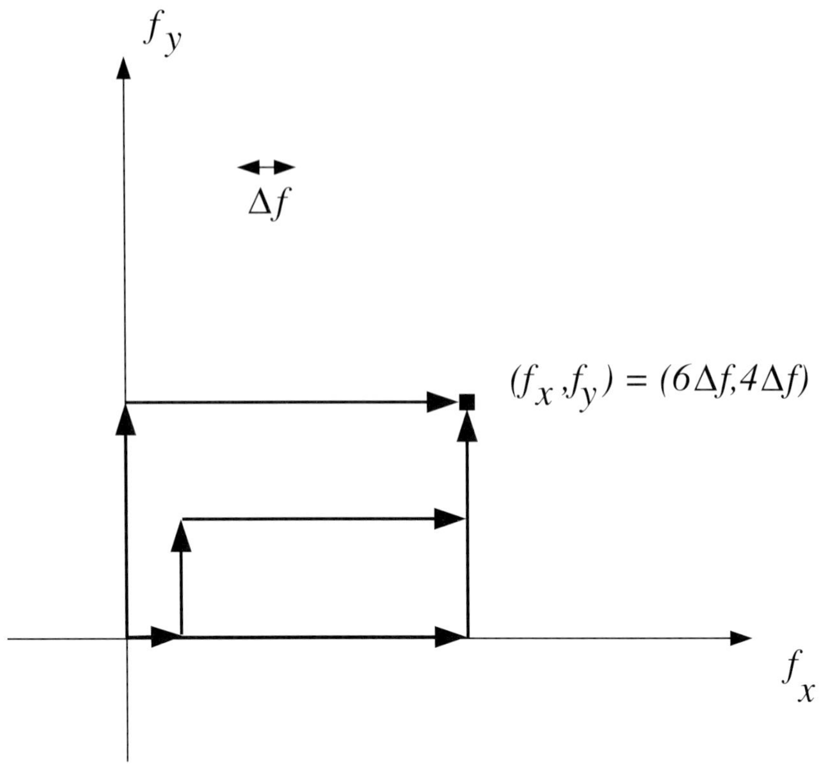

FIGURE 4.8
Three possible phase reconstruction paths to the point $\vec{f} = (6\Delta f, 4\Delta f)$.

where N_x and N_y are integers. Eq. (4.110) says that the phase at any point which can be written as an integer multiple of $(\Delta f_x, \Delta f_y)$ can be obtained by summing the phase differences from $\vec{f} = (0, 0)$ to the point of interest.

It should be noted that Eq. (4.110) indicates only one path to the point $\vec{f} = (N_x \Delta f_x, N_y \Delta f_y)$. However, a multiplicity of paths from $\vec{f} = (0, 0)$ to any point $(N_x \Delta f_x, N_y \Delta f_y)$ can be drawn, as illustrated in Fig. 4.8. In Fig. 4.8 three of a very large number of possible paths from $\vec{f} = (0, 0)$ to $\vec{f} = (6\Delta f, 4\Delta f)$ are shown. The phase arising from taking an arbitrary path can be written most generally as a path integral given by [27]

$$\phi_o(\vec{f}) = \oint \mathrm{d}\vec{s} \cdot \nabla \phi_o(\vec{f}), \tag{4.111}$$

where $\mathrm{d}\vec{s}$ is a differential vector along the path of interest. Note that in a sampled space Eq. (4.111) must be evaluated using summations. In a noise free system all of the possible paths to any point in frequency space would sum to the same phase value. However, noise effects can cause the sums of the phase differences along different paths to the same point to have different values. Averaging the sum of the phase differences along non redundant paths to a given point in frequency space is often used to reduce the effects of noise [27].

A few practical comments regarding use of the cross spectrum method are in order before proceeding to a discussion of the bispectrum technique. The offset vectors Δf_x and Δf_y are often chosen to be equal to a single sample spacing in frequency space if only a single pair of offsets are to be used. This choice permits all points in frequency space to be reached by the discrete path integration required to reconstruct the phase. However, the cross spectrum is often quite noisy due to low light level, poor seeing, or both. Under these conditions multiple offset vectors are often used. The purpose of using multiple offset vectors is to create a larger number of paths to average to obtain the object phase spectrum [13]. However, using multiple offset vectors leads to the creation of a four dimensional cross spectrum.

A detailed treatment of noise effects on the cross spectrum is beyond the scope of the present work. Evaluating the cross spectrum transfer function required the computation of a fourth order moment of the pupil plane field. Evaluating the second moment of the cross spectrum transfer function, needed to compute its variance, requires the computation of an eighth order moment of the pupil plane field. An excellent treatment of this topic is presented in Ref. [13].

4.4.2 Bispectrum technique

The bispectrum provides another technique for obtaining the phase spectrum of the object from short exposure images. Like the cross spectrum, the bispectrum uses a specialized moment of the measured image spectrum. The bispectrum of an image was defined in Eq. (4.9) as [9]

$$B(\vec{f}_1, \vec{f}_2) = I(\vec{f}_1)I(\vec{f}_2)I^*(\vec{f}_1 + \vec{f}_2). \tag{4.112}$$

The bispectrum is a four dimensional data object since it is a function of four scalar spatial variables. The bispectrum has eight-fold symmetry. These symmetries are easily obtained by using the definition of the bispectrum and noting the Hermite symmetry of the image spectrum, and are given by [9]

$$\begin{aligned} B(\vec{f}_1, \vec{f}_2) &= B(\vec{f}_2, \vec{f}_1) \\ B(\vec{f}_1, \vec{f}_2) &= B(\vec{f}_1 - \vec{f}_2, \vec{f}_2) \\ B(\vec{f}_1, \vec{f}_2) &= B^*(-\vec{f}_1, -\vec{f}_2). \end{aligned} \tag{4.113}$$

The object phase spectrum is encoded in the phases of the bispectrum. The relationship between the bispectrum phase and the object phase spectrum can be obtained by using the definition of the bispectrum and Eqs. (4.3) and (4.84),

$$\begin{aligned} B(\vec{f}_1, \vec{f}_2) = {} & |O(\vec{f}_1)||O(\vec{f}_2)||O(\vec{f}_1 + \vec{f}_2)| \\ & \times |\mathcal{H}(\vec{f}_1)||\mathcal{H}(\vec{f}_2)||\mathcal{H}(\vec{f}_1 + \vec{f}_2)| \\ & \times \exp\Big\{ j[\phi_o(\vec{f}_1) + \phi_o(\vec{f}_2) - \phi_o(\vec{f}_1 + \vec{f}_2) \\ & \quad + \phi_{\mathcal{H}}(\vec{f}_1) + \phi_{\mathcal{H}}(\vec{f}_2) - \phi_{\mathcal{H}}(\vec{f}_1 + \vec{f}_2)] \Big\}. \end{aligned} \tag{4.114}$$

The object phase spectrum information is encoded in the $\exp\{j[\phi_o(\vec{f}_1) + \phi_o(\vec{f}_2) - \phi_o(\vec{f}_1 + \vec{f}_2)]\}$ term of Eq. (4.114). In a single realization of the bispectrum the object phase spectrum information is corrupted by the random phase spectrum of the OTF, present in the $\exp\{j[\phi_{\mathcal{H}}(\vec{f}_1) + \phi_{\mathcal{H}}(\vec{f}_2) - \phi_{\mathcal{H}}(\vec{f}_1 + \vec{f}_2)]\}$ term of Eq. (4.114). This corruption is overcome by averaging. The expected value of the bispectrum is given by

$$\mathrm{E}\{B(\vec{f}_1, \vec{f}_2)\} = O(\vec{f}_1)O(\vec{f}_2)O(\vec{f}_1 + \vec{f}_2)\mathrm{E}\{\mathcal{H}(\vec{f}_1)\mathcal{H}(\vec{f}_2)\mathcal{H}^*(\vec{f}_1 + \vec{f}_2)\}. \tag{4.115}$$

The term $\mathrm{E}\{\mathcal{H}(\vec{f}_1)\mathcal{H}(\vec{f}_2)\mathcal{H}^*(\vec{f}_1 + \vec{f}_2)\}$ relates the object spectrum information to the bispectrum and is referred to as the bispectrum transfer function. An argument similar to that presented in Eqs. (4.98) to (4.102) can be used to show that the bispectrum transfer function is real. Evaluation of the bispectrum transfer function requires the calculation of a sixth order moment of the field in the pupil plane. The interested reader is referred to Refs. [9] and [13] for the details of this analysis. The key result of these analyses is that the average bispectrum is found to be nonzero out to spatial frequencies approaching the diffraction-limited cutoff frequency if the range of allowed values of $(\vec{f}_1 + \vec{f}_2)$ is suitably constrained.

Unlike the cross spectrum, the bispectrum is insensitive to random motion of the image centroid. This can be easily shown by using the definition of a shifted image given in Eq. (4.104) and the spectrum of the shifted image given in Eq. (4.105). Specifically, if the shifted version of the image $i(\vec{x})$, with spectrum given by $I(\vec{f})$, is given by $i(\vec{x} + \Delta\vec{x})$, then the spectrum of the shifted image is given by $\exp\{-j2\pi\vec{f} \cdot \Delta\vec{x}\}I(\vec{f})$. The bispectrum of the shifted image, $B_s(\vec{f}_1, \vec{f}_2)$, is given by

$$\begin{aligned} B_s(\vec{f}_1, \vec{f}_2) &= I(\vec{f}_1)I(\vec{f}_2)I^*(\vec{f}_1 + \vec{f}_2) \\ &\quad \times \exp\left\{-j2\pi\vec{f}_1 \cdot \Delta\vec{x} - j2\pi\vec{f}_2 \cdot \Delta\vec{x} + j2\pi(\vec{f}_1 + \vec{f}_2) \cdot \Delta\vec{x}\right\} \\ &= B(\vec{f}_1, \vec{f}_2), \end{aligned} \tag{4.116}$$

where $B(\vec{f}_1, \vec{f}_2)$ is the bispectrum of the unshifted image. The practical consequence of this insensitivity to turbulence-induced random motions of the image is that, unlike the cross spectrum case, images do not have to be shifted to a common centroid prior to computing the bispectrum. At low light levels and for extended objects significant errors can be present in the estimate of the location of the centroid [28], leading to an attenuation of the average cross spectrum which is not present in the average bispectrum.

It is necessary to constrain the range of allowed values of $(\vec{f}_1 + \vec{f}_2)$. This is generally accomplished by letting $\vec{f}_1$ sweep through the entire range of frequency space of interest, and letting $\vec{f}_2$ be a small offset vector which we shall denote as $\Delta\vec{f}$. Simulation studies [29, 30] and theoretical investigations [13] have shown that little useful information is obtained from the bispectrum if $|\Delta\vec{f}| > r_0/\lambda d_i$. A heuristic argument can be used to understand the reason for this limitation. Consider the bispectrum transfer function, $\mathrm{E}\{\mathcal{H}(\vec{f}_1)\mathcal{H}(\vec{f}_2)\mathcal{H}^*(\vec{f}_1 + \vec{f}_2)\}$ for the case of $|\vec{f}_1| \gg r_0/\lambda d_i$, $\vec{f}_2 = \Delta\vec{f}$, and $|\Delta\vec{f}| < r_0/\lambda d_i$. In this case the OTF at $\Delta\vec{f}$ is finite, and based on the earlier discussion of the cross spectrum transfer function, approximately uncorrelated with the OTF at $\vec{f}_1$ and $\vec{f}_1 + \Delta\vec{f}$ so that

$$\mathrm{E}\{\mathcal{H}(\vec{f}_1)\mathcal{H}(\Delta\vec{f})\mathcal{H}^*(\vec{f}_1 + \Delta\vec{f})\} \approx \mathrm{E}\{\mathcal{H}(\Delta\vec{f})\} \times \mathrm{E}\{\mathcal{H}(\vec{f}_1)\mathcal{H}^*(\vec{f}_1 + \Delta\vec{f})\}. \tag{4.117}$$

The second term on the right side of Eq. (4.117), $\mathrm{E}\{\mathcal{H}(\vec{f}_1)\mathcal{H}^*(\vec{f}_1 + \Delta\vec{f})\}$, is recognized as the cross spectrum transfer function presented in Eq. (4.100). As shown in the discussion of Eq. (4.100) the $\mathrm{E}\{\mathcal{H}(\vec{f}_1)\mathcal{H}^*(\vec{f}_1 + \Delta\vec{f})\}$ term has significant value only for $|\Delta\vec{f}| < r_0/\lambda d_i$. Thus, the bispectrum has significant value for bispectrum elements in the range $|\Delta\vec{f}| < r_0/\lambda d_i$.

It is also important to limit the *size* of the bispectrum data object. To obtain an appreciation for the magnitude of the problem of computing the bispectrum, consider the total bispectrum of a 128 × 128 pixel image. The image spectrum also contains 128 × 128 samples. For each point in frequency space, f_1, there are $(128^2 - 1)$ possible f_2 vectors. Hence, in the total bispectrum of this image there are $128^2 \times (128^2 - 1)$ = 264,257,536 points. Due to the eight-fold symmetry of the bispectrum [9] it is only necessary to compute one eighth of the bispectrum, reducing the number of points in the complete bispectrum to 33,032,192. This is still a staggering data object for widely available computers, and steps to reduce its size are very much in order. Constraining the range of values which $(\vec{f}_1 + \vec{f}_2)$ is allowed to take, as discussed above, provides a useful means for reducing the size of the bispectrum data object.

Given the four dimensional nature of the bispectrum, a brief discussion of practical techniques for computing the bispectrum is in order. One such method is to let $\vec{f}_1$ sweep through all the appropriate points in the sampled frequency space, and to let $\vec{f}_2$ be defined as a small offset vector, $\Delta\vec{f}$ [13], restricted to the range $|\Delta\vec{f}| \leq r_0/(\lambda d_i)$. For any given $\vec{f}_1$ the set of $\Delta\vec{f}$ such that $|\Delta\vec{f}| \leq r_0/(\lambda d_i)$ describes a circular set of points about $\vec{f}_1$. In this scheme, the bispectrum is computed for all triplets of points consisting of $\vec{f}_1$, one of the allowed values of $(\vec{f}_1 + \Delta\vec{f})$ in the circular region, and $\Delta\vec{f}$. It is also possible to restrict $\Delta\vec{f}$ to a square region about $\vec{f}_1$ defined by $-r_0/(\lambda d_i) \leq \Delta f_x, \Delta f_y \leq r_0/(\lambda d_i)$, where Δf_x and Δf_y are the orthogonal components of $\Delta\vec{f}$. Many points of the bispectrum contain redundant information, as indicated in the symmetries stated in Eq. (4.113). Computing the bispectrum for redundant points can be avoided by use of conditional statements within the iteration loops. Lookup tables can be used to store the values $\vec{f}_1$ and $\Delta\vec{f}$ for which the bispectrum is to be computed. As shown in the next subsection, it is necessary to remove a photon noise bias from the bispectrum of each detected image. Each realization of the bias-removed bispectrum is added to the accumulator for eventual computation of the sample-based estimate of the average bispectrum.

The remainder of this section provides a treatment of using the bispectrum for reconstructing the object spectrum. In the next subsection the problem of obtaining an unbiased estimator for the photon and additive noise effects is presented. In the following subsection phase spectrum reconstruction from the bispectrum phase is discussed.

Unbiased estimator for the bispectrum

When the bispectrum is computed directly from photon-limited images a photon noise bias arises similar to the photon noise bias present in speckle interferometry and the cross spectrum. As a result of this bias the bispectrum of the detected image is not directly proportional to the bispectrum of the object. Rather, undesired additive terms arising from the photon-limited detection process are present. These undesired terms must be removed during the calculation of the bispectrum.

The nature and origin of these bias terms can be derived by using the definition of the Fourier transform of a photon-limited image, given in Eq. (4.24), in the definition of the bispectrum, given in Eq. (4.9). Performing this operation yields

$$\begin{aligned} B(\vec{f}_1, \vec{f}_2) &= D(\vec{f}_1)D(\vec{f}_2)D^*(\vec{f}_1 + \vec{f}_2) \\ &= \sum_{m=1}^{K} \exp\{-j2\pi\vec{f}_1 \cdot \vec{x}_m\} \sum_{n=1}^{K} \exp\{-j2\pi\vec{f}_2 \cdot \vec{x}_n\} \sum_{l=1}^{K} \exp\{j2\pi(\vec{f}_1 + \vec{f}_2) \cdot \vec{x}_l\} \\ &= \sum_{m=1}^{K}\sum_{n=1}^{K}\sum_{l=1}^{K} \exp\left\{-j2\pi\left[\vec{f}_1 \cdot \vec{x}_m + \vec{f}_2 \cdot \vec{x}_n - \vec{f}_1 \cdot \vec{x}_l - \vec{f}_2 \cdot \vec{x}_l\right]\right\}, \end{aligned} \tag{4.118}$$

where K is the number of photoevents in the image, and x_q is the location of the qth photoevent. As stated in the discussion of Eqs. (4.27) and (4.29), photon biases arise when photoevent locations coincide in evaluating Eq. (4.118).

To apply this observation to Eq. (4.118) we must set $m = n \neq l$, $m = l \neq n$, $n = l \neq m$, and $m = n = l$, and evaluate the terms which arise. As an example, consider the bias term which arises when $m = n$. Evaluating Eq. (4.118) for the $m = n$ case yields

$$B(\vec{f}_1, \vec{f}_2)_{m=n\neq l} = \sum_{m=1}^{K}\sum_{l=1}^{K}{}_{m\neq l} \exp\left\{-j2\pi\left[(\vec{f}_1 + \vec{f}_2) \cdot \vec{x}_m - (\vec{f}_1 + \vec{f}_2) \cdot \vec{x}_l\right]\right\}. \tag{4.119}$$

Eq. (4.119) can be interpreted as the modulus squared of the detected image spectrum evaluated at $(\vec{f}_1 + \vec{f}_2)$ minus the terms which would arise from $m = l$, which sum to K. Hence, one term which must be subtracted from $B(\vec{f}_1, \vec{f}_2)$ to remove the photon bias is

$$|D(\vec{f}_1 + \vec{f}_2)|^2 - K. \tag{4.120}$$

Comparing this result to Eq. (4.29) shows that Eq. (4.120) is identical to the unbiased speckle interferometry estimator, $Q(\vec{f}_1 + \vec{f}_2)$.

To evaluate the total photon-induced noise bias in the bispectrum this process must be performed for all of the terms giving rise to photon biases in the bispectrum. The remaining terms which give rise to photon biases in the bispectrum have been evaluated [12, 13, 19, 21]. For the photon-limited detection case the unbiased estimator for the bispectrum, $B_U(\vec{f}_1, \vec{f}_2)$, is given by [21]

$$B_U(\vec{f}_1, \vec{f}_2) = D(\vec{f}_1)D(\vec{f}_2)D^*(\vec{f}_1 + \vec{f}_2) - |D(\vec{f}_1)|^2$$
$$-|D(\vec{f}_2)|^2 - |D(\vec{f}_1 + \vec{f}_2)|^2 + 2K. \tag{4.121}$$

It is sometimes more convenient to express Eq. (4.121) in terms of the unbiased speckle interferometry estimator, $Q(\vec{f})$, which is often available during the computation since the speckle interferometry and bispectrum computations are generally combined in a single program. The result of expressing $B_U(\vec{f}_1, \vec{f}_2)$ in terms of $Q(\vec{f})$ is [21]

$$B_U(\vec{f}_1, \vec{f}_2) = D(\vec{f}_1)D(\vec{f}_2)D^*(\vec{f}_1 + \vec{f}_2) - Q(\vec{f}_1) - Q(\vec{f}_2)$$
$$-Q(\vec{f}_1 + \vec{f}_2) - K. \tag{4.122}$$

Note that the photon bias will vary from image to image, and hence the photon bias must be removed from the bispectrum of each detected image before averaging the bispectrum.

Equation (4.121) is valid only for the photon-limited detection case. If additive noise is present in the image measurement then an additional term is needed in the unbiased estimator for the bispectrum to remove a bias which arises from the additive noise. This is in contrast to the cross spectrum case, where it was shown that the cross spectrum is not biased by a term arising from additive noise. The analysis of the bias in the bispectrum arising from the additive noise can be conducted using the model for additive noise given in Eq. (4.55). The result of this analysis is that in the presence of additive noise the unbiased estimator for the bispectrum is

$$B_U(\vec{f}_1, \vec{f}_2) = D(\vec{f}_1)D(\vec{f}_2)D^*(\vec{f}_1 + \vec{f}_2) - |D(\vec{f}_1)|^2$$
$$-|D(\vec{f}_2)|^2 - |D(\vec{f}_1 + \vec{f}_2)|^2 + 2K + 3P\sigma_n^2, \tag{4.123}$$

where P is the number of pixels in the image, and σ_n is the standard deviation of the additive noise expressed in photo electrons.

After bias removal, each realization of the bispectrum is added to an accumulator for eventual calculation of the sample-based estimate of the mean bispectrum. The mean bispectrum forms the input to the phase spectrum reconstruction algorithm. The problem of reconstructing the object phase spectrum from the bispectrum is treated in the next subsection.

Phase spectrum reconstruction from the bispectrum

In this subsection the problem of reconstructing the object phase spectrum from the phase of the bispectrum is addressed. The most widely used technique is recursive in nature, and is similar to the cross spectrum reconstruction technique presented in Section 4.4.1. The recursive reconstruction process uses the fact that the object phase spectrum at $(\vec{f}_1 + \vec{f}_2)$ can be expressed as

$$\phi_o(\vec{f}_1 + \vec{f}_2) = \phi_o(\vec{f}_1) + \phi_o(\vec{f}_2) - \phi_B(\vec{f}_1, \vec{f}_2), \tag{4.124}$$

where $\phi_B(\vec{f}_1, \vec{f}_2)$ is the phase of the mean bispectrum. Thus, if the object spectrum at $\vec{f}_1$ and $\vec{f}_2$ is known, the object phase spectrum at $(\vec{f}_1 + \vec{f}_2)$ can be computed directly. Of course, two starting points are required for this process: the object phase spectrum at some initial values of $\vec{f}_1$ and $\vec{f}_2$ must be determined. The object phase spectrum at $\vec{f} = 0$ is identically zero so that

$$\phi_o(\vec{f} = 0) = 0. \tag{4.125}$$

While this is a sufficient starting point for object phase spectrum reconstruction from the cross spectrum, an additional known phase is required for phase reconstruction from the bispectrum. The additional starting point typically used is to choose

$$\begin{aligned}\phi_o(\pm\Delta f, 0) &= 0,\\ \phi_o(0, \pm\Delta f) &= 0,\end{aligned} \tag{4.126}$$

where Δf is the sample spacing in the frequency domain. In words, Eq. (4.126) says that the phase of the four points closest to $\vec{f} = (0, 0)$ are set equal to zero. It is easy to show that any nonzero choice of $\phi_o(\pm\Delta f, 0)$ or $\phi_o(0, \pm\Delta f)$ causes a term linear in Δf to appear in the reconstructed phase. This linear phase term corresponds to a shift of the image in space. Since we are typically indifferent to the absolute location of the object within the image, no information of value is lost by making the choice in Eq. (4.126).

Within the bispectrum there are many possible combinations of $\vec{f_1}$ and $\vec{f_2}$ which sum to $(\vec{f_1} + \vec{f_2})$. Under noise free conditions the reconstructed phase at $(\vec{f_1} + \vec{f_2})$ would be the same for all paths. However, noise effects corrupt the phase estimation process so that the reconstructed phase at $(\vec{f_1} + \vec{f_2})$ is a function of the path taken. To reduce these noise effects the phase estimates from several unique paths to $(\vec{f_1} + \vec{f_2})$ are generally averaged [21].

It should be noted that the bispectrum phases are only known modulo 2π. Hence, the recursive reconstructor in Eq. (4.124) could, and in practice often does, lead to 2π phase mismatches between the phase spectrum values computed along different paths to the same point in frequency space. Phases from different paths to the same point cannot be averaged to reduce noise under this condition [31]. To overcome this problem a variation of Eq. (4.124) is often used. In this variation the recursive phase reconstruction is performed using unit amplitude phasors. Mathematically, the unit amplitude phasor recursive reconstructor is given by [32]

$$\exp\{j\phi_o(\vec{f_1} + \vec{f_2})\} = \exp\{j\phi_o(\vec{f_1})\}\exp\{j\phi_o(\vec{f_2})\}\exp\{-j\phi_B(\vec{f_1}, \vec{f_2})\}. \tag{4.127}$$

The object phase spectrum is obtained by computing the argument of the left-hand side of Eq. (4.127). Phase spectrum values obtained using Eq. (4.127) are indifferent to the 2π phase ambiguities. Hence, phase spectrum values obtained from multiple paths to the same point may be averaged to reduce noise effects. The unit amplitude phasor recursive reconstructor has been used successfully in several studies of the bispectrum technique [13, 21, 30, 31, 32].

While the present treatment of phase reconstruction from the bispectrum has been limited to recursive phase reconstruction, it should be noted that this problem has received a great deal of attention in the literature. Meng *et al.* [32] have developed a least squares formulation of the phase reconstruction problem. Northcott *et al.* [31] have developed a reconstruction technique based on the projection slice theorem of tomography and the Radon transform. Matson [30, 33] has developed two weighted least-squares estimation formulations of the phase reconstruction problem. Ayers *et al.* [13] and Nakajima [34] have performed extensive studies of the signal-to-noise ratio behavior of the bispectrum, and Matson [33] has studied phase spectrum reconstruction errors when using noisy bispectra. These analyses have consistently shown that the bispectrum phase spectrum reconstruction technique can be used to obtain high quality phase spectrum estimates for point source objects in the mid frequency regime of 40 to 60% of the diffraction limited cutoff frequency of the telescope at low light levels. This performance improves as the light level increases [29].

Interestingly, if the squared phase spectrum error $\phi_\epsilon^2(\vec{f})$, for the estimated phase $\tilde{\phi}_o(\vec{f})$, is defined by

$$\phi_\epsilon^2(\vec{f}) = \left(\phi_o(\vec{f}) - \tilde{\phi}_o(\vec{f})\right)^2, \tag{4.128}$$

Matson [33] has found through simulation that the spatial frequency at which $\phi_\epsilon^2(\vec{f})$ becomes greater than 1 rad^2 is approximately the same spatial frequency at which the signal-to-noise ratio of the unbiased speckle interferometry estimator, $Q(\vec{f})$, becomes less than unity. This trend has been observed in other studies of speckle imaging phase and modulus estimation [14]. The practical value of this observation is that the signal-to-noise ratio of the unbiased speckle interferometry estimator, $Q(\vec{f})$, which can be easily calculated using Eq. (4.52), can be used to estimate the highest spatial frequency for which good phase spectrum estimation will be obtained.

The treatment of speckle imaging to this point has been mainly theoretical. We have not yet shown how to obtain a reconstructed image from speckle imaging measurements and computations. In the next section we provide the details of image reconstruction for speckle imaging. Some representative results of speckle imaging are provided to demonstrate the expected level of performance which can be obtained with speckle imaging techniques.

4.5 Image reconstruction for speckle imaging

In this section the problem of reconstructing an image from speckle imaging measurements and computations is discussed. Representative results of speckle imaging are provided to demonstrate the level of imaging performance which can be expected for speckle imaging. All of the results presented were obtained using a simulation of speckle imaging like that shown in Fig. 4.1, with the addition of phase spectrum reconstruction using bispectrum technique. The required modification to the simulation is shown in Fig. 4.9. Specifically, after the photon-limited image is created in the simulation both the unbiased speckle interferometry estimator $Q(\vec{f})$, and the unbiased bispectrum estimator, $B_U(\vec{f}_1, \vec{f}_2)$, must be computed for each frame, and the necessary quantities required to compute the mean and standard deviation of $Q(\vec{f})$ and $B_U(\vec{f}_1, \vec{f}_2)$ must be accumulated. If the cross spectrum is to be used $C_U(\vec{f}, \Delta\vec{f})$ should be substituted for $B_U(\vec{f}_1, \vec{f}_2)$.

Once the required quantities have been computed from the measurements, reconstructed images can be computed in the following manner. The modulus squared of the object spectrum may be obtained using a variation of the inverse filter stated in Eq. (4.76), specifically

$$|\tilde{O}_n(\vec{f})|^2 = \frac{Q(\vec{f})}{Q_R(\vec{f}) + \alpha/SNR_Q(\vec{f})}, \tag{4.129}$$

where $SNR_Q(\vec{f})$ is the signal-to-noise ratio of $Q(\vec{f})$ computed from sample-based estimates of the mean and standard deviation of $Q(\vec{f})$, and α is a constant selected by the user. The purpose of the term $\alpha/SNR_Q(\vec{f})$ in Eq. (4.129) is to reduce the effects of noise at high spatial frequencies similar to the Wiener filter [24, 35]. The Wiener filter is a minimum error reconstruction [24], which is not true for the filter in Eq. (4.129). However, it should be noted that the information required to create a Wiener filter is not in general available in speckle imaging. Equation (4.129) is often referred to as the pseudo-Wiener filter. At low spatial frequencies, where $SNR_Q(\vec{f})$ is high, the term $\alpha/SNR_Q(\vec{f})$ is small, and has little effect on the calibration. However, at high spatial frequencies $SNR_Q(\vec{f})$ can be small so that the term $\alpha/SNR_Q(\vec{f})$ dominates the denominator of Eq. (4.129), essentially tapering the high spatial frequency components of $|\tilde{O}_n(\vec{f})|^2$ to reduce the effects of high spatial frequency noise in the reconstructed image.

Next, the square root of $|\tilde{O}_n(\vec{f})|^2$ is computed, and the phase spectrum reconstructed from the bispectrum phase is associated with $|\tilde{O}_n(\vec{f})|$ to obtain $\tilde{O}_n(\vec{f})$. After the average unbiased bispectrum is obtained the phase spectrum was reconstructed using the recursive technique [31] for all examples shown here. An image can be obtained by applying the inverse Fourier transform operation to $\tilde{O}_n(\vec{f})$. However, there can still be high frequency noise effects present in image space, particularly if a relatively small value of α was chosen.

One way to mitigate this effect is to apply an additional filter to $\tilde{O}_n(\vec{f})$ before the inverse Fourier transform operation. One useful filter is a conical filter defined by

$$H(\vec{f}) = \begin{cases} 1 - \frac{|\vec{f}|}{f_c} & |\vec{f}| \le f_c \\ 0 & |\vec{f}| > f_c \end{cases}, \tag{4.130}$$

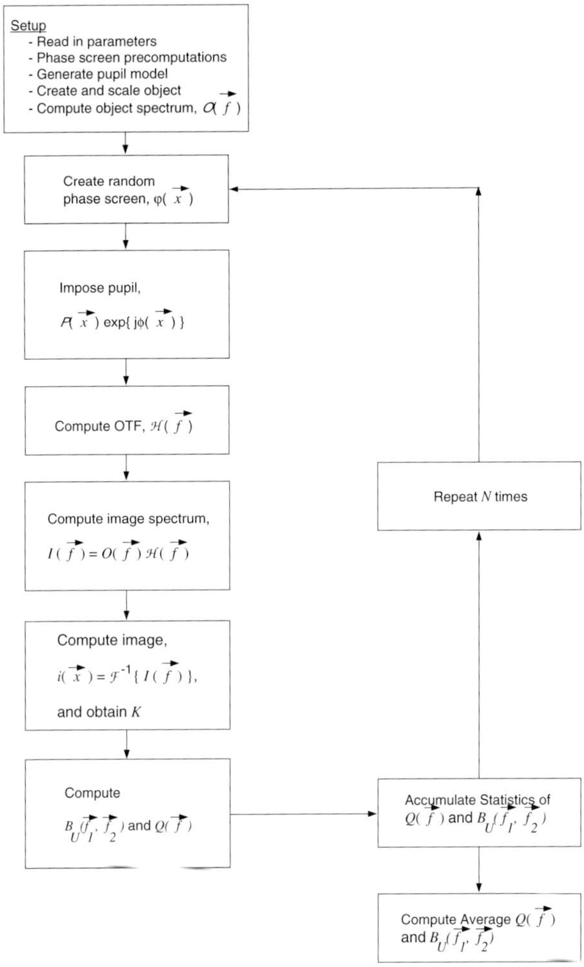

FIGURE 4.9
Block diagram for speckle imaging simulation.

where f_c is a user-selected cutoff frequency [35, 36]. Another useful filter is the tophat filter defined by

$$H(\vec{f}) = \begin{cases} 1 & |\vec{f}| \leq f_c \\ 0 & |\vec{f}| > f_c \end{cases}, \tag{4.131}$$

where f_c is a user-selected cutoff frequency.

The first example provided is the reconstructed image of the binary star used as an example in the discussion of simulating speckle interferometry, shown in Fig. 4.6. Recall that the components of the binary in this example are separated by 1.5 μrad while the diffraction-limited resolution of the telescope is approximately 0.5 μrad. In Fig. 4.6 the long exposure image, the unbiased speckle interferometry estimator, $Q(\vec{f})$, and the estimate of the modulus squared of the object spectrum are shown. In Fig. 4.10 the reconstructed image of this binary star is shown. For this example, the bispectrum was computed for offsets $|\Delta\vec{f}| \leq r_0/(\lambda d_i)$. The phase spectrum was reconstructed using the recursive technique of Northcott *et al.* [31]. For the image shown in Fig. 4.10 the value $\alpha = 0.001$ was used. Observe that in the long exposure image in Fig. 4.6(a) there was no indication that the object was a binary star, while in

FIGURE 4.10
Reconstructed image of binary star from the example in Fig. 4.6. A negative image is shown for clarity.

Fig. 4.10 the components of the binary star can be clearly resolved.

Several examples are now presented to illustrate the level of performance which can be provided by speckle imaging. All of the results were obtained using the simulation previously described. A one meter diameter telescope was modeled for these examples. Two objects were simulated: (1) a binary star with separation of 1.5 μrad, and brightness ratio of $d_1/d_2 = 2$; and (2) a computer rendering of a satellite object shown in Fig. 4.11. The Fried seeing parameter was set at two different levels, $r_0 = 0.1$ m and $r_0 = 0.05$ m. For the binary star example the average number of photoevents per image was set at three different levels using the visual magnitudes m_v = 4, 8, and 12. For the satellite object example the average number of photoevents per image was set using m_v = 2, 4, and 6. Exposure time for the image measurements was fixed at 10 ms, and the mean wavelength was chosen to be 500 nm with a bandwidth of 50 nm centered on the mean wavelength. These choices yield the following average numbers of photoevents per image for the binary star object: for m_v = 4, $\overline{K}$ = 45,177; for m_v = 4, $\overline{K}$ = 1,135; and for m_v = 12, $\overline{K}$ = 180. For the satellite object the following average numbers of photoevents per image resulted: for m_v = 2, $\overline{K}$ =280,034; for m_v = 4, $\overline{K}$ = 45,177; and for m_v = 6, $\overline{K}$ = 7,160. The satellite object was assumed to be at a range of 600 km and was assumed to be 10 m across. Two hundred frames were used to obtain all of the results shown.

Radially averaged plots of $\text{SNR}_Q(\vec{f})$ for the binary star are shown in Fig. 4.12. Figure 4.12(a) contains the results for $r_0 = 0.1$ m, and Fig. 4.12(b) contains the results for $r_0 = 0.05$ m. These plots were obtained by computing the sample-based estimates of the mean and the variance of $Q(\vec{f})$ for every spatial frequency inside the region of frequency space where the diffraction-limited OTF is non-zero. The values of the two dimensional function $Q(\vec{f})$ were then averaged along circles of constant radius to obtain the plots shown in Fig. 4.12, which are a function of only a single spatial frequency variable. Note that the plots shown in Fig. 4.12 *have not* been multiplied by $\sqrt{N}$. Also note that, consistent with the intuition obtained from Eq. (4.52), $\text{SNR}_Q(\vec{f})$ monotonically decreases with decreasing light level and decreasing r_0. The signal-to-noise ratio limited cutoff frequency, f_{SNR}, can be estimated from the plots in Fig. 4.12 by choosing the spatial frequency at which $\sqrt{N}\text{SNR}_Q(\vec{f})$ takes some arbitrary value, chosen here to be $\sqrt{N}\text{SNR}_Q(\vec{f}) = \sqrt{2}$. This choice for defining f_{SNR} allows f_{SNR} to be obtained directly from Fig. 4.12 as the point at which $\text{SNR}_Q(\vec{f}) = 0.1$.

The reconstructed images for the binary star are shown in Fig. 4.13. Six images are shown in Fig. 4.13. These images were reconstructed using the tophat filter of Eq. (4.131) with cutoff frequency given by $f_c = f_{\text{SNR}}$. For clarity, the central 100×100 pixel region of the images are shown in Fig. 4.13. The images in the left column are for $D/r_0 = 10$, and the images in the right column are for $D/r_0 = 20$. Reconstructed images for m_v = 4 are in the top row; reconstructed images for m_v = 8 are in the middle row; and reconstructed images for m_v = 12 are in the bottom row. Note that for the higher signal levels and better seeing conditions the binary star is easily resolved. However, as the overall $\text{SNR}_Q(\vec{f})$ becomes lower, and f_{SNR} decreases, two effects become apparent: (1) the resolution, as indicated by

FIGURE 4.11
Computer generated satellite model used to simulate speckle imaging of an extended object.

the width of one of the star images, becomes degraded; and (2) the images become noisier, as indicated by the energy in the "halo" surrounding the reconstructed stars. This linkage between resolution and signal-to-noise ratio applies in a very general sense to all image reconstruction techniques which are applied to noisy data.

Radially averaged plots of $\mathrm{SNR}_Q(\vec{f})$ for the satellite object are shown in Fig. 4.14. Figure 4.14(a) contains the results for $r_0 = 0.1$ m, and Fig. 4.14(b) contains the results for $r_0 = 0.05$ m. These plots were obtained using the same method as used to obtain Fig. 4.12, and the plots shown in Fig. 4.14 have not been multiplied by $\sqrt{N}$. Note that, as in the case of the binary star object, $\mathrm{SNR}_Q(\vec{f})$ monotonically decreases with decreasing light level and decreasing r_0. However, the signal-to-noise ratio limited cutoff frequencies, f_{SNR}, for these examples are much lower than for the equivalent binary star case. This reduced $\mathrm{SNR}_Q(\vec{f})$ results from the fact that, unlike the spectrum of the binary star, the spectrum of the satellite object rolls off to small values at low spatial frequencies, reducing $\mathrm{SNR}_Q(\vec{f})$, consistent with the result in Eq. (4.52). Thus, more frames of data are generally required to obtain a specified value of $\mathrm{SNR}_Q(\vec{f})$ at some spatial frequency for an extended object than for an object consisting of a single point or a few point-like objects. As an example of the data set size needed to reconstruct extended objects, in Ref. [29] several thousand frames of data were measured and processed to reconstruct images of the planets in this solar system.

The reconstructed images for the satellite object are shown in Fig. 4.15. Figure 4.15 is organized similar to Fig. 4.13. The images in the left column are for $D/r_0 = 10$, and the images in the right column are for $D/r_0 = 20$. Reconstructed images for $m_v = 2$ are in the top row; reconstructed images for $m_v = 4$ are in the middle row; and reconstructed images for $m_v = 6$ are in the bottom row. All of the images were reconstructed using the tophat filter of Eq. (4.131) with cutoff frequency given by $f_c = f_{\mathrm{SNR}}$. Note that as the light level decreases and the seeing parameter r_0 gets smaller the resolution, as indicated by the ability to resolve fine details on the object, becomes degraded. This fact is consistent with the observation made for the binary star example that as the overall $\mathrm{SNR}_Q(\vec{f})$ becomes lower, and

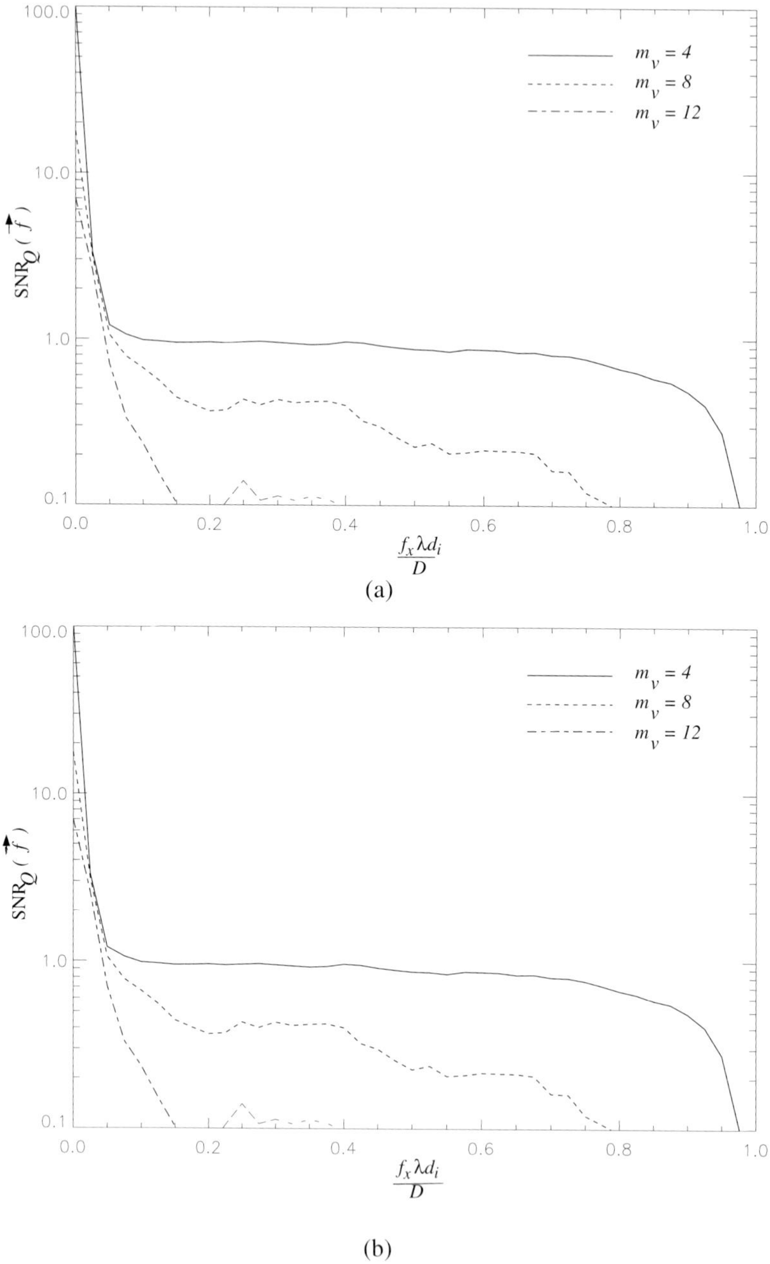

FIGURE 4.12
$\text{SNR}_Q(\vec{f})$ for binary star example: (a) r_0 = 0.1 m; (b) r_0 = 0.05 m.

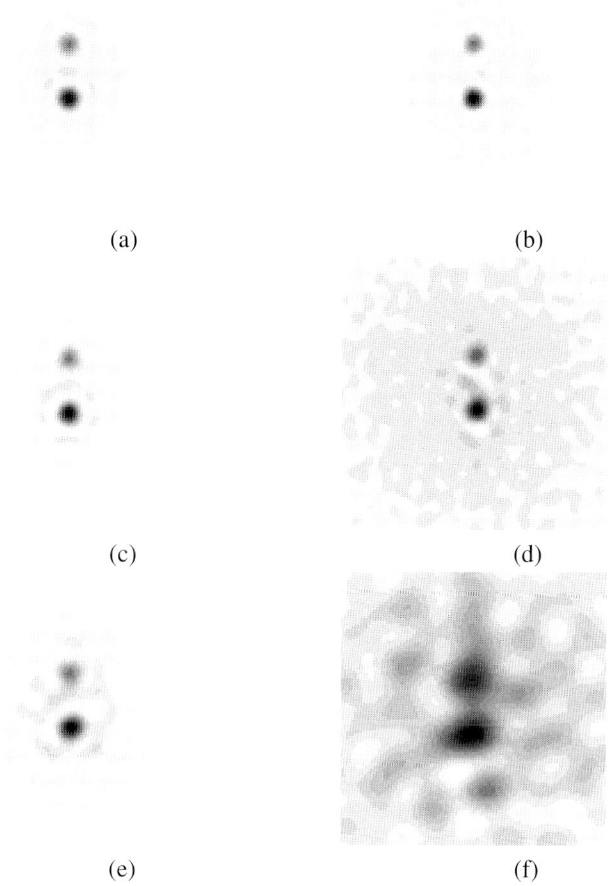

FIGURE 4.13
Reconstructed binary star images: (a) $D/r_0 = 10, m_v = 4$**; (b)** $D/r_0 = 20, m_v = 4$**; (c)** $D/r_0 = 10, m_v = 8$**; (d)** $D/r_0 = 20, m_v = 8$**; (e)** $D/r_0 = 10, m_v = 12$**; (f)** $D/r_0 = 20, m_v = 12$**. Angular separation of the binary components = 1.5** μ**rad, telescope diameter = 1.0 m, and ratio of brightnesses = 2:1. The central** 100×100 **pixels of** 256×256 **pixel images are presented. Negative images are displayed for clarity.**

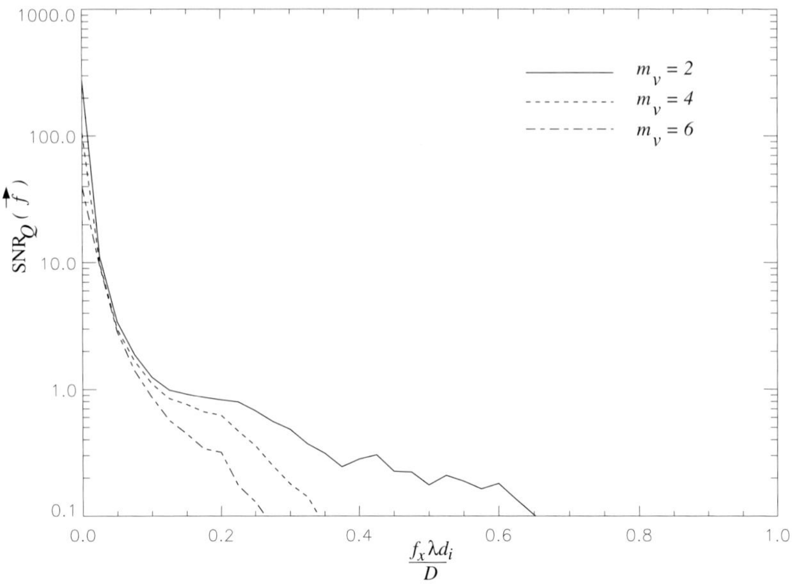

(a)

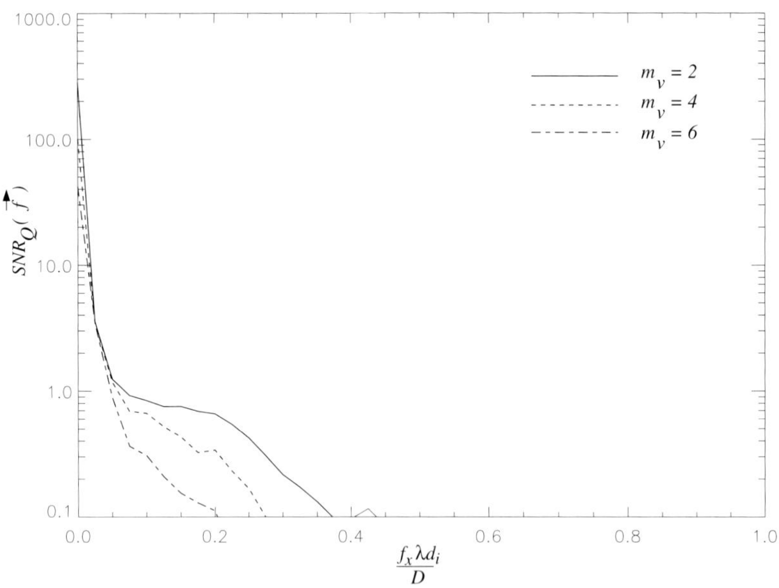

(b)

FIGURE 4.14
$\mathrm{SNR}_Q(\vec{f})$ for satellite example: (a) $r_0 = 0.1$ m; (b) $r_0 = 0.05$ m.

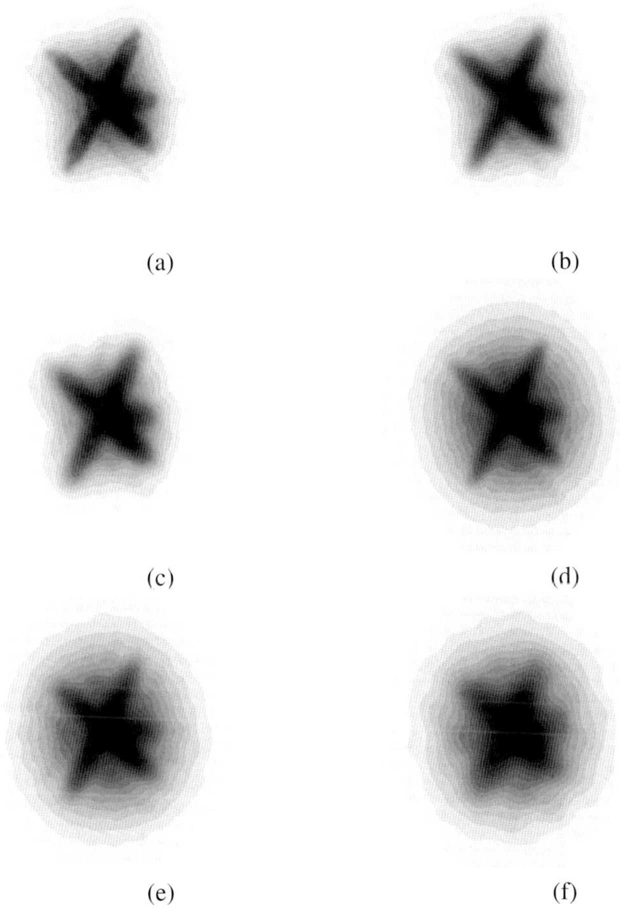

FIGURE 4.15
Reconstructed satellite images: (a) $D/r_0 = 10, m_v = 2$**; (b)** $D/r_0 = 20, m_v = 2$**; (c)** $D/r_0 = 10, m_v = 4$**; (d)** $D/r_0 = 20, m_v = 4$**; (e)** $D/r_0 = 20, m_v = 6$**; (f)** $D/r_0 = 20, m_v = 6$**. Negative images are displayed for clarity.**

f_{SNR} decreases, the resolution degrades, and the images become noisier.

4.6 Conclusion.

This chapter has provided a discussion of pure post-detection image processing techniques for overcoming the effects of atmospheric turbulence on astronomical imaging systems. As presented in technical discussion and illustrated visually in the examples, considerations of signal and noise control the quality of the reconstructed image and the resolution which can be obtained through the atmosphere using speckle imaging. While we have attempted to cover all of the relevant topics, it is impossible to give a fully comprehensive review of all of the relevant literature here. Detailed presentations on specific related topics are available in the literature. For more detailed discussions of the signal-to-noise ratio of $Q(\vec{f})$ the reader is referred to Refs. [11], [14], [15], and [22]. Phase spectrum reconstruction from the cross spectrum is discussed in Refs. [8], [13], and [27]. Phase spectrum reconstruction from the bispectrum is discussed in Refs. [9], [13], and [30]. A treatment of photon noise effects on the bispectrum is given in Refs. [13] and [34], and the effects of photon noise on the quality of the phase spectrum reconstructed from the bispectrum are discussed in Refs. [13], [14], [30], and [33]. Also, an interesting comparison of the cross spectrum and bispectrum techniques is provided in Ref. [13].

Bibliography

[1] I. Newton, "Optics," in *Great Books of the Western World* (R. M. Hutchins, ed.), vol. 34, Chicago: Encyclopedia of Brittanica, 1952.

[2] F. Roddier, "The effects of atmospheric turbulence in optical astronomy," in *Progress in Optics* (E. Wolf, ed.), vol. XIX, New York: North-Holland, 1981.

[3] V. I. Tatarskii and V. U. Zavorotny, "Atmospheric turbulence and the resolution limits of large ground-based telescopes: comment," *J. Opt. Soc. Am. Communications*, vol. 10, pp. 2410–2414, 1993.

[4] A. N. Kolmogorov, "The local structure of turbulence in incompressible viscous fluids for very large reynolds' numbers," in *Turbulence, Classic Papers on Statistical Theory* (S. K. Friedlander and L. Topper, eds.), pp. 151–155, New York: Wiley-Interscience, 1961.

[5] D. L. Fried, "Statistics of a geometric representation of wavefront distortion," *J. Opt. Soc. Am.*, vol. 55, pp. 1427–1435, 1965.

[6] D. L. Fried, "Optical resolution through a randomly inhomogeneous medium for very long and very short exposures," *J. Opt. Soc. Am.*, vol. 56, pp. 1372–1379, 1966.

[7] A. Labeyrie, "Attainment of diffraction limited resolution in large telescopes by fourier analyzing speckle patterns in star images," *Astron. Astrophys.*, vol. 6, p. 85, 1970.

[8] K. T. Knox and B. J. Thompson, "Recovery of images from atmospherically degraded short exposure images," *Astrophys. J.*, vol. 193, pp. L45–L48, 1974.

[9] A. W. Lohmann, G. Weigelt, and B. Wirnitzer, "Speckle masking in astronomy: triple correlation theory and applications," *Appl. Opt.*, vol. 22, pp. 4028–4037, 1983.

[10] D. Korff, "Analysis of a method for obtaining near-diffraction-limited information in the presence of atmospheric turbulence," *J. Opt. Soc. Am.*, vol. 63, pp. 971–980, 1973.

[11] J. W. Goodman, *Statistical Optics*. New York: John Wiley & Sons, 1985.

[12] W. I. Beavers, D. E. Dudgeon, J. W. Beletic, and M. T. Lane, "Speckle imaging through the atmosphere," *The Lincoln Laboratory Journal*, vol. 2, pp. 207–228, 1989.

[13] G. R. Ayers, M. J. Northcott, and J. C. Dainty, "Knox-thompson and triple correlation imaging through atmospheric turbulence," *J. Opt. Soc. Am. A*, vol. 5, pp. 963–985, 1988.

[14] M. C. Roggemann and C. L. Matson, "Power spectrum and fourier phase spectrum estimation by using fully and partially compensating adaptive optics and bispectrum postprocessing," *J. Opt. Soc. Am. A*, vol. 9, pp. 1525–1535, 1992.

[15] J. C. Dainty and A. H. Greenaway, "Estimation of spatial power spectra in speckle interferometry," *J. Opt. Soc. Am.*, vol. 5, pp. 786–790, 1979.

[16] P. S. Idell and A. Webster, "Resolution limits for coherent optical imaging: signal-to-noise analysis in the spatial-frequency domain," *J. Opt. Soc. Am. A*, vol. 9, pp. 43–56, 1992.

[17] M. C. Roggemann, "Limited degree-of-freedom adaptive optics and image reconstruction," *Appl. Opt.*, vol. 30, pp. 4227–4233, 1991.

[18] J. W. Goodman and J. F. Belsher, "Photon limitations in imaging and image restoration," contract, Rome Air Development Center, Griffiss AFB, NY, 1977.

[19] J. W. Beletic, "Deterministic photon bias in speckle imaging," *Opt. Commun.*, vol. 71, pp. 130–133, 1989.

[20] A. Papoulis, *Probability, Random Variables, and Stochastic Processes*. New York: McGraw-Hill, third ed., 1991.

[21] T. W. Lawrence, D. M. Goodman, E. M. Johansson, and J. P. Fitch, "Speckle imaging of satellites

at the u.s. air force maui optical station," *Appl. Opt.*, vol. 31, pp. 6307–6321, 1992.

[22] M. G. Miller, "Noise considerations in stellar speckle interferometry," *J. Opt. Soc. Am.*, vol. 67, pp. 1176–1184, 1977.

[23] J. W. Goodman, *Introduction to Fourier Optics*. New York: McGraw-Hill Book Co., 1968.

[24] R. C. Gonzalez and R. E. Woods, *Digital Image Processing*. Reading, Massachusetts: Addison-Wesley, 1993.

[25] W. Press, B. Flannery, S. Teukolsky, and W. Vetterling, *Numerical Recipes - The Art of Scientific Computing*. Cambridge, UK: Cambridge University Press, 1986.

[26] D. Goldsmith and N. Cohen, *Mysteries of the Milky Way*. Chicago: Contemporary Books Co., 1991.

[27] K. T. Knox, "Image retrieval from astronomical speckle patterns," *J. Opt. Soc. Am.*, vol. 66, pp. 1236–1239, 1976.

[28] H. T. Yura and M. T. Tavis, "Centroid anisoplanatism," *J. Opt. Soc. Am. A*, vol. 2, pp. 765–773, 1985.

[29] J. W. Beletic and R. M. Goody, "Recovery of planetary images by speckle imaging," *Appl. Opt.*, vol. 31, pp. 6909–6921, 1992.

[30] C. L. Matson, "Weighted-least-squares phase reconstruction from the bispectrum," *J. Opt. Soc. Am. A*, vol. 8, pp. 1905–1913, 1991.

[31] M. J. Northcott, G. R. Ayers, and J. C. Dainty, "Algorithms for image reconstruction from photon-limited data using the triple correlation," *J. Opt. Soc. Am. A*, vol. 5, pp. 986–992, 1988.

[32] J. Meng, G. J. M. Aitken, E. K. Hege, and J. S. Morgan, "Triple-correlation subplane reconstruction of photon-address images," *J. Opt. Soc. Am. A*, vol. 7, pp. 1243–1250, 1990.

[33] C. L. Matson, I. A. DeLarue, T. M. Gray, and I. E. Drunzer, "Optimal fourier spectrum estimation from the bispectrum," *Comp. & Elec. Eng.*, vol. 18, pp. 485–497, 1992.

[34] T. Nakajima, "Signal-to-noise ratio of the bispectral analysis of speckle interferometry," *J. Opt. Soc. Am. A*, vol. 5, pp. 1477–1491, 1988.

[35] M. C. Roggemann, E. L. Caudill, D. W. Tyler, M. J. Fox, M. A. Von Bokern, and C. L. Matson, "Compensated speckle imaging: theory and experimental results," *Appl. Opt.*, vol. 33, pp. 3099–3110, 1994.

[36] M. C. Roggemann, D. W. Tyler, and M. F. Bilmont, "Linear reconstruction of compensated images: theory and experimental results," *Appl. Opt.*, vol. 31, pp. 7429–7441, 1992.

5

Adaptive Optical Imaging Systems

5.1 Introduction

It is clear from the discussion presented in Chapters 1 and 3 that the wave front field aberrations induced by atmospheric turbulence can severely degrade the performance of an optical imaging system. This degradation can be visualized as a de-phasing of the fringes that make up the intensity distribution in the image. The fringes no longer add up in an in-phase manner and, as a result, the image is blurred. Atmospheric turbulence so limits ground-based imaging that the best observatories in the world only achieve resolutions associated with diffraction limited apertures of 10 to 30 cm. Babcock [1] was the first to propose that these de-phased fringes could be put back into their proper relationship by flattening the perturbed wave front in the pupil of the imaging system. Babcock's idea is the foundation for all adaptive optics used to compensate for atmosphere turbulence. Systems using adaptive optics are commonly called adaptive optical telescopes or adaptive optical imaging (AOI) systems. AOI systems must basically perform two main functions: 1) sense the wave front deformations and 2) compensate in real time for the deformation.

The components required to perform the two functions of an AOI system are shown in Fig. 5.1. These include the deformable mirror (DM), the wave front sensor (WFS), and an actuator command computer. Aberrated light entering the telescope is first reflected from the deformable mirror. Some of this light is focused to form an image, and some is diverted to the WFS. The entire process, from the acquisition of a WFS measurement to the positioning of the surface of the DM, must be performed at speeds commensurate with the rate of change of the turbulence-induced wave front phase errors. The goal in the positioning of the DM surface is to present an approximation of the conjugate of the turbulence-induced field so that the field reflected from the DM will more closely approximate the field that would exist in the absence of atmospheric turbulence.

The DM is a flexible mirror whose surface shape can be modified at high speed in response to applied electrical signals. The DM surface is controlled in such a way as to cancel the atmospheric induced wave front phase aberration in the pupil. The mirror surface or figure is controlled by a number of actuators that push and pull on the mirror surface to cause it to deform in a manner that approximates the turbulence induced aberration. The WFS measures the wave front phase deformations; however it does not do so directly. Rather, wave front sensors are generally sensitive to the wave front phase gradient. The wave front phase is obtained by using the measured wave front gradient in a phase reconstruction algorithm. The DM and WFS are connected through the actuator command computer. The purpose of the control computer is to take measurements from the WFS and map them into real time control signals for the DM. The command computer basically strives to achieve wave front reconstruction by controlling the surface of the DM.

If the adaptive optical concepts discussed above worked perfectly, the AOI system would achieve the ideal performance limit: diffraction limited imaging. In this case the imaging system would operate as if in an aberration free environment and only the diffraction effects discussed in Chapter 2 would limit performance. In principle, if an AOI system achieves the full potential of diffraction limited performance

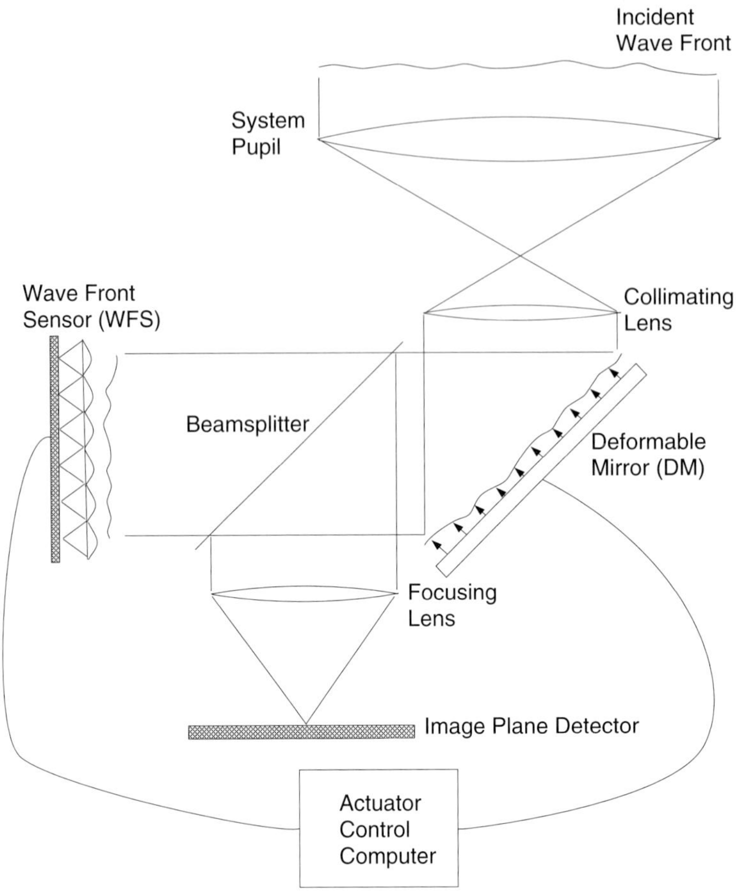

FIGURE 5.1
Simplified optical configuration of an AOI system.

or even nearly diffraction limited performance, it will provide imaging performance superior to the post-processing techniques discussed in Chapter 4. The dynamic range and sensitivity of the imaging system are dramatically improved for an AOI system as compared to post-processing. Unfortunately theoretical and experimental [2] evidence have shown that true diffraction limited performance is impossible to achieve. The actual performance of an AOI system may encompass the entire range from nearly no improvement to performance having characteristics of nearly diffraction limited imagery. Evaluating the expected performance of AOI systems has been the subject of a large amount of research over the past 20 years [3]–[46]. All of this past research has given us a rich set of tools to perform almost any performance or design analysis imaginable. It is our hope that the material presented here will be directly transferable to the adaptive optical systems that are of interest to the reader. In the sections to follow we present a generalized framework for analyzing the performance of any AOI system. In addition to presenting the analysis tools and examples of their use, we present a number of previously calculated results that are significant and general enough to be of interest to a wide range of readers.

Before jumping into the detailed discussions of the components and performance analysis of AOI systems we first present in Section 5.2 a qualitative discussion of the factors that fundamentally limit AOI system performance. In Section 5.3 we present a brief introduction to the components that make up the typical AOI system. This introduction includes discussions of the practical limitations of these components and also introduces the system models that will be used in the performance modeling of Section 5.4. These system models are defined in a generalized manner such that nearly any AOI system can be modeled. AOI system performance modeling is presented in Section 5.4. This section begins with a discussion of the metrics that are used to measure performance. The required statistical description of the residual wave front phase is then presented. Section 5.4 ends with a discussion of a number of fundamental and system performance results using the models presented in this chapter. The chapter is concluded in Section 5.5.

5.2 Factors that degrade AOI system performance

An ideal AOI system is one that perfectly senses and compensates for the atmospherically induced wave front deformations. A large number of factors can be attributed to the limitations that keep AOI systems from achieving this ideal performance. To discuss all of them is beyond the scope of what can be reasonably presented here. The factors that are included in our discussion are those that may well represent the most fundamental and limiting factors in any AOI system. These fundamental factors include

- finite light levels in the WFS,
- differences between the sensed and object wave fronts associated with anisoplanatism,
- finite spatial sampling of the WFS,
- finite number of degrees-of-freedom available from the DM,
- and the finite temporal response of the AOI system.

5.2.1 WFS light levels and anisoplanatism

The finite amount of light in the WFS and anisoplanatism may well represent the most severe of all factors that degrade AOI system performance. These two factors are discussed together because they are inextricably tied to each other in the case of telescopes used for astronomical purposes. In most of the adaptive optics systems built to date light from either the object being imaged or a nearby "beacon" source is used to provide light for the WFS. In either case the finite light level available to the wavefront sensor can severely limit the accuracy of the wave front sensing. Inaccurate wave front sensing leads to inaccurate wave front reconstruction which degrades AOI system performance significantly below the theoretical limit. The problem of obtaining enough light for accurate wave front sensing has been the most significant limiting factor in the application of adaptive optics technology to ground-based astronomy. Over the past twenty years adaptive optics have generally not been applied to ground-based astronomical telescopes because of the inability to obtain bright beacon sources over any significant portion of the night sky. The problem lies in the fact that in most instances the object being imaged is itself not bright enough to be used by the WFS. As a result, a nearby beacon source must be found to provide the light for the WFS. In the case of astronomy this nearby beacon is generally a nearby bright star, sometimes referred to as a guide star. If finding the closest bright beacon was the only problem to overcome then adaptive optics would have been successfully used in astronomy years ago. Simply finding the nearest bright beacon star is not adequate unless it lies within the isoplanatic angle of the object being imaged.

The isoplanatic angle is qualitatively the maximum angular separation between the object and the beacon such that the turbulence induced wave front deformations for the object and beacon wave fronts are still reasonably similar. Later in the chapter we quantify exactly what "reasonably similar" means. In order to understand the implication of the isoplanatic angle, recall that the light from the beacon is used to sense the perturbed wave front and this information is in turn used to compensate for the

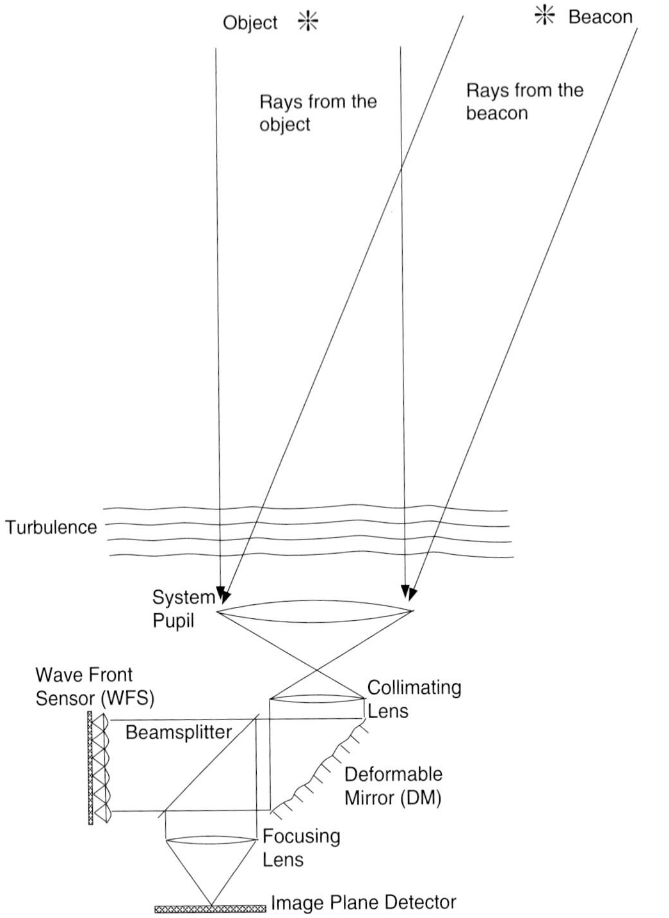

FIGURE 5.2
Angular anisoplanatism.

distortions in the object wave front. If the beacon is separated from the object, the two wave fronts propagate through slightly different regions in the atmosphere, as shown in Fig. 5.2. This difference in propagation path results in slightly different wave front deformations induced on the two wave fronts. This effect is referred to as an anisoplanatic effect. Besides finite light levels in the WFS, anisoplanatism represents the other major factor that causes the AOI system to perform in a less than ideal way. In fact, as seen in the discussion above, light levels in the WFS and anisoplanatism are inextricably tied together since it is anisoplanatism that prevents the AOI system from using nearby bright guide stars.

The isoplanatic angle has been defined quantitatively as the largest angle separation between the object and beacon such that some measure of performance is kept within a specified level. The isoplanatic angle is very small for astronomical observations in the visible wavelength range. At visible wavelengths the object being imaged has to be within 5 to 10 μrad of the beacon. This restriction on the angular separation of the object and beacon implies that the percentage of the night sky accessible using the naturally occurring, bright, natural guide stars is essentially zero [17]. There simply are not enough bright stars to view a significant portion of the night sky using an adaptive optical telescope. The outlook is more optimistic at infrared wavelengths where the isoplanatic angle is considerably larger (100's of μrad). This trend with respect to wavelength is understood when considering the isoplanatic angle defined by Fried [10]:

$$\theta_0 = 58.1 \times 10^{-3} \lambda^{6/5} \left[\int_0^L \mathrm{d}z C_n^2(z) z^{5/3} \right]^{-3/5}, \tag{5.1}$$

where $C_n^2(z)$ is the structure constant of the turbulence induced index of refraction fluctuations, L is the path length through turbulence, and λ is the optical wave length. In this definition the isoplanatic angle has a wavelength dependence that is identical to the dependence found for the Fried parameter, r_0. The isoplanatic angle and r_0 are proportional to $\lambda^{6/5}$ (see Eq. (3.61)). In applications involving long wave lengths (such as infrared (IR) astronomy) r_0 increases from the small value it takes at visible wavelengths. This increase in r_0 in turn allows the WFS subapertures to be larger than those required for visible wavelengths. Dimmer beacons can be used as the WFS subapertures become larger. This decreasing brightness requirement in combination with an increased isoplanatic angle size gives rise to a situation in which IR astronomy using natural stars is possible. Studies of the applications of adaptive optics to infrared astronomy have shown that there are enough bright guide stars to access nearly 100% of the night sky. The fact that the first applications of adaptive optics to astronomy have been for infrared systems is a testament to this fact [47]. The isoplanatic angle defined above is derived for near field atmospheric conditions. The effects of anisoplanatism have also been investigated for far field turbulence effects using a more rigorous diffraction analysis [48, 49].

A possible solution to the problems associated with natural guide stars is the creation of an artificial beacon using a laser to project a spot in the atmosphere above the telescope [17]. In 1985 Foy and Labeyrie [50] were the first to suggest, in the open literature, that it may be possible to create a suitable artificial guide star in the upper atmosphere with a laser. Within the United States' Defense Advanced Research Projects Agency the idea was being considered as early as 1982 [51]. Artificial guide stars allow access to nearly the entire sky even at visible wavelengths since the laser can be pointed in any direction. The transmitted laser beam is focused to a spot in the center of the field of view of the telescope at a finite altitude above the telescope. Within the region where the beam is focused some of the light is scattered from the atmospheric constituents, and some of this scattered light is returned in the direction of the telescope. The backscattered light is used by the WFS to measure the turbulence-induced aberrations. The laser spot is referred to as a laser beacon or artificial guide star. Ideally the use of laser beacons removes the fundamental problem of low light levels in the WFS since the brightness of the beacon is controlled by the laser power.

The problem associated with the angular anisoplanatic effects discussed above is seemingly solved with the use of laser beacons. The laser beacon can be placed in exactly the same angular direction as the object, thereby forcing the beacon and object wave fronts to arise from the same angular point in the sky. Another anisoplanatic effect arises with the use of laser beacons that is slightly different in nature from the angular anisoplanatism discussed above. With the use of laser beacons the AOI system still suffers from a fundamental anisoplanatic effect known as focus anisoplanatism. This type of anisoplanatism is qualitatively the result of the same phenomenon as discussed above. In the case of a laser beacon, anisoplanatic effects arise from the fact that the light emitted from the laser beacon is a spherical wave emitted from a finite altitude source. Figure 5.3 shows the geometry of interest. Since the object wave front generally arises from a source a very large distance away (as is the case for celestial bodies) it is well modeled as a plane wave. The difference in plane and spherical wave propagation results in an anisoplanatic effect that is again due to the two wave fronts propagating through different regions of the turbulence as shown in Fig. 5.3. This type of anisoplanatism is referred to as focus anisoplanatism. So even with the use of a bright laser beacon the AOI system will still be fundamentally limited by the effects of focus anisoplanatism. The effects of focus anisoplanatism can be lessened by placing the laser beacon at the highest altitude possible. The two main approaches for the generation of laser beacons involve the use of Rayleigh scattering in the stratosphere and the use of resonance fluorescence of sodium (Na) atoms in the mesosphere [17]. Beacons employing Rayleigh scattering are called Rayleigh beacons, while guide stars employing Na resonance fluorescence are called Na beacons.

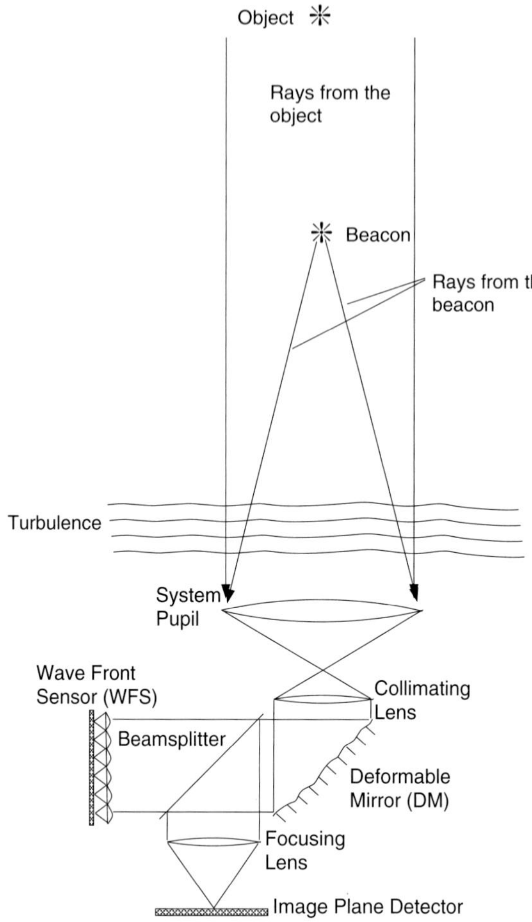

FIGURE 5.3
Focus anisoplanatism.

The advantage of using a Rayleigh beacon is that the lasers can be purchased today with many of the required characteristics. In particular the power required to create sufficiently bright Rayleigh guide stars is available with "off-the-shelf" lasers [17, 47]. Rayleigh guide stars are most efficiently created at relatively low altitudes, ranging from 10 to 20 km. Na beacons, on the other hand, are formed in a naturally occurring layer of Na in the mesosphere. The Na layer is located at an altitude of approximately 90 km [52]. Since Na beacons are created at a much higher altitude than Rayleigh beacons, the effects of focus anisoplanatism are much less severe for Na beacons than for Rayleigh beacons. The effect of focus anisoplanatism is significant enough that Na laser beacons are the preferred approach assuming availability of a laser to produce them. However, to date only expensive, special purpose lasers have succeeded in creating sufficiently bright Na beacons [53, 54].

In addition to focus anisoplanatism, an AOI system employing an artificial beacon also suffers from a form of angular anisoplanatism. This fact may seem counterintuitive since the whole point of creating a laser beacon is that it can be placed at the same angular position as the object thereby avoiding the angular anisoplanatism discussed earlier. Unfortunately, though, in the upward projection of the laser beacon the exact position of the focused spot can not be determined. The laser beacon optical field propagates up through turbulence before being focused into a spot. The upward propagation through the turbulence causes the angular position of the beacon to shift randomly. The random motion denies the AOI system exact knowledge of the position of the beacon. Without this knowledge the AOI system cannot reliably use the laser beacon to predict the wave front tilt. Wave front tilt refers to the portion of the wave front deformation that corresponds to the best fit plane over the aperture of the system. The Zernike polynomials corresponding to wave front tilt for a circular aperture are the 2nd and 3rd modes listed in Table 3.2. The wave front tilt induced on the beacon field is caused by both the turbulence for the downward propagation and the motion of the beacon (i.e., the turbulence in the upward propagation). Since the beacon motion is random and unknown it is impossible to measure the portion of the wave front tilt induced by the downward propagation through the atmosphere. The insensitivity to wave front tilt for an artificial beacon implies the AOI system must employ another beacon with a fixed location in the sky (i.e., a natural star) for sensing and compensation of overall wave front tilt.

This last statement may make it appear that we are right back where we started when we introduced the use of laser beacons to solve the angular anisoplanatism problem. Fortunately the natural star is only needed for sensing the portion of the perturbed wave front associated with wave front tilt. The isoplanatic angle for tilt sensing is much larger than the isoplanatic angle associated with the higher order deformations [17]. In other words the angular separation of the natural beacon for tilt sensing and the object can be much larger. This angular limit is called the tilt isoplanatic angle. Since the tilt isoplanatic angle is larger than the isoplanatic angle associated with the higher order deformations, obtaining natural stars for tilt sensing is not nearly as a severe a limitation as obtaining natural stars for use in compensating the higher order wave front deformations. The bottom line is that tilt anisoplanatism represents another fundamental limitation in the performance of AOI systems as long as the natural guide star used for tilt sensing is separated from the angular direction of the object. The effects of tilt anisoplanatism can be decreased by decreasing the angular separation of the natural guide star and the direction to the object.

Finally, it is also worth noting that the performance of the overall tilt correction has a profound effect on the performance of the system [23]. From Fig. 3.13 and Table 3.4 in Chapter 3 it is apparent that 87% of the power in the aperture averaged mean square wave front fluctuations is due to the full aperture wave front tilt. Since such a large portion of the wave front aberrations is associated with wave front tilt, tilt correction is critical for effective AOI system performance. Fig. 3.13 shows that the higher order Zernike aberrations have nearly an order of magnitude less power than the combined tilt terms. Any correction of these higher order aberrations will be swamped by uncorrected tilt if the system suffers from poor tilt correction. The brightness and angular separation of the tilt guide star contributes significantly to the limiting performance of the AOI system.

5.2.2 Finite pupil sampling and system temporal response

The latter three performance limitations presented in the list on page 171 are grouped together since they arise from fundamental limitations of the components of the AOI system itself. Both the finite spatial sampling of the WFS and the finite degrees-of-freedom of the DM result in the same general type of system performance degradation. The finite sampling and the finite degrees-of-freedom can both be viewed as degrading the system's ability to sense and compensate for the high spatial frequencies of the wave front deformations. The inherent finite sampling of real world wave front sensors makes them unsuitable for sensing deformations having spatial frequencies beyond some upper frequency cutoff. The cutoff frequency of the WFS is inversely proportional to the WFS sample spacing. Aliasing of high spatial frequencies phase deformations into lower frequencies is a fundamental problem encountered with the finite sampling. The sampling interval of a WFS is defined by how the system pupil is segmented into subapertures. Each WFS subaperture represents a sample of the wave front deformation. Assuming the subapertures fill the entire pupil, it is the dimension of the single subaperture, d, that defines the sampling interval of the WFS. The finite number of degrees-of-freedom of the DM acts like a spatial filter that limits the spatial frequency content of the reconstructed wave front phase. Like the WFS, the cutoff frequency of the DM is inversely proportional to the spacing of the actuators that deform the surface of the mirror.

The performance degradation imposed by the DM and WFS can be lessened by using wave front sensors with smaller sample spacing and deformable mirrors with more actuators (i.e., more closely space actuators). The general rule of thumb for "good" wave front compensation is to design the AOI system with subaperture and actuator spacings equal to r_0. An AOI system satisfying this condition is generally referred to as a fully compensated system. Recall that r_0 is the Fried coherence cell size defined in Chapter 3. To roughly quantify what "good" wave front compensation means we can refer to Table 3.4. Table 3.4 quantifies the mean square pupil averaged residual wave front phase after compensation of succeeding higher order Zernike modes. The residual phase fluctuation given on the third line of the table corresponds to wave front compensation of modes 1, 2, and 3. The first three modes are the piston, and the x and y directed tilt over an aperture of diameter of D. If the AOI system is expected to at least correct for the tilt over an area corresponding to the WFS subapertures and all other lower order deformations, then we can use Table 3.4 to roughly estimate the level of wave front phase fluctuation after wave front compensation. For this calculation the parameter D in Table 3.4 is equated to the subaperture diameter, d. From the table the residual wave front fluctuation is expected to be $0.134(d/r_0)^{5/3}$. Equating d to r_0 for "good" compensation results in a residual wave front phase fluctuation of 0.134 rad^2. This mean square residual phase fluctuation corresponds to a RMS fluctuation of 1/17 waves.

Improving upon the performance limitations imposed by the WFS subaperture size, the DM inter-actuator spacing involves using smaller subapertures and smaller inter-actuator spacings. This approach is generally expensive in both hardware complexity and cost. The controller in an AOI system must map many WFS measurements to many DM actuator control signals. Decreasing the subaperture size and actuator spacing increases the complexity and speed requirement of the system controller. Recall the AOI system must sense and compensate for the wave front deformations at speeds commensurate with the coherence time of the atmosphere. At good observatory sites the coherence time will be on the order of 10 milliseconds [17]. The large number of subapertures and DM actuators for densely sampled pupils may well push the temporal bandwidth limits of a multiple input/output controller. Adaptive optical designs for 2 meter class telescopes and larger have generally avoided designs involving subaperture sizes and inter-actuator spacings of r_0. In general large adaptive telescopes have been designed with subaperture sizes and inter-actuator spacings larger than r_0. The motivation for adaptive optical systems designed for less than full compensation is primarily a cost and complexity issue. As an example, consider the adaptive optical system required for full compensation for a 3.5 meter telescope. At a good observatory site the value of r_0 might have an average value of 10 cm [17]. For full compensation the subaperture sizes and inter-actuator spacings are equated to r_0 and over 900 subapertures and actuators will be required.

Adaptive optical systems having subaperture sizes and inter-actuator spacings larger than r_0 are generally referred to as partially compensated systems. These systems have the advantage that they are less expensive and less complex than fully compensated systems, but they suffer in performance due to the sampling issues discussed above. However it has been shown recently [21] that partially compensated AOI systems do not suffer as much in a signal-to-noise ratio (SNR) sense as might be expected. Even though average performance measures, such as the mean square residual phase fluctuation, degrade for systems employing partial compensation it has been shown that the image spectrum SNR does not degrade as rapidly [21]. This observation has opened up possibilities of using the post processing image reconstruction techniques discussed in Chapter 6 to achieve close to the same image qualities as a fully compensated AOI system.

Another possible reason for employing a partially compensated AOI system is the effect of finite light levels on the WFS. For a fixed photon flux level there will be an optimal WFS subaperture size that minimizes the residual wave front phase fluctuations [19]. Depending on the light level expected for the WFS this optimum subaperture size may result in a partially compensated system. The optimum subaperture size exists because of the competing trends in the two main contributions to the wave front compensation errors. As the WFS subaperture size decreases the errors associated with finite sampling decrease. On the other hand, the decreasing size causes the subaperture to intercept less light, and as a result the slope measurements become noisier, and in turn the residual wave front error increases.

The temporal evolution of the turbulence induced wave front deformations and the AOI system's finite temporal response, fundamentally limits performance. In simple terms, the AOI system takes a finite amount of time from the time of wave front sensing to the time of wave front correction. This time delay causes a degradation in performance since there will exist differences between the sensed wave front and the wave front arising from the turbulence. The atmospheric correlation time, τ_c, is generally used to characterize the rate at which the wave front deformations are changing. The inverse of the τ_c is commonly used to specify the required AOI system temporal bandwidth. The temporal bandwidth requirement and light levels in the WFS are inextricably tied together from a overall performance perspective. The temporal bandwidth required to keep up with the temporally evolving wave front deformations control the rate at which WFS measurements must be taken. A WFS measurement rate implies a maximum integration time for each WFS measurement. For a fixed photon flux the integration time controls the average number of photo-events detected per measurement cycle. Just as we saw for the subaperture size, we see that competing trends will result in an optimum control bandwidth. Increasing system bandwidth will decrease the wave front errors associated with the temporally evolving wave fronts. At the same time, though, an increasing bandwidth decreases the number of photo-events detected per WFS measurement. This decreased WFS signal level results in noisier wave front slope measurements for non-ideal detectors, and in turn increased residual wave front errors.

5.2.3 Summary

As seen from the discussion above there are a large number of factors that degrade AOI system performance. Many previous analyses of these effects have considered each of these factors separately and summed their effects to get an overall system performance level. The following references provide excellent introductions to this type of performance analysis [24, 25, 55, 56, 57]. This analysis approach has the main advantage of being conceptually simple and easy to implement. The major drawback is the fact that for this simple analysis to be accurate the effects of the separate factors must be independent of each other. The factors are not independent of each other, and as a result the simple approach of adding supposedly independent performance degradation factors can result in misleading performance predictions. In fact the relationship between the various factors is quite complex and extremely difficult to summarize with simple analytical expressions. Hence, there is ample motivation for the development of an analysis framework that accurately incorporates all of the factors discussed above while at the same time is straightforward and general enough to be applied to a wide range of AOI system configurations. In subsequent sections of this chapter the goal is to present such an analysis framework.

5.3 Adaptive optical system components and models

In order to present a generalized framework for analyzing the performance of AOI systems we first present an introductory discussion of the three main components of a system: the deformable mirror (DM), the wave front sensor (WFS), and the controller. The purpose of this introduction is to provide an overall view of the performance issues associated with the components and an appreciation for the complexity of the accurate system modeling required for overall system performance predictions. These models are subsequently used in the mathematical framework presented in Section 5.4 to predict the performance of AOI systems. This section is not intended to be a comprehensive review of the detailed technologies of these three components. Others have presented this type of comprehensive review and references for these publications are cited throughout the following subsections.

5.3.1 Deformable mirrors

The deformable mirror (DM) is the system component that performs the wave front correction. In principle this component can be any device that is capable of altering the shape of the perturbed wave front. We focus on reflective devices and refer to these devices as deformable mirrors. We also note that transmissive devices such as liquid crystals and non linear optical elements can also be used for wave front control [58, 59]. The shape of the DM surface (or figure) is controlled such that deformations of the incident wavefront are cancelled (or nearly so) as the optical field reflects from the surface. If $\exp[j\phi(\vec{x},t)]$ represents the perturbed optical field at the surface of the DM just before reflection, the wave front deformation just after reflection is represented mathematically by

$$\begin{aligned}\exp[j\epsilon(\vec{x},t)] &= \exp[j\phi(\vec{x},t)]\exp[-j\hat{\phi}(\vec{x},t)], \\ &= \exp\left[j\left(\phi(\vec{x},t)-\hat{\phi}(\vec{x},t)\right)\right], \end{aligned} \tag{5.2}$$

where $\vec{x}$ is a spatial position in the plane containing the DM, t is time, $\phi(\vec{x},t)$ is the piston removed wave front phase before reflection, $\hat{\phi}(\vec{x},t)$ is the piston removed figure of the DM, and $\epsilon(\vec{x},t)$ is the residual wave front phase after reflection. The reason for considering piston removed phases in Eq. (5.2) is that any residual phase due to a non-zero aperture averaged phase (i.e., piston) has no effect on the performance of an imaging system. Since the piston has no effect on imaging system performance the wave front phase quantities considered throughout this chapter will be piston removed quantities. We also note that by considering piston removed phase we avoid the mathematical problems associated with the infinite variance of the piston term for atmospheric turbulence described by the Kolmogorov power spectrum (see Section 3.6). Following the notation established by Eq. (3.156), the piston removed phase, $\phi(\vec{x},t)$, is expressed as

$$\phi(\vec{x},t) = \psi(\vec{x},t) - \int d\vec{x}' W(\vec{x}')\psi(\vec{x}',t), \tag{5.3}$$

where $\psi(\vec{x},t)$ is the incident wave front phase and $W(\vec{x})$ is the pupil weighting function. Note that $W(\vec{x})$ is normalized to have a unit volume as specified by Eq. (3.139).

In an ideal AOI system the figure of the DM, $\hat{\phi}(\vec{x},t)$, and the perturbed wave front, $\phi(\vec{x},t)$, exactly cancel and the residual wave front phase $\epsilon(\vec{x},t) = 0$. In Section 5.2 we discussed the factors that prevent AOI systems from achieving ideal performance. A number of these factors are associated with the DM. In particular, the performance of the DM is limited by the finite degrees of freedom associated with controlling the figure of the mirror and the finite temporal response. The finite degrees of freedom restrict the range of surface shapes that can be placed on the DM. The degrees of freedom are closely related to the number of actuators used to control the surface of the DM. The deformable mirror's surface is controlled by actuators that push, pull and/or bend the surface. As the number of actuators increases, the range of shapes that can be placed on the DM surface increases and as a result the degrees of freedom increase. Deformable mirrors with increasing numbers of actuators are generally increasingly capable of matching $\hat{\phi}(\vec{x},t)$ to $\phi(\vec{x},t)$. The trade-off though is the increased complexity of the DM and the resulting increased through-put required of the DM controller.

The temporal response of the DM characterizes how the DM surface responds temporally to an instantaneous command. Both the speed of the response (i.e., how fast the surface moves to the desired shape) and resonances are of interest. Fortunately the state of current DM design is such that these temporal concerns are not major considerations for imaging through the atmosphere. Deformable mirrors are being designed to have bandwidths on the order of a kilohertz or greater and resonances well outside the operation range of interest [26, 60]. This bandwidth easily meets the requirements associated with the temporal rates required for atmospheric compensation: several 100 Hz [26]. For more comprehensive discussions of the characteristics and limitations of the various types of deformable mirrors see references [26], [55] and [61].

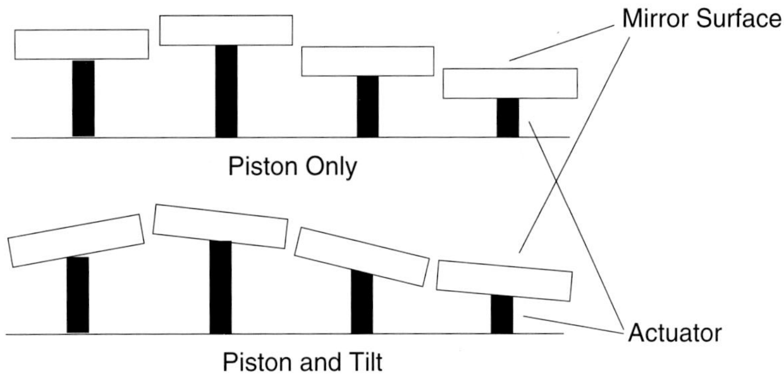

FIGURE 5.4
Cross-section of two types of segmented deformable mirrors.

DM types: segmented and continuous

There are two broad classes of deformable mirrors. These two classes are categorized by the nature of the mirror surface: segmented or continuous. The surface of a segmented DM is controlled by positioning a number of independently controlled, closely packed, flat mirror segments. Figure 5.4 illustrates a cross section of the two types of the segmented deformable mirrors. The individual segments of the DM can either be piston only or piston plus tilt elements. The piston only element is a mirror segment that can only be moved perpendicular to the mirror plane. The piston plus tilt mirror has the added degree of freedom for tilt control in two orthogonal directions. The advantages associated with segmented deformable mirrors are the modularity and simplicity of the design. Modularity is associated with the ability to combine rectangular arrays of mirror segments to form larger deformable mirrors. The simplicity is associated with designs that allow individual segments to be easily replaced. Another advantage associated with segmented mirrors is the conceptually simple control requirements [61]. Since the segmented elements operate completely independently of each other a complex interaction matrix describing inter-actuator coupling is not required for optimal control. Disadvantages include diffraction effects from the individual segment edges and the difficulty of achieving inter-segment alignment. An example of a AOI system using a segmented mirror is described in reference [62]. In this work a 19 element segmented mirror is used in an AOI system for solar observations.

The disadvantages associated with segmented deformable mirrors are largely addressed by continuous deformable mirrors. The continuous DM has a single continuous mirror surface. The mirror surface is usually called the face plate or face sheet. Figure 5.5 shows a cross section of a representative continuous face plate DM. The shape of the mirror surface is controlled by an array of actuators that push, pull and bend the face sheet to obtain the desired figure. Continuous deformable mirrors provide the advantage of avoiding the diffraction effects associated with segmented deformable mirrors. A major disadvantage of continuous mirrors is the complicated algorithm required to control continuous deformable mirrors. The face sheet deformation associated with a particular actuator is not generally independent of the other actuators responses. Even if the actuator responses are independent the face sheet deformation associated with a particular actuator generally overlaps the response of neighboring actuators. The coupling and overlap of the responses of each actuator give rise to a complicated coupling matrix that makes control more difficult than for the segmented mirror. Another disadvantage of the continuous DM is the practical problem of repairing bad actuators [61]. Examples of continuous deformable mirror designs are discussed in references [26], [60], [61], [63], [64] and [65].

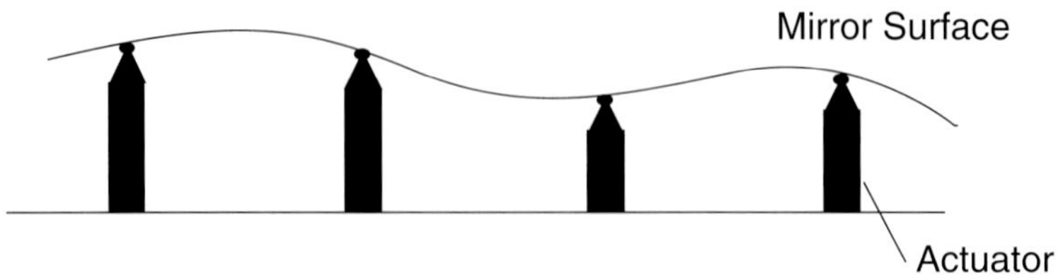

FIGURE 5.5
Cross-section of a continuous DM.

Influence functions and DM modeling

In order to model the DM in the performance analysis presented in Section 5.4 it is necessary to quantify how the signal applied to a particular actuator changes the surface of the DM. In the model presented below we ignore the temporal response of the DM by assuming that it responds instantaneously. Generally the temporal bandwidth of currently available deformable mirrors is high enough to justify this assumption [26, 60]. The change in the surface shape due to the application of a signal to an actuator is called the influence function. The ith influence function describes how the surface of the DM is deformed with an unit signal applied to the ith actuator. The influence function for a segmented DM is localized to the region of the surface defined by a particular mirror segment. The influence function for the continuous DM is not restricted to a region around the actuator location as in the segmented DM. In the continuous DM case influence functions have been modeled with polynomials, trigonometric functions and Gaussian functions [6, 11, 15, 26, 66]. Let the ith influence function be designated $r_i(\vec{x})$ where $\vec{x}$ is a position on the DM surface. By assuming linearity of the responses of all the actuators we can model the surface of the DM with the following relationship

$$\hat{\phi}(\vec{x}, t) = \sum_i c_i(t) r_i(\vec{x}), \tag{5.4}$$

where $c_i(t)$ is the control signal applied to the ith actuator at time t. The key assumption affecting the validity of the model presented by Eq. (5.4) is the linearity of the actuator influence functions. Two factors that prevent linearity from being assumed is the dependence of the influence functions on neighboring actuator responses and hysteresis. For the performance analysis presented in Section 5.4 we assume the DM model represented by Eq. (5.4) is accurate.

Tilt compensation mirrors and models

Before moving to the WFS discussion we finally note that the first types of deformable mirrors used to compensate for the effects of atmospheric turbulence were tilt mirrors. These mirrors are sometimes called tip-tilt mirrors, and represent the simplest type of wave front compensation. For tilt compensation the surface of the DM is simply a flat mirror that can be adjusted to tilt in two orthogonal directions. Early adaptive optics research concentrated on tilt correction systems since the random tilt induced by the atmosphere contains 87% of the power of the wave front phase fluctuations (see Table 3.4). Modern AOI systems typically have a tilt correction mirror as the first stage of wave front compensation. Performing tilt correction separately from the higher order corrections lowers the dynamic range requirements of the DM. The "dynamic range" refers here to the magnitude of the DM surface deformations. Performance issues associated with tilt correction are the speed requirement of the mirror, temporal resonances and the accuracy of the tilt control.

The bandwidth requirement for tilt correction for a circular aperture can be calculated from the temporal correlation function of the Zernike polynomial coefficients, $a_2(t)$ and $a_3(t)$. The Zernike polynomials are discussed in Section 3.6.1 and the polynomial coefficients $a_2(t)$ and $a_3(t)$ are associated with the x and y-directed tilts over the aperture (see Table 3.2). The computation of the correlation function is a straightforward application of Eq. (3.142) generalized to account for temporal effects by using the "frozen flow" concepts presented in Section 3.4.3. Tyler [27] has derived the necessary correlation function and the associated temporal power spectrum. Assuming the use of a first order controller for tilt compensation he finds that the residual angular tilt variance after tilt compensation can be expressed as

$$\sigma_\theta^2 = \left(\frac{f_T}{f_{3\mathrm{dB}}}\right)^2 \left(\frac{\lambda}{D}\right)^2, \tag{5.5}$$

where $f_{3\mathrm{dB}}$ is the temporal frequency at which the tilt compensator response is down 3 dB from the maximum value, λ is the optical wavelength, D is the diameter of the system pupil and

$$f_T = 0.368 D^{-1/6} \lambda^{-1} \left[\int \mathrm{d}z C_n^2(z) v^2(z)\right]^{1/2}. \tag{5.6}$$

The parameter $f_{3\mathrm{dB}}$ is also referred to as the bandwidth of the tilt compensator. In Eq. (5.6) $C_n^2(z)$ is the structure constant of the index-of-refraction fluctuations, z is distance along the optical path from the aperture, and $v(z)$ is the path dependent turbulence velocity profile. In the special case that the velocity profile is constant with respect to z, Eq. (5.6) can be written in terms of the Fried parameter given in Eq. (3.61):

$$f_T = 0.0902 \left(\frac{r_0}{D}\right)^{1/6} \left(\frac{v}{r_0}\right). \tag{5.7}$$

Good tilt correction should result in an angular RMS tilt variation that is 1/4th the angular resolution of the diffraction limited pupil, λ/D [27]. To obtain this level of tilt correction the bandwidth of the controller, $f_{3\mathrm{dB}}$, must be greater than $4 f_T$. As an example, consider an aperture diameter of $D = 20 r_0$ and a wind velocity of $v = 100 r_0$/s. For these assumptions we find the controller bandwidth must be 22 Hz.

The mathematical model for wave front tilt correction is expressed as

$$\phi_{tr}(\vec{x}, t) = \phi(\vec{x}, t) - \vec{a} \cdot \vec{x}, \tag{5.8}$$

where $\phi_{tr}(\vec{x}, t)$ designates the tilt-removed phase and $\vec{a}$ is the vector slope. The vector slope $\vec{a}$ is the slope of a tilted plane fitted to $\phi(\vec{x}, t)$. The value of $\vec{a}$ can be found by performing a least squares fit of a plane to $\phi(\vec{x}, t)$ over the pupil. For a circular aperture the x and y-directed components of $\vec{a}$ correspond to one half the Zernike coefficients a_2 and a_3, respectively (see Section 3.6.1). For an arbitrarily defined aperture the vector slope $\vec{a}$ is given by Eq. (3.127), which, for convenience, is repeated here:

$$\vec{a} = \hat{x} K_x \int \mathrm{d}\vec{x}\, W(\vec{x}) \left[\vec{x} \cdot \hat{x}\right] \psi(\vec{x}) + \hat{y} K_y \int \mathrm{d}\vec{x}\, W(\vec{x}) \left[\vec{x} \cdot \hat{y}\right] \psi(\vec{x}), \tag{5.9}$$

where $\hat{x}$ and $\hat{y}$ are x- and y-directed unit vectors in the pupil plane, $W(\vec{x})$ is the pupil aperture function, and K_x and K_y are normalization factors required to satisfy the equality in Eq. (5.9) if $\psi(\vec{x})$ is replaced by $\vec{a} \cdot \vec{x}$. These normalization factors are given by

$$\begin{aligned} K_x &= \left[\int \mathrm{d}\vec{x}\, W(\vec{x}) (\vec{x} \cdot \hat{x})^2\right]^{-1}, \\ K_y &= \left[\int \mathrm{d}\vec{x}\, W(\vec{x}) (\vec{x} \cdot \hat{y})^2\right]^{-1}. \end{aligned} \tag{5.10}$$

Summary

The main contribution of this section are the DM models presented in Eqs. (5.4) and (5.8). The mathematical models only characterize the spatial response of the DM surface for steady state actuator commands. In Section 5.4 the DM models are incorporated into a generalized analysis framework for predicting the performance of AOI systems.

5.3.2 Wavefront sensors

The wave front sensor (WFS) component of an AOI system provides the means of measuring the perturbed wave front phase which is in turn used to control the DM to achieve wave front compensation. Wave front sensors used in AOI systems do not directly measure the wave front phase, $\phi(\vec{x}, t)$. Instead the wave front sensors measure the spatial gradient or Laplacian of $\phi(\vec{x}, t)$. The spatial gradient is commonly referred to as the wave front slope while the Laplacian is referred to as the wave front curvature. An estimate of $\phi(\vec{x}, t)$ is computed from wave front derivative information using a phase reconstruction algorithm. The controller in an AOI system performs this phase reconstruction by mapping the wave front derivative information from the WFS to the control signals, $c_j(t)$, for the DM. The various algorithms that can be used to perform the phase reconstruction are discussed in Section 5.3.3.

One of the fundamental performance issues associated with the fidelity of phase reconstruction is the accuracy of the measured derivatives given finite light levels in the WFS and detector noise. The other fundamental issue affecting phase reconstruction is the wave front sample spacing. We introduce the basic types of wave front sensors used in AOI systems. We also introduce mathematical models for two of the most common types of wave front sensors used in adaptive optics and present expressions that quantify the effects of photon and detector noise. These models will be used in the performance analysis presented in Section 5.4.

WFS types and models

The two main types of wave front sensors used in AOI systems are derived from wave front sensors used in optical testing [67]. Both of these wave front sensors are sensitive to the slope of the wave front phase. They both have the advantage of being broad band, or white light sensors; an important consideration for astronomical imaging.

Hartmann WFS: The first sensor we discuss is based on the Hartmann test. A diagram of the Hartmann WFS (H–WFS) is depicted in the left side of Fig. 5.6. The H–WFS spatially segments the incident wave front with a lenslet array. Each array element is referred to as a subaperture and it focuses a spot onto an array of detectors in the lenslet focal plane. The right portion of Figure 5.6 shows that the location of the spot focused on the detector, $\vec{x}_s$, is related to the wavefront slope $\vec{s}$ over the subaperture by

$$\vec{s} = k\frac{|\vec{x}_s|}{f_l}, \tag{5.11}$$

where $k = 2\pi/\lambda$ and f_l is the focal length of the subaperture lens. The spot location, $\vec{x}_s$, is estimated by calculating the centroid of the spot on the detector plane. One of the most common detector array configurations is 4 detectors forming the quadrants of a square detector area. This type of detector is called a quad cell. The main drawbacks associated with quad cells are the limited dynamic range of the tilt measurement and the effect of the spot size on the algorithm that computes the spot centroid. Hartmann wave front sensors with larger arrays of detector elements (4×4 to 8×8 detector array sizes) circumvent these problems but at the cost of increasing electronic noise effects. A number of

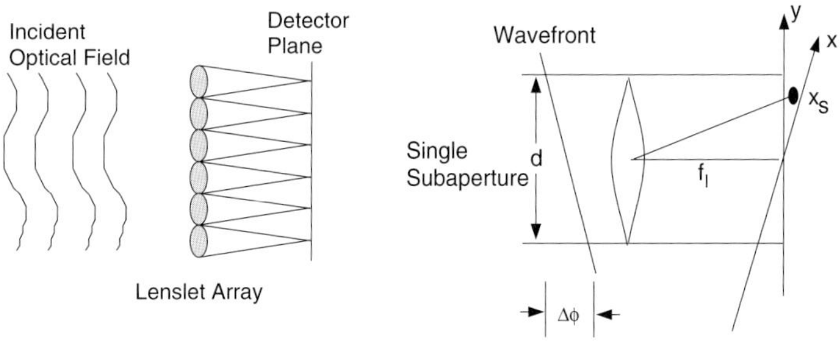

FIGURE 5.6
Diagram of the Hartmann Wavefront Sensor (H–WFS).

authors have investigated the performance of different detector array configurations as well as computing optimal detector element sizes [68, 69, 70, 71, 72].

H–WFS Model: The spot centroid computation associated with the H–WFS and its relationship to the wave front slope via Eq. (5.11) can be shown to be equivalent to computing the average wave front phase gradient over the WFS subaperture [73]:

$$\vec{s}(j) = \int d\vec{x}\, W_{sj}(\vec{x}) \nabla \phi(\vec{x}, t), \tag{5.12}$$

where $\vec{s}(j)$ is the wave front slope associated with the jth subaperture, $\nabla\phi(\vec{x}, t)$ is the spatial gradient of $\phi(\vec{x}, t)$ and $W_{sj}(\vec{x})$ is the jth subaperture weighting function. Note that the subaperture function is normalized such that

$$\int d\vec{x}\, W_{sj}(\vec{x}) = 1. \tag{5.13}$$

The H–WFS measurement model is obtained by adding a term to Eq. (5.12) to account for the measurement noise:

$$\begin{aligned} \vec{s}_m(j) &= \vec{s}(j) + \vec{s}_n(j), \\ &= \int d\vec{x}\, W_{sj}(\vec{x}) \nabla\phi(\vec{x}, t) + \vec{s}_n(j) \quad \text{(H–WFS Model)}, \end{aligned} \tag{5.14}$$

where $\vec{s}_m(j)$ is the measured slope for the jth subaperture and $\vec{s}_n(j)$ is a random vector that accounts for both photon and read noise. The phase $\phi(\vec{x}, t)$ and the noise $\vec{s}_n(j)$ are assumed independent. In the next subsection the variance of $\vec{s}_n(j)$ is quantified as a function of the light level in the WFS, the read noise levels in the detector and atmospheric effects. How the variance of $\vec{s}_n(j)$ depends on the size of the beacon source is also investigated. We find it useful to write the H–WFS model given by Eq. (5.14) in terms of $\phi(\vec{x}, t)$ rather than $\nabla\phi(\vec{x}, t)$ by integrating the first term by parts [11]:

$$\vec{s}_m(j) = -\int d\vec{x}\, \nabla W_{sj}(\vec{x}) \phi(\vec{x}, t) + \vec{s}_n(j) \quad \text{(H–WFS Model)}, \tag{5.15}$$

where $\nabla W_{sj}(\vec{x}, t)$ is the gradient of the jth subaperture aperture function.

Shearing Interferometer WFS: The other common WFS used in adaptive optical imaging is the lateral shearing interferometer (SI–WFS) [74]. A sketch of the SI–WFS is shown in Figure 5.7. Our discussion of the SI–WFS is more involved than the H–WFS discussion due to the less intuitive operation of this sensor. The SI–WFS splits the incident field into two optical beams: an x-leg and y-leg beam. The most common approach used to achieve the shear involves the use of gratings [75, 76]. Another approach makes use of beam separation by polarization [77]. Regardless of the beam separation approach, the

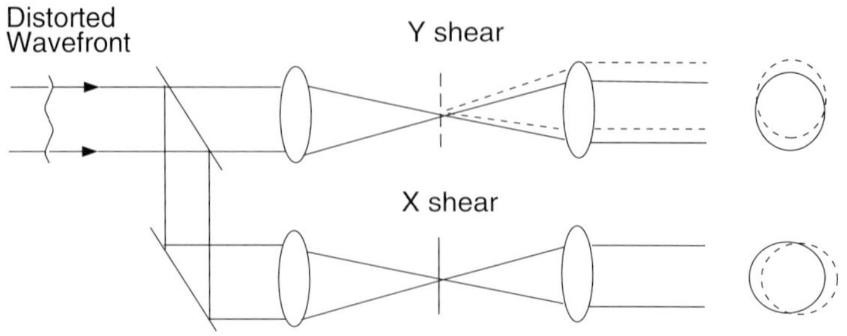

FIGURE 5.7
Simple optical diagram of the shearing interferometer WFS [26].

x-and y-directed wavefront slopes are sensed separately in each leg. Depending on the specific type of SI–WFS, the optical field in each leg is split a number of times [78]. Before optical detection the beams are split one last time and laterally shifted (sheared) with respect to each other. The amount of shear is an adjustable parameter. The sheared beams are superimposed on an optical detector to form an interference pattern. For the H–WFS the lenslet array establishes the subapertures. In the case of the SI–WFS, the subapertures are established by the pixels on the detector. The detector plane is conjugated to the pupil plane of the optical system and each detector element defines a single subaperture.

The particular algorithm used to transform the pixel irradiance measurements into wavefront slope estimates depends on the type of SI–WFS. For the AC SI–WFS a time varying phase modulation is applied to one of the sheared beams [74, 76]. The detected signal in this case is a sinusoidal time varying signal. The phase of the sinusoidal signal is related to the slope of the incident wavefront. In the case of the static 3- and 4-bin (sometimes referred to as the 3 and 4 point) shearing interferometer wave front sensors, the detected irradiances are constant signals that can be combined together to estimate the phase slope [78].

To understand the relationship between these detected intensities and the wavefront phase slope, consider a beacon giving rise to an incident field $u_b(\vec{x})e^{j\phi(\vec{x})}$. The quantity $u_b(\vec{x})$ is the complex optical field that would exist in the pupil in the absence of atmospheric turbulence and $\phi(\vec{x})$ is the phase deformation caused by propagation through the atmosphere. Now consider detecting the superposition of laterally sheared versions of $u_b(\vec{x})e^{j\phi(\vec{x})}$ as described above. The detected irradiance over the pixel defined by the jth subaperture, $W_{sj}(\vec{x})$, is given by

$$i_d(j) = \int d^2\vec{x}\, W_{sj}(\vec{x}) \left\langle \left| u_b\left(\vec{x} + \frac{\Delta\vec{x}}{2}\right) \exp\left\{ j\phi\left(\vec{x} + \frac{\Delta\vec{x}}{2}\right)\right\} + u_b\left(\vec{x} - \frac{\Delta\vec{x}}{2}\right) \exp\left\{ j\phi\left(\vec{x} - \frac{\Delta\vec{x}}{2}\right)\right\}\right|^2 \right\rangle, \tag{5.16}$$

where $\Delta\vec{x}$ is the lateral shear. Expanding Eq. (5.16) gives

$$i_d(j) = \int d^2\vec{x}\, W_{sj}(\vec{x}) \Bigg\{ \left\langle \left| u_b\left(\vec{x} + \frac{\Delta\vec{x}}{2}\right)\right|^2 \right\rangle + \left\langle \left| u_b\left(\vec{x} - \frac{\Delta\vec{x}}{2}\right)\right|^2 \right\rangle + 2\left\langle u_b\left(\vec{x} + \frac{\Delta\vec{x}}{2}\right) u_b^*\left(\vec{x} - \frac{\Delta\vec{x}}{2}\right)\right\rangle \cos\left(\phi\left(\vec{x} + \frac{\Delta\vec{x}}{2}\right) - \phi\left(\vec{x} - \frac{\Delta\vec{x}}{2}\right)\right)\Bigg\}. \tag{5.17}$$

To simplify the notation we designate the phase difference

$$\phi\left(\vec{x}-\frac{\Delta\vec{x}}{2}\right)-\phi\left(\vec{x}+\frac{\Delta\vec{x}}{2}\right)=\Delta\phi(\vec{x},\Delta\vec{x}), \tag{5.18}$$

and the mutual intensity of the beacon field as

$$\mathrm{E}\left\{u_b(\vec{x})u_b^*(\vec{x}')\right\}=i_b\mu_b(\vec{x}-\vec{x}'), \tag{5.19}$$

where μ_b is the normalized mutual intensity, or visibility of the beacon field, and i_b is the irradiance of each of the sheared beams. The Van Cittert-Zernike theorem given in Eq. (2.71) has been used to justify the second order stationary properties of $u_b(\vec{x})$ assumed in writing Eq. (5.19). The subscript b in Eq. (5.19) indicates that the quantities depend on the properties of the beacon. Using Eqs. (5.18) and (5.19), the detector irradiance given by Eq. (5.17) now reduces to the following simplified form:

$$i_d(j)=2i_b\int \mathrm{d}^2\vec{x}\,W_{sj}(\vec{x})\left[1+\mu_b(\Delta\vec{x})\cos\left(\Delta\phi(\vec{x},\Delta\vec{x})\right)\right]. \tag{5.20}$$

The key realization to make from Eq. (5.20) is that the phase difference $\Delta\phi(\vec{x},\Delta\vec{x})$ over the jth subaperture is encoded in the measured irradiance $i_d(j)$.

This phase difference, $\Delta\phi(\vec{x},\Delta\vec{x})$, is decoded from the irradiance measurements either by processing the time varying signal of the AC SI–WFS or processing the multiple, static measurements of the 3- and 4-bin SI–WFS measurements [74, 78]. For the AC SI–WFS the argument of the cosine term has a additional, linear time varying term:

$$i_d(j,t)=2i_b\int \mathrm{d}^2\vec{x}\,W_{sj}(\vec{x})\left[1+\mu_b(\Delta\vec{x})\cos\left(\omega t+\Delta\phi(\vec{x},\Delta\vec{x})\right)\right], \tag{5.21}$$

where ω is the angular frequency of the time varying phase delay induced in one of the two sheared beams. In this case the phase difference $\Delta\phi$ is sensed by detecting the phase of the sinusoidally varying signal $i_d(j,t)$. For the 3- and 4-bin SI-WFS, fixed delays are induced in one of the sheared beams before detection. In these cases, the argument of the cosine contains an additional term equal to the phase delay. For the ith phase delay we have

$$i_d(j,i)=2i_b\int \mathrm{d}^2\vec{x}\,W_{sj}(\vec{x})\left[1+\mu_b(\Delta\vec{x})\cos\left(\Phi_i+\Delta\phi(\vec{x},\Delta\vec{x})\right)\right], \tag{5.22}$$

where Φ_i takes on values of 0, $2\pi/3$ and $4\pi/3$ radians for the 3-bin SI–WFS and values of 0, $\pi/2$, π and $3\pi/2$ radians for the 4-bin SI–WFS. For the 3-bin SI–WFS the phase difference $\Delta\phi$ is estimated from $i_d(j,1)$, $i_d(j,2)$ and $i_d(j,3)$ using [78]

$$\Delta\phi(\vec{x},\Delta\vec{x})=-\tan^{-1}\left[\frac{\sqrt{3}(i_d(j,3)-i_d(j,2))}{2i_d(j,1)-i_d(j,2)-i_d(j,3)}\right]. \tag{5.23}$$

For the 4-bin SI–WFS the phase difference $\Delta\phi$ is estimated from $i_d(j,1)$, $i_d(j,2)$, $i_d(j,3)$ and $i_d(j,4)$ using [78]

$$\Delta\phi(\vec{x},\Delta\vec{x})=-\tan^{-1}\left[\frac{i_d(j,2)-i_d(j,4)}{i_d(j,3)-i_d(j,1)}\right]. \tag{5.24}$$

Equations (5.23) and (5.24) can be shown to give an exact relationship by substituting in Eq. (5.22) and assuming that $\Delta\phi(\vec{x},\Delta\vec{x})$ is constant within the subaperture.

SI–WFS Model: Regardless of the specific type of SI-WFS discussed above, for modeling purposes the measured slope signal for jth subaperture is given by

$$\vec{s}_m(j)=\int \mathrm{d}\vec{x}\,W_{sj}(\vec{x})\left[\frac{\Delta\phi(\vec{x},\Delta\vec{x}_x)}{|\Delta\vec{x}_x|}\hat{x}+\frac{\Delta\phi(\vec{x},\Delta\vec{x}_y)}{|\Delta\vec{x}_y|}\hat{y}\right]+\vec{s}_n(j) \quad \text{(SI–WFS Model)}, \tag{5.25}$$

where $\hat{x}$ and $\hat{y}$ are unit vectors in orthogonal directions in the sensing plane, $\Delta\vec{x}_x$ and $\Delta\vec{x}_y$ are shear vectors in the $\hat{x}$ and $\hat{y}$ directions and $\vec{s}_n(j)$ again represents the measurement noise. The first term in Eq. (5.25) is the error free portion of the measurement. The variance of the measurement error term, $\vec{s}_n$, depends on light level, beacon size, atmospheric effects and shear distances, $\Delta\vec{x}_x$ and $\Delta\vec{x}_y$. In the next subsection we quantify how the measurement error depends on these parameters.

Curvature Sensing: The other type of WFS that has found use in AOI systems is the curvature sensor [67, 79]. The curvature sensor measures a signal proportional to the Laplacian or second derivative of the wave front phase. The Laplacian measurements are combined with wave front tilt measurements around the edge of the system pupil to reconstruct the wave front by solving the Poisson equation. One of the advantages of using a curvature sensor is that solving the Poisson equation can be accomplished directly by applying the Laplacian measurements to a membrane or bimorph mirror [65, 67, 79]. Since curvature sensors have not found wide spread use in AOI systems we only introduce the reader to the curvature sensing concept and refer the more interested reader to the references cited above. In particular Rousset [67] provides an excellent comparison of the curvature sensor to the slope sensors discussed above. We do not present a curvature sensor model or any performance computations for an AOI system using a curvature sensor. These types of performance calculations are presented in references [80], [81], and [82].

Measurement errors

Having established how the two slope sensors measure the wavefront slope, we turn our attention to quantifying the variance of the noise term $\vec{s}_n$ in Eqs. (5.15) and (5.25). The variance of $\vec{s}_n$ is designated σ_n^2. Since $\vec{s}_n$ is a 2-dimensional vector, the variance σ_n^2 is the sum of the variance of the two orthogonal components of $\vec{s}_n$.

H–WFS: The root mean square error of the slope for a H–WFS is related to the standard deviation of the spot location measurement, σ_c. Both shot noise and read noise contribute to the spot measurement error. In the case of a H–WFS using a quad cell, the contribution to σ_n due to shot or photon noise has been found to be [68, 83]

$$\sigma_{ns} = k\left(\frac{\sigma_c}{f_l}\right) = \frac{\sqrt{2}\pi}{d(\overline{K})^{1/2}\int_{-1}^{1} df_x I_b(f_x,0)\mathcal{H}_{tr}(f_x,0)} \text{ (radians/m)}, \tag{5.26}$$

where $k = 2\pi/\lambda$, d is the subaperture dimension, f_l is the subaperture lens focal length, and $\overline{K}$ is the total average detected photon count per subaperture measurement. The subscript ns designates the noise contribution due to shot noise. The function $I_b(f_x, f_y)$ is the Fourier transform of the beacon intensity distribution projected into the subaperture lens focal plane and $\mathcal{H}_{tr}(f_x, f_y)$ is the tilt-removed optical transfer function (OTF) of the subaperture. The spectrum $I_b(f_x, f_y)$ is normalized such that $I_b(0,0) = 1$. The tilt-removed OTF is the same as the short exposure OTF discussed in Section 3.5.4. In Eq. (5.26) the spatial frequency variables f_x and f_y are normalized to the diffraction limit of the subaperture, $d/\lambda f_l$. Let the beacon intensity distribution projected into the focal plane of the lenslet array be designated $i_b(\vec{x})$. This intensity distribution is related to $I_b(\vec{f})$ through the following Fourier transform relationship:

$$I_b(\vec{f}) = \frac{\mathcal{F}\left[i_b\left(\frac{\vec{\alpha}\lambda f_l}{d}\right)\right]}{\mathcal{F}\left[i_b\left(\frac{\vec{\alpha}\lambda f_l}{d}\right)\right]\Big|_{\vec{f}=0}} \tag{5.27}$$

where $\mathcal{F}$ is the 2-dimensional Fourier transform operator defined by Eq. (2.5) and $\vec{\alpha}$ is the focal plane coordinate normalized by $\lambda f_l/d$. Equation (5.26) is the basic shot noise result that defines the performance limit for the Hartmann WFS. It accounts for increases in the noise variance due to the spreading of the spot caused by atmospheric turbulence and the finite size of the beacon.

In the special case that the beacon intensity distribution is well modeled by a Gaussian like function we can write

$$i_b(\vec{x}) \propto \exp\left[-\frac{|\vec{x}|^2}{\left(\frac{f_l}{z_b}\sigma_b\right)^2}\right], \tag{5.28}$$

where $i_b(\vec{x})$ is the intensity projected onto the focal plane of the subaperture lenslet and σ_b is the RMS size of the beacon at the beacon altitude z_b. Substituting this intensity distribution into Eq. (5.27) gives the normalized beacon spectrum

$$I_b(\vec{f}) = \exp\left\{-\pi^2|\vec{f}|^2\beta^2\left(\frac{d}{r_0}\right)^2\right\}, \tag{5.29}$$

where β is the ratio of the angular size of the beacon to the angular seeing limit: $\beta = (\sigma_b/z_b)/(\lambda/r_o)$. Substituting Eq. (5.29) into Eq. (5.26) gives

$$\sigma_{ns} = k\left(\frac{\sigma_c}{f_l}\right) = \frac{\sqrt{2}\pi}{d(\overline{K})^{1/2}\int_{-1}^{1} df_x \exp\left\{-\pi^2|f_x|^2\beta^2\left(\frac{d}{r_0}\right)^2\right\}\mathcal{H}_{tr}(f_x,0)}\ (\text{radians/m}), \tag{5.30}$$

The dependency of Eq. (5.30) on a number of key factors is evident. The finite light level is characterized by the average photon count per measurement $\overline{K}$. The effect of the beacon size on the accuracy of the measurement is characterized by the parameter β (angular size of the source normalized by the diffraction limited angular resolution). Finally the broadening of the beacon spot in the lenslet focal plane due to atmosphere turbulence is characterized by the tilt-removed OTF, $\mathcal{H}_{tr}$. The slope error given in Eq. (5.30) cannot be reduced further without an expression for $\mathcal{H}_{tr}(\vec{f})$. Using the short exposure OTF given in Eq. (3.135) of Chapter 3 it is possible to numerically evaluate Eq. (5.30). Figure 5.8 is a plot of the σ_{ns} versus d/r_0 for β ranging from 0 to 2. The parameter d is the subaperture dimension. It is apparent from Fig. 5.8 that increasing the beacon size causes a significant worsening of the shot noise error. A decreasing value of r_0 (or equivalently an increasing value of the ratio d/r_0) is also seen to increase the measurement error.

A lower bound on the RMS noise expression given in Eq. (5.30) or Eq. (5.26) results when we consider the situation in which $\mathcal{H}_{l_i}(\vec{f})$ is equal to the diffraction limited OTF of the subaperture and the beacon is a point source. This case arises when $d \ll r_0$ and $\beta \ll 1$ (i.e., minimal atmospheric turbulence effects and the angular beacon size are small compared to λ/r_0). In this case Eq. (5.26) reduces to

$$\sigma_{ns} = \frac{\sqrt{2}\pi}{d(\overline{K})^{1/2}}\ (\text{radians/m}). \tag{5.31}$$

The RMS noise given in Eq. (5.30) or Eq. (5.26) accounts for the contributions to the measurement error due to shot or photon noise. The other main source of error is attributed to the noise added in the process of reading the detector. This noise is referred to as sensor noise or, in the special case of a charge-coupled device (CCD) detector, as read noise. In the case that $r_0 \geq d$ and $\beta \ll 1$ Parenti [24] has shown the contribution to σ_n due to read noise for a quad cell is

$$\sigma_{nr} = \frac{\sqrt{2}\pi}{d\sqrt{\frac{\overline{K}^2}{4\sigma_e^2}}}, \tag{5.32}$$

where σ_e^2 is the variance of the read or electronic noise for each pixel in the detector array expressed in units of photoelectrons squared (a 2×2 array for the quad cell). More general expressions for the

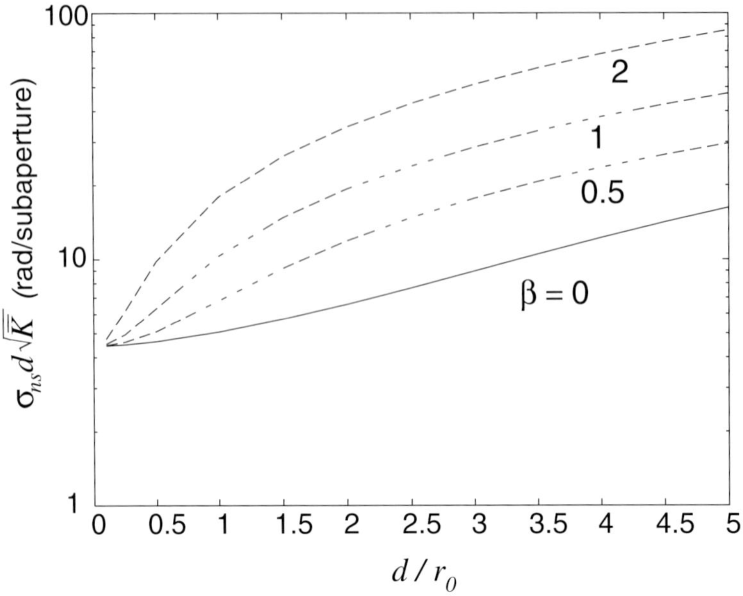

FIGURE 5.8
Plot of σ_{ns} versus d/r_0 for the H–WFS and β ranging from 0 to 2. The parameter d is the subaperture dimension, $\sqrt{K}$ is the average detected photon-count per measurement and β is the ratio of the angular beacon size to the angular seeing limit, $\beta = (\sigma_b/z_b)/(\lambda/r_0)$.

read noise contribution to σ_n have been developed by Parenti [24] for larger detector arrays and for spot broadening due to atmospheric turbulence.

In summary, to obtain the total slope measurement error, the shot noise and read noise errors must be combined to give the total measurement error σ_n^2:

$$\sigma_n^2 = \sigma_{ns}^2 + \sigma_{nr}^2. \tag{5.33}$$

SI–WFS: The measurement error due to shot noise for the SI–WFS has been found to be [83]

$$\sigma_{ns} = \frac{2\sqrt{2}}{r_0(\overline{K})^{1/2} s \mu_a(s)\mu_b(s)} \text{(radians/m)}, \tag{5.34}$$

where $\overline{K}$ is the total average photon count detected per subaperture measurement, $\mu_a(s)$ is the visibility reduction due to atmospheric seeing effects, $\mu_b(s)$ is the visibility reduction due to the size of the beacon, and s is the normalized shear, $s = |\Delta\vec{x}_x|/r_o$ or $|\Delta\vec{x}_y|/r_o$. The shot noise given by Eq. (5.34) is applicable for the SI–WFS that employs equal shears in the x and y directions: $|\Delta\vec{x}_x| = |\Delta\vec{x}_y|$. The atmospheric visibility factor $\mu_a(s)$ accounts for the reduction in the contrast of the detected irradiance caused by the variation of the wave front phase over the subaperture. This visibility factor has been computed in Ref. [83] and is given by

$$\mu_a(s) = \int \mathrm{d}^2\vec{x}\, W_s(\vec{x}) \exp\left\{-\frac{1}{2}\mathrm{E}\left\{\Delta\tilde{\phi}^2(\vec{x}, r_0 s)\right\}\right\}, \tag{5.35}$$

where

$$\Delta\tilde{\phi}(\vec{x}, \Delta\vec{x}) = \Delta\phi(\vec{x}, \Delta\vec{x}) - \int \mathrm{d}\vec{x}'\, W_s(\vec{x}')\Delta\phi(\vec{x}', \Delta\vec{x}), \tag{5.36}$$

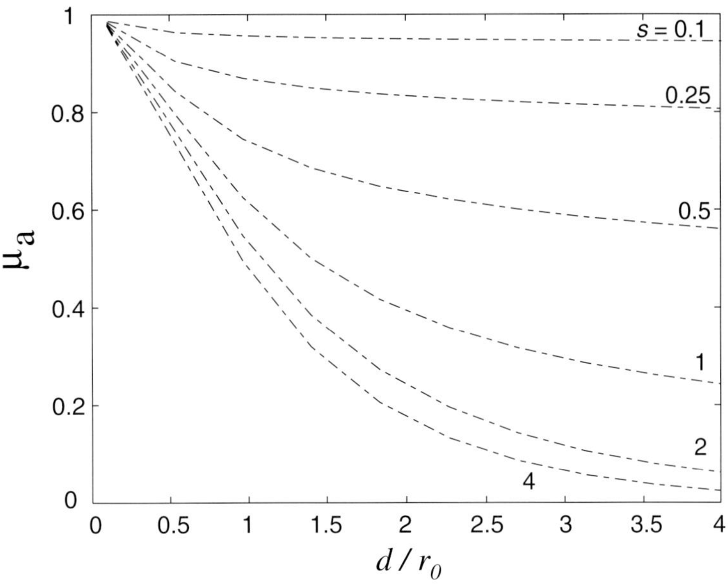

FIGURE 5.9
Plot of the atmospheric visibility μ_a versus normalized subaperture size d/r_0 for the SI-WFS. These results were calculated for a square subaperture of dimension d. The parameter s is the normalized shear, $|\Delta\vec{x}_x|/r_0$ or $|\Delta\vec{x}_y|/r_0$.

and $W_s(\vec{x})$ is a subaperture weighting function. The other visibility factor, $\mu_b(s)$, is related to the beacon irradiance distribution. It is mathematically equal to the normalized mutual intensity of the beacon field as given by the Van Cittert-Zernike theorem. This term accounts for the reduction in the visibility of the interference pattern on the detector surface due to the finite size of the beacon. For the specific case in which the beacon is modeled by a Gaussian irradiance distribution having an RMS width of σ_b and an altitude of z_b as given by Eq. (5.28), we find, through application of the Van Cittert-Zernike theorem, that

$$\mu_b(s) = \exp\left[-\pi^2\beta^2 s^2\right], \tag{5.37}$$

where β is angular size of the beacon relative to the seeing limited angle ($\beta = (\sigma_b/z_b)/(\lambda/r_0)$). Substituting Eq. (5.37) into Eq. (5.34) gives

$$\sigma_{ns} = \frac{2\sqrt{2}}{r_0(\overline{K})^{1/2} s\mu_a(s)} \exp\left[\pi^2\beta^2 s^2\right] \text{ (radians/m)}. \tag{5.38}$$

The atmospheric visibility term $\mu_a(s)$ has been investigated for the case of atmospheric turbulence having the Kolmogorov index-of-refraction power spectral density given in Eq. (3.8) [83]. Figure 5.9 illustrates a plot of the $\mu_a(s)$ versus d/r_0 for a square subaperture of dimension d and various values of the normalized shear, s. The atmospheric visibility term obviously decreases for increasing shear values as well as decreasing values of r_0. Using the results presented in Fig. 5.9, the shot noise given in Eq. (5.38) is evaluated and presented in the graph shown in Fig. 5.10. Figure 5.10 is a plot of σ_{ns} versus d/r_0 for β ranging from 0 to 2. Just as we saw for the H–WFS, the error increases dramatically for increasing beacon size and decreasing seeing conditions (i.e., decreasing r_0).

Read noise also contributes to the slope error for the SI–WFS. Sandler [78] has investigated the read

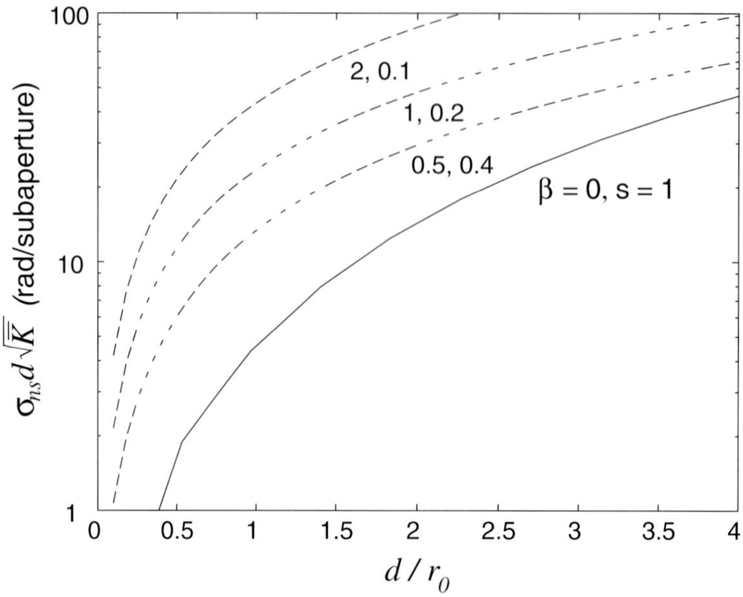

FIGURE 5.10
Plot of σ_{ns} versus d/r_0 for the SI–WFS and β ranging from 0 to 2. The parameter d is the subaperture dimension, $\sqrt{\overline{K}}$ is the average detected photon-count per measurement and β is the ratio of the angular beacon size to the angular seeing limit, $\beta = (\sigma_b/z_b)/(\lambda/r_0)$. For each value of β the optimum value of shear was found and it is this optimum value that is used to compute the shot noise error presented in the graph.

noise contribution to the slope error and found that

$$\sigma_{nr} = \frac{\sqrt{8}\sigma_e}{r_0 \overline{K} s \mu_a(s)\mu_b(s)} \quad \text{(rad/m)} \tag{5.39}$$

for the 4-bin interferometer and

$$\sigma_{nr} = \frac{\sqrt{6}\sigma_e}{r_0 \overline{K} s \mu_a(s)\mu_b(s)} \quad \text{(rad/m)} \tag{5.40}$$

for the 3-bin interferometer. To obtain the total slope measurement error, σ_n, the shot noise and read noise errors are combined as in Eq. (5.33).

Summary

The main contribution of this section is the presentation of the two WFS models applicable for the H–WFS and SI–WFS. These two models are given in Eqs. (5.15) and (5.25). Along with these models we presented shot and read noise expressions that characterize the variance of the additive slope measurement error $\vec{s}_n$. These measurement error models are presented and discussed in Eqs. (5.26) through (5.40). The numerical results shown in Figs. 5.8 and 5.10 indicate that the SI–WFS has better performance for small subapertures ($d/r_o < 1$) and point source beacons ($\beta = 0$). On the other hand for larger subapertures ($d/r_o > 1$) and finite size beacons ($\beta > 0$), the H–WFS can have significantly lower measurement errors. Both the WFS and measurement error models will be incorporated within the mathematical framework to be established in Section 5.4 for the AOI system performance predictions.

5.3.3 Controllers

In Section 5.3.1 we introduced a model to characterize how the control signals, $c_i(t)$, are mapped into the mirror surface, $\hat{\phi}(\vec{x}, t)$. This model is given in Eq. (5.4) and is repeated here:

$$\hat{\phi}(\vec{x}, t) = \sum_i c_i(t) r_i(\vec{x}). \tag{5.41}$$

The summation is over all deformable mirrors actuators, and $r_i(\vec{x})$ is the influence function of the ith actuator. In an AOI system the information provided by the WFS is used to control the signals sent to the actuators. The problem at hand is finding a mapping from WFS measurements to control signals that results in the desired system performance. The phrase "desired performance" has different meanings depending on the performance metric being minimized or maximized. Qualitatively the goal is to achieve a mapping that results in $\phi(\vec{x}, t) \approx \hat{\phi}(\vec{x}, t)$. The degree to which the reconstructed phase matches the actual phase depends on the noise in the WFS measurements, anisoplanatism and the spatial sampling of the WFS subapertures and spacing of the DM actuators. In the following two subsections we present two basic approaches for achieving the mapping of the WFS measurement to DM control signals. In both cases it is assumed that we desire a linear mapping from WFS measurements to actuator control signals. In other words we want to derive a control matrix M so that

$$\mathbf{c} = M\mathbf{s}_m, \tag{5.42}$$

where $\mathbf{c}$ is a column vector of the control signals and $\mathbf{s}_m$ is a column vector of the measured subaperture slopes. In writing Eq. (5.42) we have dropped the explicit time dependence of $\mathbf{c}$ and $\mathbf{s}_m$. Note that the orthogonal components of the vector slope expressions given previously represent individual scalar elements of the column vector $\mathbf{s}_m$. In the development to follow the slope and control signal time dependence is suppressed for the sake of simplicity. We return to the temporal nature of the control problem in the last subsection of this section. The two basic approaches for deriving the control matrix M are distinguished from each other by the quantity that is minimized in the solution process. In the first approach M is derived by minimizing the difference between the measured wave front slopes and the slopes associated with the the DM surface. This minimization falls within the formalism of "maximum a posteriori estimation" and "least squares" estimation [84]. In the second approach statistical knowledge of both the wave front statistics and the WFS noise characteristics are used to derive a control matrix that minimizes the aperture average mean square residual phase error.

Maximum a posteriori (MAP) estimator

For the maximum a posteriori (MAP) estimation development to follow the slope measurement model is assumed to take the following form

$$\mathbf{s}_m = \mathbf{s}_{dm} + \mathbf{s}_n, \tag{5.43}$$

where $\mathbf{s}_{dm}$ is a column vector of the slopes associated with the surface of the DM, and $\mathbf{s}_n$ is the column vector of associated noise realizations due to shot and read noise. In the MAP estimation approach we attempt to minimize $|\mathbf{s}_m - \mathbf{s}_{dm}|^2$, where the notation $|\mathbf{a}|^2 = \mathbf{a}^T\mathbf{a}$. This approach is well suited for closed loop adaptive optical systems in which the difference, $\mathbf{s}_m - \mathbf{s}_{dm}$, is the closed loop measurement used to update the mirror control signals. The elements of $\mathbf{s}_{dm}$ are given by the WFS models of Eqs. (5.15) or (5.25) with $\phi(\vec{x}, t)$ replaced by $\hat{\phi}(\vec{x}, t)$ and the noise assumed to be zero. As an example, consider the H–WFS model of Eq. (5.15):

$$s_{dm}(j) = -\int d\vec{x}\,(\nabla W_{sj}(\vec{x}) \cdot \vec{d}_j)\hat{\phi}(\vec{x}), \tag{5.44}$$

where $s_{dm}(j)$ is the jth slope measurement element of the vector $\mathbf{s}_{dm}$ and $\vec{d}_j$ is a unit vector specifying the direction of sensitivity of the jth slope measurement. Recall that each subaperture generates a vector slope having two components: one in the $\vec{d}_j$ direction and another in a direction perpendicular to $\vec{d}_j$. As such $\vec{d}_j$ takes on one of two orthogonal directions for each subaperture. To proceed with the goal of deriving the optimal control vector $\mathbf{c}$, a relationship must be established between $\mathbf{s}_{dm}$ and the control vector $\mathbf{c}$. Substituting Eq. (5.41) into Eq. (5.44) gives this relationship:

$$s_{dm}(j) = -\sum_i c_i \int d\vec{x}\,(\nabla W_{sj}(\vec{x}) \cdot \vec{d}_j) r_i(\vec{x}). \tag{5.45}$$

By defining the jith element of the matrix H as

$$H_{ji} = -\int d\vec{x}\,(\nabla W_{sj}(\vec{x}) \cdot \vec{d}_j) r_i(\vec{x}), \tag{5.46}$$

we can write the following matrix result from Eq. (5.45)

$$\mathbf{s}_{dm} = H\mathbf{c}. \tag{5.47}$$

The matrix H is sometimes called the Jacobian matrix. Substituting Eq. (5.47) into the WFS measurement model of Eq. (5.43) gives the relationship between $\mathbf{s}_m$ and $\mathbf{c}$:

$$\mathbf{s}_m = H\mathbf{c} + \mathbf{s}_n. \tag{5.48}$$

With the slope measurement model given in Eq. (5.48) the control problem becomes choosing the estimate of $\mathbf{c}$ in a way that maximizes the likelihood of the measurement $\mathbf{s}_m$. In other words we want to find the estimate of $\mathbf{c}$ that maximizes the probability $p(\mathbf{s}_m|\mathbf{c})p(\mathbf{c})$ where $p(\mathbf{s}_m|\mathbf{c})$ is the conditional probability density function of the measurement $\mathbf{s}_m$ given the control vector $\mathbf{c}$ and $p(\mathbf{c})$ is the a priori probability density function of control vector $\mathbf{c}$. This approach results in what is called the MAP estimate of $\mathbf{c}$. The result of this derivation will also be shown to reduce to the standard least squares approach when the noise is assumed to be zero.

In order to proceed the probability density functions $p(\mathbf{s}_m|\mathbf{c})$ and $p(\mathbf{c})$ must be known. From the WFS measurement model given in Eq. (5.48) it is apparent that $p(\mathbf{s}_m|\mathbf{c})$ is essentially the distribution of the measurement noise, $\mathbf{s}_n$. The noise $\mathbf{s}_n$ is assumed to be a zero mean random Gaussian vector [11, 15]. Considering the a-prior probability $p(\mathbf{c})$, we first note that the control vector $\mathbf{c}$ is linearly related to the measured wave front slopes, $\mathbf{s}_m$, as given by Eq. (5.42). This linear relationship establishes that $\mathbf{c}$ is also a zero mean Gaussian vector since $\mathbf{s}_m$ is assumed to be zero mean and Gaussian [85, 86]. We also note that $\mathbf{c}$ and $\mathbf{s}_n$ are assumed to be independent. Given these assumptions the only quantity needed to completely characterize the statistics of $\mathbf{c}$ and $\mathbf{s}_n$ are their respective covariance matrices. The covariance matrices for $\mathbf{c}$ and $\mathbf{s}_n$ are designated with the following notation:

$$\begin{aligned} C_c &= \mathrm{E}\{\mathbf{c}\mathbf{c}^T\}, \\ C_n &= \mathrm{E}\{\mathbf{s}_n\mathbf{s}_n^T\}, \end{aligned} \tag{5.49}$$

where the superscript T designates vector transpose.

MAP estimate: Due to the assumption of zero mean Gaussian statistics for $\mathbf{s}_m$ and $\mathbf{c}$ and the use of the measurement model in Eq. (5.48) the MAP estimate is equivalent to estimating $\mathbf{c}$ such that $|\mathbf{s}_{dm} - \mathbf{s}_m|^2$ is minimized. The result of the minimization is found in Ref. [84] with the resulting optimum control vector being

$$\mathbf{c} = \left(H^T C_n^{-1} H + C_c^{-1}\right)^{-1} H^T C_n^{-1}\mathbf{s}. \tag{5.50}$$

Matching the terms in Eqs. (5.42) and (5.50) we see that the MAP control matrix is given by

$$M_{MAP} = \left(H^T C_n^{-1} H + C_c^{-1}\right)^{-1} H^T C_n^{-1}. \tag{5.51}$$

The WFS slope measurement noise covariance matrix, C_n, will in many cases be diagonal since the photon noise and the read noise are independent for each WFS measurement. In the case that the variance of each subaperture measurement is the same and equal to σ_n^2, Eq. (5.51) reduces to

$$M_{MAP} = \left(H^T H + \sigma_n^2 C_c^{-1}\right)^{-1} H^T. \tag{5.52}$$

This control matrix has been derived by others [28, 87]. The noise variance, σ_n^2, can be calculated using the measurement error models for the H–WFS and SI–WFS given in Section 5.3.2.

To compute the control matrix given in Eqs. (5.52), the control signal covariance matrix, C_c, is required. The covariance matrix C_c is calculated in a manner similar to the approach used to compute the covariance of the Zernike coefficients presented in Section 3.6. If the actuator influence functions, $r_i(\vec{x})$, satisfy the orthonormal conditions given in Eqs. (3.137) through (3.139), Eq. (3.142) can be used to compute the covariance matrix C_c. In the notation of Eq. (3.142) the coefficient a_i is equated to the control signal c_i.

However, in nearly all practical situations the DM actuator influence functions are not orthonormal. In this case finding the covariance C_c is more difficult. To find the covariance matrix C_c we must first establish a relationship between the wave front phase $\phi(\vec{x})$ and the control vector $\mathbf{c}$. For orthonormal influence functions this relationship is established by Eq. (3.140). To establish the relationship for non-orthonormal influence functions we choose to relate $\mathbf{c}$ to $\phi(\vec{x})$ by minimizing the difference between $\phi(\vec{x})$ and $\sum c_i r_i(\vec{x})$. Gaffard and Boyer [12] have derived such a relationship by minimizing the following pupil averaged mean square phase difference:

$$\epsilon^2 = \int d\vec{x}\, W(\vec{x}) \mathrm{E}\left[\left(\phi(\vec{x}) - \sum_i c_i r_i(\vec{x})\right)^2\right], \tag{5.53}$$

where $W(\vec{x})$ is the pupil weighting function. Minimization of Eq. (5.53) results in a control vector $\mathbf{c}$ that must satisfy [12]

$$\mathbf{c} = R^{-1}\mathbf{p}, \tag{5.54}$$

where R is a matrix with elements defined by

$$r_{ij} = \int d\vec{x}\, W(\vec{x}) r_i(\vec{x}) r_j(\vec{x}), \tag{5.55}$$

and $\mathbf{p}$ is a column vector with elements defined by

$$p_i = \int d\vec{x}\, W(\vec{x}) \phi(\vec{x}) r_i(\vec{x}). \tag{5.56}$$

Using Eq. (5.54) we can now write the covariance matrix:

$$\begin{aligned} C_c &= \mathrm{E}\{\mathbf{c}\mathbf{c}^T\} \\ &= R^{-1}\mathrm{E}\{\mathbf{p}\mathbf{p}^T\}R^{-1} \\ &= R^{-1} P R^{-1} \end{aligned} \tag{5.57}$$

where we have used the fact that R is symmetric to obtain the second line of Eq. (5.57), and the matrix $P = \mathrm{E}\{\mathbf{p}\mathbf{p}^T\}$ with elements given by

$$p_{ij} = \int\int d\vec{x} d\vec{x}'\, W(\vec{x}) W(\vec{x}') r_i(\vec{x}) r_j(\vec{x}) \mathrm{E}\{\phi(\vec{x}, t)\phi(\vec{x}', t)\}. \tag{5.58}$$

We finally note that Eq. (5.57) reduces to Eq. (3.142) when the set of influence functions $r_i(\vec{x})$ are orthonormal. In this special case the matrix R is an identity matrix and the elements of C_c given by Eq. (5.58) reduce exactly to the covariance elements given by Eq. (3.142).

Least squares estimate: We wrap up this section with a presentation of the least squares control matrix. It turns out that the MAP reconstructor given above in Eq. (5.52) reduces to the well known least squares solution as the measurement noise variance decreases to zero. The defining characteristic of the least squares solution is that it does not make use of either the atmospheric or noise statistics. The minimization problem is treated as being deterministic in that the measurements $\mathbf{s}_m$ are assumed to be noiseless. As such, the least squares control matrix is the simplest controller to compute. Neither the noise or atmospheric statistics are required. Due to the fact the least squares controller is so simple to compute it is used most often in current AOI system implementations. Letting the noise variance go to zero in Eq. (5.52) gives the least squares control matrix:

$$M_{LS} = \left(H^T H\right)^{-1} H^T. \tag{5.59}$$

The result in Eq. (5.59) can also be derived by using standard least squares methodology in the minimization of $|\mathbf{s}_m - H\mathbf{c}|$ [84]. In fact this is exactly the approach presented in Section 6.2.2 of Chapter 6 for the wave front reconstruction required in deconvolution from wave front sensing (DWFS). We note that M_{LS} corresponds to the Morre-Penrose generalized inverse or pseudo inverse of H. A large number of authors have presented the least squares development for the control of the AOI system DM [11, 21, 28, 55, 67].

Minimum variance controller

In the development of MAP estimate presented above it was asserted that the measurement model of Eq. (5.48) represented reality. Using this model, the approach for finding the control vector $\mathbf{c}$ was straightforward. The estimate was found by maximizing the likelihood that a particular command generated the observed slope measurements $\mathbf{s}_m$. The major problem with this approach is validity of the measurement model of Eq. (5.48). In this model the noiseless slope vector, $\mathbf{s}_{dm} = H\mathbf{c}$, is modeled using only the range of wave front surfaces capable of being produced by the DM. The implicit assumption of Eq. (5.48) is that the DM modes or actuator influence functions are capable of exactly reproducing the wave front phase of the incident wave front, $\phi(\vec{x})$.

Since the DM has a finite number of degrees of freedom, it is not capable of reproducing $\phi(\vec{x})$ exactly, and as a result the WFS slope measurement $\mathbf{s}_m$ is only partially modeled by the expression given in Eq. (5.48). Another term must be added to Eq. (5.48) to account for the portion of the wave front outside the range of possible surface deformations of the DM. Not accounting for this extra term gives rise to aliasing. The wave front phase deformations outside the range of the DM are aliased onto those deformations within the range of the DM. If this extra term were added to the WFS model of Eq. (5.48) then the minimization of $|\mathbf{s}_{dm} - \mathbf{s}_m|^2$ would not be the optimal approach for finding $\mathbf{c}$.

An approach that circumvents the drawback associated with the MAP estimate (and the least squares estimate) is a derivation that results in what has been called the "minimum variance controller." The difference between this derivation and the MAP derivation is the quantity minimized. For the minimum variance controller the aperture averaged residual mean square phase is minimized rather than $|\mathbf{s}_{dm} - \mathbf{s}_m|^2$. To define the aperture averaged mean square residual phase recall that the residual phase is designated $\epsilon(\vec{x})$ and is defined as

$$\epsilon(\vec{x}) = \phi(\vec{x}) - \hat{\phi}(\vec{x}), \tag{5.60}$$

where $\phi(\vec{x})$ is the piston removed wave front phase deformation and $\hat{\phi}(\vec{x})$ is the piston removed wave front phase correction applied by the DM. The aperture averaged mean square residual phase is defined by

$$\mathrm{E}\{\epsilon^2\} = \mathrm{E}\left\{\int \mathrm{d}\vec{x}\, W(\vec{x})\epsilon^2(\vec{x})\right\},$$
$$= \int \mathrm{d}\vec{x}\, W(\vec{x})\mathrm{E}\{\epsilon^2(\vec{x})\}, \tag{5.61}$$

where $W(\vec{x})$ is the AOI system pupil weighting function. The weighting function is normalized to unit area in the same manner as defined in Eq. (3.139). Minimization of Eq. (5.61) avoids the aliasing problem associated with the MAP estimation approach since the actual wave front phase $\phi(\vec{x})$ is an integral component of the quantity being minimized rather than the reduced order model represented by Eq. (5.48). The minimum variance controller will result in optimal performance since its derivation factors in statistical information concerning both the incident wave front phase $\phi(\vec{x})$ and the noise on the WFS measurements. Additionally it has been shown that minimizing $\mathrm{E}\{\epsilon^2\}$ is equivalent to maximizing the AOI system's Strehl ratio [29]. The Strehl ratio is defined as the ratio of the peak of the system PSF to the peak of the PSF of an unaberrated system. The PSF of an incoherent imaging is the Fourier transform of the OTF as discussed in Section 2.2.5.

Using Eqs. (5.41) and (5.60) we can explicitly show the dependence of $\mathrm{E}\{\epsilon^2\}$ on $\phi(\vec{x})$ and the DM control signals:

$$\mathrm{E}\{\epsilon^2\} = \int \mathrm{d}\vec{x}\, W(\vec{x})\mathrm{E}\left\{\left(\phi(\vec{x}) - \sum_i c_i r_i(\vec{x})\right)^2\right\}. \tag{5.62}$$

This expression is identical to the quantity minimized in Eq. (5.53). In the previous minimization the control vector $\mathbf{c}$ was derived assuming perfect knowledge of the wave front phase $\phi(\vec{x})$. In the minimization presented here we only have a set of noisy WFS measurements of $\phi(\vec{x})$. Recall that we are restricting our attention to the class of linear controllers as specified in Eq. (5.42). Substituting Eq. (5.42) into Eq. (5.62) shows the dependence of $\mathrm{E}\{\epsilon^2\}$ on the slope measurements $\mathbf{s}_m$ and the control matrix M:

$$\mathrm{E}\{\epsilon^2\} = \int \mathrm{d}\vec{x}\, W(\vec{x})\mathrm{E}\left\{\left(\phi(\vec{x}) - \sum_i \sum_j M_{ij} s_m(j) r_i(\vec{x})\right)^2\right\}, \tag{5.63}$$

where M_{ij} is the ijth element of M and $s_m(j)$ is the j WFS slope measurement. The slope measurement model used here is given by either Eq. (5.15) for the H–WFS or Eq. (5.25) for the SI–WFS. Wallner [11] first considered the problem of minimizing Eq. (5.63) and many others have since studied AOI system performance using the minimum variance controller [15, 18, 19, 20, 21, 22, 30, 32]. In particular both Roggemann [21] and Wallner [11] have shown the minimum variance controller provides better performance than the least squares controller of Eq. (5.59).

Expanding the square inside the expected value operator in Eq. (5.63) allows us to write $\mathrm{E}\{\epsilon^2\}$ in the following way [11, 15]

$$\mathrm{E}\{\epsilon^2\} = \mathrm{E}\{\epsilon_o^2\} + \sum_i \sum_{i'} \sum_j \sum_{j'} M_{ij} M_{i'j'} S_{jj'} R_{ii'} - 2\sum_i \sum_j M_{ij} A_{ij}, \tag{5.64}$$

where $S_{jj'}$, $R_{ii'}$ and A_{ij} are elements of the matrices S, R and A, respectively, and $\mathrm{E}\{\epsilon_o^2\}$ is the mean square uncorrected wave front phase. Before explicitly defining these matrices it is notationally convenient to define an inner product. Consider the notation for the inner product of the 2-D functions $f(\vec{x})$ and $g(\vec{x})$:

$$[f(\vec{x}), g(\vec{x})] = \int \mathrm{d}\vec{x}\, W(\vec{x}) f(\vec{x}) g(\vec{x}). \tag{5.65}$$

Using this notation the matrices S, R, and A, and $\mathrm{E}\{\epsilon_o^2\}$ are compactly defined as

$$S = \mathrm{E}\left\{\mathbf{s}_m \mathbf{s}_m^T\right\}, \tag{5.66}$$

$$R = \left[\mathbf{r}(\vec{x}), \mathbf{r}^T(\vec{x})\right], \tag{5.67}$$

$$A = \left[\mathbf{r}(\vec{x}), \mathrm{E}\left\{\phi(\vec{x})\mathbf{s}_m^T\right\}\right], \tag{5.68}$$

$$\mathrm{E}\{\epsilon_o^2\} = [\phi(\vec{x}), \phi(\vec{x})], \tag{5.69}$$

where $\mathbf{s}_m$ is a column vector composed of the subaperture WFS slope measurements and $\mathbf{r}(\vec{x})$ is a column vector composed of the deformable mirror actuator influence functions. Noting that the multiple summations in Eq. (5.64) can be written using the matrix trace operator we can finally write $\mathrm{E}\{\epsilon^2\}$ as [30]

$$\mathrm{E}\{\epsilon^2\} = \mathrm{E}\{\epsilon_o^2\} - tr(2AM^T - MSM^T R^T), \tag{5.70}$$

where $tr(B)$ denotes the trace of the square matrix B. The trace of a square matrix is the sum of the diagonal elements. Equation (5.70) can be minimized by taking the derivative of the expression with respect to the control matrix M and equating the result to zero. Using standard rules for taking the derivative of the trace we can show [88]

$$\frac{\partial \mathrm{E}\{\epsilon^2\}}{\partial M} = 2A - 2RMS. \tag{5.71}$$

Setting Eq. (5.71) equal to zero and solving for the minimum variance (MV) controller gives

$$M_{MV} = R^{-1}AS^{-1}. \tag{5.72}$$

Finally substituting the MV control matrix back into the expression for $\mathrm{E}\{\epsilon^2\}$ gives the minimum error:

$$\mathrm{E}\{\epsilon^2\} = \mathrm{E}\{\epsilon_o^2\} - tr(R^{-1}AS^{-1}A^T). \tag{5.73}$$

The minimum variance controller given by Eq. (5.72) requires detailed knowledge of both the wave front phase and WFS noise statistics as reflected in the requirement to compute the matrices S and A. The computation of R only requires knowledge of the deformable mirror influence functions. The practical drawback associated with implementing the minimum variance controller is accurately characterizing the wave front phase and WFS noise statistics so that the appropriate S and A can be computed. In most situations this prior knowledge is not available. If accurate prior knowledge of the required statistics is available then it has been shown that the minimum variance controller outperforms the least squares solution of Eq. (5.59) [11, 21].

Closed loop, finite temporal response controllers

In the derivations of the controller presented in the previous two subsections we have ignored the finite temporal response of the controller. Implicit in the derivations was the assumption that the controller responded instantaneously to the slope measurement $\mathbf{s}_m(t)$. In reality the controller will have finite temporal response and will be configured in a closed loop system. The closed loop arrangement is used in practical control applications for stability reasons [33]. In the closed loop arrangement the WFS measures the wave front after compensation by the DM. In such an arrangement the measured wave front slope is the difference between the incident wave front slopes and the slopes due to the DM surface:

$$\begin{aligned} \Delta\mathbf{s}(t) &= \mathbf{s}_m(t) - \mathbf{s}_{dm}(t), \\ &= \mathbf{s}_m(t) - H\mathbf{c}(t), \end{aligned} \tag{5.74}$$

where $\mathbf{s}_m(t)$ is the instantaneous slope vector associated with the incident wave front phase, H is the Jacobian matrix defined in Eq. (5.46) and $\mathbf{c}(t)$ is the DM control vector. The closed loop WFS measurement $\Delta\mathbf{s}(t)$ is mapped to a vector of control signals $\mathbf{c}(t)$ via a servo-loop compensator. The finite temporal response of the loop compensator can be modeled with a filter having the following representative first order differential description [30]

$$\frac{\mathrm{d}\mathbf{c}}{\mathrm{d}t} = k_g M \Delta\mathbf{s}(t), \tag{5.75}$$

where k_g is the gain of the filter in units of radians per second and M is the control matrix that maps the slope signals $\Delta\mathbf{s}(t)$ into the control signals. The filter gain is related to the 3 dB bandwidth of the filter by $f_{3\mathrm{dB}} = k_g/2\pi$. Substituting Eq. (5.74) into Eq. (5.75) and solving for the steady state solution of $\mathbf{c}(t)$ gives [30]

$$\mathbf{c}(t) = \int_0^\infty \mathrm{d}\tau k_g \exp\left(-k_g \tau M H\right) M \mathbf{s}_m(t-\tau). \tag{5.76}$$

In the special case that the control matrix M is given by the least squares solution of Eq. (5.59), Eq. (5.76) reduces to [30]

$$\begin{aligned} \mathbf{c}(t) &= M_{LS} \int_0^\infty \mathrm{d}\tau k_g \exp(-k_g \tau) \mathbf{s}_m(t-\tau), \\ &= M_{LS} \tilde{\mathbf{s}}_m(t), \end{aligned} \tag{5.77}$$

where $\tilde{\mathbf{s}}_m(t)$ is the temporally filtered version of the instantaneous slope $\mathbf{s}_m(t)$. Equation (5.77) is a significant result since it establishes a simple relationship between $\mathbf{c}(t)$ and the filtered measurement $\tilde{\mathbf{s}}_m(t)$ that is identical to the simple linear relationship established by Eq. (5.42). It turns out this relationship is not unique to the loop compensator given by Eq. (5.75). The linear relationship in the last line of Eq. (5.77) can be established for any temporal response characterized by an ordinary differential equation with constant coefficients [30].

The least squares control matrix M_{LS} is required to obtain the linear relationship between $\mathbf{c}(t)$ and the filtered measurement $\tilde{\mathbf{s}}_m(t)$ given by Eq. (5.77). Ellerbroek [30] has examined the problem of deriving a minimum variance control matrix for the closed loop arrangement described above. By imposing two linear constraints on the control matrix M, Ellerbroek shows that Eq. (5.77) can also be used to relate $\mathbf{c}(t)$ and $\tilde{\mathbf{s}}_m(t)$ for the minimum variance controller. With this simple linear relationship Ellerbroek derives a minimum variance control matrix that is structurally similar to the minimum variance solution given in Eq. (5.72) except for the additional complexity resulting from the constraints.

Summary

The control matrices presented in this section have established the required relationship between the WFS measurements and the DM actuator signals. These control matrices are sometimes referred to as the phase reconstruction matrices. The approaches used to derive the matrices include the maximum a posteri and the minimum variance methodologies. The maximum a posteri methodology encompasses the derivation of the most common phase reconstruction matrix: the least squares reconstructor. The minimum variance and the least squares reconstructors are both used in the numerical examples presented in Section 5.4.8.

5.4 AOI system performance modeling

In the following section we present the framework for using the AOI system component models presented in the previous sections to predict imaging performance. These models include the WFS measurement models of Eqs. (5.15) and (5.25), the DM model of Eq. (5.4) and the controller model of Eq. (5.42). The WFS measurement models cited above must be used in conjunction with a noise model. The WFS noise models are given in Section 5.3.2. These component models have been presented within a generalized mathematical framework that allows nearly any AOI system to be modeled. The goal of the following section is to demonstrate the type of AOI system performance modeling that is possible.

In Subsection 5.4.1 we present the imaging metrics commonly used to quantify AOI system performance. The performance metrics will, in turn, define the quantities that must be computed and illustrate what statistics are required. In Subsection 5.4.2 the wave front phase statistics required to compute the performance metrics are reviewed. Most of these statistics have been introduced in Chapter 3 and are reintroduced here in the context of AOI system modeling. With the performance metrics and the required system component models in hand the next consideration is the method used to actually conduct the performance modeling. Subsection 5.4.6 presents an overview of the two numerical approaches that can be used to accomplish the performance modeling. These two approaches are numerical evaluation of expressions describing the statistical performance metrics and Monte Carlo simulation of the AOI system. Finally we demonstrate the use of the performance analysis methods presented here with a number of performance calculations. These calculations are presented in Subsections 5.4.7 and 5.4.8, and represent fundamental performance results that will generally be of interest to anyone interested in understanding the performance limitations of AOI systems.

5.4.1 Performance metrics

In order to accomplish the goal of developing a comprehensive performance analysis, we first need to define a set of metrics used to quantify performance. These metrics include classic measures that are well known to the optics community. They include the squared residual wave front phase after wave front correction, $\epsilon^2(\vec{x})$, the optical transfer function (OTF), the point spread function (PSF), and the Strehl ratio. In the context of AOI system modeling we will be considering the statistical nature of these performance metrics. In particular we will be computing the expected value or ensemble average of the metrics and, in some cases, the variance of the metric.

It is fairly obvious why we might be interested in the average value of a particular performance metric. The mean indicates how the AOI system performs on average for a large number of atmospheric turbulence realizations. Due to the temporal evolution of atmospheric turbulence, a large number of wave front phase realizations can generally be thought of as a long exposure image. This type of average performance calculation has already been performed for the uncompensated imaging system operating in the presence of the atmosphere. The long exposure OTF, $\mathcal{H}_{LE}$, derived in Section 3.5.3 is an example of the average OTF.

The motivation for computing the variance of a performance metric may not be as obvious. To achieve the level of performance indicated by the mean value of a particular performance metric requires an ensemble average of an infinite number of realizations. In practice we never conduct experiments in an ensemble average sense. Measurements are always conducted in such a way that a finite number of observations or realizations are measured to determine the quantity of interest. In many situations it is of interest whether a small or large number of observations are required to achieve the level of performance predicted by the mean. To answer this question the variation of the performance metric becomes the quantity of interest. A large variation of the metric about the mean indicates that a large number of observations will be required to achieve the performance level predicted by the mean. On an instantaneous basis we would expect the performance to vary widely about the mean. On the other hand, a small variation of the performance metric indicates that it will take only a small number of observations to achieve the performance predicted by the mean. In summary, the variance of a performance metric in conjunction with its mean value can be used together to quantify the performance fluctuation of the system. Mathematically this fluctuation is quantified with the signal-to-noise ratio (SNR). The SNR has been previously introduced in Chapters 2 and 4 and is defined

$$\mathrm{SNR} = \frac{|\overline{\mathrm{metric}}|}{\sqrt{\mathrm{var}[\mathrm{metric}]}}, \tag{5.78}$$

where $\overline{a}$ and var$[a]$ are mean and variance of the random variable a, respectively. High values of the SNR indicate the performance fluctuation of the metric is small compared to the average value. A low SNR, on the other hand, indicates the performance fluctuation is large compared to the average value. The SNR is particularly useful for characterizing the fluctuation of the optical transfer function (OTF).

To compute any of the performance metrics cited above the wave front phase in the pupil after wave front correction is required. The phase of the corrected wave front is referred to as the residual phase and it is designated with the notation $\epsilon(\vec{x})$. The residual phase $\epsilon(\vec{x})$ is the difference between the incident wave front phase deformation, $\phi(\vec{x})$, and the wave front correction applied by the DM, $\hat{\phi}(\vec{x})$:

$$\epsilon(\vec{x}) = \phi(\vec{x}) - \hat{\phi}(\vec{x}). \tag{5.79}$$

Note that both $\phi(\vec{x})$ and $\hat{\phi}(\vec{x})$ are piston removed quantities as defined in Eq. (5.3). It is through Eq. (5.79) that the AOI system models of Section 5.3 and the performance metrics of this section are tied together. In the ideal situation, the residual phase $\epsilon(\vec{x})$ is zero but, as discussed in Section 5.2, there are many factors that prevent the AOI system from achieving perfect wave front correction. It is also important to note that in concentrating on the residual phase $\epsilon(\vec{x})$ we are explicitly assuming that the AOI system is dominated by wave front phase errors rather than amplitude or scintillation effects. This assumption is commonly invoked when the turbulence is in the near field. Near field conditions refer to the situation in which most of the turbulence effects are relatively close to the aperture plane of the AOI system. In the case of astronomical imaging this assumption is accurate under almost all situations of interest [89].

One of the most common and easiest system performance metrics to compute is the aperture-average, squared, residual phase. This quantity was defined in Eq. (5.61) during the development of the minimum variance controller and is repeated here

$$\epsilon^2 = \int \mathrm{d}\vec{x}\; W(\vec{x})\epsilon^2(\vec{x}), \tag{5.80}$$

where $W(\vec{x})$ is a weighting function defining the pupil of the AOI system. The weighting function is normalized such that $\int \mathrm{d}\vec{x}\; W(\vec{x}) = 1$. One of the advantages of this metric is that it summarizes the AOI system performance with a single number. This single valued metric was particularly valuable in the derivation of the minimum variance controller of the previous section and in the case where $\epsilon^2 < (2\pi/10)^2$ rad^2 it can be used to estimate the Strehl ratio [26].

In many cases a single valued performance metric, such as ϵ^2, is not adequate to fully understand how the AOI system performs. For incoherent imaging the optical transfer function (OTF) and the point spread function (PSF) are two extremely useful metrics. The OTF and PSF retain a large amount of information about the spatial correlation properties of $\epsilon(\vec{x})$. The OTF is defined in Eq. (3.108) of Chapter 3, and in the context of the residual phase $\epsilon(\vec{x})$ is rewritten here

$$\mathcal{H}(\vec{f}) = N_F^{-1}\Big(W(\vec{f}\lambda d_i)\exp[j\epsilon(\vec{f}\lambda d_i)] \star W(\vec{f}\lambda d_i)\exp[j\epsilon(\vec{f}\lambda d_i)]\Big), \tag{5.81}$$

where $\vec{f}$ is the spatial frequency, λ is the optical wavelength, d_i is the distance between the image and exit pupil planes, and N_F is a normalization constant that enforces $\mathcal{H}(0) = 1$. The OTF clearly illustrates the response of the AOI system to the whole range of spatial frequencies. High spatial frequencies are of particular interest, since it is in this region that the magnitude of the OTF indicates the achievable resolution. The PSF is derived directly from the OTF via a 2-dimensional inverse Fourier transform (see Chap 2, Section 2.2.5):

$$s(\vec{x}) = \mathcal{F}^{-1}\left[\mathcal{H}(\vec{f})\right], \tag{5.82}$$

where $\vec{x}$ designates a point in the image plane. The PSF is interpreted as the image-plane intensity distribution that results from imaging a point source. This intensity distribution is a useful performance measure, since the resolution of the AOI system can, in the infinite signal case, be determined directly from the width of the main lobe of the PSF.

The Strehl ratio can also be computed from the PSF by comparing the peak of the main lobe of the PSF with that of an unaberrated system. Mathematically we can express the Strehl ratio, SR, as

$$\mathrm{SR} = \frac{s(\vec{x}=0)}{s_{dl}(\vec{x}=0)}, \tag{5.83}$$

where $s_{dl}(\vec{x})$ is the diffraction limited PSF. The Strehl ratio can also be estimated. If the aperture averaged residual wave front phase ϵ^2 is less than $(2\pi/10)^2$ rad^2 then the Strehl ratio is well approximated by [26]

$$SR \approx \exp[-\epsilon^2]. \tag{5.84}$$

5.4.2 Residual wavefront phase statistics

The residual phase $\epsilon(\vec{x})$ is inherently a random process that is a function of time and space. As such, the definitions for ϵ^2, $\mathcal{H}(\vec{f})$, $s(\vec{x})$, and SR given in the previous section are for a particular temporal realization of the random process $\epsilon(\vec{x})$. To compute the average and variance of these performance metrics requires knowledge of the statistics of $\epsilon(\vec{x})$. In Chapter 3 Gaussian statistics were used to describe the atmospherically induced statistical fluctuations of the phase in the pupil, $\phi(\vec{x})$. It is straightforward to argue that the residual phase, $\epsilon(\vec{x})$, will also have Gaussian statistics since $\hat{\phi}(\vec{x})$ is linearly related to $\phi(\vec{x})$ (see the WFS and controller models of Sections 5.3.2 and 5.3.3). With $\epsilon(\vec{x})$ characterized as a Gaussian random process, the only other parameters needed to fully characterize the residual phase statistics is the mean and correlation function. In Section 3.5.3 of Chapter 3 the incident phase $\phi(\vec{x})$ is assumed to be zero mean. This zero mean assumption and the linearity of the relationship between $\epsilon(\vec{x})$ and $\phi(\vec{x})$, implies that $\epsilon(\vec{x})$ is also zero mean. This fact leaves only the requirement to specify the correlation function for $\epsilon(\vec{x})$.

Let $\Gamma_\epsilon(\vec{x},\vec{x}')$ designate the correlation function of $\epsilon(\vec{x})$:

$$\Gamma_\epsilon(\vec{x},\vec{x}') = \mathrm{E}\left\{\epsilon(\vec{x})\epsilon(\vec{x}')\right\}. \tag{5.85}$$

In the special case in which the AOI system applies no wave front compensation $\hat{\phi}(\vec{x})$ is zero and $\epsilon(\vec{x})$ reduces to the uncorrected wave front $\phi(\vec{x})$. This case corresponds to uncompensated imaging, and the correlation properties (as characterized by the phase structure function $D_\phi(\vec{x}-\vec{x}')$) of $\epsilon(\vec{x})$ match those discussed in Section 3.4 of Chapter 3. If on the other hand the AOI system does apply wave front compensation the correlation properties of $\epsilon(\vec{x})$ become more complicated. We will show that these correlation properties still depend on the phase structure function $D_\phi(\vec{x}-\vec{x}')$ but the dependence is complicated by the AOI system components. In particular we note that $\Gamma_\epsilon(\vec{x},\vec{x}')$, or equivalently, the residual phase structure function

$$\begin{aligned} D_\epsilon(\vec{x},\vec{x}') &= \mathrm{E}\left\{\left[\epsilon(\vec{x}) - \epsilon(\vec{x}')\right]^2\right\}, \\ &= \Gamma_\epsilon(\vec{x},\vec{x}) + \Gamma_\epsilon(\vec{x}',\vec{x}') - 2\Gamma_\epsilon(\vec{x},\vec{x}'), \end{aligned} \tag{5.86}$$

will not, in general, display the wide-sense stationary properties of $\phi(\vec{x})$.

5.4.3 Role of Γ_ϵ and D_ϵ in the computation of the performance metrics

To illustrate how the correlation function Γ_ϵ and the phase structure function D_ϵ are needed for the computation of the metrics defined earlier let us specifically consider the role of Γ_ϵ and D_ϵ in the calculation of $\mathrm{E}\{\epsilon^2\}$, $\mathrm{E}\{\mathcal{H}(\vec{f})\}$, and $\mathrm{E}\{|\mathcal{H}(\vec{f})|^2\}$. The simplest to consider is $\mathrm{E}\{\epsilon^2\}$. The mean squared, aperture averaged, residual phase is found by performing an ensemble average of both sides of Eq. (5.80):

$$\begin{aligned}\mathrm{E}\{\epsilon^2\} &= \int \mathrm{d}^2\vec{x}\, W(\vec{x})\mathrm{E}\{\epsilon^2(\vec{x})\},\\ &= \int \mathrm{d}^2\vec{x}\, W(\vec{x})\Gamma_\epsilon(\vec{x},\vec{x}),\end{aligned} \tag{5.87}$$

where the second line of the expression illustrates the dependence of $\mathrm{E}\{\epsilon^2\}$ on Γ_ϵ.

Next consider the computation of the average OTF, $\mathrm{E}\{H(\vec{f})\}$. The average OTF is found by taking the ensemble average of both sides of Eq. (5.81)

$$\mathrm{E}\left\{\mathcal{H}(\vec{f})\right\} = N_F^{-1}\mathrm{E}\left\{\left(W(\vec{f}\lambda d_i)\exp(j\epsilon(\vec{f}\lambda d_i)) \star W(\vec{f}\lambda d_i)\exp(j\epsilon(\vec{f}\lambda d_i))\right)\right\}. \tag{5.88}$$

Making use of the definition of the correlation operator $\star$ defined in Eq. (2.43) we can rewrite Eq. (5.88) as

$$\mathrm{E}\left\{\mathcal{H}(\vec{f})\right\} = N_F^{-1}\int \mathrm{d}^2\vec{x}\, W(\vec{x})W(\vec{x}-\vec{f}\lambda d_i)\mathrm{E}\left\{\exp\left[j\left(\epsilon(\vec{x})-\epsilon(\vec{x}-\vec{f}\lambda d_i)\right)\right]\right\}. \tag{5.89}$$

We now make use of the fact that $\epsilon(\vec{x})$ is a zero mean Gaussian random process. Equation (3.46) of Chapter 3 presents a general result for Gaussian random variables that can be used to simplify Eq. (5.89). This result states that given b is a zero mean Gaussian random variable we can state the following relationship

$$\mathrm{E}\left\{\exp[jb]\right\} = \exp\left[-\frac{1}{2}\mathrm{E}\left\{|b|^2\right\}\right]. \tag{5.90}$$

Using Eq. (5.90) we can write Eq. (5.89) as

$$\mathrm{E}\left\{\mathcal{H}(\vec{f})\right\} = N_F^{-1}\int \mathrm{d}^2\vec{x}\, W(\vec{x})W(\vec{x}-\vec{f}\lambda d_i)\exp\left[-\frac{1}{2}\mathrm{E}\left\{\left(\epsilon(\vec{x})-\epsilon(\vec{x}-\vec{f}\lambda d_i)\right)^2\right\}\right]. \tag{5.91}$$

We can now easily write Eq. (5.91) in terms of the correlation function Γ_ϵ,

$$\begin{aligned}\mathrm{E}\left\{\mathcal{H}(\vec{f})\right\} &= N_F^{-1}\int \mathrm{d}^2\vec{x}\, W(\vec{x})W(\vec{x}-\vec{f}\lambda d_i)\\ &\times\exp\left[-\frac{1}{2}\left(\Gamma_\epsilon(\vec{x},\vec{x})+\Gamma_\epsilon(\vec{x}-\vec{f}\lambda d_i,\vec{x}-\vec{f}\lambda d_i)-2\Gamma_\epsilon(\vec{x},\vec{x}-\vec{f}\lambda d_i)\right)\right],\end{aligned} \tag{5.92}$$

or in terms of the residual phase structure function D_ϵ,

$$\mathrm{E}\left\{\mathcal{H}(\vec{f})\right\} = N_F^{-1}\int \mathrm{d}^2\vec{x}\, W(\vec{x})W(\vec{x}-\vec{f}\lambda d_i)\exp\left[-\frac{1}{2}D_\epsilon(\vec{x},\vec{x}-\vec{f}\lambda d_i)\right]. \tag{5.93}$$

The second moment of the OTF can be found using a development that is identical to the development presented for the speckle transfer function presented in Section 4.3.1 of Chapter 4. It is easy to show that the speckle transfer result of Eq. (4.22) can be generalized by replacing D_ψ with D_ϵ:

$$\begin{aligned} \mathrm{E}\left\{|\mathcal{H}(\vec{f})|^2\right\} = N_F^{-2} \int \mathrm{d}^2\vec{x}\, W(\vec{x})W(\vec{x}-\vec{f}\lambda d_i) \exp\left[-\frac{1}{2}D_\epsilon(\vec{x},\vec{x}-\vec{f}\lambda d_i)\right] \\ \times \int \mathrm{d}^2\vec{x}'\, W(\vec{x}')W(\vec{x}'-\vec{f}\lambda d_i) \exp\left[-\frac{1}{2}D_\epsilon(\vec{x}',\vec{x}'-\vec{f}\lambda d_i)\right] \\ \times \exp\left[-\frac{1}{2}\left(D_\epsilon(\vec{x},\vec{x}') + D_\epsilon(\vec{x}-\vec{f}\lambda d_i,\vec{x}'-\vec{f}\lambda d_i) - D_\epsilon(\vec{x}-\vec{f}\lambda d_i,\vec{x}') - D_\epsilon(\vec{x},\vec{x}'-\vec{f}\lambda d_i)\right)\right]. \end{aligned} \tag{5.94}$$

This result is more complicated than the expression given by Eq. (4.22) since we can not assume that $D_\epsilon(\vec{x},\vec{x}')$ can be written as a function of $\vec{x}-\vec{x}$. With the second moment result of Eq. (5.94) it is straightforward to compute the variance of the OTF using Eq. (4.38):

$$\mathrm{var}\left\{\mathcal{H}(\vec{f})\right\} = \mathrm{E}\left\{|\mathcal{H}(\vec{f})|^2\right\} - \left|\mathrm{E}\left\{\mathcal{H}(\vec{f})\right\}\right|^2. \tag{5.95}$$

In some situations the statistics of the real and imaginary parts of $\mathcal{H}(\vec{f})$ are of interest. It is obvious from the result given above that $\mathrm{E}\{\mathcal{H}(\vec{f})\}$ is entirely real. This observation implies that the average real part of $\mathcal{H}(\vec{f})$ is $\mathrm{E}\{\mathcal{H}(\vec{f})\}$ and that the average of the imaginary part of $\mathcal{H}(\vec{f})$ is zero. The variance of the real and imaginary parts of $\mathcal{H}(\vec{f})$ is also of interest. In particular these second order statistics are shown to be useful in computing the fourth order moment $\mathrm{E}\{|\mathcal{H}(\vec{f})|^4\}$ in Section 4.3.2 of Chapter 4. Using a development similar to the development of the second moment $\mathrm{E}\{|\mathcal{H}(\vec{f})|^2\}$ in Section 4.3.1 of Chapter 4 it is easy to show that

$$\mathrm{var}\left\{\mathrm{Re}\left[\mathcal{H}(\vec{f})\right]\right\} = \frac{1}{2}\left[\mathrm{E}\left\{|\mathcal{H}(\vec{f})|^2\right\} + \mathrm{E}\left\{\mathcal{H}^2(\vec{f})\right\}\right] - \left|\mathrm{E}\left\{\mathcal{H}(\vec{f})\right\}\right|^2, \tag{5.96}$$

$$\mathrm{var}\left\{\mathrm{Im}\left[\mathcal{H}(\vec{f})\right]\right\} = \frac{1}{2}\left[\mathrm{E}\left\{|\mathcal{H}(\vec{f})|^2\right\} - \mathrm{E}\left\{\mathcal{H}^2(\vec{f})\right\}\right], \tag{5.97}$$

where the operators $\mathrm{Re}[\cdot]$ and $\mathrm{Im}[\cdot]$ are the real and imaginary parts of the argument and $\mathrm{E}\{\mathcal{H}^2(\vec{f})\}$ is given by an expression nearly identical to $\mathrm{E}\{|\mathcal{H}(\vec{f})|^2\}$ given in Eq. (5.94) except for a critical sign difference in the argument of the third exponential argument:

$$\begin{aligned} \mathrm{E}\left\{\mathcal{H}^2(\vec{f})\right\} = N_F^{-2} \int \mathrm{d}^2\vec{x}\, W(\vec{x})W(\vec{x}-\vec{f}\lambda d_i) \exp\left[-\frac{1}{2}D_\epsilon(\vec{x},\vec{x}-\vec{f}\lambda d_i)\right] \\ \times \int \mathrm{d}^2\vec{x}'\, W(\vec{x}')W(\vec{x}'-\vec{f}\lambda d_i) \exp\left[-\frac{1}{2}D_\epsilon(\vec{x}',\vec{x}'-\vec{f}\lambda d_i)\right] \\ \times \exp\left[\frac{1}{2}\left(D_\epsilon(\vec{x},\vec{x}') + D_\epsilon(\vec{x}-\vec{f}\lambda d_i,\vec{x}'-\vec{f}\lambda d_i) - D_\epsilon(\vec{x}-\vec{f}\lambda d_i,\vec{x}') - D_\epsilon(\vec{x},\vec{x}'-\vec{f}\lambda d_i)\right)\right]. \end{aligned} \tag{5.98}$$

As expected, the variances of the real and imaginary parts given by Eqs. (5.96) and (5.97) sum to the variance of $|\mathcal{H}(\vec{f})|$ given by Eq. (5.95).

Summary

In the development of the performance metrics described above it is clear that the correlation function and the structure function of the residual phase, Γ_ϵ and D_ϵ, plays a critical role in numerical evaluation of the derived expressions. We next turn our attention to obtaining the expressions that described the required second order statistical correlation properties of the residual phase.

5.4.4 Computing Γ_ϵ

Since the phase structure function for the residual phase, D_ϵ, is related to the correlation function Γ_ϵ in a straightforward manner (see Eq. (5.86)) we concentrate in this subsection on developing a generalized expression for Γ_ϵ. Substituting the expression for $\epsilon(\vec{x})$ given by Eq. (5.79) into Eq. (5.85) gives

$$\Gamma_\epsilon(\vec{x}, \vec{x}') = \mathrm{E}\left\{\left(\phi(\vec{x}) - \hat{\phi}(\vec{x})\right)\left(\phi(\vec{x}') - \hat{\phi}(\vec{x}')\right)\right\}. \tag{5.99}$$

Next substituting in the model for the wave front correction $\hat{\phi}(\vec{x})$ given by Eq. (5.4) we obtain

$$\Gamma_\epsilon(\vec{x}, \vec{x}') = \mathrm{E}\left\{\left(\phi(\vec{x}) - \mathbf{c}^T\mathbf{r}(\vec{x})\right)\left(\phi(\vec{x}') - \mathbf{c}^T\mathbf{r}(\vec{x}')\right)\right\}, \tag{5.100}$$

where we have used the vector notation introduced in Section 5.3.3. The vector $\mathbf{c}$ is a column vector consisting of actuator control signals. The ith element of $\mathbf{c}$ is the control signal associated with the ith actuator. The vector $\mathbf{r}(\vec{x})$ is composed of the deformable mirror actuator influence functions. The ith element of $\mathbf{r}(\vec{x})$ is the influence function associated with the ith actuator. The control vector $\mathbf{c}$ is related to the control matrix, M, and the WFS measurement vector $\mathbf{s}_m$ by Eq. (5.42):

$$\mathbf{c} = M\mathbf{s}_m. \tag{5.101}$$

The elements of $\mathbf{s}_m$ are discussed in Section 5.3.2 and defined by Eqs. (5.15) and (5.25) for the Hartmann and shearing interferometer wave front sensors, respectively. Derivations of a number of different control matrices, M, are presented in Section 5.3.3. Substituting the relationship between $\mathbf{c}$, M and $\mathbf{s}_m$ given by Eq. (5.101) into Eq. (5.100) gives

$$\Gamma_\epsilon(\vec{x}, \vec{x}') = \mathrm{E}\left\{\left(\phi(\vec{x}) - \mathbf{s}_m^T M^T\mathbf{r}(\vec{x})\right)\left(\phi(\vec{x}') - \mathbf{s}_m^T M^T\mathbf{r}(\vec{x}')\right)\right\}. \tag{5.102}$$

Expanding Eq. (5.102) gives

$$\begin{aligned}\Gamma_\epsilon(\vec{x}, \vec{x}') = {} & \mathrm{E}\left\{\phi(\vec{x})\phi(\vec{x}')\right\} + \mathbf{r}^T(\vec{x})M\mathrm{E}\left\{\mathbf{s}_m\mathbf{s}_m^T\right\}M^T\mathbf{r}(\vec{x}') \\ & -\mathrm{E}\left\{\phi(\vec{x}')\mathbf{s}_m^T\right\}M^T\mathbf{r}(\vec{x}) - \mathrm{E}\left\{\phi(\vec{x})\mathbf{s}_m^T\right\}M^T\mathbf{r}(\vec{x}').\end{aligned} \tag{5.103}$$

Before examining the individual terms of Eq. (5.103) we first note that anisoplanatic and temporal effects can be easily incorporated in the evaluation of the Γ_ϵ. Recall that anisoplanatic effects arise in AOI systems from differences between the sensed wave front and the object wave front. The AOI system uses the optical field from the beacon source for sensing the wave front phase deformations while applying the wave front correction to the wave front associated with the object. As discussed in Section 5.2.1 anisoplanatic effects typically arise when the WFS beacon source is spatially separated from the direction to the imaged object. Temporal effects result in the same type of degradation as anisoplanatic effects. The temporal degradation of an AOI system arises from the difference in time between wave front sensing and the application of wave front correction. Anisoplanatic and temporal effects can be incorporated into the mathematical framework established in this section by associating $\phi_o(\vec{x}, t)$ with the object wave front phase deformation and $\phi_b(\vec{x}, t)$ with the beacon wave front phase deformation. Using this notation we rewrite Eq. (5.103) as

$$\begin{aligned}\Gamma_\epsilon(\vec{x}, \vec{x}') = {} & \mathrm{E}\left\{\phi_o(\vec{x}, t)\phi_o(\vec{x}', t)\right\} + \mathbf{r}^T(\vec{x})M\mathrm{E}\left\{\mathbf{s}_m(t)\mathbf{s}_m^T(t)\right\}M^T\mathbf{r}(\vec{x}') \\ & -\mathrm{E}\left\{\phi_o(\vec{x}', t)\mathbf{s}_m^T(t)\right\}M^T\mathbf{r}(\vec{x}) - \mathrm{E}\left\{\phi_o(\vec{x}, t)\mathbf{s}_m^T(t)\right\}M^T\mathbf{r}(\vec{x}'),\end{aligned} \tag{5.104}$$

where the dependence on the beacon wave front phase, $\phi_b(\vec{x}, t)$, is embedded in the WFS slope vector $\mathbf{s}_m(t)$. To illustrate this dependence the two WFS models of Eqs. (5.15) and (5.25) are rewritten using the notation for the beacon wave front, $\phi_b(\vec{x})$, along with a generalization to account for the possibility of temporal filtering of the slope measurement as given by Eq. (5.77):

$$\vec{s}_m(t) = -\int_{-\infty}^{t} \mathrm{d}\tau \left(\int \mathrm{d}\vec{x}\, \nabla W_s(\vec{x}) \phi_b(\vec{x}, t-\tau) \right) h(\tau) + \vec{s}_n, \qquad \text{(H–WFS)}, \tag{5.105}$$

$$\vec{s}_m(t) = \int_{-\infty}^{t} \left(\int \mathrm{d}\vec{x}\, W_s(\vec{x}) \left[\Delta\phi_b(\vec{x}, \Delta\vec{x}_x, t-\tau)\hat{x} \right.\right.$$

$$\left.\left. + \Delta\phi_b(\vec{x}, \Delta\vec{x}_y, t-\tau)\hat{y} \right] \right) h(\tau) + \vec{s}_n, \qquad \text{(SI–WFS)}, \tag{5.106}$$

where $h(t)$ is an impulse response that accounts for the possibility of a temporal filtering of the WFS measurement and $\Delta\phi_b(\vec{x}, \Delta\vec{x}, t)$ is the phase difference defined by Eq. (5.18) with $\phi(\vec{x})$ replaced with $\phi_b(\vec{x}, t)$. In the simple case in which the effect of a time delay between wave front sensing and correction is being investigated the impulse response takes the form $h(t) = \delta(t-\tau)$, where τ is the time delay. A finite integration time in combination with a wave front correction delay can be modeled with $h(t) = \tau_{int}^{-1}\mathrm{rect}((t-\tau)/\tau_{int})$ where τ_{int} is the WFS integration time. Note that the orthogonal components of the vector slope expressions given above each represent individual scalar elements of the column vector $\mathbf{s}_m(t)$.

In order to derive an expression for Γ_ϵ it is apparent that a number of statistical correlations are fundamental to the calculation. From Eq. (5.104) it can be seen these correlations are

$$\mathrm{E}\left\{\phi_o(\vec{x}, t)\phi_o(\vec{x}', t)\right\} = \Gamma_o(\vec{x}, \vec{x}', t, t), \tag{5.107}$$

$$\mathrm{E}\left\{\mathbf{s}_m(t)\mathbf{s}_m^T(t)\right\} = S, \tag{5.108}$$

$$\mathrm{E}\left\{\phi_o(\vec{x}, t)\mathbf{s}_m^T(t)\right\} = \Gamma_{os}(\vec{x}, t)., \tag{5.109}$$

where Γ_o is a scalar correlation function of $\phi_o(\vec{x}, t)$, S is for the slope correlation matrix and Γ_{os} is a correlation vector for the object phase $\phi_o(\vec{x}, t)$ and the slope measurement vector $\mathbf{s}_m(t)$. Expressions for these correlations have been presented and used by a number of researchers for AOI system performance [11, 15, 18, 21, 30]. The correlations can be found in a straightforward manner by using the expression for the piston removed wave front phase deformation, $\phi_o(\vec{x}, t)$, given by Eq. (5.3) and the WFS measurement models of Eqs. (5.105) and (5.106). Rewriting Γ_ϵ in terms of the notation given in Eqs. (5.107) through (5.109) gives

$$\Gamma_\epsilon(\vec{x}, \vec{x}') = \Gamma_o(\vec{x}, \vec{x}', t, t) + \mathbf{r}^T(\vec{x}) M S M^T \mathbf{r}(\vec{x}')$$

$$-\Gamma_{os}(\vec{x}', t) M^T \mathbf{r}(\vec{x}) - \Gamma_{os}(\vec{x}, t) M^T \mathbf{r}(\vec{x}'). \tag{5.110}$$

In the next three subsections we consider how the correlations given in Eqs. (5.107) through (5.109) are related to the object and beacon phase statistics. Ultimately we will be interested in establishing this relationship when the object and beacon phase statistics are applicable for propagation through atmospheric turbulence.

Object phase correlation function, Γ_o:

First consider the phase correlation given by Eq. (5.107). Substituting Eq. (5.3) in (5.107) gives

$$\Gamma_o(\vec{x},\vec{x}',t,t) = \mathrm{E}\left\{\psi_o(\vec{x},t)\psi_o(\vec{x}',t) - \int \mathrm{d}\vec{x}''W(\vec{x}'')\psi_o(\vec{x}'',t)\psi_o(\vec{x}',t)\right.$$
$$\left. - \int \mathrm{d}\vec{x}'''W(\vec{x}''')\psi_o(\vec{x}''',t)\psi_o(\vec{x},t) + \int\int \mathrm{d}\vec{x}''\mathrm{d}\vec{x}'''W(\vec{x}'')W(\vec{x}''')\psi_o(\vec{x}'',t)\psi_o(\vec{x}''',t)\right\} \tag{5.111}$$

where $\psi_o(\vec{x},t)$ is the object wave front phase deformation before piston removal. The phase structure function D_o is defined by

$$D_o(\vec{x}-\vec{x}',t-t') - \mathrm{E}\left\{(\psi_o(\vec{x},t) - \psi_o(\vec{x}',t'))^2\right\}. \tag{5.112}$$

Using Eq. (5.112) to rewrite Eq. (5.111) results in

$$\Gamma_o(\vec{x},\vec{x}',t,t) = \frac{1}{2}\int \mathrm{d}\vec{x}''W(\vec{x}'')\left[D_o(\vec{x}-\vec{x}'',0) + D_o(\vec{x}'-\vec{x}'',0)\right]$$
$$-\frac{1}{2}\int\int \mathrm{d}\vec{x}''\mathrm{d}\vec{x}'''W(\vec{x}'')W(\vec{x}''')\left[D_o(\vec{x}-\vec{x}',0) + D_o(\vec{x}''-\vec{x}''',0)\right]. \tag{5.113}$$

A change of variables reduces Eq. (5.113) to its simplest form:

$$\Gamma_o(\vec{x},\vec{x}',t,t) = \frac{1}{2}\int \mathrm{d}\vec{x}''W(\vec{x}'')\left[D_o(\vec{x}-\vec{x}'',0) + D_o(\vec{x}'-\vec{x}'',0)\right]$$
$$-\frac{1}{2}\int \mathrm{d}\Delta\vec{x}\left[W(\Delta\vec{x})\star W(\Delta\vec{x})\right]\left[D_o(\vec{x}-\vec{x}',0) + D_o(\Delta\vec{x},0)\right], \tag{5.114}$$

where the notation $\star$ designates autocorrelation as defined by Eq. (2.43) of Chapter 2. Since Eq. (5.114) is not strictly a function of the difference $\vec{x}-\vec{x}'$, we conclude the piston removed phase does not display the wide sense stationary properties assumed for the incident wave front phase $\psi_o(\vec{x},t)$. Even though this wide sense stationary property is lost it is easy to show $\phi_o(\vec{x},t)$ retains the "stationarity in first increments" property of $\psi_o(\vec{x},t)$. Stationarity in first increments implies that the phase structure function depends only on the difference in spatial coordinates. In fact it is easy to show that the phase structure functions for $\phi_o(\vec{x},t)$ and $\psi_o(\vec{x},t)$ are identical:

$$\mathrm{E}\left\{(\phi_o(\vec{x},t) - \phi_o(\vec{x}',t))^2\right\} = \mathrm{E}\left\{(\psi_o(\vec{x},t) - \psi_o(\vec{x}',t))^2\right\}. \tag{5.115}$$

Slope correlation matrix, S:

The slope-to-slope measurement correlation matrix of Eq. (5.108) is found by substituting in one of the slope measurement models of Eq. (5.105) or (5.106). For illustration purposes consider using the H–WFS slope measurement model for computing the ijth element of S:

$$\mathrm{E}\left\{s_m(i,t)s_m(j,t)\right\} = \int_{-\infty}^{t}\int_{-\infty}^{t} \mathrm{d}\tau\mathrm{d}\tau' \int\int \mathrm{d}\vec{x}\mathrm{d}\vec{x}'\,(\nabla W_{si}(\vec{x})\cdot\vec{d}_i)(\nabla W_{sj}(\vec{x}')\cdot\vec{d}_j)$$
$$\times\mathrm{E}\left\{\phi_b(\vec{x},t-\tau)\phi_b(\vec{x}',t-\tau')\right\}h(\tau)h(\tau') + \mathrm{E}\left\{s_n(i)s_n(j)\right\}, \tag{5.116}$$

where the notation $s_m(i,t)$ is used to designate the ith slope measurement at time t and $s_n(i)$ is the measurement noise associated with the ith measurement. The vector $\vec{d}_i$ is an unit vector designating the direction of the slope sensitivity of the ith measurement. The result given in Eq. (5.116) is obtained by recognizing the independence and zero mean nature of the measurement noise $s_n(i)$ and the beacon wave front phase $\phi_b(\vec{x}, t)$. To avoid the possibility of confusion concerning the subscript notation used in the subaperture weighting function $W_{si}(\vec{x})$, it should be noted that each subaperture yields two slope measurements (one for each orthogonal sensitivity direction). Since each slope measurement is indexed with i there will be two distinct index values, i and i', for which $W_{si}(\vec{x}) = W_{si'}(\vec{x})$ and $\vec{d}_i$ is perpendicular to $\vec{d}_{i'}$. Hence, the slope measurement index i is not an index for the WFS subaperture.

In order to compute the slope measurement correlation given by Eq. (5.116) the beacon wave front phase and measurement noise correlation functions are required. The beacon phase correlation, $\Gamma_b(\vec{x}, \vec{x}', t, t') = \mathrm{E}\{\phi_b(\vec{x}, t)\phi_b(\vec{x}', t')\}$, is derived in a manner nearly identical to the derivation of the object phase correlation function, $\Gamma_o(\vec{x}, \vec{x}', t, t) = \mathrm{E}\{\phi_o(\vec{x}, t)\phi_o(\vec{x}', t)\}$. The only difference between the two derivations is the requirement to incorporate the time difference between the beacon wave front phases. Using the space-time phase structure function defined for the incident beacon wave front $\psi_b(\vec{x}, t)$,

$$D_b(\vec{x} - \vec{x}', t - t') = \mathrm{E}\left\{(\psi_b(\vec{x}, t) - \psi_b(\vec{x}', t'))^2\right\}, \tag{5.117}$$

it is straightforward to show using a development similar to the one used to obtain Eq. (5.114) that

$$\begin{aligned}\Gamma_b(\vec{x}, \vec{x}', t, t') = {} & \frac{1}{2}\int \mathrm{d}\vec{x}'' W(\vec{x}'')\left[D_b(\vec{x} - \vec{x}'', t - t') + D_b(\vec{x}' - \vec{x}'', t - t')\right] \\ & - \frac{1}{2}\int \mathrm{d}\Delta\vec{x}\left[W(\Delta\vec{x}) \star W(\Delta\vec{x})\right]\left[D_b(\vec{x} - \vec{x}', t - t') + D_b(\Delta\vec{x}, t - t')\right].\end{aligned} \tag{5.118}$$

Substituting Eq. (5.118) into Eq. (5.116) gives

$$\begin{aligned}\mathrm{E}\{s_m(i, t)s_m(j, t)\} = {} & -\frac{1}{2}\int_{-\infty}^{t}\int_{-\infty}^{t} \mathrm{d}\tau \mathrm{d}\tau' \int\int \mathrm{d}\vec{x}\mathrm{d}\vec{x}'\ (\nabla W_{si}(\vec{x}) \cdot \vec{d}_i)(\nabla W_{sj}(\vec{x}') \cdot \vec{d}_j) \\ & \times D_b(\vec{x} - \vec{x}', \tau' - \tau)h(\tau)h(\tau') + \mathrm{E}\{s_n(i)s_n(j)\}.\end{aligned} \tag{5.119}$$

This simplified result is obtained by taking advantage of the fact that the integral of $\nabla W_{si}(\vec{x})$ over an even function of $\vec{x}$ is zero if $W_{si}(\vec{x})$ is even.

Since the slope measurement noise is typically modeled as being zero mean and independent for each subaperture, the noise correlation $\mathrm{E}\{s_n(i)s_n(j)\}$ is given by the following model:

$$\mathrm{E}\{s_n(i)s_n(j)\} = \begin{cases} \sigma_n^2 & \text{for } i = j \\ 0 & \text{for } i \neq j \end{cases}, \tag{5.120}$$

where σ_n^2 is the noise variance for the slope measurement. The magnitude of the measurement noise variance is a function of the WFS type, light level, read noise and atmospheric turbulence effects. In Section 5.3.2 we presented expressions that can be used to characterize σ_n for the H–WFS and SI–WFS.

Object phase and slope correlation, Γ_{os}:

The final correlation required to compute Γ_ϵ is fundamentally different from the previous correlation quantities. Whereas the correlations Γ_o and S ultimately depend on auto-correlations of the object and beacon phases, the correlation Γ_{os} depends on the cross correlation between the object and beacon phases. Substituting the H–WFS slope measurement model of Eq. (5.105) into Eq. (5.109) gives the ith element of the vector Γ_{os}

$$\mathrm{E}\{\phi_o(\vec{x},t)s_m(i,t)\} = -\int_{-\infty}^{t} \mathrm{d}\tau \int \mathrm{d}\vec{x}' \, (\nabla W_{si}(\vec{x}') \cdot \vec{d}_i)\mathrm{E}\{\phi_o(\vec{x},t)\phi_b(\vec{x}',t-\tau)\} \, h(\tau). \quad (5.121)$$

Just as in the derivation of the Γ_o and S the computation hinges on characterizing a phase correlation function. In this case we must characterize the cross-correlation

$$\Gamma_{ob}(\vec{x},\vec{x}',t,t') = \mathrm{E}\{\phi_o(\vec{x},t)\phi_b(\vec{x}',t')\}. \quad (5.122)$$

Substituting the respective piston removed phase expressions for $\phi_o(\vec{x},t)$ and $\phi_b(\vec{x}',t')$ into Eq. (5.122) gives

$$\begin{aligned}\Gamma_{ob}(\vec{x},\vec{x}',t,t') = \frac{1}{2} \int \mathrm{d}\vec{x}'' W(\vec{x}'') \left[D_{ob}(\vec{x},\vec{x}'',t-t') + D_{ob}(\vec{x}',\vec{x}'',t-t')\right] \\ -\frac{1}{2} \int\int \mathrm{d}\vec{x}''\mathrm{d}\vec{x}''' W(\vec{x}'')W(\vec{x}''') \left[D_{ob}(\vec{x},\vec{x}',t-t') + D_{ob}(\vec{x}'',\vec{x}''',t-t')\right], \quad (5.123)\end{aligned}$$

where D_{ob} is the cross phase structure function defined as

$$D_{ob}(\vec{x},\vec{x}',t-t') = \mathrm{E}\left\{(\psi_o(\vec{x},t) - \psi_b(\vec{x}',t'))^2\right\}. \quad (5.124)$$

Note that the cross phase structure function defined above is purposely *not* written as a function of $\vec{x} - \vec{x}'$ since its functional dependence is more complicated. In the next section we will investigate the form of the phase structure functions D_o, D_b and D_{ob} as they relate to optical wave propagation through atmospheric turbulence.

Summary

The key equations required to compute the residual pupil phase correlation function, Γ_ϵ, have been presented in this section. These equations have been written in terms of the auto- and cross-correlation functions of the object and beacon wave fronts. Equations (5.114), (5.118) and (5.123) illustrate that these wave front correlation functions ultimately depend on the auto- and cross phase structure functions D_o, D_b and D_{ob}. The next section introduces a mathematical form for these structure functions for Kolmogorov turbulence.

5.4.5 Atmospheric phase statistics

It is apparent from the expressions for Γ_o, S, and Γ_{os} that knowledge of the auto- and cross-correlation properties of the object and beacon wave front phases, $\psi_o(\vec{x},t)$ and $\psi_b(\vec{x},t)$, respectively, are required. In particular, the expressions for these phase correlation functions, Eqs. (5.114), (5.118) and (5.123), show that all that is required is knowledge of the respective phase structure functions. The required phase structure functions are

$$D_o(\vec{x},\vec{x}',t,t') = \mathrm{E}\{(\psi_o(\vec{x},t) - \psi_o(\vec{x}',t'))^2\}, \quad (5.125)$$

$$D_b(\vec{x},\vec{x}',t,t') = \mathrm{E}\{(\psi_b(\vec{x},t) - \psi_b(\vec{x}',t'))^2\}, \quad (5.126)$$

$$D_{ob}(\vec{x},\vec{x}',t,t') = \mathrm{E}\{(\psi_o(\vec{x},t) - \psi_b(\vec{x}',t'))^2\}. \quad (5.127)$$

In the special case that the phase deformations are induced by propagation through atmospheric turbulence, the development of the phase structure functions presented in Section 3.4.2 of Chapter 3 is applicable. In particular the geometry of Fig. 3.7 is directly applicable for the calculation of the phase structure functions given above. For convenience Fig. 3.7 is repeated in Fig. 5.11 using the notation applicable for the object and beacon wave fronts. In this figure the object location is designated by $\vec{r}_o$ and the beacon location is designated by $\vec{r}_b$. Two points in the pupil plane are designated with the vectors $\vec{x}$ and $\vec{x}'$. The straight ray path from the object to the pupil point $\vec{x}$ is the path traversed by the object wave front at time t. The wandering line from the beacon to the pupil point represents the difference in the transverse location of the atmosphere at some time t' later ($t' > t$). The wandering nature of this second ray path is due to the time difference between t' and t and the transverse velocity of the turbulence layers (see Figure 3.6 of Section 3.4.3).

A phase structure function accounting for anisoplanatic effects for Kolmogorov turbulence is given by Eq. (3.93) of Chapter 3. This phase structure function is easily generalized to account for time differences by using the "frozen flow" concepts of Section 3.4.3. Generalizing Eq. (3.93) to account for the time differences between the wave front phases gives

$$D_\psi(\vec{x}, \vec{x}', t, t') = 6.88 \int_0^L \mathrm{d}z \left(\frac{|\Delta\vec{p}(z, t - t')|}{r_0} \right)^{5/3} \tilde{C}_n^2(z). \tag{5.128}$$

where L is the length of the vertical propagation path to the minimum altitude source, r_0 is the plane wave Fried parameter defined in Eq. (3.61), $\tilde{C}_n^2(z)$ is a normalized index-of-refraction fluctuation structure constant defined in Eq. (3.88), and $\Delta\vec{p}$ is the transverse separation of the two ray paths shown in Fig. 5.11. The transverse separation depends on the phase structure function being computed. The three required expressions for computing D_o, D_b or D_{ob} are given by

$$\Delta\vec{p}(z, t - t') = \begin{cases} \left(1 - \frac{z}{z_o}\right)(\vec{x} - \vec{x}') - \vec{v}(z)(t - t') & \text{for computing } D_o, \\ \left(1 - \frac{z}{z_b}\right)(\vec{x} - \vec{x}') - \vec{v}(z)(t - t') & \text{for computing } D_b, \\ \vec{x} - \vec{x}' + \frac{z}{z_o}(\vec{r}_0 - \vec{x}) & \\ \qquad - \frac{z}{z_b}(\vec{r}_b - \vec{x}') - \vec{v}(z)(t - t') & \text{for computing } D_{ob}. \end{cases} \tag{5.129}$$

The z coordinates, z_o and z_b, represent the altitudes of object and beacon sources, respectively. The velocity $\vec{v}(z)$ is the transverse velocity of the turbulence layers as a function of altitude.

5.4.6 Getting results

In the presentation of Sections 5.3 and 5.4 the goal has been to present a generalized mathematical framework from which the performance of nearly any AOI system can be predicted. This mathematical framework includes models for the deformable mirror (DM), the wave front sensor (WFS) and the controller, as well as definitions of the performance metrics of interest and the required statistical quantities. In this section we discuss two approaches for actually computing the performance metrics defined in subsection 5.4.1. First we discuss the obvious approach involving numerical evaluation of the analytical expressions for the performance metrics. All of the required expressions and quantities to facilitate this type of analytical computation have been given in subsections 5.4.1 through 5.4.5. The second approach involves performing Monte Carlo simulations of the AOI system. In this latter approach the AOI system models of Section 5.3 are used in conjunction with a random wave front phase screen generator. The performance metrics defined in subsection 5.4.1 are computed for each realization of the generated wave front phase and performance statistics are accumulated over a large number of wave front realizations. In the remainder of this section we discuss the issues, advantages and disadvantages of these two approaches for computing performance results.

Numerical evaluation of the analytical performance expressions

Numerical evaluation of the analytical performance expressions given in Eqs. (5.87) through (5.98)

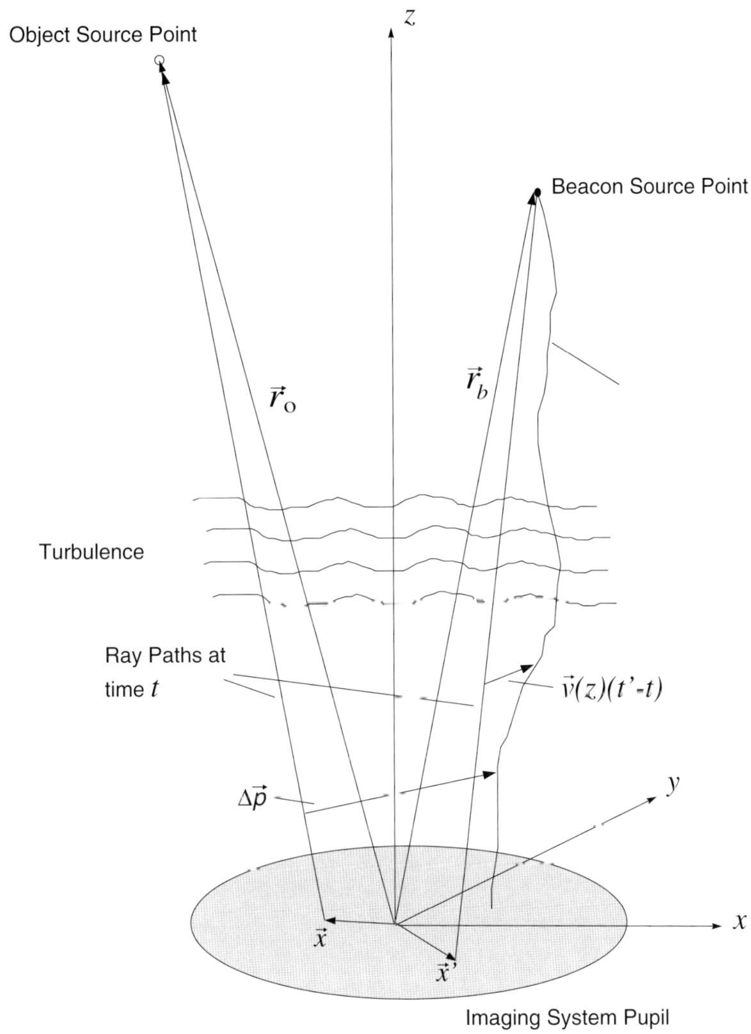

FIGURE 5.11
Geometry for deriving the structure functions for the object and beacon wave front phases.

offers a straightforward way of quantifying the statistical performance of the AOI system. The results of the numerical evaluation are inherently statistical and as a consequence are applicable for describing the average performance of the AOI system for a large number of independent realizations of the atmospheric phase deformations. In the language of atmospheric optics, a large number of independent wave front realizations is synonymous with a long exposure.

Possibly the most significant advantage associated with analytical performance computations is the ability to use the mathematical framework presented here to compute "fundamental performance limitations". When we speak of "fundamental performance limitations" we are referring to the types of computations in which generally all, except one, of the the AOI system components are idealized (i.e., assumed to perform perfectly). This type of assumption lends itself to investigations of the fundamental limitations associated with individual AOI system components or individual atmospheric turbulence effects. To emphasize how to set up and perform these fundamental performance calculations three examples are considered in Section 5.4.7.

Another advantage of computing performance using the analytical expressions in Eqs. (5.87) through (5.98) includes the fact that these expressions are exact and accuracy is only limited by the accuracy of the numerical solution of the integrals. Accuracy is achieved at the expense of computational intensity. This computational intensity is one of the major disadvantages of performing analytical computations. Even though the integrands of the integrals in question are all generally well behaved functions, all the integrals of interest are at least four dimensional. The integrals in question are like those in Eqs. (5.87), (5.93), and (5.94). Each of these integrals in turn rely on the evaluation of the residual phase correlation Γ_ϵ which involves integrals like those in Eqs. (5.114), (5.118) and (5.123). For many calculations the required computations are extremely time consuming even on the fastest widely available workstations. Even so, a number of researchers that have performed the required computations to study the performance of complete AOI systems. A partial set of examples include the papers in the following reference list: [10, 11, 15, 18, 19, 21, 30, 32, 38, 39, 40, 41, 42, 43, 44, 90]. In these papers it is typical to see that a number of approximations are made to speed up the numerical computations. The approximations normally result in one or more of the integrations being evaluated analytically. One of the most common assumptions is that the structure constant profile, $C_n^2(z)$, and the wind velocity, $\vec{v}(z)$, are constant over the optical path. These two assumptions generally result in an expression that can be analytically evaluated for the phase structure function given by Eq. (5.128). Another common simplifying assumption is modeling the distributed atmosphere as a finite collection of thin turbulent layers, as discussed in Section 3.4.1. Others have simplified the integrals involved in the H–WFS model, Eq. (5.105). In this case the gradient of the subaperture weighting function, $\nabla W_s(\vec{x})$, is represented with Dirac-delta function along the edge of the aperture. This representation is used in Chapter 6 in the development of Section 6.2.2 and leads to an easy evaluation of the two dimensional integral in the H–WFS model of Eq. (5.105). Besides these approximations it has been observed that integrals of the form given by Eqs. (5.114), (5.118) and (5.123) can be evaluated for circular apertures using a Gegenbauer expansion technique [38]. In fact the results of this expansion have already been used in the numerical evaluation of the piston-removed phase correlation function given by Eq. (3.211) of Chapter 3. The Gegenbauer expansion of the integrals results in an infinite sum that is useful for speeding up the computations since the sum converges rapidly. Tyler [38] presents an excellent summary of the use of this expansion for evaluating integrals of the type given in Eqs. (5.114), (5.118) and (5.123).

Another drawback associated with analytical performance computations is that the performance results are inherently statistical quantities. These statistical performance measures do not give a direct insight into how the AOI system performs for individual realizations of the wave front phase deformations. Finally, some performance measures simply do not lend themselves to solutions that can be written down in terms of analytically tractable expressions. A prime example is computing the statistics of the phase of the OTF for an AOI system. Understanding the phase statistics of the OTF is extremely useful for understanding how successful image reconstruction techniques might be. Obtaining an analytical expression for the statistics of the OTF phase is extremely difficult [22]. More recently we have seen the rise and acceptance of Monte Carlo simulations to obtain numerical results for the performance metrics defined earlier as well as performance metrics that defy analytical solution. The next subsection presents a discussion of the simulation approach.

Simulating AOI systems

In Section 3.7 of Chapter 3 we discussed the motivation for generating random phase screens. The motivational factors discussed in Section 3.7 were primarily directed toward the drawbacks discussed above for the analytical computations. Besides the computational intensity associated with numerically evaluating the analytical performance expressions, some performance metrics simply cannot be written in a form which can be evaluated analytically. At this point a Monte Carlo simulation of the AOI system becomes the only feasible performance calculation approach. Simulations have found widespread use for investigating the performance of optical systems operating in the presence of atmospheric turbulence [13, 14, 21, 22, 45, 91, 92, 93, 94]. In these simulations the components of an optical system are modeled using exactly the same AOI component models presented in Section 5.3 [21]. The simulations are executed by generating random realizations of the wave front phase in the pupil of the system. Using the component models of the optical system it is possible to employ the Fourier optics concepts presented in Chapter 2 to propagate each realization of the pupil field through the system. Statistical performance results are obtained by calculating the performance metrics for each realization of the pupil phase and accumulating the results. Section 3.7 contains an extensive discussion on how to generate the random phase screens with the proper first and second order statistics.

It is instructive to consider the performance measures associated with the OTF when discussing how simulations address the drawbacks discussed above for analytical computations. A single simulation can be used to compute any number of the statistical measures of the OTF: the mean and variance, mean and variance of the real and imaginary parts, the mean and variance of the phase , and any other higher order statistical quantities desired. These OTF performance measures are evaluated on a two dimensional grid of sampled spatial frequency points simultaneously. Due to the fact that simulations are computationally less intensive than analytical computations the spacing of the frequency samples is generally much smaller than the sample spacing used in analytic calculations. The computational requirements for simulations are primarily driven by the number of Monte Carlo runs required to achieve some level of convergence for the statistical quantities of interest. The other major computational driver is the number of sample points required to adequately sample the pupil of the AOI system.

Besides making feasible some performance computations that are simply not possible using analytical computations, the simulation offers the ability to observe the performance of the AOI system for individual realizations of the phase screen. This capability leads to the ability to form simulated imagery: a capability absent from analytical computations. In addition, the imagery generated by the simulation can be corrupted by noise effects attributed to photon and additive camera detector noise. Both of these noise effects have been discussed extensively in Section 2.3.2. Simulating the effects of noise in the image measurement is extremely important for investigations involving image reconstruction and speckle imaging and other more specialized techniques such as statistical frame selection [95].

Simulations do present a problem with respect to the accuracy of the finite dimensional representation of the wave front phase in the pupil of the AOI system. Since the wave front phase is a continuous random process, a representation of the phase would theoretically require an infinite dimensional basis set. Due to the fact that the simulated wave front phase, $\phi(\vec{x})$, is inherently represented using a finite dimensional representation there will be unavoidable errors in this representation. The errors can be reduced by increasing the dimension of the phase representation, but at the cost of increased computational time. If $\phi(\vec{x})$ is being represented by a finite sum of Zernike polynomials (as discussed in Section 3.6.1) the dimension of the representation is increased simply by including more polynomials in the sum. Generally the finite sum includes the lowest order polynomials (i.e., the lowest spatial frequency terms) and increasing the number of terms improves the higher order (i.e., the high spatial frequency) representation of $\phi(\vec{x})$. If $\phi(\vec{x})$ is being represented by a rectangular grid of discrete phase values (as discussed in Section 3.7.2) the dimension of the representation is increased by decreasing the spacing of the grid points which in turn leads to a larger number of total grid points. Decreasing the grid spacing results in a higher fidelity representation of higher spatial frequencies. This increased fidelity is attributed to a lessening of aliasing. Another inaccuracy associated with the discrete grid representation of the phase is associated with the fact that the array of grid points is only represented over a finite area. Usually this area is referred to as a window and corresponds to an area at least as large as the area of the imaging system pupil, $W(\vec{x})$. This "windowing" gives rise to an effect called Gibbs phenomenon. The window results in a convolution of the power spectrum of the sampled random process with the Fourier transform of the window. Generally Gibbs phenomenon is not such a problem for power spectra in which the size of the window is large and the structure in the power spectrum is relatively smooth. The problem encountered in simulating atmospheric induced phase deformations lies in the fact that, for Kolmogorov statistics, the power spectrum of $\phi(\vec{x})$ rises sharply and goes to infinity as the spatial frequency approaches zero. For this particular power spectrum, the convolution of the Fourier transform of the window has the effect of smearing the low spatial frequency portion of the power spectrum and, as a result, under representing the power in the lowest spatial frequencies. These lowest order wave front phase deformations are not accurately represented in the simulated wave front phase. Cochran [96] has addressed this problem by specifically adding in the appropriate amount of low order wave front deformations to achieve the correct power spectrum power level at low spatial frequencies. The effect of Gibbs phenomenon can also be lessened by using the Von Karman index-of-refraction power spectrum described in Section 3.3 instead of the Kolmogorov spectrum.

The block diagram of a simulation suitable for simulating the optical performance of an AOI system is shown in Fig. 5.12. The essential elements of simulating an AOI system are

- the setup block,
- the phase screen generator,
- the tilt removal system,
- the WFS model,
- the actuator command computer,
- the phase screen spatial shifter,
- the DM model, and
- the OTF computation.

Here we explain some of the important implementation details of each functional block. Note that the discussion presented here dwells on the minimum required functions of an AOI imaging simulation. Enhancements to this minimal configuration are possible, and some will be discussed in Chap 6.

Setup: The purpose of the setup block is to read in the key parameters for the simulation, and to perform one–time calculations needed later in the simulation. Specific quantities which must be read in include the pupil diameter D, WFS geometry information, DM actuator location information, the Fried seeing parameter r_0, a nominal wavelength λ_n, the mean wavelength for image formation λ_{IMG}, the mean wavelength for WFS operation λ_{WFS}, and the mean numbers of photo-events per integration time in the image plane $\overline{K}$ and in the WFS $\overline{K}_W$. Additional quantities which may be read in include the wind velocity $\vec{v}$, the time delay between sensing and correcting for the wave front τ, and a decision about whether to model a separate tilt removal system. If images are desired the object must either be created within the simulation, such as for the trivial case of a point source, or read in from outside the simulation, such as in the case of an extended object. One–time computations which must be performed include calculations needed for the phase screen generator, such as computing and factoring the phase covariance matrix as discussed in Section 3.7, computation of the reconstruction matrix M, and computation of the object spatial frequency spectrum. The reconstruction matrix M can be computed using the techniques described in Section 5.3.3, and the generalized notation M is used to represent any of the reconstruction matrices. Note that the ordering of the elements of M must be consistent with the ordering of the WFS

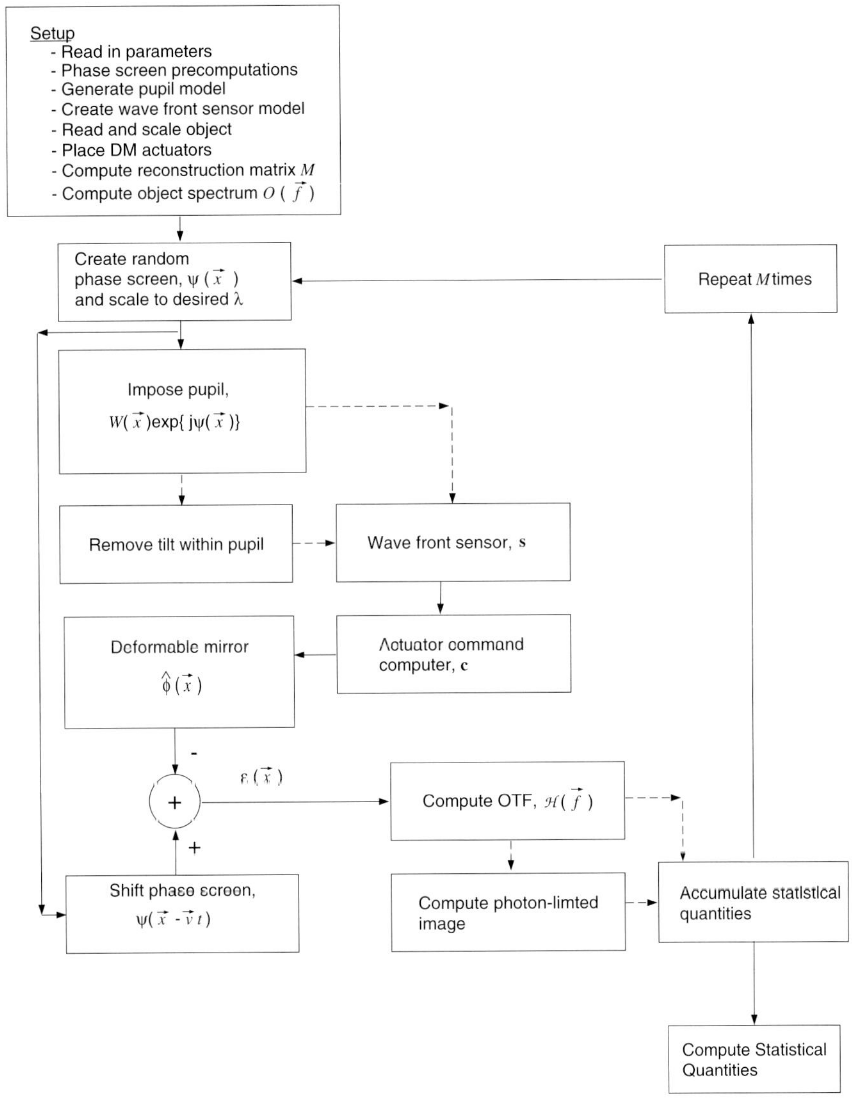

FIGURE 5.12
Block diagram for simulating the performance of a AOI system.

slope measurement vector $\mathbf{s}_m$ and the actuator control signal vector $\mathbf{c}$, defined in Eq. (5.42).

Defining the pupil geometry is one of the most important aspects of the setup block. In particular, the number of samples across the pupil N_{pup}, the sample spacing Δx, the number of samples per WFS subaperture, and the number of subapertures across the pupil diameter must be established. If it is assumed that one sample in the pupil model lies at the center of the pupil, then the sample spacing in the pupil Δx is given by

$$\Delta x = \frac{D}{N_{pup} - 1}. \tag{5.130}$$

The overall requirement that $r_0/\Delta x \geq 2$ must also be met to insure that the simulated wave front is well–sampled.

Phase screen generator: The value of the Fried seeing parameter r_0 is generally established at some nominal wavelength λ_n. The phase screen can be generated by any suitable means which produces the desired wave front statistics, including the techniques discussed in Section 3.7 and in Ref. [97]. Phase screens are generated to have the Fried seeing parameter value of r_0 at the wavelength λ_n. To serve as inputs for the WFS and DM models the phase screen must be scaled to a desired wavelength, λ_d. The appropriate relationship for this scaling is

$$\psi(\vec{x}, \lambda_d) = \frac{\lambda_n}{\lambda_d}\psi(\vec{x}, \lambda_n), \tag{5.131}$$

where $\psi(\vec{x}, \lambda)$ designates the wave front phase at the point $\vec{x}$ and the wavelength λ. The phase screen is passed to both the tilt removal subsystem and possibly the phase screen shifting routine.

Tilt removal system: While it is possible to model the performance of an AOI system where all wave front correction is provided by the DM, in practice the wave front tilt is removed separately by a tilt correction system [2]. The purpose of the tilt correction system is to remove the Zernike tilt modes, Z_2 and Z_3 (see Section 3.6) from the wave front. Motivation for separately removing the tilt component of the turbulence–induced aberration derives from two considerations: (1) tilt provides the strongest contribution to the turbulence–induced aperture averaged mean square wave front phase error and as such can easily exceed the stroke of the DM; and (2) simple flat mirrors can be used to removed the tilt component of the aberration, while the DM actuator influence functions cannot in general perfectly fit tilt. Because of the inability of a DM to fit the tilt component, the optical performance of systems which explicitly removes tilt using a separate subsystem is known to be superior to those which do not remove tilt [21]. If a tilt removal system is required the tilt component of $\psi(\vec{x}, \lambda)$ must be computed using Eq. (3.154), and subtracted from the wave front phase to create the tilt–removed phase $\phi(\vec{x}, \lambda)$ The optional nature of a separate tilt removal subsystem in the simulation is indicated in Fig. 5.12 by use of broken lines connecting the appropriate blocks. If tilt is not removed, then the wave front is passed directly to the wave front sensor block after the pupil is imposed.

WFS model: The phase $\phi(\vec{x}, \lambda_{WFS})$ is passed to the WFS model to compute the vector of WFS slopes $\mathbf{s}_m$ where λ_{WFS} is the optical wavelength used by the wave front sensor. For the Hartmann sensor, noise free wave front slopes $\mathbf{s}$ are computed most directly using Eq. (5.15). Note that if the subaperture function $W_{sj}(\vec{x})$ is constant within the subaperture and zero outside the subaperture, then the x-direction component of $\nabla W_{sj}(\vec{x})$, denoted $\nabla^x W_{sj}(\vec{x})$ is given by

$$\nabla^x W_{sj}(x, y) = \frac{1}{d^2}\mathrm{rect}\left(\frac{y - y_{cj}}{d}\right)[\delta(x - x_{cj} + d/2) - \delta(x - x_{cj} - d/2)], \tag{5.132}$$

where d is the dimension of a square subaperture, (x_{cj}, y_{cj}) is the center location of the jth subaperture, the function rect(x) is defined in Eq. (2.23) and $\delta(x)$ is a Dirac delta function. The y-directed gradient of $W_{sj}(\vec{x})$ is obtained from Eq. (5.132) by substituting y for x and x for y. In a simulation Eq. (5.15) must be implemented using discrete integration. The trapezoid rule or Simpson's extended rule provide suitable integration strategies for evaluating Eq. (5.15) [98, 99]. If WFS noise effects are to be modeled, a random draw of a Gaussian–distributed random variable with variance given by Eq. (5.33) in the case of a Hartmann WFS is added to each component of the noise free slope to create the noise–corrupted slope measurement. The measured slope vector is then passed to the actuator command computer.

Actuator command computer: The actuator command computer maps slope measurements to actuator control signals using Eq. (5.42). This operation is easily accomplished using a vector–matrix multiplication routine provided in widely available software packages [99, 100]. Actuator control signals are passed to the DM model, and are used to compute the figure of the DM.

Phase screen spatial shifter: If a time delay between sensing and correcting for the wave front aberration $\psi(\vec{x}, \lambda)$ is to be modeled it is necessary to shift the phase screen by the amount $\vec{v}\tau$, where $\vec{v}$ is the vector velocity of the wind, and τ is the time delay between sensing and correcting for the turbulence effects. This shift creates the phase screen $\psi(\vec{x} - \vec{v}\tau, \lambda)$ The phase screen must be shifted *after* the wave front sensor model, but *before* the correction is applied by the DM. Care must be exercised in performing this shift. The shift $\vec{v}\tau$ must in practice be composed of orthogonal components which can be expressed as integer multiples of the sample spacing in the pupil, Δx.

DM model: The deformable mirror model performs two functions: (1) it computes the figure of the DM $\hat{\phi}(\vec{x}, \lambda)$ using Eq. (5.4), based on the actuator command signals and the DM influence functions; and (2) it computes the residual phase error $\epsilon(\vec{x})$ using Eq. (5.79). The residual phase error $\epsilon(\vec{x})$ is then passed to the OTF computation module.

OTF computation: The residual phase error $\epsilon(\vec{x})$ is used to compute the instantaneous OTF using Eq. (5.81). It must be noted that Eq. (5.81) is the autocorrelation of the generalized pupil function. Direct numerical implementation of the required spatial integration is impractical in most cases. More frequently, fast Fourier transform (FFT) techniques are used compute the instantaneous OTF. FFT techniques can be used to evaluate the OTF using the autocorrelation theorem of Fourier transforms which states that for an arbitrary complex-valued function $f(\vec{x})$ [59]

$$\mathcal{F}\{f(\vec{x}) \star f(\vec{x})\} = |F(\vec{f})|^2, \tag{5.133}$$

where $F(\vec{f})$ is the Fourier transform of $f(\vec{x})$. The algorithm for computing the OTF based on Eqs. (5.81) and (5.133) is as follows

$$\mathcal{H}(\vec{f}) = \mathcal{F}^{-1}\left\{|\mathcal{F}\{W(\vec{x})\exp[j\epsilon(\vec{x})]\}|^2\right\}. \tag{5.134}$$

Care must be taken to perform the computation in Eq. (5.134) in an array sufficiently large to hold the outcome of the autocorrelation and avoid wrap-around error [101]. The array size requirement is that if N_{arr} is the size of the array used to implement Eq. (5.134), then $N_{arr} \geq 2N_{pup}$. Further, many FFT routines require that $N_{arr} = 2^n$, where n is a positive integer. Note that $\mathcal{H}(\vec{f})$ given in Eq. (5.134) is defined in a sampled spatial frequency domain. The sample spacing in the frequency domain can be obtained by noting that the radius of the OTF in samples N_{OTF} is

$$N_{OTF} = N_{pup} - 1. \tag{5.135}$$

The angular cutoff frequency for the OTF f_c is given by

$$f_c = D/\lambda_{IMG}, \tag{5.136}$$

so that the sample spacing in the OTF Δf in cycles per radian is given by

$$\Delta f = \frac{f_c}{N_{OTF}} = \frac{D}{\lambda_{IMG}(N_{pup} - 1)}. \tag{5.137}$$

If the instantaneous point spread function $s(\vec{x})$ is computed by applying the inverse FFT to $\mathcal{H}(\vec{f})$ then the sample spacing in $s(\vec{x})$, Δx_s, is given by

$$\Delta x_s = \frac{1}{N_{arr}\Delta f} = \frac{\lambda_{IMG}(N_{pup} - 1)}{DN_{arr}}, \tag{5.138}$$

where Δx_s is an angle with units of radians. An instantaneous realization of $\mathcal{H}(\vec{f})$ is the desired outcome of the operations described to this point.

Realizations of $\mathcal{H}(\vec{f})$ can be stored or passed to a routine which accumulates quantities required to compute sample-based estimates of desired quantities, such as the mean and the variance of $\mathcal{H}(\vec{f})$. This step is indicated by use of a broken line connecting the OTF computation block to the statistical quantity accumulation block in Fig. 5.12. If adaptive optics images are to be simulated the steps described in Section 4 to create a photon–limited image using the object spectrum and $\mathcal{H}(\vec{f})$ can be followed. The broken lines in Fig. 5.12 connecting the OTF computation block, image computation block, and the statistical quantity accumulation block indicate this possible simulation path.

In summary, a simulation that is validated against analytical computations is an extremely valuable tool for predicting the performance of complicated AOI systems. Simulations address all of the drawbacks discussed for the analytical computation approach. Most importantly, an accurate simulation of a complex AOI system generally requires much less computational resources than analytical performance calculations. This fact, in conjunction with the fact simulations can be used to compute some performance metrics that are analytically intractable, makes AOI system simulations an indispensable tool for both system design as well as basic research.

5.4.7 Fundamental AOI system performance results

In this section we present a number of fundamental AOI system results. The term "fundamental" refers to the fact that the results focus on a single factor that degrades performance while idealizing all other aspects of the AOI system.

Temporal Delay: First consider an investigation of the effect of a time delay between wave front sensing and correction. Suppose that we desire to understand the most fundamental performance limitation associated with this delay without factoring in the limitations of the WFS, the DM, anisoplanatism, and anything else that limits performance but is not connected with the time delay. This goal is easily accomplished by expressing the residual phase as

$$\epsilon(\vec{x}, t) = \phi_o(\vec{x}, t) - \phi_o(\vec{x}, t - \tau), \tag{5.139}$$

and proceeding with the computation of the residual phase correlation function Γ_ϵ in the same manner as described in Section 5.4.4. This type calculation can be thought of in the context of a AOI system that works perfectly except for the fact that a delay τ exists between wave front sensing and correction. In this paradigm the estimated wave front $\hat{\phi}(\vec{x}, t)$, which is normally given by the surface of the DM, is, instead, simply the time delayed object phase, $\phi_o(\vec{x}, t - \tau)$. Since the system WFS and DM are being idealized, the computation of Γ_ϵ is not complicated by the WFS, DM or controller models and as a result it is easy to show using Eq. (5.139) that Γ_ϵ is given by

$$\Gamma_\epsilon(\vec{x}, \vec{x}) = \Gamma_o(\vec{x}, \vec{x}, t, t) + \Gamma_o(\vec{x}, \vec{x}, t - \tau, t - \tau) - 2\Gamma_o(\vec{x}, \vec{x}, t, t - \tau), \tag{5.140}$$

where Γ_o is given by Eq. (5.114). Assume the aperture average mean square phase error, $\mathrm{E}\{\epsilon^2\}$, is the performance metric of interest. By substituting Eq. (5.140) into the expression for $\mathrm{E}\{\epsilon^2\}$ given by Eq. (5.87) and using Eq. (5.114) it can be shown that $\mathrm{E}\{\epsilon^2\}$ reduces to

$$\mathrm{E}\{\epsilon^2\} = \int \mathrm{d}\Delta\vec{x}\, [W(\Delta\vec{x}) \star W(\Delta\vec{x})]\, [D_o(0, \tau) + D_o(\Delta\vec{x}, 0) - D_o(\Delta\vec{x}, \tau)], \tag{5.141}$$

where the phase structure function D_o is given by Eq. (5.128). Carrying the example to completion we have evaluated Eq. (5.141) and plotted it as a function of time delay τ. Figure 5.13 is a plot of $\mathrm{E}\{\epsilon^2\}/(D/r_0)^{5/3}$ versus normalized time delay $v\tau/D$, where v is the transverse velocity of the turbulent atmosphere and D is the dimension of the telescope aperture. To obtain the result shown in Fig. 5.13 the atmospheric velocity profile $\vec{v}(z)$ was assumed to be constant with respect

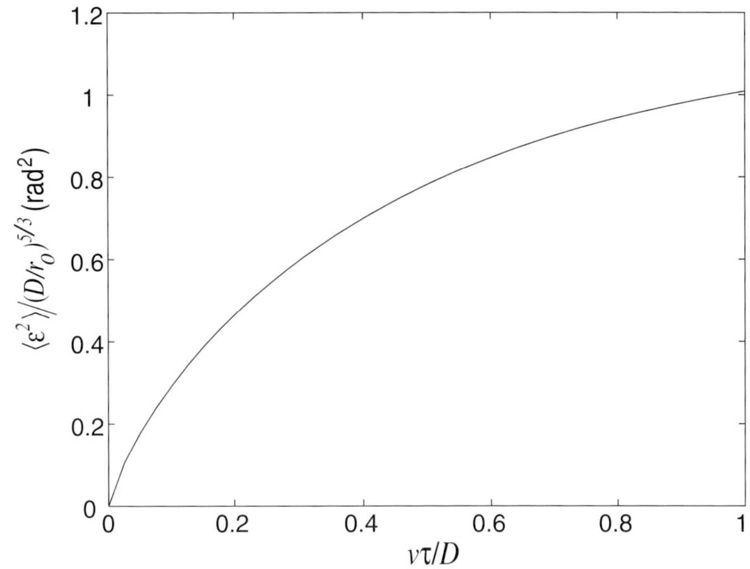

FIGURE 5.13
Plot of the aperture average, mean square residual phase, $E\{\epsilon^2\}/(D/r_0)^{5/3}$ versus $v\tau/D$, where D is the dimension of the telescope aperture and v is the transverse wind speed of the turbulence layer.

to altitude z. Fried [37] and Karr [102] have also investigated $E\{\epsilon^2\}$ as a function of time delay but have done so for $\psi(\vec{x}, t) - \psi(\vec{x}, t - \tau)$ instead of the piston-removed quantity given by Eq. (5.139). Welsh [36] has investigated the effect of a time delay for the case in which the overall wave front tilt is also removed.

Focus Anisoplanatism: Another example of a fundamental computation is the investigation of focus anisoplanatism. Recall that focus anisoplanatism is a fundamental limitation that effects AOI systems employing finite altitude beacons. The applicable geometry is shown in Fig. 5.3. The residual phase in this case can be expressed as

$$\epsilon(\vec{x}) = \phi_o(\vec{x}) - \phi_b(\vec{x}), \tag{5.142}$$

where $\phi_o(\vec{x})$ is the object wave front and $\phi_b(\vec{x})$ is the beacon wave front and we have dropped the time dependence. The object is modeled as an on axis, celestial source and the beacon is an on axis, finite altitude source. Using Eq. (5.142) the residual phase correlation function Γ_ϵ can be written

$$\Gamma_\epsilon(\vec{x}, \vec{x}) = \Gamma_o(\vec{x}, \vec{x}) + \Gamma_b(\vec{x}, \vec{x}) - 2\Gamma_{ob}(\vec{x}, \vec{x}), \tag{5.143}$$

where Γ_o, Γ_b and Γ_{ob} are given by Eqs. (5.114), (5.118) and (5.123), respectively. Again assuming the aperture average mean square phase error, $E\{\epsilon^2\}$, is the performance metric of interest, it can be shown that

$$\begin{aligned} E\{\epsilon^2\} = \frac{1}{2} \int d\Delta\vec{x}\, [W(\Delta\vec{x}) \star W(\Delta\vec{x})]\, [D_o(\Delta\vec{x}) + D_b(\Delta\vec{x})] \\ - \int\int d\vec{x}'\vec{x}'' W(\vec{x}')W(\vec{x}'')\, [D_{ob}(\vec{x}', \vec{x}'') + D_{ob}(\vec{x}', \vec{x}')], \end{aligned} \tag{5.144}$$

where the phase structure functions D_o, D_b and D_{ob} are given by Eq. (5.128). Equation (5.144) has been evaluated for the four layer C_n^2 models, HV-21, and SLC-N, SLC-D, described in Section

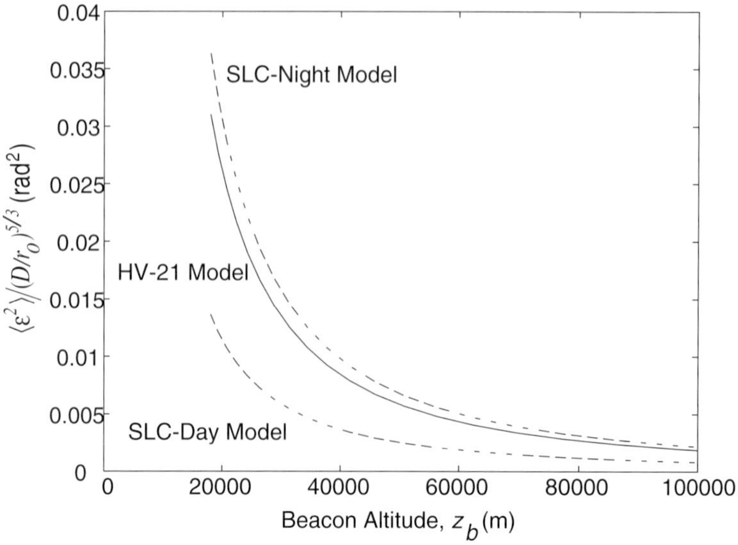

FIGURE 5.14
Plot of the aperture average, mean square residual phase, $E\{\epsilon^2\}/(D/r_0)^{5/3}$ due to focus anisoplanatism versus beacon altitude, z_b. The results are applicable for a circular aperture, an on axis laser beacon and the four layer C_n^2 models, HV-2, SLC-N and SLC-D, described in Section 3.4.1.

3.4.1. The mean square wave front is evaluated as a function of the beacon altitude z_b. Figure 5.14 illustrates a plot of $E\{\epsilon^2\}/(D/r_0)^{5/3}$ versus z_b. As expected, the degradation caused by focus anisoplanatism decreases as the beacon altitude is increased. It is also evident from Fig. 5.14 that C_n^2 models with proportionally more of the turbulence located at low altitudes are less susceptible to focus anisoplanatism. Table 3.1 of Chapter 3 gives the relative weights and altitudes of the turbulence layers for the three C_n^2 models used in the computation of the results for Fig. 5.14. Fried and Belsher [37] have also investigated focal anisoplanatism. In their analysis they argue that the overall tilt should also be removed since artificial beacons provide no information for sensing overall tilt. As a result their analysis is considerably more involved than the simple analysis presented above, and their results also show lower residual mean square errors.

Angular Anisoplanatism: Equation (5.144) is also applicable for investigating angular anisoplanatism. The applicable geometry in this case is shown in Fig. 5.2. Substituting the appropriate expressions from Eq. (5.128) for the phase structure functions for this geometry into Eq. (5.144) gives the mean square wave front error as a function of the angular separation of the beacon and object sources. Figure 5.15 illustrates a plot of $E\{\epsilon^2\}/(D/r_0)^{5/3}$ as a function of object-beacon angular separation, θ. Again we see that the turbulence profiles having proportionally more of the turbulence at low altitudes are less susceptible to angular anisoplanatism. Fried [10], Sasiela [103], and Stone [104] have also investigated the fundamental limitations associated with angular anisoplanatism.

As these three examples illustrate, fundamental performance calculations are easily accomplished within the mathematical models and framework presented in this chapter. A number of authors have performed investigations of the fundamental nature as described above. An example includes Greenwood's [34] investigation of the effects of finite bandwidth controllers on AOI system performance. This investigation resulted in the definition of a characteristic atmospheric frequency: the Greenwood frequency f_g. For the first order loop compensator described by Eq. (5.75) the mean square residual

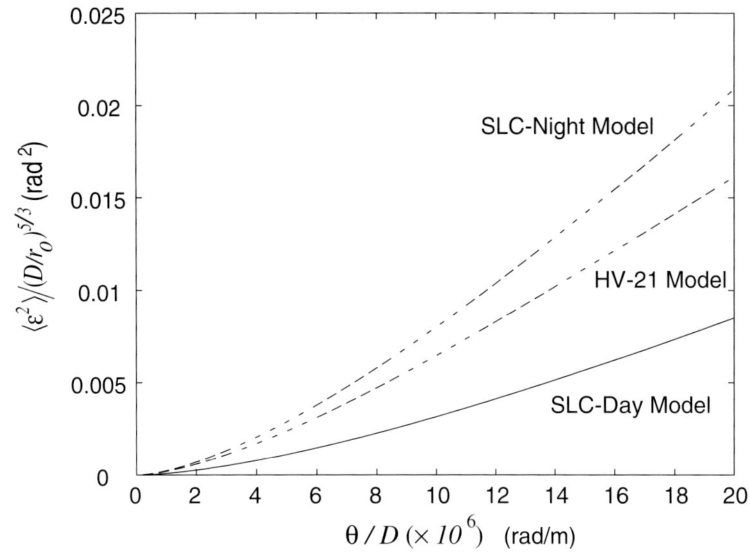

FIGURE 5.15
Plot of the aperture average, mean square residual phase, $E\{\epsilon^2\}/(D/r_0)^{5/3}$ versus the normalized angular separation of the object and beacon, θ/D. The parameter D is the diameter of the circular telescope aperture. The results are applicable for the four layer C_n^2 models, HV-2, SLC-N and SLC-D, described in Section 3.4.1.

phase error is related to the Greenwood frequency by [37]

$$E\{\epsilon^2\} = \left(\frac{f_g}{f_{3\text{dB}}}\right)^{5/3}, \tag{5.145}$$

where $f_{3\text{dB}}$ is the 3 dB bandwidth of the compensator. For the case in which the atmosphere is modeled with a single atmospheric turbulence layer, f_g is related to the Fried parameter by

$$f_g = 0.426\frac{|\vec{v}|}{r_0}, \tag{5.146}$$

where $\vec{v}$ is the transverse velocity of the layer. Another example is Tyler's [27, 38] investigation of the bandwidth requirements for tilt compensation as well as the limiting wave front correction performance possible using multiple laser beacon sources.

5.4.8 Examples of AOI system performance calculations

In the following subsection we present a series of representative AOI system performance calculation results. The purpose here is to illustrate both the utility of the analysis approach described in the previous sections and to also demonstrate some basic results that will be of interest to a broad range of readers. The AOI system considered in the following subsections has the following general characteristics:

- *Telescope pupil:* The pupil is annular with an obscuration ratio of $D_o/D = 0.4$, where D is the pupil diameter and D_o is the diameter of the obscuration. The pupil diameter is assumed to be 3 m. The pupil of the AOI system is shown in Fig. 5.16.
- *Wave Front Sensor (WFS):* The WFS is a Hartmann sensor with square subapertures of side length d. The number of subapertures spanning the pupil diameter ranges from 15 to 16. Figure 5.16

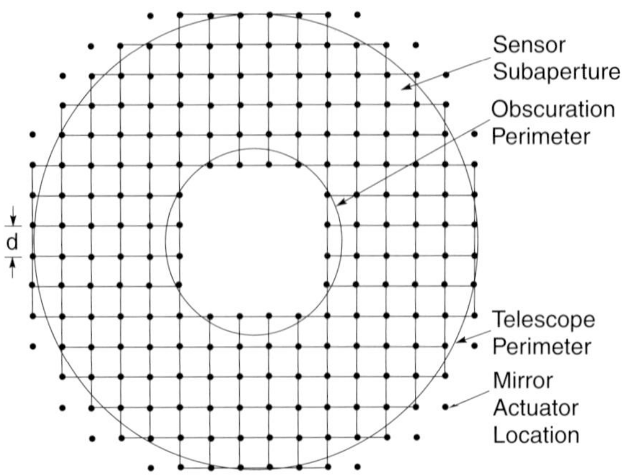

FIGURE 5.16
Telescope pupil, deformable mirror and wave front sensor geometry.

shows the H–WFS subaperture geometry for the case in which 15 subapertures span the diameter. The slope measurements are corrupted by shot noise with a standard deviation given by [15]

$$\sigma_n = \begin{cases} \dfrac{0.86\pi\eta}{(\overline{K}_W)^{1/2} r_0} \text{ rad/m} & d > r_0 \\ \dfrac{0.74\pi\eta}{(\overline{K}_W)^{1/2} d} \text{ rad/m} & d \leq r_0 \end{cases}, \tag{5.147}$$

where η is a parameter accounting for imperfections of the detector array used in the focal plane of the WFS (i.e., the effects of finite size detector elements and dead space between elements) and $\overline{K}_W$ is the average number of photon events per WFS slope measurement. We note that the shot noise expression given by Eq. (5.147) is nearly identical to the H–WFS quad cell expression given by Eq. (5.31). The expression given above was derived for an array of detectors as opposed to a quad cell and employs a simplified approach for accounting for the degradations caused by atmospheric effects by introducing r_0 in the denominator for $d > r_0$. The factor η is greater than or equal to unity and η is unity only in the ideal case of a perfect detector array (infinitely small pixels and zero dead-space). For the computational results of the following subsections we assume $\eta = 1.35$. This value is typical of the minimum achievable value for currently available intensified CCD detector arrays [70].

- *Deformable Mirror (DM):* The mirror is assumed to be of monolithic design with an actuator spacing equal to the subaperture dimensions d. The influence function for the ith actuator, $r_i(\vec{x})$, is modeled by a Gaussian function:

$$r_i(\vec{x}) = \exp\left(\frac{-(\vec{x} - \vec{x}_i)^2}{d_a^2}\right), \tag{5.148}$$

where $\vec{x}_i$ specifies the location of the ith actuator and d_a is the influence radius. The Gaussian response is often used to model piezoelectric or membrane deformable mirrors [15]. For the results presented here we assume $d_a = d$.

- *Beacon:* We consider cases in which the AOI system uses both a finite altitude laser beacon as well as a natural guide star. In order to incorporate the effects of anisoplanatism we account for the altitude of the guide star as well as the altitude distribution of the turbulence. When we refer to the effects of anisoplanatism we restrict our attention to the degradation of the on axis wave front compensation caused by the difference in the optical ray path from the guide star and the

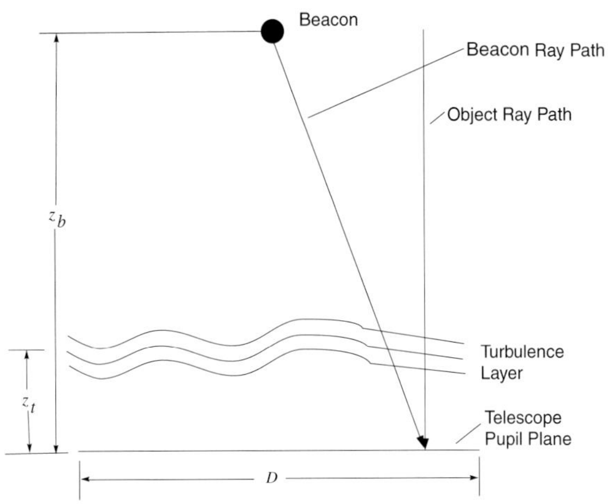

FIGURE 5.17
Laser Beacon-Pupil Geometry

ray path from the celestial object of interest. The difference between these ray paths is shown in Figure 5.17. The height of the guide star and the height of the turbulence in conjunction with the telescope diameter are factors that determine the magnitude of the anisoplanatic degradation. The effects of anisoplanatism can also be considered in terms of the limitations imposed on the field-of-view (FOV) of the adaptive telescope system. This limitation is not considered here, but has been treated elsewhere [10, 18, 31, 32].

Gardner *et al.* [17] describe two possible approaches to creating the laser guide star. They consider guide stars created at a relatively low altitude within the stratosphere using Rayleigh scattering from air molecules. They also consider guide stars created using resonance fluorescence scattering at a relatively high altitude within the mesospheric sodium (Na) layer. For the results presented here we restrict our attention to guide stars created in the Na layer. The Na layer provides an excellent scattering medium for creation of laser guide stars because of its high altitude and its large resonant backscattering coefficient [52]. The layer has a mean altitude of 92 km, and its average thickness is about 10 km FWHM. The high altitude of the layer offers considerable advantages in terms of the number of guide stars required to fully sample the telescope pupil [17]. Due to the finite altitude of the guide star and the small size of the isoplanatic angle at visible wavelengths, multiple laser guide stars are required to compensate a large telescope pupil. This was first pointed out by Foy and Labeyrie [50]. For the results presented here we assume that a single Na guide star at an altitude of $z_b = 92$ km is created on the axis of the telescope. This geometry is shown in Fig. 5.17. We also consider the case in which the reference beacon is a single, on axis, natural star (i.e., $z_b = \infty$).

- C_n^2 *Altitude Profile:* The turbulence altitude profile is characterized by the refractive index structure constant of the refractive index fluctuations, C_n^2. Reference [18] presents an analysis of the effects of anisoplanatism on laser-guided telescopes. In this previous analysis, the authors consider a single layer model of the turbulence altitude profile, as well as the more realistic Hufnagel-Valley continuous C_n^2 model [105]. For the results presented here we use the simpler, single layer model. The single layer is assumed to be located at the altitude z_t. This model is used instead of the more realistic continuous models of Section 3.2 primarily because of the prohibitive amount of computation time required to obtain results using the continuous models.

  In the following subsection we assume the turbulence altitude is $z_t = 10$ km. Assuming the turbulence is concentrated at 10 km will, in most cases, result in a conservative prediction of the

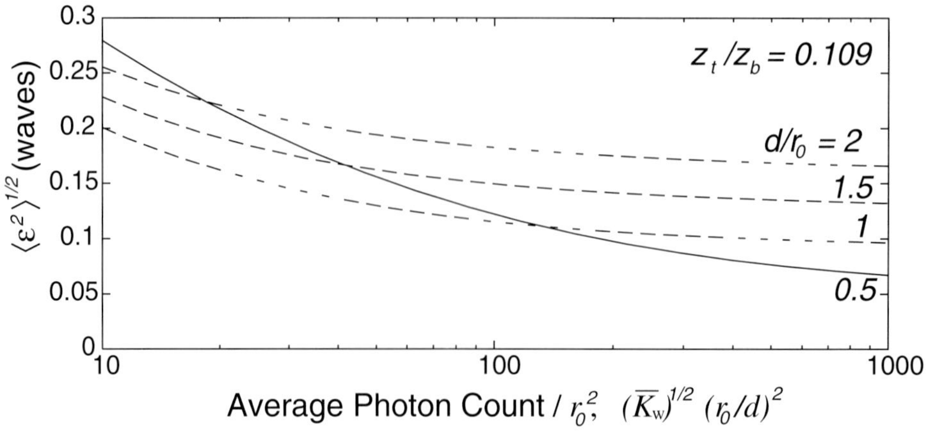

FIGURE 5.18
RMS wave front error, $E\{\epsilon^2\}^{1/2}$ plotted versus photon flux for the AOI system described on pages 219 to 222. The ratio of the turbulence altitude to the guide star altitude is $z_t/z_b = 0.109$. The family of curves are for d/r_0 ranging from 0.5 to 2.

anisoplanatic effects. Generally it would be more appropriate to model the turbulence at a lower altitude.

The results presented in the following subsections have been computed by numerically evaluating the analytical expressions for the performance metrics and Monte Carlo simulation. The analytical expressions are the same as the expressions described in the sections leading up to this point. The simulations are based on the AOI system models described in Section 5.3 and a phase screen generator similar to those described in Section 3.7. The analytical results are largely drawn from Ref. [19] while the simulation results are generated from a family of simulation computer programs that has been used to predict the performance predictions for a number of investigations [20, 21, 22, 45, 46, 91, 92, 93, 94, 106].

The assumptions listed above do not completely specify the characteristics of the AOI system or the atmosphere. These previous assumptions serve only to lay the basic framework for the calculations. The system and atmospheric parameters that have yet to be specified include the WFS element and DM actuator spacing relative to the Fried coherence cell size r_0 and beacon brightness. Many of the results to be presented have applicability to a range of system parameters. For example, the ratio of the subaperture dimension to the Fried cell size, d/r_0, is varied from a minimum value of 0.5 to a maximum value of 2. This range of values allows us to investigate AOI system performance over a performance regime covering "fully" compensated ($d/r_0 \leq 1$) to "partially" compensated ($d/r_0 > 1$). Some of the results will also show how the performance varies with light level in the WFS.

Analytical results: mean square residual phase error, $E\{\epsilon^2\}$

We start by considering the simplest and easiest to compute performance metric: the aperture averaged, RMS residual phase error, $E\{\epsilon^2\}$. Figure 5.18 is a plot of $(E\{\epsilon^2\})^{1/2}$ versus photon flux incident on the WFS for the case of $z_t/z_b = 0.109$. This particular ratio approximately matches the geometry in which the guide star is located at an altitude of 92 km (nominal height of the mesospheric Na layer) and the turbulence is at a height of 10 km. These results were computed by evaluating the analytical expression for $E\{\epsilon^2\}$ given by Eq. (5.87) and assuming the AOI system employs the minimum variance

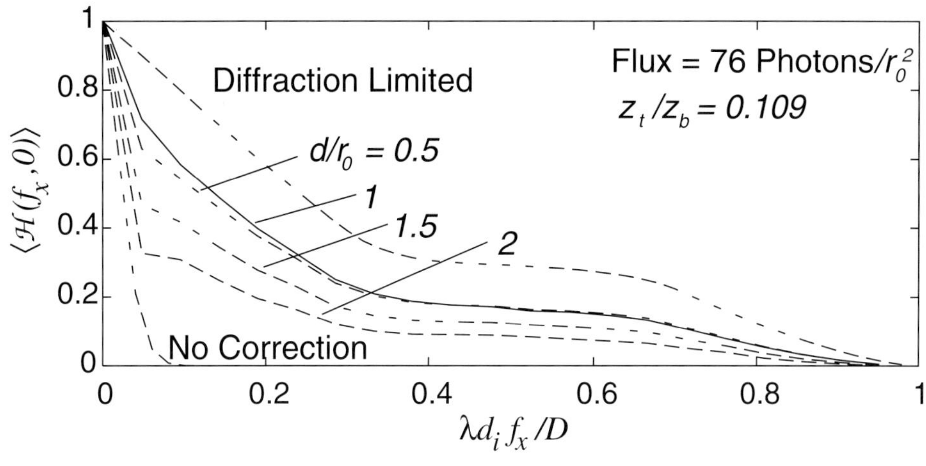

FIGURE 5.19
Ensemble average OTF, E$\{\mathcal{H}(f_x, 0)\}$ versus spatial frequency f_x for the AOI system described on pages 219 to 222. Also shown are OTF's for the case of diffraction limited performance and the case of no wave front compensation. For each curve the flux is 76 photons/r_0^2 and the ratio $z_t/z_b = 0.109$.

(MV) controller described in Section 5.3.3. It is assumed that the AOI system does not remove overall aperture tilt before compensation with the DM. The wave front statistics are characterized using the phase structure function given by Eq. (5.128). As expected, $(\mathrm{E}\{\epsilon^2\})^{1/2}$ decreases with increasing flux. This trend simply indicates the slope measurements become less noisy as the guide star brightness is increased. The asymptotic behavior of each curve as the flux increases is indicative of the limitations imposed on performance by the WFS and mirror actuator sampling intervals. Notice that at the higher flux levels, $(\mathrm{E}\{\epsilon^2\})^{1/2}$ decreases with decreasing subaperture and actuator spacing (i.e., decreasing d/r_0). This trend is expected since decreasing d/r_0 results in smaller sampling intervals in both the WFS and DM, which in turn results in improved sampling and reconstruction of the higher spatial frequency phase deformations. Recall that $d/r_0 \leq 1$ falls within the category of AOI system that we characterize as a fully compensated system. Figure 5.18 indicates that even systems labeled as being fully compensated are fundamentally limited by either the light level or the finite sampling of the incident wave front phase.

In contrast to the trends observed for high light levels, at the lower flux levels we observe performance can actually be better for larger values of d/r_0. Systems characterized by $d/r_0 > 1$ are characterized as partially compensated systems. This reversal in performance is explained by realizing that at the low flux levels, noise is the predominant cause of wave front error. For a given flux level the measurement noise effects will increase for decreasing values of d/r_0. The measurement noise increases because the light collected by each WFS subaperture is proportional to d^2. At low flux levels, decreasing values of d/r_0 increase the effect of measurement noise at a faster rate than the beneficial effects expected from smaller subapertures. These two opposing trends indicate that for a given flux level there will exist an optimum value of d/r_0 that minimizes the wave front error.

Analytical results: OTF and PSF, E$\{\mathcal{H}(\vec{f})\}$ and E$\{s(\vec{x})\}$

The graph in Fig. 5.19 shows the average OTF plotted versus normalized spatial frequency. The OTF is plotted along the x-directed component of $\vec{f}$ and the frequency axis is normalized to the diffraction

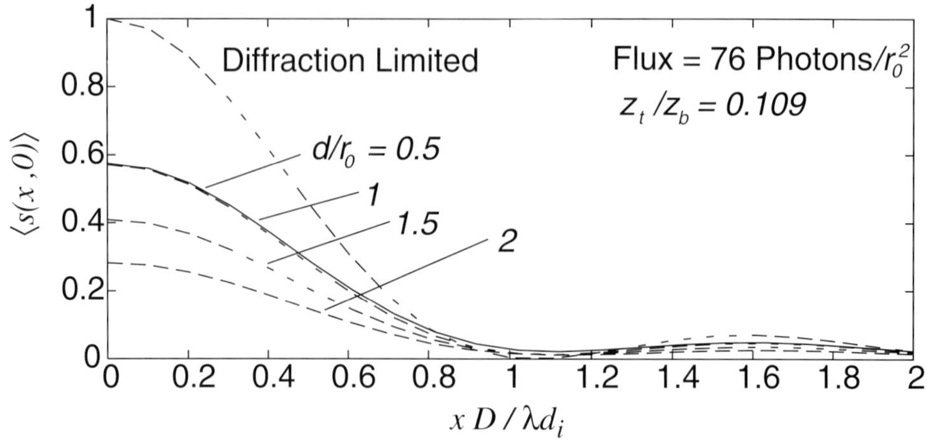

FIGURE 5.20
Ensemble average PSF, E$\{s(x,0)\}$ versus image plane coordinate x for the AOI system described on pages 219 to 222. Also shown is the ensemble PSF for the case of diffraction limited performance. For each curve the flux is 76 photons/r_0^2 and the ratio $z_t/z_b = 0.109$.

limited cut-off of the pupil. The WFS flux is held constant at 76 photons/r_0^2. This level corresponds to the level specified by Gardner *et al.* [17] in their study of AOI systems employing laser beacons. The OTF curves for the case of no correction and for the case of diffraction limited seeing are also shown. Between these two limiting cases are the OTF curves for d/r_0 ranging from 0.5 to 2. These results were computed by numerically evaluating the analytical expression for the OTF given by Eq. (5.93). Notice the significant improvement in the response of the OTF for decreasing values of d/r_0. This improvement is most noticeable in going from the partially compensated system with $d/r_0 = 2$ to the fully compensated system with $d/r_0 = 1$. It is also interesting to notice that for each of the values of d/r_0 shown, the AOI system preserves spatial frequencies out to the diffraction limit. This is in contrast to the case of no wave front correction, where the spatial frequencies are cutoff at $\sim$ 10 percent of the maximum spatial frequency for diffraction limited imaging.

The graph shown in Fig. 5.20 illustrates the average PSF curves derived from the OTF curves shown in Fig. 5.19. Recall that the PSF and OTF are related by a Fourier transform relationship. The figure shows a cross-section of the average PSF plotted versus a normalized image plane coordinate. The image plane coordinate x is normalized by $\lambda d_i/D$. Note how the peak of the PSF curves decrease with decreasing levels of wave front compensation (i.e., increasing d/r_0). We finally note that both the Strehl ratio and the angular resolution of the AOI system can be obtained from the PSF data. The Strehl ratio is the value of the PSF at the origin ($xD/\lambda d_i = 0$) normalized by the diffraction limited value of the PSF at the origin. For the normalization of Fig. 5.20 the Strehl is obtained directly from the graph by identifying the intersection point of the PSF with vertical axis. The angular resolution is derived from some predefined measure of the width of the PSF. The full width, half maximum (FWHM) is a common measure of angular resolution. The interesting point to note from the PSF data shown in Fig. 5.20 is that angular resolution as determined by the FWHM metric. does not significantly degrade as d/r_0 increases. This observation is typical for AOI systems. It is well known that the performance of AOI systems degrade in such a way that the width of the PSF core remains constant as the PSF peak decreases. The decrease of the PSF peak (or equivalently the Strehl ratio) causes more light to be spread

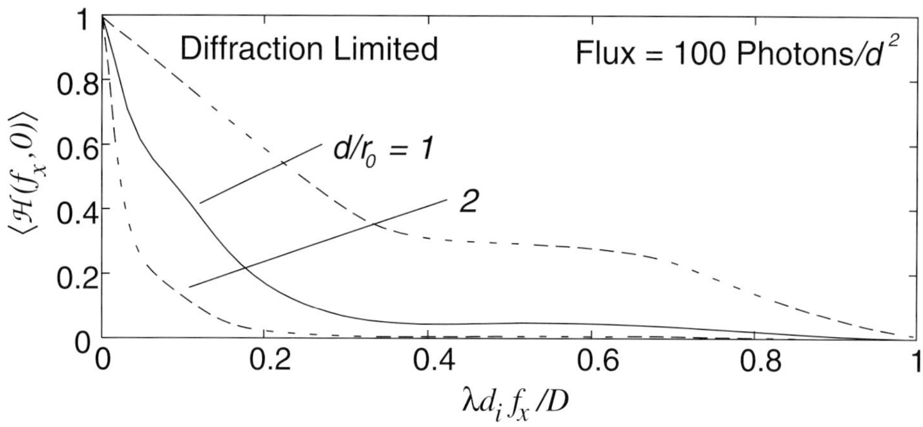

FIGURE 5.21
Simulation result: Ensemble average OTF, E$\{\mathcal{H}(f_x, 0)\}$ versus spatial frequency f_x for the AOI system described on pages 219 to 222. Also shown is the OTF for the case of diffraction limited performance. The WFS light level corresponds to 100 photons/d^2 and the ratio $z_t/z_b = 0.0$ (natural guide star).

into a halo outside the central core.

Simulation results: OTF mean and variance, E$\{\mathcal{H}(\vec{f})\}$ and var$\{\mathcal{H}(\vec{f})\}$

In many situations the variance of $\mathcal{H}(\vec{f})$ is also of interest. Recall that the variance is used together with the mean to quantify the SNR of a particular random quantity. The SNR is defined by Eq. (5.78). To compute the variance requires computation of the second moment of the OTF as given in Eq. (5.94). As seen in Eq.(5.94) the computation of E$\{|\mathcal{H}(\vec{f})|^2\}$ is a four-fold integral over space, in addition to the integrals involved in evaluating the residual phase structure function D_ϵ. In our research we have found it more computationally efficient to use the simulation to evaluate second order quantities such as var$\{\mathcal{H}(\vec{f})\}$, var$\{\text{Re}[\mathcal{H}(\vec{f})]\}$, and var$\{\text{Im}[\mathcal{H}(\vec{f})]\}$. Before showing any second order results we first show the average OTF computed using a simulation like that described in pages 211 to 216. For the following results the simulation parameters have been chosen to closely match the AOI system described at the beginning of this section. In the simulation the overall wave front tilt was removed prior to wave front compensation by the DM. The major reason there is a slight difference between the simulation results and the analytical results presented above is the difference in the wave front reconstruction matrix. For the analytical results presented in Figs. 5.18 through 5.20 the reconstruction matrix is the MV control matrix described by Eq. (5.72). In contrast, for the simulation results the least squares (LS) reconstruction matrix described by Eq. (5.59) is used. Figure 5.21 shows a cross section of E$\{\mathcal{H}(f_x, 0)\}$ obtained from the simulation. For these results the beacon is assumed to be a natural guide star and the light level in the WFS is 100 photo-events/subaperture. The ratio d/r_0 takes on values of 1 and 2. Comparison of this result with the average OTF results of Fig. 5.19 show qualitative similarities between the simulation and analytical computation approaches. It is apparent from the comparison of the levels of the OTF curves in the two figures that the MV controller (results shown in Fig. 5.19) gives better performance than the LS controller (results shown in Fig. 5.21). Roggemann presents a thorough comparison of the minimum variance and least squares wave front reconstruction approaches as well as a demonstration that simulations and the corresponding analytical expressions give identical results

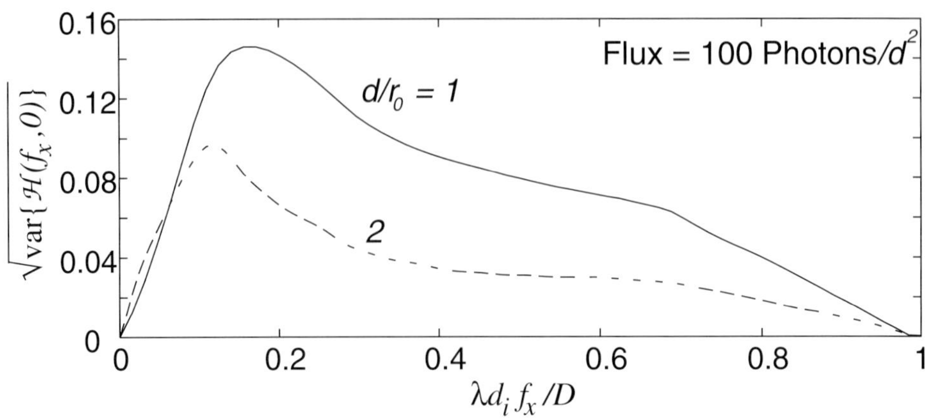

FIGURE 5.22
Standard deviation of $|\mathcal{H}(f_x, 0)|$ versus spatial frequency f_x for the AOI system described on pages 219 to 222. The WFS light level corresponds to 100 photons/d^2 and the ratio $z_t/z_b = 0.0$ (natural guide star).

[21].

Figure 5.22 shows a plot of the standard deviation of the OTF plotted versus spatial frequency. Figures 5.23 and 5.24 show the corresponding standard deviations for the real and imaginary parts of the OTF, respectively. These three graphs show that the OTF fluctuates most for normalized spatial frequencies from 0.1 to 0.3. The graphs also show that the fluctuation of the real part of the OTF is significantly larger than the fluctuation of the imaginary part. Further interpretation of the variance data is difficult without also considering the average value of the OTF.

Putting together the variance data of Fig. 5.22 and the average OTF data in Fig. 5.21 into the expression for the SNR given by Eq. (5.78) results in the SNR plot shown in Fig. 5.25. A SNR plot like this is extremely valuable for investigating the use of short exposure imagery for image reconstruction. The image reconstruction approaches of interest include the speckle imaging techniques discussed in Chapter 4 and the hybrid image reconstruction techniques discussed in Chapter 6. In particular, the SNR indicate how strongly the OTF is expected to fluctuate relative to the average value. Large values of the OTF SNR indicates that the OTF attenuation is known with a large degree of certainty, and as a result image reconstruction can be used to compensate for the attenuation. Low SNRs, on the other hand, indicate that the OTF attenuation is not known with a large degree of certainty and as a result image reconstruction is difficult. The point at which the SNR falls below unity has been used in the past to quantify the cut-off frequency for reliable image reconstruction [22]. Figure 5.25 shows that the cut-off frequency is approximately 0.23 for $d/r_0 = 1$ and 0.12 for $d/r_0 = 2$. Obviously decreasing levels of compensation (i.e., d/r_0 increasing) can significantly affect the SNR. It should also be noted that even though the SNR decreases below unity at relatively low normalized frequencies, the SNR does not continue to significantly decrease for higher frequencies. This trend is fortunate since it indicates that spatial frequencies out to nearly the diffraction limit can be recovered by averaging a number of independent estimates of the image. In other words the SNR can be improved by averaging. It has been noted in Chapter 4 that the SNR improves by a factor of $\sqrt{N}$, where N is the number of independent measured images used in the image reconstruction.

One very tangible benefit of performing a simulation is the ability to investigate statistical expressions for quantities that are analytically very difficult, if not intractable, to compute. One of these quantities is the phase of the OTF. Since the average of the imaginary part of the OTF is identically zero, the phase of the average OTF is zero. A quantity of interest is the variance of the phase, or equivalently the mean square phase error. Figure 5.26 illustrates a plot of the OTF RMS phase error as a function of spatial

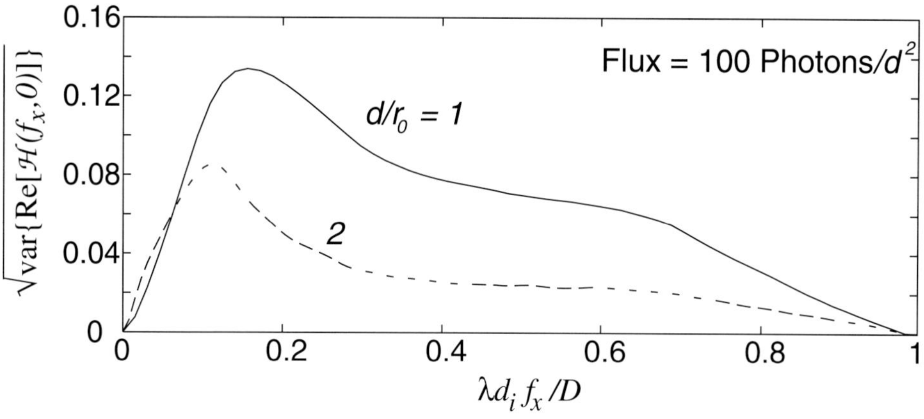

FIGURE 5.23
Standard deviation of the real part of $\mathcal{H}(f_x, 0)$ versus spatial frequency f_x for the AOI system described on pages 219 to 222. The WFS light level corresponds to 100 photons/d^2 and the ratio $z_t/z_b = 0.0$ (natural guide star).

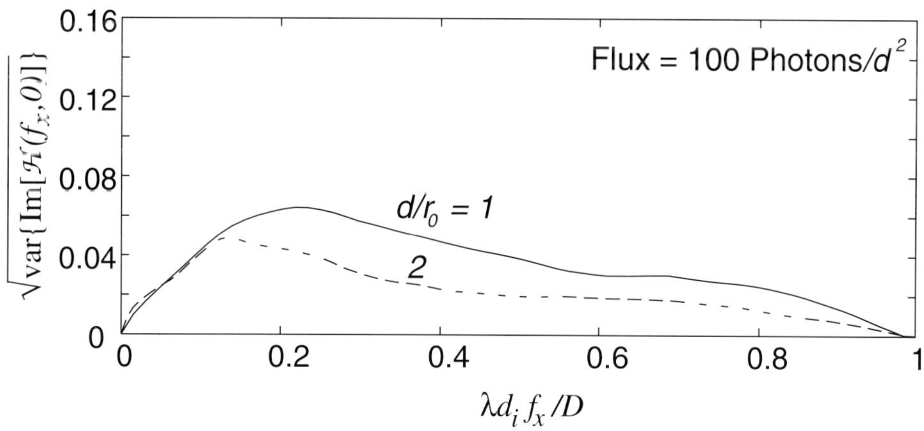

FIGURE 5.24
Standard deviation of the imaginary part of $\mathcal{H}(f_x, 0)$ versus spatial frequency f_x for the AOI system described on pages 219 to 222. The WFS light level corresponds to 100 photons/d^2 and the ratio $z_t/z_b = 0.0$ (natural guide star).

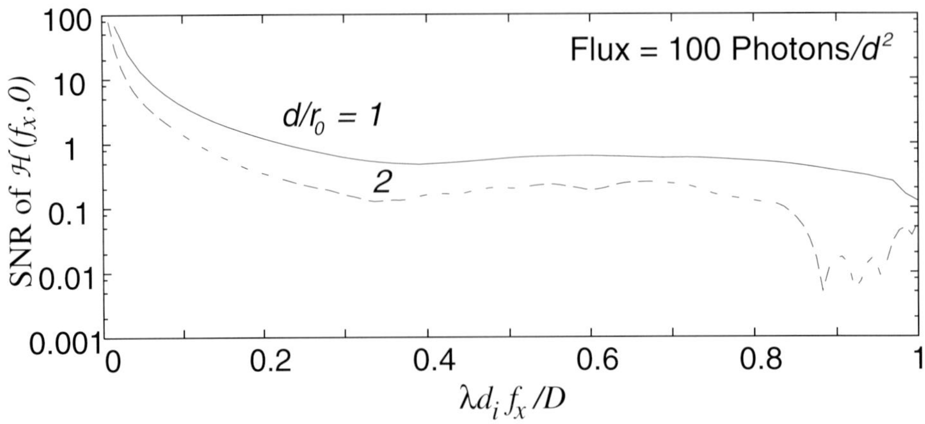

FIGURE 5.25
SNR of $\mathcal{H}(f_x, 0)$ plotted versus normalized spatial frequency f_x for the AOI system described on pages 219 to 222. The WFS light level corresponds to 100 photons/d^2 and the ratio $z_t/z_b = 0.0$ (natural guide star).

frequency. The two curves shown in the graph are for the AOI system described on pages 219 to 222 and for $d/r_0 = 1$ and 2. It is interesting to note from a comparison of Figs. 5.25 and 5.26 that the RMS phase error exceeds 1 rad at about the same point for which the SNR of $|\mathcal{H}|$ drops below unity. This correlation between the drop of the SNR below unity and the phase error exceeding 1 rad has also been observed for speckling imaging [107].

Another advantage of performing a simulation, as opposed to an analytical computation, is that the simulation results in numerical values for the OTF statistics over the entire frequency domain grid simultaneously. To obtain the same OTF statistics over a dense frequency domain grid using analytical computations is beyond the reasonable capability of today's computers. In particular we find that obtaining the average PSF from a simulation is as easy as applying the Fast Fourier Transform (FFT) to the computed average OTF. Figures 5.27 and 5.28 show two dimensional surface maps of the average OTF and PSF for the AOI system described on pages 219 to 222.

Simulation results: single realization quantities

One of the key advantages of using simulations to compute the performance of AOI systems is the ability to investigate how the simulated system is performing on a realization-to-realization basis. A single realization corresponds in practice to the short exposure performance of the system. In particular it may be useful to investigate the instantaneous wave front reconstruction achieved by the DM. Another area of interest is the investigation of speckle imaging techniques used in conjunction with either partially or fully compensated AOI systems. In this case the quantities of interest are the short exposure images as discussed in Chapter 4. The instantaneous short exposure images of the object can also be obtained from a simulation. These short exposure images are obtained by simply multiplying an individual realization of the OTF by the spectrum of the object being imaged. This product is then Fourier transformed to obtain a noise free version of the short exposure image. Shot noise and detector noise effects can be easily added by appropriately randomizing the irradiance levels. The details of forming simulated imagery corrupted by photon noise are discussed in the development presented from Eqs. (4.68) to (4.70) of Chapter 4. Examples of simulated long and short exposure imagery of a binary star pair are presented

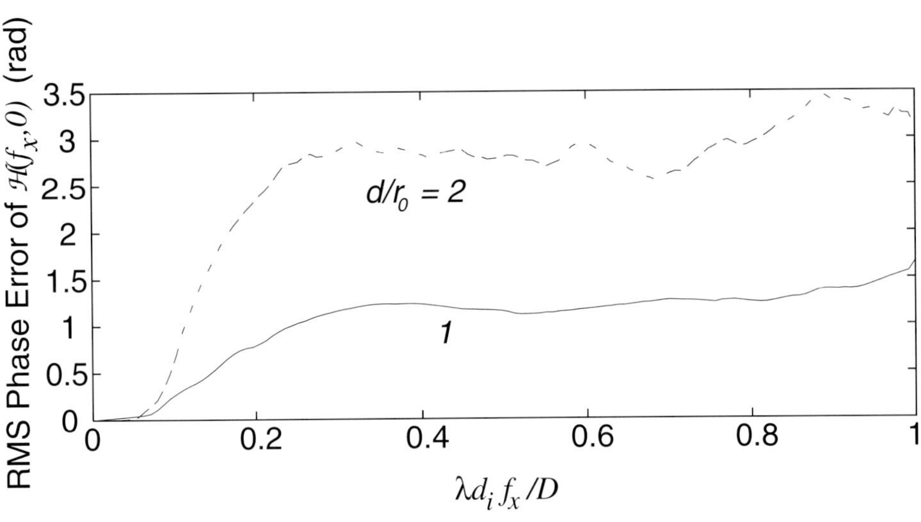

FIGURE 5.26
RMS phase error of $\mathcal{H}(f_x, 0)$ plotted versus normalized spatial frequency f_x for the AOI system described on pages 219 to 222. The WFS light level corresponds to 100 photons/d^2 and the ratio $z_t/z_b = 0.0$ (natural guide star).

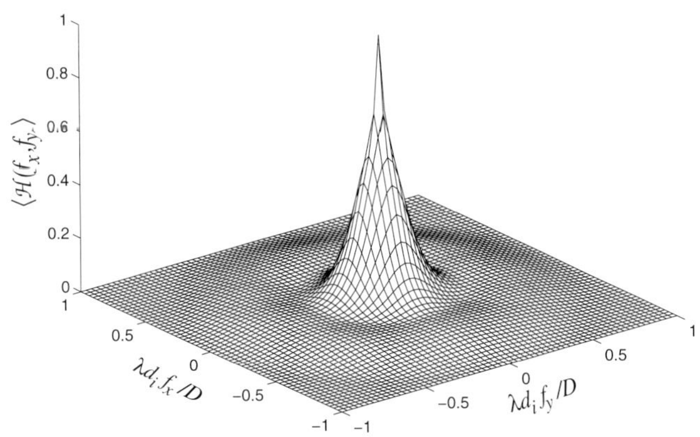

FIGURE 5.27
Surface plot of the OTF, E$\{\mathcal{H}(\vec{f})\}$, for the AOI system described on pages 219 to 222. This result is for $d/r_0 = 1$. The WFS light level corresponds to 100 photons/d^2 and the ratio $z_t/z_b = 0.0$ (natural guide star).

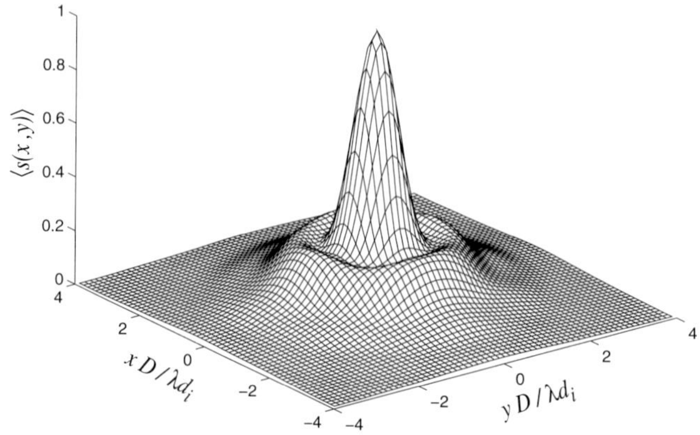

FIGURE 5.28
Surface plot of the PSF, E$\{s(\vec{x})\}$ for the AOI system described on pages 219 to 222. This result is for $d/r_0 = 1$. The WFS light level corresponds to 100 photons/d^2 and the ratio $z_t/z_b = 0.0$ (natural guide star).

in the following section.

Figures 5.29 through 5.31 show a series of single realization results picked from the large data set of realizations used to compute the results shown in Figs. 5.21 through 5.28. Figures 5.29(a) and 5.29(b) show a single realization of the incident perturbed wave front and the wave front correction placed on the DM. Figure 5.30 shows the residual wave front after compensation (i.e., the difference between the wave fronts shown in Fig. 5.29(a) and (b)). In this particular realization the AOI system is obviously effective in sensing and compensating for the incident wave front phase deformations. The aperture averaged mean square phase for the incident wave front shown in Fig. 5.29(a) is 18.48 rad^2. After compensation the aperture averaged mean square phase for the residual wave front shown in Fig. 5.30 is reduced to a value of 0.65 rad^2.

Figures 5.31(a) and 5.31(b) show the instantaneous OTF and PSF corresponding to the residual wave front in Fig. 5.30. It can be seen that the instantaneous OTF and PSF do not match the average values shown in Figs. 5.27 and 5.28. The PSF shown in Fig. 5.31(b) can be thought of as a short exposure image from the AOI system for a single bright star. We finally note that other instantaneous realizations of the OTF and PSF will exhibit characteristics that are both better and worse than those exhibited in Figs. 5.29 to 5.31.

Simulation imagery of a binary star pair

Figures 5.32 through 5.34 show simulated imagery of a binary star pair for the 3 m AOI system used in the previous examples. High light levels are assumed for all of the imagery shown in these figures. Before showing how the AOI system performs we first show in Fig. 5.32 the diffraction limited image of the binary star pair. The stars are separated by $2\lambda/D$ radians which corresponds to 0.33 μrad (0.069 arcsec) for $\lambda = 0.5\ \mu$m. The brightness of the star on the right is one half that of the star on the

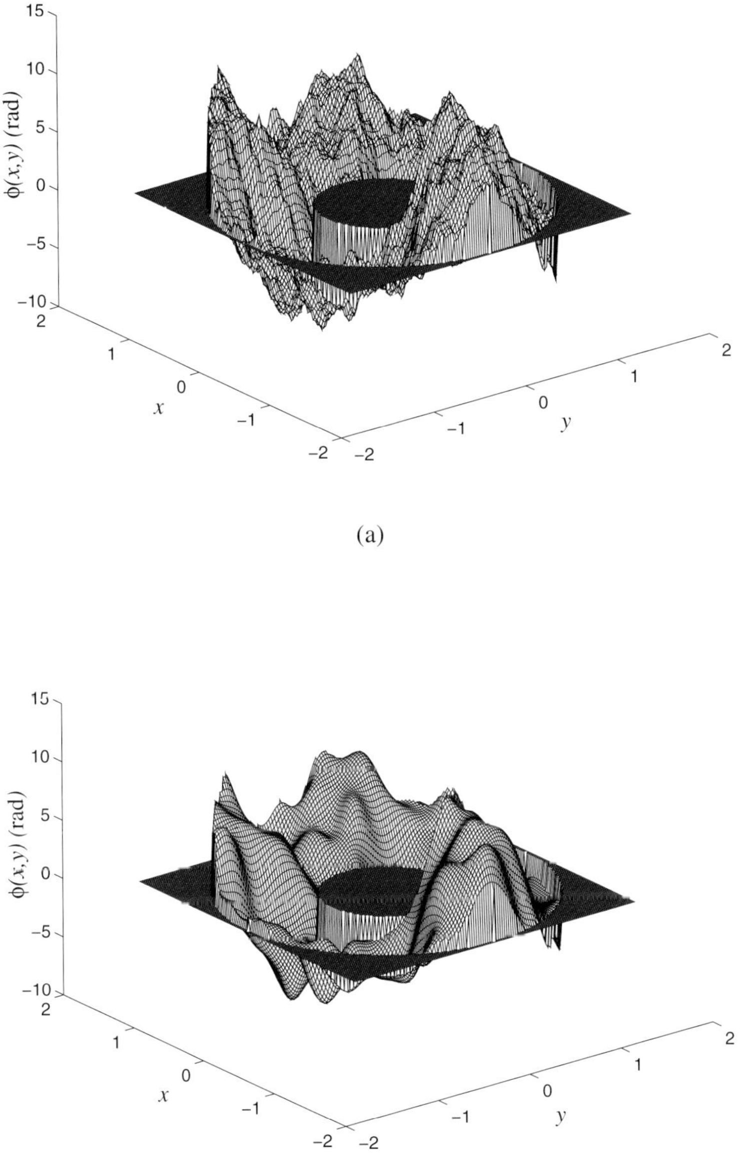

FIGURE 5.29
Pupil wave front phase maps corresponding to a) the incident perturbed wave front and b) the correction wave front applied to the DM for the AOI system described on pages 219 to 222. This result is for $d/r_0 = 1$. The WFS light level corresponds to 100 photons/d^2 and the ratio $z_t/z_b = 0.0$ (natural guide star).

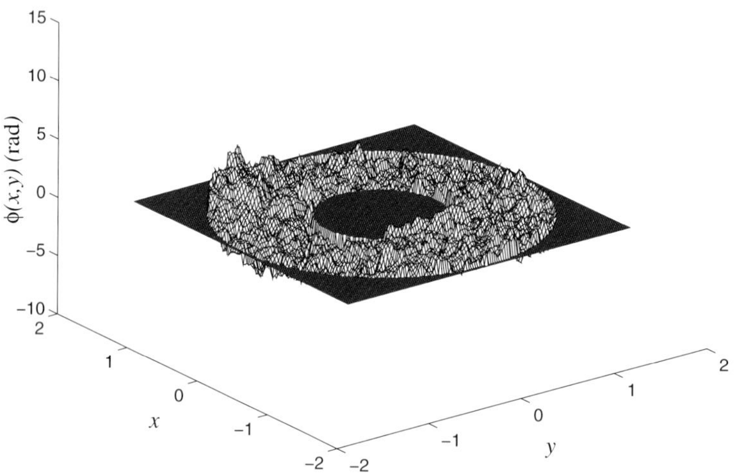

FIGURE 5.30
Residual phase after wave front compensation for the AOI system described on pages 219 to 222. The residual phase is the difference between the wave fronts shown in Fig. 5.29. This result is for $d/r_0 = 1$. The WFS light level corresponds to 100 photons/d^2 and the ratio $z_t/z_b = 0.0$ (natural guide star).

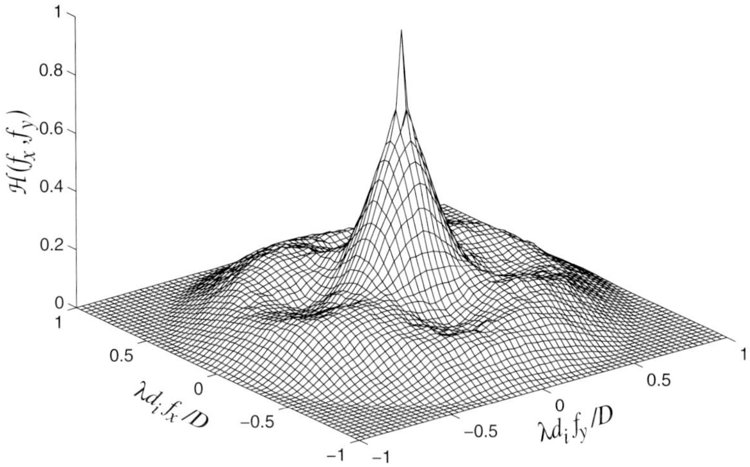

(a)

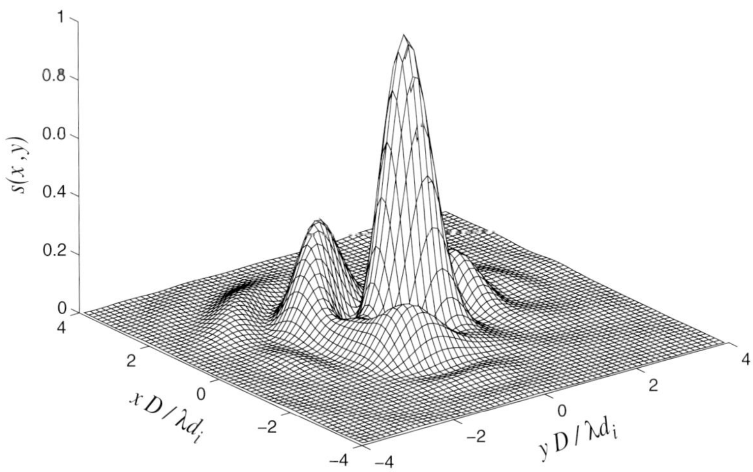

(b)

FIGURE 5.31
Instantaneous a) OTF, $\mathcal{H}(\vec{f})$, and b) PSF, $s(\vec{x})$, corresponding to the residual wave front in Fig. 5.30.

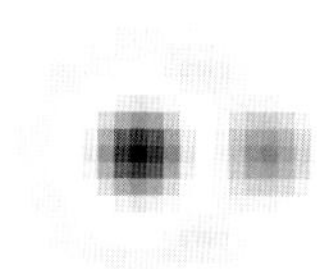

FIGURE 5.32
Diffraction limited image of a binary star pair for the 3 m telescope described on page 219. The stars are separated by 0.33 μrad (0.069 arcsec) and the simulated imaging wavelength is 0.5 μm. The brightness of the star on the right is one half that of the star on the left.

left. The three images shown in Fig. 5.33 illustrate the effects of atmospheric turbulence. The three images correspond to (a) a long exposure image for an uncompensated telescope, (b) a long exposure for a compensated telescope using a single natural guide star with $d/r_0 = 1$ and (c) a long exposure for a compensated telescope using a single natural guide star with $d/r_0 = 2$. Comparison of the uncompensated image shown in Fig. 5.33(a) and the diffraction limited image shown in Fig. 5.32 clearly demonstrates the severe performance degradation caused by the atmosphere. The compensated images of 5.33(b) and 5.33(c) illustrate the ability of the AOI system to dramatically improve performance. Both of the compensated images reveal that the components of the binary star pair can be resolved. Notice the broader halo of light for the case of $d/r_0 = 2$ as compared to the case of $d/r_0 = 1$. Figure 5.34 illustrates a series of short exposure images of the same binary star pair (i.e., a single realization from the simulation). Figure 5.34(a) shows a short exposure image for the uncompensated telescope while Figs. 5.34(b) and 5.34(c) show compensated imagery for the cases of $d/r_0 = 1$ and $d/r_0 = 2$, respectively. These latter three images can be used in investigations involving the speckle and hybrid imaging techniques described in Chapters 4 and 6.

The imagery shown in Figures 5.32 through 5.34 is applicable for high light conditions ($\sim$ 1 million photon events per short exposure image). Under lower light conditions a more realistic image is obtained by randomizing the irradiance levels in a way that accounts for photon and read noise. The steps for generating low light level images are described in the development presented from Eqs. (4.68) to (4.70) of Chapter 4.

5.5 Summary

The goal of this chapter has been to present a generalized mathematical framework that allows the reader to investigate nearly any performance aspect of an AOI system. To achieve this goal we first presented a qualitative discussion of the major factors that degrade the performance of AOI systems. These include WFS sampling, the finite degrees of freedom of the DM, the finite light levels incident on the WFS, anisoplanatism (both focus and angular) and the finite temporal response of the system. Generalized system component models were next introduced as well as the mathematical expressions required to characterize the second order statistics of the incident and residual wave front phase. Using these models and statistics the resulting analysis framework can be tailored to investigate fundamental performance issues such as anisoplanatism and finite temporal responses, as well as more system oriented investigations in which the entire AOI system is accurately characterized. The system can be modeled with varying levels of detail ranging from the investigation where most of the components are idealized, as in the fundamental computations, to detailed modeling where all of the major system components are accurately characterized. The chapter is summarized by presenting two practical approaches for conducting these performance investigations. These two approaches include numerical computations involving the analytical expressions presented in the chapter and Monte Carlo simulation using the system models. Examples of both computation approaches show the types of computations that can be accomplished as well as the flexibility of the overall analysis framework to investigate a wide range of performance metrics.

FIGURE 5.33
Simulated long exposure imagery of a binary star pair for (a) an uncompensated telescope, (b) a compensated telescope with $d/r_0 = 1$ and (c) a compensated telescope with $d/r_0 = 2$. The results were computed for the adaptive optical telescope described on pages 219 to 222. The stars are separated by 0.33 μrad (0.069 arcsec) and the imaging wavelength is 0.5 μm. The brightness of the star on the right is one half that of the star on the left.

FIGURE 5.34
Simulated short exposure imagery of a binary star pair for (a) an uncompensated telescope, (b) a compensated telescope with $d/r_0 = 1$ and (c) a compensated telescope with $d/r_0 = 2$. The results were computed for the adaptive optical telescope described on pages 219 to 222. The stars are separated by 0.33 μrad (0.069 arcsec) and the imaging wavelength is 0.5 μm. The brightness of the star on the right is one half that of the star on the left.

Bibliography

[1] H. W. Babcock, "The possibility of compensating astronomical seeing," *Publ. Astron. Soc. Pac.*, vol. 65, pp. 229–236, 1953.

[2] R. Q. Fugate, B. L. Ellerbroek, C. H. Higgins, M. P. Jelonek, W. J. Lange, A. C. Slavin, W. J. Wild, D. M. Winker, J. M. Wynia, J. M. Spinhirne, B. R. Boeke, R. E. Ruane, J. F. Moroney, M. D. Oliker, D. W. Sindle, and R. A. Cleis, "Two generations of laser-guide-star adaptive-optics experiments at the starfire optical range," *J. Opt. Soc. Am. A*, vol. 11, pp. 310–314, 1994.

[3] R. J. Noll, "Zernike polynomials and atmospheric turbulence," *J. Opt. Soc. Am.*, vol. 66, pp. 207–211, 1976.

[4] D. L. Fried, "Least-square fitting a wave front distortion estimate to an array of phase-difference measurements," *J. Opt. Soc. Am.*, vol. 67, pp. 370–375, 1977.

[5] R. H. Hudgin, "Wave front reconstruction for compensated imaging," *J. Opt. Soc. Am.*, vol. 67, pp. 375–377, 1977.

[6] R. H. Hudgin, "Wave front compensation error due to finite corrector-element size," *J. Opt. Soc. Am.*, vol. 67, pp. 393–395, 1977.

[7] J. Y. Wang and J. K. Markey, "Modal compensation of atmospheric turbulence phase distortion," *J. Opt. Soc. Am.*, vol. 68, pp. 78–87, 1978.

[8] D. P. Greenwood, "Mutual coherence function of a wave front corrected by zonal adaptive optics," *J. Opt. Soc. Am.*, vol. 69, pp. 549–554, 1979.

[9] W. H. Southwell, "Wave-front estimation from wave-front slope measurements," *J. Opt. Soc. Am.*, vol. 70, pp. 998–1006, 1980.

[10] D. L. Fried, "Anisoplanatism in adaptive optics," *J. Opt. Soc. Am.*, vol. 72, pp. 52–61, 1982.

[11] E. P. Wallner, "Optimal wave front correction using slope measurements," *J. Opt. Soc. Am.*, vol. 73, pp. 1771–1776, 1983.

[12] J. P. Gaffard and C. Boyer, "Adaptive optics for optimization of image resolution," *Appl. Opt.*, vol. 26, pp. 3772–3777, 1987.

[13] P. Nisenson and R. Barakat, "Partial atmospheric correction with adaptive optics," *J. Opt. Soc. Am. A*, vol. 4, pp. 2249–2253, 1987.

[14] R. C. Smithson *et al.*, "Quantitative simulation of image correction for astronomy with a segmented actice mirror," *Appl. Opt.*, vol. 27, pp. 1615–1620, 1988.

[15] B. M. Welsh and C. S. Gardner, "Performance analysis of adaptive optics systems using slope sensors," *J. Opt. Soc. Am. A*, vol. 6, pp. 1913–1923, 1989.

[16] R. C. Smithson and M. L. Peri, "Partial correction of astronomical images with active mirrors," *J. Opt. Soc. Am. A*, vol. 6, pp. 92–97, 1989.

[17] C. S. Gardner, B. M. Welsh, and L. A. Thompson, "Design and performance analysis of adaptive optical telescopes using laser guide stars," *Proc. IEEE*, vol. 78, pp. 1721–1743, 1990.

[18] B. M. Welsh and C. S. Gardner, "Effects of turbulence induced anisoplanatism on the imaging performance of adaptive astronomical telescopes using laser guide stars," *J. Opt. Soc. Am. A*, vol. 8, pp. 69–80, 1991.

[19] B. M. Welsh, "Imaging performance analysis of adaptive telescopes using laser guide stars," *Appl. Opt.*, vol. 30, pp. 5021–5030, 1991.

[20] M. C. Roggemann, "Limited degree-of-freedom adaptive optics and image reconstruction," *Appl. Opt.*, vol. 30, pp. 4227–4233, 1991.

[21] M. C. Roggemann, "Optical performance of fully and partially compensated adaptive optics

systems using least-squares and minimum variance phase reconstruction," *Comp. & Elec. Eng.*, vol. 18, pp. 451–466, 1992.

[22] M. C. Roggemann and C. L. Matson, "Power spectrum and fourier phase spectrum estimation by using fully and partially compensating adaptive optics and bispectrum postprocessing," *J. Opt. Soc. Am. A*, vol. 9, pp. 1525–1535, 1992.

[23] F. Rigaut and E. Gendron, "Laser guide star adaptive optics: the tilt determination problem," *Astron. Astrophys.*, vol. 261, pp. 677–684, 1993.

[24] R. R. Parenti and R. J. Sasiella, "Laser-guide-star systems for astronomical applications," *J. Opt. Soc. Am. A*, vol. 11, pp. 288–309, 1994.

[25] D. T. Gavel, J. R. Morris, and R. G. Vernon, "Systematic design and analysis of laser-guide-star adaptive-optics systems for large telescopes," *J. Opt. Soc. Am. A*, vol. 11, pp. 914–924, 1994.

[26] J. W. Hardy, "Active optics: A new technology for the control of light," *Proc. IEEE*, vol. 66, pp. 651–697, 1978.

[27] G. A. Tyler, "Bandwidth considerations for tracking through turbulence," *J. Opt. Soc. Am. A*, vol. 11, pp. 358–367, 1994.

[28] D. L. Fried, "Atmospheric turbulence optical effects: understanding the adaptive-optics implications," in *Adaptive Optics for Astronomy* (D. Alloin and J.-M. Mariotti, eds.), vol. C423 of *NATO Advanced Study Institude Series*, pp. 25–57, Kluwer Academic Publishers, 1994.

[29] J. Herrmann, "Phase variance and strehl ratio in adaptive optics," *J. Opt. Soc. Am. A*, vol. 9, pp. 2257–2258, 1992.

[30] B. L. Ellerbroek, "First-order performance evaluation of adaptive optics systems for atmospheric turbulence compensation in extended field-of-view astronomical telescopes," *J. Opt. Soc. Am. A*, vol. 11, pp. 783–805, 1994.

[31] D. C. Johnston and B. M. Welsh, "Estimating the contribution of different parts of the atmosphere to optical wavefront aberration," *Comp. & Elec. Eng.*, vol. 18, pp. 467–484, 1992.

[32] D. C. Johnston and B. M. Welsh, "Analysis of multiconjugate adaptive optics," *J. Opt. Soc. Am. A*, vol. 11, pp. 394–408, 1994.

[33] M. Demerlé, P. Y. Madec, and G. Rousset, "Servo-loop analysis for adaptive optics," in *Adaptive Optics for Astronomy* (D. Alloin and J.-M. Mariotti, eds.), vol. C423 of *NATO Advanced Study Institude Series*, pp. 73–88, Kluwer Academic Publishers, 1994.

[34] D. P. Greenwood, "Bandwidth specifications for adaptive optics systems," *J. Opt. Soc. Am.*, vol. 67, pp. 390–392, 1977.

[35] D. L. Fried, "Limiting resolution looking down through the atmosphere," *J. Opt. Soc. Am.*, vol. 56, pp. 1380–1384, 1966.

[36] B. M. Welsh, "Effect of an adaptive optical system's temporal response on imaging performance," in *SPIE Proceedings on Atmospheric Propagation and Remote Sensing*, vol. 1688, pp. 536–547, 1992.

[37] D. L. Fried, "Time-delay-induced mean-square error in adaptive optics," *J. Opt. Soc. Am. A*, vol. 7, pp. 1224–1225, 1990.

[38] G. A. Tyler, "Merging: a new method for tomography through random media," *J. Opt. Soc. Am. A*, vol. 10, pp. 409–425, 1993.

[39] B. M. Welsh and M. C. Roggemann, "Evaluating the performance of adaptive optical telescopes," in *Adaptive Optics for Astronomy* (D. Alloin and J.-M. Mariotti, eds.), vol. C423 of *NATO Advanced Study Institude Series*, Kluwer Academic Publishers, 1994.

[40] J. D. Downie and J. W. Goodman, "Optimal wavefront control for adaptive segmented mirrors," *Appl. Opt.*, vol. 28, pp. 5326–5332, 1989.

[41] D. P. Petersen and K. H. Cho, "Sampling and reconstruction of a turbulence-distorted wave front," *J. Opt. Soc. Am. A*, vol. 3, pp. 818–825, 1986.

[42] D. L. Fried, "Analysis of fundamental limits to artificial guide star adaptive optics system performance for astronomical imaging," *J. Opt. Soc. Am. A*, vol. 11, pp. 277–287, 1994.

[43] B. L. Ellerbroek, C. V. Loan, N. P. Pitsianis, and R. J. Plemmons, "Optimizing closed-loop adaptive-optics performance with use of multiple control bandwidths," *J. Opt. Soc. Am. A*, vol. 11, pp. 2871–2886, 1994.

[44] G. A. Tyler, "Wave-front compensation for imaging with off-axis guide stars," *J. Opt. Soc. Am. A*, vol. 11, pp. 339–346, 1994.

[45] M. C. Roggemann and J. A. Meinhardt, "Image reconstruction by means of wave-front sensor measurements in closed-loop adaptive-optics systems," *J. Opt. Soc. Am. A*, vol. 10, pp. 1996–2007, 1993.

[46] M. C. Roggemann, D. W. Tyler, and M. F. Bilmont, "Linear reconstruction of compensated images: theory and experimental results," *Appl. Opt.*, vol. 31, pp. 7429–7441, 1992.

[47] L. A. Thompson, "Adaptive optics in astronomy," *Physics Today*, pp. 24–31, December 1994.

[48] S. E. Troxel, B. M. Welsh, and M. C. Roggemann, "Off-axis optical transfer function calculations in an adaptive-optics system by means of a diffraction calculation for weak index fluctuations," *J. Opt. Soc. Am. A*, vol. 11, pp. 2100–2111, 1994.

[49] S. E. Troxel, B. M. Welsh, and M. C. Roggemann, "Anisoplanatism effects on signal-to-noise ratio performance of adaptive optical systems," *J. Opt. Soc. Am. A*, vol. 12, pp. 570–577, March 1995.

[50] R. Foy and A. Labeyrie, "Feasibility of adaptive telescopes with laser probe," *Astr. Astrophys.*, vol. 152, pp. 129–131, 1985.

[51] W. Happer, G. MacDonald, C. Max, and F. J. Dyson, "Atmospheric-turbulence compensation by resonant optical backscattering from the sodium layer in the upper atmosphere," *J. Opt. Soc. Am. A*, vol. 11, pp. 263–275, 1994.

[52] C. S. Gardner, "Sodium resonance fluorescence lidar applications in atmospheric science and astronomy," *Proc. IEEE*, vol. 77, pp. 408–418, 1989.

[53] R. Q. Fugate, "Laser beacon adaptive optics," *Optics and Photonic News*, vol. 4, pp. 14–19, 1993.

[54] G. P. Collins, "Making stars to see stars: Dod adaptive optics work is declassified," *Physics Today*, vol. 45, pp. 17–21, 1992.

[55] R. K. Tyson, *Principles of Adaptive Optics*. San Diego, California: Academic Press, Inc., 1991.

[56] D. Sandler, S. Stahl, J. R. P. Angel, M. Lloyd-Hart, and D. McCarthy, "Adaptive optics for diffraction-limited infrared imaging with 8-m telescopes," *J. Opt. Soc. Am. A*, vol. 11, pp. 925–945, 1994.

[57] R. K. Tyson, "Adaptive optics system performance approximations for atmospheric turbulence correction," *Opt. Eng.*, vol. 29, pp. 1165–1173, 1990.

[58] G. D. Love, J. S. Fender, and S. R. Restaino, "Adaptive wavefront shaping with liquid crystals," *Optics & Photonics News*, vol. 6, pp. 16–21, 1995.

[59] J. W. Goodman, *Introduction to Fourier Optics*. New York: McGraw-Hill Book Co., 1968.

[60] M. A. Ealey and J. F. Washeba, "Continuous facesheet low voltabe deformable mirrors," *Opt. Eng.*, vol. 29, pp. 1191–1198, 1990.

[61] E. N. Ribak, "Deformable mirrors," in *Adaptive Optics for Astronomy* (D. Alloin and J.-M. Mariotti, eds.), vol. C423 of *NATO Advanced Study Institude Series*, pp. 149–161, Kluwer Academic Publishers, 1994.

[62] D. S. Acton and R. C. Smithson, "Solar imaging with a segmented adaptive mirror," *Appl. Opt.*,

vol. 31, pp. 3161–3169, 1992.

[63] M. A. Ealey and J. A. Wellman, "Deformable mirrors: design fundamentals, key performance specifications, and parametric trades," in *Proc. SPIE on Active and Adaptive Optical Components*, vol. 1543, pp. 36–51, 1991.

[64] B. Hulburd and D. Sandler, "Segmented mirrors for atmospheric compensation," *Opt. Eng.*, vol. 29, pp. 1186–1190, 1990.

[65] C. Schwartz, E. Ribak, and S. G. Lipson, "Bimorph adaptive mirrors and curvature sensing," *J. Opt. Soc. Am. A*, vol. 11, pp. 895–902, 1994.

[66] K. E. Moore and G. N. Lawrence, "Zonal model of an adaptive mirror," *Appl. Opt.*, vol. 29, pp. 4622–4628, 1990.

[67] G. Rousset, "Wavefront sensing," in *Adaptive Optics for Astronomy* (D. Alloin and J.-M. Mariotti, eds.), vol. C423 of *NATO Advanced Study Institude Series*, pp. 115–137, Kluwer Academic Publishers, 1994.

[68] G. A. Tyler and D. L. Fried, "Image position error associated with a quadrant detector," *J. Opt. Soc. Am.*, vol. 72, pp. 804–808, 1982.

[69] K. A. Winnick, "Cramer-rao lower bounds on the performance of charge-coupled-device optical position estimators," *J. Opt. Soc. Am. A*, vol. 3, pp. 1809–1815, 1986.

[70] T. J. Kane, B. M. Welsh, C. S. Gardner, and L. A. Thompson, "Wave front detector optimization for laser guided adaptive telescopes," in *SPIE Proceedings on Active Telescope Systems*, vol. 1114, pp. 160–171, 1989.

[71] G. Cao and X. Yu, "Accuracy analysis of a hartmann-shack wavefront sensor operated with a faint source," *Opt. Eng.*, vol. 33, pp. 2331–2335, 1994.

[72] C. Witthoft, "Wavefront sensor noise reduction and dynamic range expansion by means of optical image intensification," *Opt. Eng.*, vol. 29, pp. 1233–1238, 1990.

[73] H. T. Yura and M. T. Tavis, "Centroid anisoplanatism," *J. Opt. Soc. Am. A*, vol. 2, pp. 765–773, 1985.

[74] J. C. Wyant, "Use of an ac hetrodyne lateral shear interferometer with real-time wavefront correction systems," *Appl. Opt.*, vol. 14, pp. 2622–2626, 1975.

[75] J. C. Wyant, "White light extended source shearing interferometer," *Appl. Opt.*, vol. 13, pp. 200–202, 1974.

[76] B. A. Horwitz, "Multiplex techniques for real-time shearing interferometry," *Opt. Eng.*, vol. 29, pp. 1223–1232, 1990.

[77] J. W. Hardy and A. J. MacGovern, "Shearing interferometry: a flexible technique for wavefront measurement," in *SPIE Proceedings on Interferometric Metrology*, vol. 816, pp. 180–195, 1987.

[78] D. Sandler, L. Cuellar, M. Lefebvre, T. Barrett, R. Arnold, P. Johnson, A. Rego, G. Smith, G. Taylor, and B. Spivey, "Shearing interferometry for laser-guide-star atmospheric correction at large d/r_o," *J. Opt. Soc. Am. A*, vol. 11, pp. 858–873, 1994.

[79] F. Roddier, "Curvature sensing and compensation: a new concept in adaptive optics," *Appl. Opt.*, vol. 27, pp. 1223–1225, 1988.

[80] F. Roddier, M. Northcott, and J. E. Graves, "A simple low-order adaptive optics system for near-infrared applications," *Pub. Astr. Soc. Pac.*, pp. 131–149, January 1991.

[81] C. Roddier and F. Roddier, "Wave-front reconstruction from defocused images and the testing of ground-based optical telescopes," *J. Opt. Soc. Am. A*, vol. 10, pp. 2277–2287, 1993.

[82] D. C. Johnston, B. L. Ellerbroek, and S. M. Pompea, "Curvature sensing analysis," in *SPIE Proceedings on Adaptive Optics in Astronomy*, vol. 2201, pp. 528–538, 1994.

[83] B. M. Welsh, B. L. Ellerbroek, M. C. Roggemann, and T. L. Pennington, "Fundamental per-

formance limitations and comparison of the hartmann and shearing interferometer wavefront sensors," *Appl. Opt.*, vol. 34, pp. 4186–4195, 1995.

[84] J. L. Melsa and D. L. Cohn, *Decision and Estimation Theory*. New York: McGraw-Hill Book Company, 1978.

[85] J. W. Goodman, *Statistical Optics*. New York: John Wiley & Sons, 1985.

[86] V. I. Tatarskii, *Wave Propagation in a Turbulent Medium*. New York: Dover Publications, 1967.

[87] R. J. Sasiela and J. G. Mooney, "An optical phase reconstructor based on using a multiplier-accumulator approach," in *SPIE Proceedings on Adaptive Optics*, vol. 551, pp. 170–176, 1985.

[88] W. H. Beyer, ed., *CRC Standard Mathematical Tables*. Boca Raton, Florida: CRC Press, Inc., 25 ed., 1978.

[89] A. T. Young, "Seeing: Its cause and cure," *Astrophys. J.*, vol. 189, pp. 587–604, 1974.

[90] B. M. Welsh and R. N. Von Niederhausern, "Performance analysis of the self-referenced speckle holography image reconstruction technique," *Appl. Opt.*, vol. 32, pp. 5071–5078, 1993.

[91] M. C. Roggemann, B. M. Welsh, and J. Devey, "Biased estimators and object-spectrum estimation in the method of deconvolution from wavefront sensing," *Appl. Opt.*, vol. 33, pp. 5754–5763, 1994.

[92] M. C. Roggemann and B. M. Welsh, "Signal-to-noise ratio for astronomical imaging by deconvolution from wavefront sensing," *Appl. Opt.*, vol. 33, pp. 5400–5414, 1994.

[93] M. C. Roggemann, B. L. Ellerbroek, and T. A. Rhoadarmer, "Widening the effective field-of-view of adaptive optics telescopes using deconvolution from wavefront sensing: average and signal-to-noise ratio performance," *Appl. Opt.*, vol. 34, pp. 1432–1444, 1995.

[94] M. C. Roggemann, C. A. Stoudt, and B. M. Welsh, "Image spectrum signal-to-noise ratio improvements by statistical frame selection for adaptive optics imaging through atmospheric turbulence," *Opt. Eng.*, vol. 33, pp. 3254–3264, October 1994.

[95] S. D. Ford, M. C. Roggemann, and B. M. Welsh, "Frame selection performance limits for statistical image reconstruction of adaptive optics compensated images," in *SPIE Proceedings on Adaptive Optical Systems and Applications*, vol. 2534, 1995.

[96] G. Cochran, "Phase screen generation," Tech. Rep. TR-663, The Optical Sciences Company, Placentia, California, 1985.

[97] G. M. Cochran, B. L. Ellerbroek, and D. L. Fried, "Spectral and temporal windows in knox-thompson imaging," Tech. Rep. TR-819R, The Optical Sciences Company, 1988.

[98] R. L. Bruden, J. D. Faires, and A. C. Reynolds, *Numerical Analysis*. Boston: Prindle, Weber, and Schmidt, 1980.

[99] W. Press, B. Flannery, S. Teukolsky, and W. Vetterling, *Numerical Recipes - The Art of Scientific Computing*. Cambridge, UK: Cambridge University Press, 1986.

[100] "Math library/special functions," in *IMSL FORTRAN Subroutines for Mathematical Applications*, 1991.

[101] R. C. Gonzalez and R. E. Woods, *Digital Image Processing*. Reading, Massachusetts: Addison-Wesley, 1993.

[102] T. J. Karr, "Temporal response of atmospheric turbulence compensation," *Appl. Opt.*, vol. 30, pp. 363–364, 1991.

[103] R. J. Sasiela, "Strehl ratios with various types of anisoplanatism," *J. Opt. Soc. Am. A*, vol. 9, pp. 1398–1405, 1992.

[104] J. Stone, P. H. Hu, S. P. Mills, and S. Ma, "Anisoplanatic effects in finite-aperture optical systems," *J. Opt. Soc. Am. A*, vol. 11, pp. 347–357, 1994.

[105] R. E. Hufnagel, "Variations of atmospheric turbulence," in *Digest of Technical Papers, Topical*

Meeting on Optical Propagation through Turbulence, Optical Society of America, 1974.

[106] M. C. Roggemann, E. L. Caudill, D. W. Tyler, M. J. Fox, M. A. Von Bokern, and C. L. Matson, "Compensated speckle imaging: theory and experimental results," *Appl. Opt.*, vol. 33, pp. 3099–3110, 1994.

[107] C. L. Matson, I. A. DeLarue, T. M. Gray, and I. E. Drunzer, "Optimal fourier spectrum estimation from the bispectrum," *Comp. & Elec. Eng.*, vol. 18, pp. 485–497, 1992.

6

Hybrid Imaging Techniques

6.1 Introduction

As we have seen in Chapters 4 and 5, neither speckle imaging nor adaptive optics can completely overcome the effects of atmospheric turbulence on imaging systems. Speckle imaging techniques are limited by the signal-to-noise ratio performance of the estimators used. Adaptive optics are limited by the finite spatial sampling of the wave front by the wave front sensor, a finite number of degrees-of-freedom in the deformable mirror, finite signal levels available to the wave front sensor, and time delays between sensing and correcting the turbulence-induced aberration. Further, adaptive optics for large telescopes operating at visible wavelengths can require several hundred to a few thousand wave front sensor subapertures and deformable mirror actuators to correct for turbulence effects. As a result, adaptive optics systems can be quite complicated and expensive to obtain and operate. Finally, both adaptive optics and speckle imaging have limited correctable field of view due to anisoplanatic effects.

These observations provide motivation for the study of imaging techniques which combine elements of adaptive optics and post detection image processing to obtain high resolution images through turbulence. Imaging techniques which combine elements of adaptive optics and post detection image processing are often referred to as *hybrid imaging* techniques. Hybrid imaging techniques have been shown to provide a tradeoff between correcting for turbulence effects before light is detected using adaptive optics, and correcting for turbulence effects after light is detected using image post processing concepts. That is, high resolution images can be obtained through turbulence by performing some correction using adaptive optics, and the remainder of the correction using image processing. In this chapter we present hybrid imaging techniques, derive mathematical models for the performance of the techniques, and provide representative results.

There are two broad classes of hybrid imaging techniques: (1) those which do not require the use of predetection wave front correction; and (2) those which do require some degree of predetection wave front correction. The category which does not require predetection correction of the wave front consists of a single imaging technique: deconvolution from wavefront sensing (DWFS). In DWFS light entering the telescope is split between an imaging camera and a wave front sensor, and simultaneous image and wave front sensor measurements are made. The wave front sensor data is used to estimate the instantaneous state of the generalized pupil function (GPF) [1], which is in turn used in image post processing to recover a high resolution image. DWFS is considered to be a hybrid imaging technique because wave front sensor measurements are required. Because DWFS does not require predetection correction of the aberrated wavefront, and because of mathematical similarities between the DWFS transfer function and the speckle transfer function introduced in Section 4.3.1, DWFS is an alternative to speckle imaging. An important aspect of DWFS is that the phase of the object spectrum is obtained directly. Unlike speckle imaging, no step corresponding to computing and processing the cross spectrum or bispectrum is required in DWFS.

Before proceeding with the discussion of hybrid techniques requiring adaptive optics, two definitions regarding adaptive optics systems presented in Section 5.2.2 are reviewed. A so-called *fully compensated* adaptive optics system is defined as an adaptive optics system which has wave front sensor subapertures of dimension r_0, and deformable mirror actuators which are separated by r_0, where r_0 is the Fried seeing parameter [2]. A *partially compensated* adaptive optics system is one in which fewer wave front sensor subapertures or deformable mirror actuators are present than in a fully compensated system. These definitions have found usage in the literature on hybrid imaging [3, 4, 5, 6], and will be used here. Some care should be exercised in applying these definitions. For example, a system which is fully compensated for $r_0 = 10$ cm would, by this definition, be partially compensated for $r_0 = 7$ cm.

The motivations for partially compensated adaptive optics systems are clear. For example, reducing the number of wave front sensor and deformable mirror elements reduces the hardware complexity and expense of the adaptive optics system. Also, if fewer subapertures are used in the wave front sensor the subapertures can have greater area in the telescope pupil, allowing more photons to be collected and improving the signal-to-noise ratio of the wave front sensor measurement. Alternatively, greater wavefront sensor subaperture size allows dimmer reference sources to be used as a beacon for the wave front sensor. There is a penalty for using partially compensated adaptive optics systems: the average performance of the imaging system, as measured by either the average optical transfer function (OTF) or the average point spread function (PSF), is degraded. As a result of this degradation images measured with partially compensated adaptive optics will appear more blurred than images measured with fully compensated systems. We shall see later in this chapter that the signal-to-noise ratio of the estimators used in the various image reconstruction techniques is greatly improved by even partially compensated adaptive optics systems, allowing high resolution images to be obtained from the blurred image measurements. Finally, it is possible for the seeing to deviate significantly from its mean value at any site through the course of a given night. Thus, a system designed to be fully compensated under nominal operating conditions may become partially compensated during some periods. Hybrid imaging techniques offer a means of recovering lost adaptive optical performance under suboptimal conditions.

Several hybrid imaging techniques have been studied which require use of adaptive optics. These include deconvolution of compensated images, compensated speckle imaging, compensated deconvolution from wavefront sensing, and frame selection. Deconvolution of compensated images is based on the observation that, even though images measured with partially compensated adaptive optics systems appear more blurred than diffraction-limited images, high spatial frequency information with good signal-to-noise ratio is present. The measured images appear blurred because the average OTF of the partially compensated system attenuates the high spatial frequency components, but does not eliminate them. The deconvolution process boosts the high spatial frequency components of the image to their appropriate values, thereby sharpening the measured images and providing higher resolution. Compensated speckle imaging consists of applying speckle imaging post processing, as presented in Chapter 4, to partially compensated images to improve the resolution of the images. Similarly, compensated DWFS is an extension of the DWFS concept to images measured using adaptive optics. Compensated DWFS has a somewhat wider applicability than other hybrid imaging techniques since it has been shown that compensated DWFS can be used to overcome some of the effects of both time delays and anisoplanatism in adaptive optics systems [6, 7]. Finally, the method of frame selection uses an image quality metric to select the "best" images from a data set, and then averages only the selected images to improve the signal-to-noise ratio of the resulting image spectrum. Deconvolution can then be used to sharpen the images resulting from frame selection.

Throughout this chapter the linear shift-invariant model for an imaging system is used [1]. This model, introduced in Chapter 2, states that the noise free image irradiance distribution $i(\vec{x})$ is given by the convolution of the object irradiance distribution $o(\vec{x})$ and the point spread function $h(\vec{x})$:

$$i(\vec{x}) = o(\vec{x}) * h(\vec{x}). \tag{6.1}$$

Fourier transforming both sides of Eq. (6.1) yields the relationship between the image spectrum $I(\vec{f})$, the object spectrum $O(\vec{f})$, and the OTF $\mathcal{H}(\vec{f})$:

$$I(\vec{f}) = O(\vec{f})\mathcal{H}(\vec{f}) \tag{6.2}$$

which was introduced in Eq. (2.41). Under conditions of finite signal we must consider the photon-limited image, represented by $d(\vec{x})$, and defined in Eq. (2.83). The detected image spectrum is the Fourier transform of $d(\vec{x})$, and is denoted by $D(\vec{f})$.

The concept of a *short exposure image* will be used throughout this chapter. A short exposure image was defined in Section 3.5 as an image measured using an exposure time sufficiently short to freeze a single realization of the turbulence-induced aberration during the image measurement. All of the techniques presented here take short exposure images, and possibly wavefront information as inputs. The short exposure image measurements are processed by the various algorithms presented here to create high resolution images.

The remainder of this chapter is organized as follows. In Section 6.2 the DWFS technique is introduced, and its performance is analyzed. The basic DWFS technique discussed in Section 6.2 does not require the use of adaptive optics. Hybrid imaging methods requiring the use of adaptive optics are presented in Section 6.3. Techniques requiring adaptive optics include linear deconvolution, compensated DWFS, compensated speckle imaging, and frame selection. The mathematical essentials of each imaging technique are presented, and representative performance results are provided.

6.2 Deconvolution from wavefront sensing

DWFS is a hybrid imaging technique which uses simultaneous measurements of a short exposure image and a wave front sensor. A block diagram of the DWFS method is shown in Fig. 6.1. Light entering the telescope is divided using a beam splitter between a wave front sensor and an imaging camera. An image of the pupil field is presented at the input to the wave front sensor, which makes a measurement of the turbulence-induced phase aberration $\psi(\vec{x})$, where $\vec{x}$ is a spatial location in the pupil. Image and wave front sensor measurements are made simultaneously and are recorded for later processing. Wave front sensor measurements are processed in the computer to reconstruct an estimate of $\psi(\vec{x})$, denoted $\tilde{\psi}(\vec{x})$, which is combined with a pupil model to form an estimate of the generalized pupil function given by $W(\vec{x})\exp\{j\tilde{\psi}(\vec{x})\}$, where $W(\vec{x})$ is the pupil function. The estimate of the generalized pupil function is processed to compute an estimate of the OTF, denoted $\tilde{\mathcal{H}}(\vec{f})$, and this estimate of the OTF is used in a spatial frequency domain deconvolution procedure described below. Averaging of many realizations of the estimator is used to improve the signal-to-noise ratio of the spectral data.

The basic concept for DWFS was first proposed by Fried [8]. This concept was later extended by Primot *et al.*, who used a different estimator than Fried originally proposed [9]. Gonglewski *et al.* provided the first experimental validation of DWFS on real astronomical images [10]. Welsh and Von Niederhausern provided an analysis of the DWFS method using an optimal estimator for the wave front phase [11]. Roggemann *et al.* showed that the Primot estimator is biased, and proposed an alternative measurement and processing technique which is immune to the effects of this bias [12]. An expression for the signal-to-noise ratio of DWFS was also derived by Roggemann and Welsh [13]. The unbiased DWFS technique proposed by Roggemann *et al.* is analyzed here. It has also been shown that the DWFS estimator proposed by Fried is appropriate for use in conjunction with adaptive optics compensation, and is presented in the compensated DWFS section later in this chapter [6, 7].

The remainder of this section is organized as follows. In the next subsection we present the estimators used for DWFS. This discussion is followed by a presentation of the wave front phase reconstructor used for DWFS. Differences in the structure functions of $\psi(\vec{x})$ and $\tilde{\psi}(\vec{x})$ are derived and the consequences are discussed. Due to the analytic complexity of the required statistical quantities, a simulation of DWFS was used to evaluate performance. The simulation is described, and representative results for the DWFS transfer function are provided. Signal-to-noise ratio considerations are presented in the subsection which follows, and the DWFS signal-to-noise ratio is compared to the signal-to-noise ratio of the speckle interferometry estimator $Q(\vec{f})$. Examples of DWFS imaging performance are provided

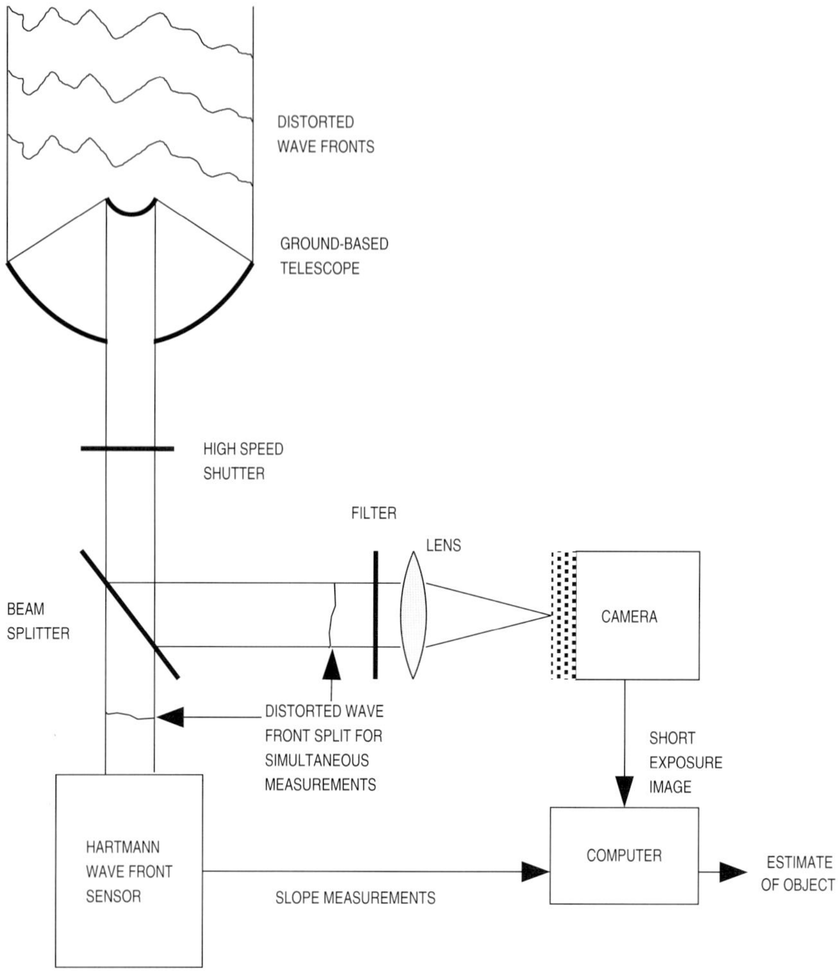

FIGURE 6.1
Block diagram of the DWFS method.

in the final subsection.

6.2.1 DWFS estimators

In this subsection we examine the estimators used to reconstruct images in the DWFS method. The estimator for the OTF is based on the expression for the OTF given in Eq. (2.44), and is given by

$$\tilde{\mathcal{H}}(\vec{f}) = N_F^{-1} \int d\vec{x}\, W(\vec{x}) W(\vec{x} - \vec{f}\lambda d_i) \exp\{j[\tilde{\psi}(\vec{x}) - \tilde{\psi}(\vec{x} - \vec{f}\lambda d_i)]\}, \tag{6.3}$$

where N_F is a normalizing factor given by

$$N_F = \int d\vec{x} |W(\vec{x})|^2, \tag{6.4}$$

λ is the mean imaging wavelength, and d_i is the distance between the exit pupil and the image plane. In the absence of noise in the image measurement the estimator for a single realization of the object spectrum $\tilde{O}(\vec{f})$ proposed by Primot *et al.* is given by [9]

$$\tilde{O}(\vec{f}) = \frac{I(\vec{f})\tilde{\mathcal{H}}^*(\vec{f})}{|\tilde{\mathcal{H}}(\vec{f})|^2}. \tag{6.5}$$

Even in the absence of measurement noise in the image plane a single realization of $\tilde{O}(\vec{f})$ is random due to turbulence effects. Hence, averaging is used to boost the signal-to-noise ratio of $\tilde{O}(\vec{f})$. The averaging technique proposed by Primot *et al.* is given by

$$\mathrm{E}\{\tilde{O}(\vec{f})\} = \frac{\mathrm{E}\{I(\vec{f})\tilde{\mathcal{H}}^*(\vec{f})\}}{\mathrm{E}\{|\tilde{\mathcal{H}}(\vec{f})|^2\}}. \tag{6.6}$$

Substituting Eq. (6.2) into Eq. (6.6) yields

$$\mathrm{E}\{\tilde{O}(\vec{f})\} = \frac{O(\vec{f})\mathrm{E}\{\mathcal{H}(\vec{f})\tilde{\mathcal{H}}^*(\vec{f})\}}{\mathrm{E}\{|\tilde{\mathcal{H}}(\vec{f})|^2\}}, \tag{6.7}$$

where the fact that $O(\vec{f})$ is deterministic has been used in writing Eq. (6.7). An estimate of the object irradiance distribution $o(\vec{x})$ is obtained by applying the inverse Fourier transform to $\mathrm{E}\{\tilde{O}(\vec{f})\}$. The form of Eq. (6.7) leads to the definition of the average DWFS transfer function $\mathrm{E}\{S(\vec{f})\}$ as

$$\mathrm{E}\{S(\vec{f})\} = \frac{\mathrm{E}\{\mathcal{H}(\vec{f})\tilde{\mathcal{H}}^*(\vec{f})\}}{\mathrm{E}\{|\tilde{\mathcal{H}}(\vec{f})|^2\}}. \tag{6.8}$$

Note that in practical applications of DWFS a photon-limited image is the best possible image available for processing. Thus, the detected image spectrum $D(\vec{f})$ must be substituted for $I(\vec{f})$ in Eq. (6.6) so that the average object spectrum estimate of interest is

$$\mathrm{E}\{\tilde{O}(\vec{f})\} = \frac{\mathrm{E}\{D(\vec{f})\tilde{\mathcal{H}}^*(\vec{f})\}}{\mathrm{E}\{|\tilde{\mathcal{H}}(\vec{f})|^2\}}. \tag{6.9}$$

One motivation for the form of the numerator of Eq. (6.6) can be seen by comparing the numerator of Eq. (6.8) to the speckle transfer function $\mathrm{E}\{|\mathcal{H}(\vec{f})|^2\}$, introduced in Eq. (4.17). In the limit of perfect wave front estimation the result

$$\mathrm{E}\{\mathcal{H}(\vec{f})\tilde{\mathcal{H}}^*(\vec{f})\} \to \mathrm{E}\{|\mathcal{H}(\vec{f})|^2\}, \tag{6.10}$$

is obtained. The right hand side of Eq. (6.10) is exactly the speckle transfer function, which was shown to be non-zero out to spatial frequencies approaching the diffraction-limited cutoff frequency in Section 4.3. Thus, if excellent wave front reconstruction is obtained the numerator of Eq. (6.7) will contain nearly diffraction limited object spectrum information.

Another motivation for the estimator of Eq. (6.7) is that DWFS provides both the *magnitude* and the *phase* of the object spectrum directly. No intermediate Fourier phase spectrum reconstruction step, such as computing and processing the cross spectrum or bispectrum required in speckle imaging, is needed in DWFS. Hence, DWFS offers a computational advantage over speckle imaging, but at the expense of making wave front sensor measurements and post processing wave front sensor data.

Note that the mid and high spatial frequencies of $O(\vec{f})$ are highly attenuated by multiplication with $\mathrm{E}\{\mathcal{H}(\vec{f})\tilde{\mathcal{H}}^*(\vec{f})\}$, and hence an image formed by applying the inverse Fourier transformation to the numerator of Eq. (6.7) would be badly blurred. The intended purpose of the denominator of Eq. (6.7) is to boost the attenuated spatial frequency components of $O(\vec{f})\mathrm{E}\{\mathcal{H}(\vec{f})\tilde{\mathcal{H}}^*(\vec{f})\}$ to their proper values. This goal is accomplished if $\mathrm{E}\{\mathcal{H}(\vec{f})\tilde{\mathcal{H}}^*(\vec{f})\}$ and $\mathrm{E}\{|\tilde{\mathcal{H}}^*(\vec{f})|^2\}$ have the same value. Unfortunately, detailed analysis has shown that in general

$$\mathrm{E}\{|\tilde{\mathcal{H}}^*(\vec{f})|^2\} \neq \mathrm{E}\{\mathcal{H}(\vec{f})\tilde{\mathcal{H}}^*(\vec{f})\}, \tag{6.11}$$

so that the DWFS transfer function given in Eq. (6.8) does not provide object spectrum estimates $\tilde{O}(\vec{f})$ with the proper amplitudes [12]. The reason for the inequality in Eq. (6.11) is that the atmospheric phase $\psi(\vec{x})$ and the estimated phase $\tilde{\psi}(\vec{x})$ have different correlation properties, as shown later. Equation (6.11) implies that undesired results will be obtained when the object spectrum estimate of Eq. (6.7) is used. For example, it is possible for $|\mathrm{E}\{S(\vec{f})\}| > 1$, which results in undesired *amplification* of some spatial frequency components [12].

Roggemann *et al.* proposed a remedy to this problem which is analogous to the reference source calibration used in speckle interferometry. Rather than use the estimated OTF $\tilde{\mathcal{H}}(\vec{f})$ in the denominator of Eq. (6.7), measurements obtained from a reference star are used. Note that the spectrum of the reference star image normalized to have the value of unity at $\vec{f} = 0$ gives the OTF of the atmosphere-telescope system at any instant, which we denote by $\mathcal{H}^{\mathrm{ref}}(\vec{f})$. A wave front sensor measurement made simultaneously with the star image measurement can be used to obtain an estimate of the OTF, denoted by $\tilde{\mathcal{H}}^{\mathrm{ref}}(\vec{f})$. Finally, a new definition of the denominator of Eq. (6.7), which has the same mean as the numerator of Eq. (6.7) if atmospheric and light level conditions are constant, is given by $\mathrm{E}\{\mathcal{H}^{\mathrm{ref}}(\vec{f})\tilde{\mathcal{H}}^{\mathrm{ref}}(\vec{f})^*\}$. The final form for a single realization of the DWFS estimator is thus given by

$$\tilde{O}(\vec{f}) = \frac{D(\vec{f})\tilde{\mathcal{H}}^*(\vec{f})}{\mathcal{H}^{\mathrm{ref}}(\vec{f})\tilde{\mathcal{H}}^{\mathrm{ref}}(\vec{f})^*}. \tag{6.12}$$

Averaging will in general be required to improve the signal-to-noise ratio behavior of the estimators so that

$$\mathrm{E}\{\tilde{O}(\vec{f})\} = \frac{\mathrm{E}\{I(\vec{f})\tilde{\mathcal{H}}^*(\vec{f})\}}{\mathrm{E}\{\mathcal{H}^{\mathrm{ref}}(\vec{f})\tilde{\mathcal{H}}^{\mathrm{ref}}(\vec{f})^*\}} = \frac{O(\vec{f})\mathrm{E}\{\mathcal{H}(\vec{f})\tilde{\mathcal{H}}^*(\vec{f})\}}{\mathrm{E}\{\mathcal{H}^{\mathrm{ref}}(\vec{f})\tilde{\mathcal{H}}^{\mathrm{ref}}(\vec{f})^*\}}. \tag{6.13}$$

Equation (6.13) motivates the definition of a new average DWFS transfer function $\mathrm{E}\{S_1(\vec{f})\}$ given by

$$\mathrm{E}\{S_1(\vec{f})\} = \frac{\mathrm{E}\{\mathcal{H}(\vec{f})\tilde{\mathcal{H}}^*(\vec{f})\}}{\mathrm{E}\{\mathcal{H}^{\mathrm{ref}}(\vec{f})\tilde{\mathcal{H}}^{\mathrm{ref}}(\vec{f})^*\}}. \tag{6.14}$$

Analytic expressions for the numerator and denominator of $\mathrm{E}\{S_1(\vec{f})\}$ given in Eq. (6.14) can be obtained by using Eqs. (2.44) and (6.3). Since the numerator and denominator of Eq. (6.14) have identical form, it is sufficient to analyze only the numerator. The numerator of $\mathrm{E}\{S_1(\vec{f})\}$, represented by $\mathrm{Num}_1(\vec{f})$, is given by

$$\begin{aligned}\mathrm{Num}_1(\vec{f}) = \mathrm{E}\Bigg\{ N_F^{-2} \int \mathrm{d}\vec{x}\, W(\vec{x})W(\vec{x} - \lambda d_i \vec{f}) \exp\left[j\left(\psi(\vec{x}) - \psi(\vec{x} - \lambda d_i \vec{f})\right)\right] \\ \times \int \mathrm{d}\vec{x}'\, W(\vec{x}')W(\vec{x}' - \lambda d_i \vec{f}) \exp\left[-j\left(\tilde{\psi}(\vec{x}') - \tilde{\psi}(\vec{x}' - \lambda d_i \vec{f})\right)\right]\Bigg\}. \end{aligned} \tag{6.15}$$

Equation (6.15) can be simplified by combining the integrals and using Eq. (3.46) to obtain

$$\begin{aligned}\mathrm{Num}_1(\vec{f}) = N_F^{-2} \int \mathrm{d}\vec{x} \int \mathrm{d}\vec{x}'\, W(\vec{x})W(\vec{x} - \lambda d_i \vec{f})W(\vec{x}')W(\vec{x}' - \lambda d_i \vec{f}) \\ \times \exp\left[-\frac{1}{2}\mathrm{E}\left\{\left(\psi(\vec{x}) - \psi(\vec{x} - \lambda d_i \vec{f}) - \tilde{\psi}(\vec{x}') + \tilde{\psi}(\vec{x}' - \lambda d_i \vec{f})\right)^2\right\}\right]. \end{aligned} \tag{6.16}$$

Equation (6.16) can be solved analytically [11]. However, due to the complexity of the argument of the exponential term in Eq. (6.16), the analytic calculations are very time consuming. As a result, simulations have also been used to evaluate $\mathrm{Num}_1(\vec{f})$ [12, 13]. $\mathrm{Num}_1(\vec{f})$ was evaluated using a simulation described later in this section.

Noise effects can prevent accurate estimation of both the numerator and the denominator of Eq. (6.13). However, errors in the denominator of Eq. (6.13) have a greater impact on the object spectrum estimate because errors in the denominator can amplify noise in the numerator. An object spectrum estimator which provides some immunity to noise amplification is given by

$$\mathrm{E}\{\tilde{O}(\vec{f})\} = \frac{\mathrm{E}\{D(\vec{f})\tilde{\mathcal{H}}^{*}(\vec{f})\}}{\mathrm{E}\{\mathcal{H}^{\mathrm{ref}}(\vec{f})\tilde{\mathcal{H}}^{\mathrm{ref}*}(\vec{f})\} + \gamma(\vec{f})}, \tag{6.17}$$

where the parameter $\gamma(\vec{f})$ is a function of spatial frequency which serves to attenuate the noisy high spatial components of $\tilde{O}(\vec{f})$, reducing noise in the reconstructed image.

In the next subsection the problem of obtaining the estimated wave front phase $\tilde{\psi}(\vec{f})$ is discussed. The spatial correlation properties of $\tilde{\psi}(\vec{f})$, as represented in the structure function, are also presented to motivate the object spectrum estimator given in Eq. (6.13).

6.2.2 Wave front phase reconstruction for DWFS

In this subsection the problem of wave front phase reconstruction is addressed. The analysis presented here is a direct extension of the analysis used to derive the least squares technique for obtaining actuator commands presented in Section 5.3.3. The telescope pupil is presumed to be fully populated with wave front sensor subapertures. The wave front slope measured by the jth subaperture in the x or y direction is represented by $s_m^q(j)$, where q indicates either the x or the y direction. An expression for $s_m^q(j)$ is given in Eqs. (5.14) and (5.15) for the Hartmann wave front sensor as

$$\begin{aligned} s_m^q(j) &= s^q(j) + s_n^q(j) \\ &= \int \mathrm{d}\vec{x}\, W_{sj}(\vec{x})\nabla^q\psi(\vec{x}) + s_n^q(j) \\ &= -\int \mathrm{d}\vec{x}\, \nabla^q W_{sj}(\vec{x})\psi(\vec{x}) + s_n^q(j), \end{aligned} \tag{6.18}$$

where $s^q(j)$ is the noise free slope measurement, ∇^q represents the q component of the spatial gradient operator, $W_{sj}(\vec{x})$ is the phase weighting applied by the wave front sensor subaperture, and $s_n^q(j)$ is an additive noise term. The noise term $s_n^q(j)$ is assumed to be a zero mean Gaussian distributed random variable with variance given in Eqs. (5.26) through (5.33). Integration by parts was applied to the second line of Eq. (6.18) to obtain the third line of that equation. The phase weighting function of the jth subaperture $W_{sj}(\vec{x})$ is defined so that $W_{sj}(\vec{x})$ is zero outside the subaperture, and inside the subaperture it is normalized so that

$$\int \mathrm{d}\vec{x}\, W_{sj}(\vec{x}) = 1. \tag{6.19}$$

The gradient of the phase weighting function in the x direction $\nabla^x W_{sj}(\vec{x})$ for square subapertures is most easily expressed in terms of the orthogonal components $\vec{x} = (x, y)$ as

$$\nabla^x W_{sj}(\vec{x}) = [\delta(x - x_{cj} + d/2) - \delta(x - x_{cj} - d/2)]\,\mathrm{rect}\left(\frac{(y - y_{cn})}{d}\right), \tag{6.20}$$

where (x_{cj}, y_{cj}) is the center location of the jth subaperture, and d is the side length of the subaperture. The y direction gradient $\nabla^y W_{sj}(\vec{y})$ is obtained by exchanging y for x in Eq. (6.20). Note that two measurements are provided from each wave front sensor subaperture: an x direction slope measurement and a y direction slope measurement. Thus, a wave front sensor containing J subapertures provides $2J$ measurements.

The estimator for the wave front phase $\tilde{\psi}(\vec{x})$ is given by the weighted sum of I elementary functions $e_i(\vec{x})$ as

$$\tilde{\psi}(\vec{x}) = \sum_{i=1}^{I} c_i e_i(\vec{x}), \tag{6.21}$$

where c_i is the weight associated with the ith elementary function. The weights c_i are obtained by

$$c_i = \sum_{j=1}^{2J} M_{ij} s_j. \tag{6.22}$$

In writing Eq. (6.22) it has been assumed that the wave front sensor contains J subapertures, and that the slope measurements are indexed from 1 to $2J$ to form a column vector of slope measurements $\mathbf{s}$ with jth entry s_j. One possible ordering of the slope measurements is as follows:

$$\begin{aligned} s_1 &= s_m^x(1), \\ s_2 &= s_m^y(1), \\ s_3 &= s_m^x(2), \\ s_4 &= s_m^y(2), \\ &\cdot \\ &\cdot \\ &\cdot \\ s_{2J-1} &= s_m^x(J), \\ s_{2J} &= s_m^y(J). \end{aligned} \tag{6.23}$$

The vector of slope measurements $\mathbf{s}$ is then written as

$$\mathbf{s} = \{s_1, s_2, ..., s_{2J}\}^T, \tag{6.24}$$

where the superscript T indicates the matrix transpose operator. Equation (6.22) can thus be written as a vector matrix multiplication by

$$\mathbf{c} = M\mathbf{s}, \tag{6.25}$$

where $\mathbf{c}$ is the column vector of elementary function weights formed by

$$\mathbf{c} = \{c_1, c_2, ..., c_I\}^T. \tag{6.26}$$

The matrix M, with ith row and jth column given by M_{ij}, is referred to as the reconstruction matrix. The matrix M maps wave front sensor slope measurements to elementary function weights. In this paradigm for wave front phase reconstruction the key problem is the choice of M.

The elementary functions $e_i(\vec{x})$ can be any suitable set of functions defined in the pupil of the telescope. Zernike polynomials, introduced in Section 3.6, are one example of a useful set of elementary functions which have been used in the literature [9, 10, 12, 13, 7]. All of the results presented later in this subsection were generated using Zernike polynomials. The ordering scheme for Zernike polynomials presented in Section 3.6 is used to index the Zernike polynomials. Note that wavefront sensors are insensitive to the first Zernike polynomial, referred to as piston. Hence the piston term will be ignored throughout the remainder of this section.

The least squares reconstruction technique was first mentioned in Section 5.3.3 in the context of controlling a deformable mirror. There is no fundamental difference between the problem of obtaining weights for deformable mirror actuator commands and the wave front reconstruction paradigm given in Eq. (6.21). Here we extend the derivation of the least squares reconstruction matrix M. Note that the maximum likelihood and minimum variance reconstruction techniques discussed in Section 5.3.3 can also be applied to the problem of wave front reconstruction for DWFS. The least squares reconstruction matrix M is obtained by minimizing a measure of squared error between a set of noise free wave front sensor slopes and the slopes which would be measured from a wave front phase composed of a linear combination of elementary functions. The appropriate error Δ is defined by

$$\Delta = \mathbf{s} - H\mathbf{c}, \tag{6.27}$$

where the matrix H is a Jacobian matrix which contains the sensitivities of the slope measurements to changes in the elementary function weights, and the noise free vector of slopes $\mathbf{s}$ must be used. The elements of H are given by

$$H_{ji} = \frac{\partial s_j}{\partial c_i}, \tag{6.28}$$

where H_{ji} is the entry in the jth row and ith column of the H matrix, and s_j is the jth element of the $\mathbf{s}$ vector defined in Eq. (6.24) under noise free conditions. An expression for H_{ji} was given in Eq. (5.46). The squared error minimized by the least squares reconstructor e^2 is given by

$$e^2 = \Delta^T \Delta. \tag{6.29}$$

This minimization is accomplished by expanding e^2 using Eq. (6.27), differentiating the resulting expression with respect to c_i, setting the derivative equal to zero, and solving for the optimal set of weights c_i^{opt}. Performing this operation yields the result for the optimal vector of elementary function weights [14]

$$\mathbf{c}^{\text{opt}} = \left(H^T H\right)^{-1} H^T \mathbf{s}. \tag{6.30}$$

Hence, with reference to Eq. (6.25), the least squares reconstruction matrix is given by

$$M = \left(H^T H\right)^{-1} H^T, \tag{6.31}$$

which is the same result given in Eq. (5.59). It is easy to show that the error between the input wave front phase $\psi(\vec{x})$ and the estimated phase obtained with the least squares reconstructor $\tilde{\psi}(\vec{x})$ is zero mean.

In the least squares phase reconstruction paradigm there is an upper bound on the number of elementary functions which can be used to estimate $\psi(\vec{x})$. One bound is imposed by the requirement that there be more measurements than unknowns, or, mathematically, $2J \geq I$. Another bound on the number of Zernike polynomials which may be used arises from the fact that as the order of the Zernike polynomial increases the spatial scale of the phase fluctuation of the Zernike polynomial becomes smaller. Beyond some maximum order the wave front sensor cannot sense structure in the phase on such a small spatial scale. The mathematical consequence is that $\left(H^T H\right)^{-1}$ fails to exist when an attempt is made to use too many Zernike polynomials to obtain the phase estimate $\tilde{\psi}(\vec{x})$. It should be noted that the maximum likelihood and minimum variance reconstruction techniques presented in Section 5.3.3 are not affected by these considerations since the R and S matrices discussed in that section are always invertible [15, 16].

It is useful to compute the structure function of the estimated phase $\tilde{\psi}(\vec{x})$, $D_{\tilde{\psi}}(\vec{x}_1, \vec{x}_2)$. The motivation for this step is to examine whether the estimator for the OTF given in Eq. (6.3) is unbiased. If $D_{\tilde{\psi}}(\vec{x}_1, \vec{x}_2)$ is very similar in character to the structure function of the atmospheric turbulence-induced phase $D_\psi(\vec{x}_1, \vec{x}_2)$ then we would expect that the estimate of the OTF $\tilde{\mathcal{H}}(\vec{f})$ would be very similar to the actual OTF $\mathcal{H}(\vec{f})$. On the other hand, if $D_{\tilde{\psi}}(\vec{x}_1, \vec{x}_2)$ has significantly different character than $D_\psi(\vec{x}_1, \vec{x}_2)$ we expect a non-zero mean error, or bias in the estimated OTF $\tilde{\mathcal{H}}(\vec{f})$.

The structure function of the estimated wave front phase is defined as [12]

$$D_{\tilde{\psi}}(\vec{x}_1, \vec{x}_2) = \mathrm{E}\left\{\left[\tilde{\psi}(\vec{x}_1) - \tilde{\psi}(\vec{x}_2)\right]^2\right\}. \tag{6.32}$$

Equation (6.32) can be expanded using Eqs. (6.21) and (6.22) to obtain

$$D_{\tilde{\psi}}(\vec{x}_1,\vec{x}_2) = \mathrm{E}\left\{\left(\sum_{i=1}^{I}\sum_{j=1}^{2J} M_{ij}s_j\left[e_i(\vec{x}_1)-e_i(\vec{x}_2)\right]\right)\right.$$
$$\left.\times\left(\sum_{i'=1}^{I}\sum_{j'=1}^{2J} M_{i'j'}s_{j'}\left[e_{i'}(\vec{x}_1)-e_{i'}(\vec{x}_2)\right]\right)\right\}. \tag{6.33}$$

Equation (6.33) can be simplified by combining terms to obtain

$$D_{\tilde{\psi}}(\vec{x}_1,\vec{x}_2) = \sum_{i=1}^{I}\sum_{j=1}^{2J}\sum_{i'=1}^{I}\sum_{j'=1}^{2J} M_{ij}M_{i'j'}S_{jj'}\left[e_i(\vec{x}_1)-e_i(\vec{x}_2)\right]\left[e_{i'}(\vec{x}_1)-e_{i'}(\vec{x}_2)\right], \tag{6.34}$$

where $S_{jj'}$ is the (j,j') element of the covariance of the wave front sensor slope measurements S defined by

$$S = \mathrm{E}\{\mathbf{s}\mathbf{s}^T\}. \tag{6.35}$$

An expression for the elements of the matrix $S_{jj'}$ is given in Eq. (5.119) as [14, 17]

$$\mathrm{E}\{s_m^q(j)s_m^r(j')\} = -\frac{1}{2}\int d\vec{x}\int d\vec{x}'W_{sj}^q(\vec{x})W_{sj'}^r(\vec{x}')D_{\psi}(\vec{x},\vec{x}') + \mathrm{E}\{s_n^q(j)s_n^r(j')\}. \tag{6.36}$$

Note that the structure function of the estimated wave front phase in Eq. (6.34) $D_{\tilde{\psi}}(\vec{x}_1,\vec{x}_2)$ is not in general spatially stationary due to the presence of the elementary functions in the expression. The fact that $D_{\tilde{\psi}}(\vec{x}_1,\vec{x}_2)$ is not spatially stationary is one key difference between the statistics of the turbulence-induced phase and the statistics of the phase estimate.

Equation (6.34) can be evaluated numerically [12]. The results of this calculation for a 1 meter diameter aperture are shown in Fig. 6.2, where $D_{\tilde{\psi}}(\vec{x}_1,\vec{x}_2)$ and $D_{\psi}(\vec{x}_1,\vec{x}_2)$ are plotted for $D/r_0 = 2,\ 10$, and $\mathrm{SNR}_W = \infty,\ 5,\ 3,\ 2$. The wave front sensor for Fig. 6.2 consisted of 10 cm square Hartmann sensor subapertures placed on a Cartesian grid within the 1 meter diameter pupil. This wave front sensor geometry resulted in 64 subapertures being present in the telescope pupil. The maximum number of Zernike polynomial elementary functions supported by this wave front sensor geometry, excluding the piston term, is 61. When more than 61 Zernike polynomials are used the matrix inverse $(H^TH)^{-1}$ in Eq. (6.31) fails to exist. The first 61 Zernike polynomials were used to obtain Fig. 6.2. Note also that since $D_{\tilde{\psi}}(\vec{x}_1,\vec{x}_2)$ is not stationary some strategy is required to choose the points $\vec{x}_1$ and $\vec{x}_2$ where $D_{\tilde{\psi}}(\vec{x}_1,\vec{x}_2)$ is to be evaluated. The approach used to create Fig. 6.2 is as follows. To achieve the maximum possible separation of 1 meter, $\vec{x}_1$ and $\vec{x}_2$ were chosen to lie on the x axis passing through the pupil center.

Figure 6.2 shows that $D_{\tilde{\psi}}(\vec{x}_1,\vec{x}_2)$ and $D_{\psi}(\vec{x}_1,\vec{x}_2)$ are not in general equal. This is true even for the case of high density spatial sampling of the wave front shown in Fig. 6.2(a). Recall from Eqs. (3.112) and (3.124) that the average OTF of the atmosphere-telescope system $\mathrm{E}\{\mathcal{H}(\vec{f})\}$ is given by

$$\mathrm{E}\{\mathcal{H}(\vec{f})\} = \mathcal{H}_0(\vec{f})\exp\left\{-\frac{1}{2}D_{\psi}(\lambda d_i\vec{f})\right\}, \tag{6.37}$$

where $\mathcal{H}_0(\vec{f})$ is the diffraction-limited OTF given in Eq. (2.44). Since $D_{\tilde{\psi}}$ is not spatially stationary we cannot rigorously substitute $D_{\tilde{\psi}}$ for D_{ψ} in Eq. (6.37). However, it is possible to conclude that in general [12]

$$\mathrm{E}\{\tilde{\mathcal{H}}(\vec{f}) - \mathcal{H}(\vec{f})\} \neq 0, \tag{6.38}$$

which in words means that $\tilde{\mathcal{H}}(\vec{f})$ is a biased estimator of $\mathcal{H}(\vec{f})$.

The implication of Eq. (6.38) for the DWFS method is that the numerator and the denominator of Eq. (6.8) are not in general equal, even under conditions of excellent wave front sensor spatial sampling

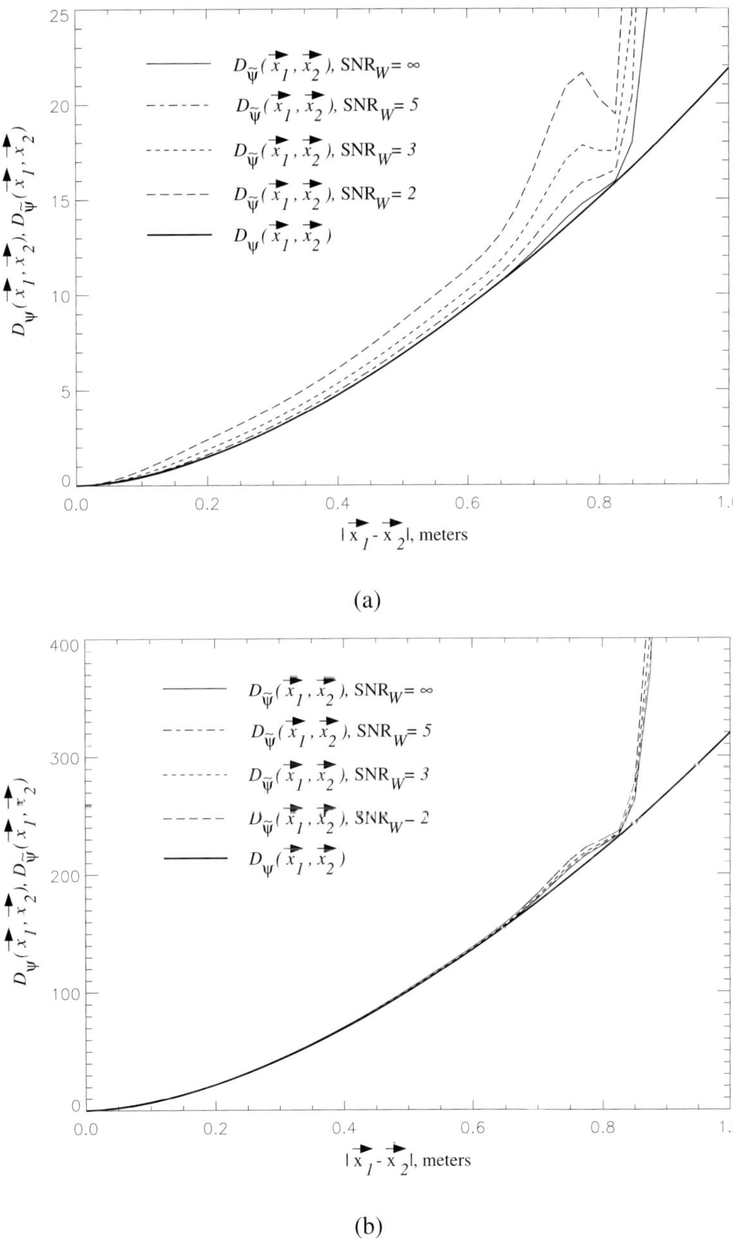

FIGURE 6.2
$D_{\tilde{\psi}}(\vec{x}_1, \vec{x}_2)$ and $D_{\psi}(\vec{x}_1, \vec{x}_2)$ for SNR$_W = \infty$, 5, 3, and 2, and first 61 Zernike polynomials used to reconstruct the phase: (a) for $D/r_0 = 2$; and (b) $D/r_0 = 10$.

and high wave front sensor signal-to-noise ratio. This can lead to unexpected and undesirable effects including both excessive attenuation and amplification of the spectral components of $O(\vec{f})$ in $E\{\tilde{O}(\vec{f})\}$ [12]. The alternative estimator presented in Eq. (6.13) overcomes this problem, but at the expense of observations of a reference star and the associated processing of the reference star data.

As previously discussed, analytic calculations of the performance of DWFS are quite complicated. Due to this analytic complexity, simulations have been used to predict the performance of the DWFS technique. In the next subsection we digress briefly and describe a simulation used to obtain the results presented later in this section.

6.2.3 Simulating DWFS

In this subsection we address the problem of simulating DWFS. The primary motivation for simulating DWFS is the complexity of the expressions which must be solved to perform analytic predictions of performance. For example, the mean of the numerator of the DWFS transfer function, given in Eq. (6.16), requires the solution of a four dimensional integral which cannot be evaluated in closed form. Hence numerical integration techniques must be used to evaluate Eq. (6.16). At each step of the numerical integration the moments and cross moments of $\psi(\vec{x})$ and $\tilde{\psi}(\vec{x})$ must be calculated. To calculate the variance of the numerator of the DWFS transfer function the second moment of $\mathcal{H}(\vec{f})\tilde{\mathcal{H}}^*(\vec{f})$ must be computed, requiring solution of an *eight* dimensional integral. These steps are very time consuming, and at best the mean and second moment of $\mathcal{H}(\vec{f})\tilde{\mathcal{H}}^*(\vec{f})$ can only be evaluated on a very coarse grid of points. Accurate simulations offer an alternative to analytic calculations of DWFS performance. The advantages of simulation are speed, the ability to evaluate performance on a dense grid of points, and the ability to actually create a simulated image of an object for those cases where a visual assessment of performance is desired.

A simulation of DWFS is shown in block diagram form in Fig. 6.3. Our discussion of simulating DWFS rests heavily on the results presented in Sections 4.3.2 and 5.4.6, where the problems of simulating speckle interferometry and adaptive optics were discussed. The specifics of the phase screen generation, pupil model, sample spacing, wave front sensor model, OTF calculation, and photon-limited image creation are identical to the approaches described in Sections 3.7, 4.3.2, and 5.4.6, respectively. Hence, the discussion dwells on the unique aspects of simulating DWFS.

The unique aspects of simulating DWFS begin after the pupil is imposed to obtain the GPF. As before, the GPF is used to create the OTF associated with each wave front. However, in the DWFS simulation the wave front is also input to a wave front sensor model. Noise-free slope measurements are generated by the wave front sensor model using the first term in Eq. (6.18). The noise-free slopes are corrupted by adding a random draw of a zero mean Gaussian random variate with standard deviation given by Eqs. (5.26) through (5.33). The measured slopes are then arranged into a column vector $\mathbf{s}$ as indicated in Eq. (6.23). Weights for the elementary functions are then obtained using Eq. (6.25), and the estimated wave front phase is obtained from Eq. (6.21). The estimated OTF $\tilde{\mathcal{H}}(\vec{f})$ is obtained by using Fourier methods described in Section 4.3.2.

Once $\tilde{\mathcal{H}}(\vec{f})$ is obtained there are two possibilities. If a simulated image is desired, a photon-limited image and its spectrum must be computed using the steps described in Section 4.3.2. These steps are indicated by the solid line leaving the box labeled "Compute OTF" in Fig. 6.3. The product of the photon-limited image spectrum $D(\vec{f})$ and the estimated OTF $\tilde{\mathcal{H}}^*(\vec{f})$ is then computed for each realization, and statistical quantities required to compute the mean and variance of this product are accumulated. However, if it is only necessary to study the DWFS transfer function the path indicated by the broken line leaving the box labeled "Compute OTF" in Fig. 6.3 is followed. Under this circumstance the product $\mathcal{H}(\vec{f})\tilde{\mathcal{H}}^*(\vec{f})$ is computed for each realization, and statistical quantities required to compute the mean and variance of this product are accumulated. In either case this process is repeated many times, where the number of iterations is sufficiently large to insure good estimation of both the mean and the variance of the quantities of interest.

In the next subsection an evaluation of the numerator of the DWFS transfer function $\text{Num}_1(\vec{f})$ given

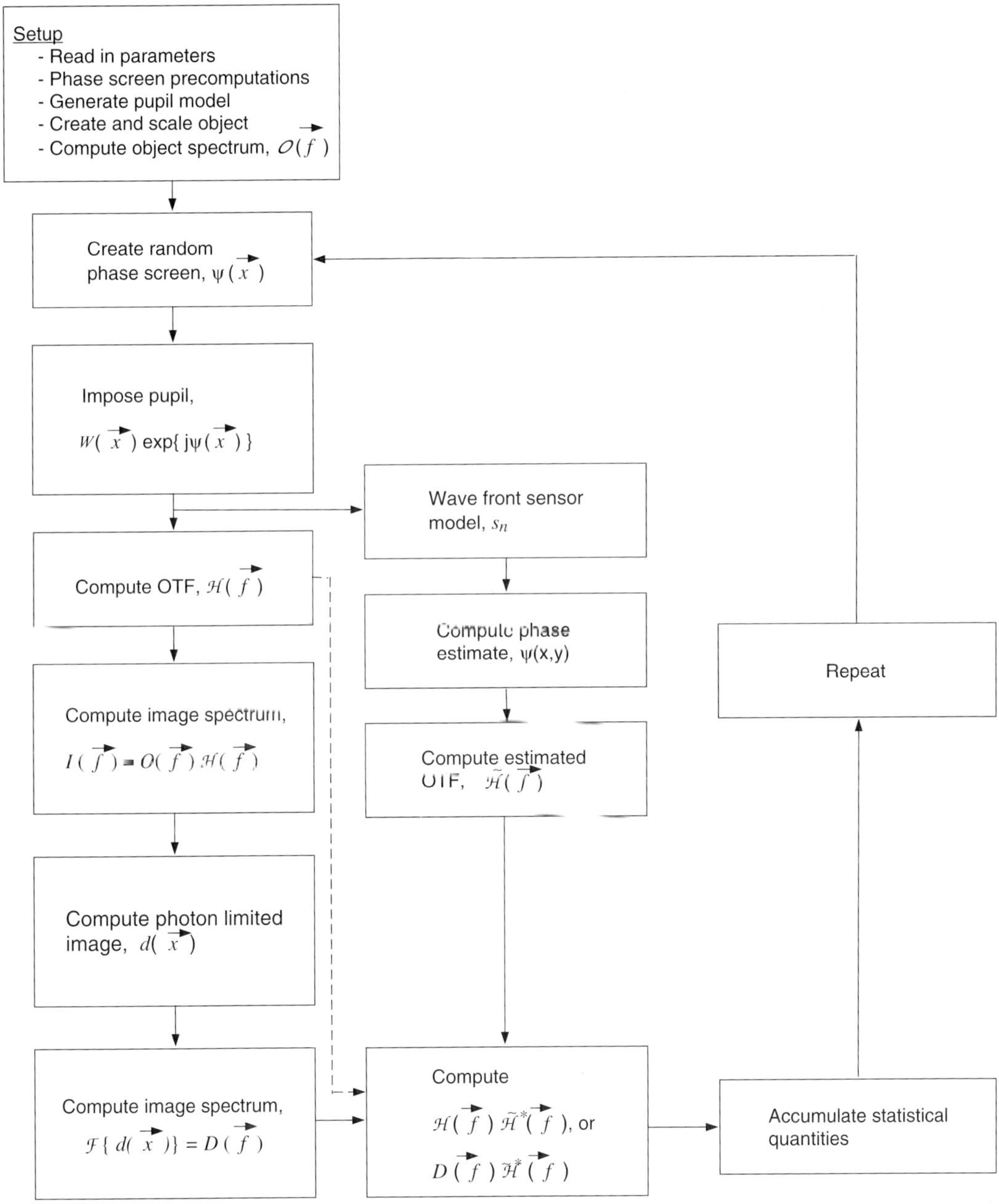

FIGURE 6.3
Block diagram for simulating DWFS.

TABLE 6.1
Input parameters for DWFS transfer function results.

Parameter	Value
Pupil diameter, D	1 m
Subaperture side length, d	0.1 m
λ_W	500 nm
λ_I	600 nm
# Zernike polynomials	61
r_0	0.5 m, 0.1 m at $\lambda = 500$ nm
SNR_W	∞, 5, 3, 2

in Eq. (6.16) is presented. The simulation described in this subsection was used to obtain all of the results presented.

6.2.4 Evaluation of the DWFS transfer function

In this subsection the numerator of the DWFS transfer function $\mathrm{Num}_1(\vec{f})$ is evaluated. The quantity $\mathrm{Num}_1(\vec{f})$ was introduced in Eq. (6.14). As noted in the discussion of Eq. (6.14), both the numerator and the denominator of $S_1(\vec{f})$ have the same mean if the wave front sensor signal level and the seeing do not change between measurements of the object and the reference data. Hence, it is sufficient to analyze either the numerator or the denominator of Eq. (6.14). All results presented here were obtained with the simulation described in Section 6.2.3.

Representative results are presented for $\mathrm{Num}_1(\vec{f}) = \mathrm{E}\{\mathcal{H}(\vec{f})\tilde{\mathcal{H}}^*(\vec{f})\}$. The speckle transfer function $\mathrm{E}\{|\mathcal{H}(\vec{f})|^2\}$ for the appropriate seeing condition is also presented for purposes of comparison. The parameters of the system simulated are given in Table 6.1. The telescope had a one meter diameter unobscured aperture. The wave front sensor had square Hartmann-type subapertures with side length equal to 10 cm, for a total of 64 fully illuminated subapertures contained inside the telescope pupil. The geometry of the wave front sensor is similar to that shown in Fig. 5.16 with the exceptions that no central obscuration is present in the present case, and a smaller aperture is modeled here. The wavefront sensor was assumed to operate at a mean wavelength of 500 nm, and the the imaging camera was assumed to operate at the mean wavelength of 600 nm. The first 61 Zernike polynomials, excluding the piston term $Z_1(\rho, \theta)$, were used to compute the phase estimate $\hat{\psi}(\vec{x})$.

Simulation results are presented in Fig. 6.4. Both $\mathrm{E}\{\mathcal{H}(\vec{f})\tilde{\mathcal{H}}^*(\vec{f})\}$ and $\mathrm{E}\{|\mathcal{H}(\vec{f})|^2\}$ are plotted in Fig. 6.4. The variable f_x in Fig. 6.4 is one of the orthogonal components of the spatial frequency variable $\vec{f}$, so that with the normalization the condition $0 \leq f_x \lambda d_i / D \leq 1$ is obtained. In Fig. 6.4(a) $\mathrm{E}\{\mathcal{H}(\vec{f})\tilde{\mathcal{H}}^*(\vec{f})\}$ and $\mathrm{E}\{|\mathcal{H}(\vec{f})|^2\}$ are shown for $r_0 = 50$ cm at $\lambda = 500$ nm, and $\mathrm{SNR}_W = \infty$, 5, 3, and 2. Fig. 6.4(b) shows a similar family of curves for the $r_0 = 10$ cm at $\lambda = 500$ nm case.

Inspection of Figs. 6.4(a) and (b) provides insight into the limits of performance of DWFS, and provides qualitative comparison of DWFS and speckle imaging. For $r_0 = 50$ cm, which gives $D/r_0 = 2$, and high SNR_W, $\mathrm{E}\{\mathcal{H}(\vec{f})\tilde{\mathcal{H}}^*(\vec{f})\}$ is reasonably close to the speckle transfer function $\mathrm{E}\{|\mathcal{H}(\vec{f})|^2\}$. The condition $\mathrm{E}\{\mathcal{H}(\vec{f})\tilde{\mathcal{H}}^*(\vec{f})\} \leq \mathrm{E}\{|\mathcal{H}(\vec{f})|^2\}$ exists since the phase of the OTF is imperfectly estimated in DWFS. As expected, $\mathrm{E}\{\mathcal{H}(\vec{f})\tilde{\mathcal{H}}^*(\vec{f})\}$ decreases with decreasing SNR_W. It must be noted that the $D/r_0 = 2$ case represents extremely good seeing, which would rarely, if ever, be encountered in practice at visible wavelengths on a one meter diameter telescope.

The more realistic seeing case of $D/r_0 = 10$ shown in Fig. 6.4(b) illustrates some of the limits of the DWFS method. Observe that for $D/r_0 = 10$ at all values of SNR_W $\mathrm{E}\{\mathcal{H}(\vec{f})\tilde{\mathcal{H}}^*(\vec{f})\}$ is significantly less than $\mathrm{E}\{|\mathcal{H}(\vec{f})|^2\}$ for $f_x \lambda d_i / D \geq 0.15$, and is less than 10^{-4} for $f_x \lambda d_i / D \geq 0.55$. By comparison, $\mathrm{E}\{|\mathcal{H}(\vec{f})|^2\}$ does not fall below 10^{-4} until $f_x \lambda d_i / D \geq 0.93$. Hence, speckle interferometry will on average provide measurements with higher spatial frequency content, though it must be noted that the

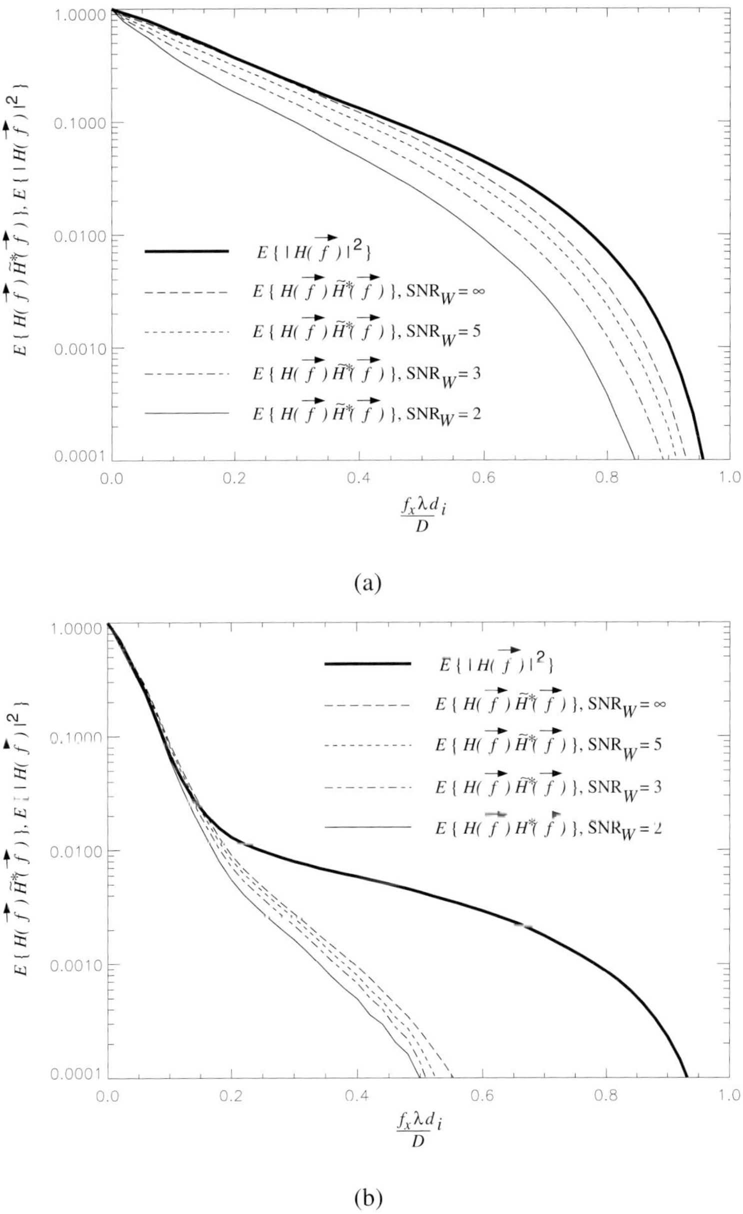

FIGURE 6.4
Speckle transfer function and numerator of DWFS transfer function for SNR$_W = \infty$, 5, 3, and 2 for: (a) $D/r_0 = 2$; and (b) $D/r_0 = 10$.

speckle interferometry measurements will not contain any object phase spectrum information.

While the speckle transfer function $\mathrm{E}\{|\mathcal{H}(\vec{f})|^2\}$ has been shown to be greater than $\mathrm{E}\{\mathcal{H}(\vec{f})\tilde{\mathcal{H}}^*(\vec{f})\}$, we cannot conclude that speckle imaging is in every case superior to DWFS. Such a conclusion must be based on signal-to-noise ratio considerations which include the effects of the object spectrum, measurement noise, and randomness in the appropriate transfer function, $|\mathcal{H}(\vec{f})|^2$ in the case of speckle imaging and $\mathcal{H}(\vec{f})\tilde{\mathcal{H}}^*(\vec{f})$ in the case of DWFS. In the next subsection the DWFS signal-to-noise ratio is derived and compared to the signal-to-noise ratio of the unbiased speckle interferometry estimator $Q(\vec{f})$.

6.2.5 Signal-to-noise ratio considerations

In this subsection the spatial frequency domain signal-to-noise ratio of the DWFS method is derived for the case of photon-limited detection. The signal-to-noise ratio analysis must account for all sources of randomness in the DWFS process. In the DWFS case the sources of randomness include photon noise in the detected image, randomness in the OTF, and randomness in the estimated OTF. We shall see that the DWFS signal-to-noise ratio expression is similar to the image spectrum signal-to-noise ratio obtained in Eq. (2.125) for the photon-limited detection case. However, the DWFS signal-to-noise ratio will be expressed in terms of the mean and variance of $\mathcal{H}(\vec{f})\tilde{\mathcal{H}}^*(\vec{f})$ and the second moment of $\tilde{\mathcal{H}}(\vec{f})$ rather than just the mean and variance of $\mathcal{H}(\vec{f})$. The signal-to-noise ratio for DWFS is then compared to the signal-to-noise ratio for the unbiased speckle imaging estimator $Q(\vec{f})$. The relevant signal-to-noise ratio for DWFS, $\mathrm{SNR}_{DW}(\vec{f})$, is the signal-to-noise ratio of the numerator of the object spectrum estimate $\hat{O}(\vec{f})$ given by

$$\mathrm{SNR}_{DW}(\vec{f}) = \frac{|\mathrm{E}\{D(\vec{f})\tilde{\mathcal{H}}^*(\vec{f})\}|}{\sqrt{\mathrm{var}\{D(\vec{f})\tilde{\mathcal{H}}^*(\vec{f})\}}}, \tag{6.39}$$

where $D(\vec{f})$ represents the spectrum of the detected photon-limited image, and $\mathrm{var}\{\cdot\}$ represents the variance of a random quantity. The analysis which follows was first presented in Ref. [13]. The models, analysis techniques, and results presented in Section 2.3.2 are used to present a shorter analysis here.

The numerator of Eq. (6.39) is evaluated by adopting the model for a photon-limited detected image $d(\vec{x})$ given in Eq. (2.83)

$$d(\vec{x}) = \sum_{n=1}^{K} \delta(\vec{x} - \vec{x}_n), \tag{6.40}$$

where $\vec{x}_n$ is the location of the nth photoevent in the image plane, and K is the total number of photoevents composing the image. The spectrum of a photon-limited image was given in Eq. (2.85) as

$$D(\vec{f}) = \sum_{n=1}^{K} \exp\{-j2\pi\vec{f}\cdot\vec{x}_n\}. \tag{6.41}$$

The numerator of Eq. (6.39) may now be evaluated using an extension of the method established in Eq. (2.115) as

$$\mathrm{E}\{D(\vec{f})\tilde{\mathcal{H}}^*(\vec{f})\} = \mathrm{E}_{\mathcal{H},\tilde{\mathcal{H}}}\left\{\mathrm{E}_{K|\mathcal{H},\tilde{\mathcal{H}}}\left\{\mathrm{E}_{x|K,\mathcal{H},\tilde{\mathcal{H}}}\{D(\vec{f})\tilde{\mathcal{H}}^*(\vec{f})\}\right\}\right\}. \tag{6.42}$$

The inner expectation in Eq. (6.42) is interpreted as the expectation over the photoevent locations $\vec{x}_n$, conditioned upon K and specific realizations of $\mathcal{H}(\vec{f})$ and $\tilde{\mathcal{H}}^*(\vec{f})$. The middle expectation is over the number of photoevents, K, conditioned upon $\mathcal{H}(\vec{f})$ and $\tilde{\mathcal{H}}^*(\vec{f})$. Finally, the outer expectation is the expectation over all possible realizations of $\mathcal{H}(\vec{f})$ and $\tilde{\mathcal{H}}^*(\vec{f})$. Repeating the steps given in Eqs. (2.116) through (2.118) yields

$$\mathrm{E}\{D(\vec{f})\tilde{\mathcal{H}}^*(\vec{f})\} = \overline{K}O_n(\vec{f})\mathrm{E}\{\mathcal{H}(\vec{f})\tilde{\mathcal{H}}^*(\vec{f})\}, \tag{6.43}$$

where $\overline{K}$ is the mean number of photoevents per image, and $O_n(\vec{f})$ is the object spectrum normalized to have a value of unity at $\vec{f} = 0$.

Comparison on Eqs. (2.118) and (6.43) shows that the term $\mathrm{E}\{\mathcal{H}(\vec{f})\}$ in Eq. (2.118) has been replaced by the term $\mathrm{E}\{\mathcal{H}(\vec{f})\tilde{\mathcal{H}}^*(\vec{f})\}$ in Eq. (6.43). The presence of $\mathrm{E}\{\mathcal{H}(\vec{f})\tilde{\mathcal{H}}^*(\vec{f})\}$ in Eq. (6.43) arises from the fact that the object spectrum can be considered to have passed through two correlated random filters: one filter given by the instantaneous OTF, and the other given by the conjugate of the estimated OTF.

Evaluation of the term $\mathrm{var}\{D(\vec{f})\tilde{\mathcal{H}}^*(\vec{f})\}$ in Eq. (6.39) is performed by first computing the second moment of $D(\vec{f})\tilde{\mathcal{H}}^*(\vec{f})$, and then subtracting the square of the mean of $D(\vec{f})\tilde{\mathcal{H}}^*(\vec{f})$. The analysis of the second moment of $D(\vec{f})\tilde{\mathcal{H}}^*(\vec{f})$ proceeds from Eq. (2.119) as

$$\mathrm{E}\{|D(\vec{f})\tilde{\mathcal{H}}^*(\vec{f})|^2\} = \mathrm{E}_{\mathcal{H},\tilde{\mathcal{H}}}\left\{\mathrm{E}_{K|\mathcal{H},\tilde{\mathcal{H}}}\left\{\mathrm{E}_{x|K,\mathcal{H},\tilde{\mathcal{H}}}\{|D(\vec{f})\tilde{\mathcal{H}}^*(\vec{f})|^2\}\right\}\right\}, \tag{6.44}$$

where the conditional expectations have the same interpretation as for Eq. (6.42). The key to evaluating Eq. (6.44) is the realization that the innermost expectation may be written as

$$\mathrm{E}_{x|K,\mathcal{H},\tilde{\mathcal{H}}}\{|D(\vec{f})\tilde{\mathcal{H}}^*(\vec{f})|^2\} = \mathrm{E}_{x|K,\mathcal{H},\tilde{\mathcal{H}}}\{|D(\vec{f})|^2\}|\tilde{\mathcal{H}}^*(\vec{f})|^2, \tag{6.45}$$

since the locations of the photoevents are independent of the estimated OTF $\tilde{\mathcal{H}}^*(\vec{f})$ [13]. The steps in Eqs. (2.119) to (2.124) can be repeated to obtain the expression for $\mathrm{var}\{D(\vec{f})\tilde{\mathcal{H}}^*(\vec{f})\}$ as

$$\mathrm{var}\{D(\vec{f})\tilde{\mathcal{H}}^*(\vec{f})\} = \overline{K}\mathrm{E}\{|\tilde{\mathcal{H}}(\vec{f})|^2\} + (\overline{K})^2|O_n(\vec{f})|^2\mathrm{var}\{\mathcal{H}(\vec{f})\tilde{\mathcal{H}}^*(\vec{f})\}. \tag{6.46}$$

The final expression for the signal-to-noise ratio of DWFS is thus

$$\mathrm{SNR}_{DW}(\vec{f}) = \frac{\overline{K}|O_n(\vec{f})\mathrm{E}\{\mathcal{H}(\vec{f})\tilde{\mathcal{H}}^*(\vec{f})\}|}{\left(\overline{K}\mathrm{E}\{|\tilde{\mathcal{H}}(\vec{f})|^2\} + (\overline{K})^2|O_n(\vec{f})|^2\mathrm{var}\{\mathcal{H}(\vec{f})\tilde{\mathcal{H}}^*(\vec{f})\}\right)^{1/2}}. \tag{6.47}$$

Comparing Eq. (6.47) to the general expression for the signal-to-noise ratio of the unbiased speckle interferometry estimator $\mathrm{SNR}_Q(\vec{f})$ given in Eq. (4.52) shows that $\mathrm{SNR}_{DW}(\vec{f})$ is *linear* with respect to the modulus of the normalized object spectrum $|O_n(\vec{f})|$, while $\mathrm{SNR}_Q(\vec{f})$ is a function of $|O_n(\vec{f})|^2$. One consequence of this different dependence on the object spectrum is that DWFS can have a signal-to-noise ratio comparable to speckle interferometry for measuring extended objects. Analytic evaluation of Eq. (6.47) would require numerical evaluation of an eight dimensional integral to obtain $\mathrm{var}\{\mathcal{H}(\vec{f})\tilde{\mathcal{H}}^*(\vec{f})\}$. This calculation is impractical on widely available computers. Hence, simulations like the one described in Section 6.2.3 have been used to evaluate Eq. (6.47) [13].

Numerical results for $\mathrm{SNR}_{DW}(\vec{f})$ are now presented to illustrate the performance of DWFS. The simulation described in Section 6.2.3 was used to obtain numerical values for $\mathrm{E}\{\mathcal{H}(\vec{f})\tilde{\mathcal{H}}^*(\vec{f})\}$, $\mathrm{E}\{|\tilde{\mathcal{H}}(\vec{f})|^2\}$, and $\mathrm{var}\{\mathcal{H}(\vec{f})\tilde{\mathcal{H}}^*(\vec{f})\}$ for a range of seeing conditions and wave front sensor light levels. These statistical quantities were then used to evaluate Eq. (6.47) for a set of objects and light levels of interest. The simulation was also used to obtain $\mathrm{E}\{|\mathcal{H}(\vec{f})|^2\}$, $\mathrm{E}\{|\mathcal{H}(2\vec{f})|^2\}$, and $\mathrm{E}\{|\mathcal{H}(\vec{f})|^4\}$ so that the signal-to-noise ratio $\mathrm{SNR}_Q(\vec{f})$ of the unbiased speckle interferometry estimator $Q(\vec{f})$ could be calculated using Eq. (4.46). The optical system parameters are summarized in Table 6.2. The optical system simulated consisted of a one meter diameter unobscured aperture. This aperture was fully populated with Hartmann-type wave front sensor subapertures with side length $d = 10$ cm which provided a total of 64 fully illuminated subapertures in the pupil. The first 40 Zernike polynomials, excluding the piston term $Z_1(\rho, \theta)$, were used to reconstruct the wave front phase for the results presented. The choice of using 40 Zernike polynomials for wave front phase reconstruction was made because of the good match between D_ψ and $D_{\tilde{\psi}}$ obtained for a wide range of SNR_W conditions for this case [12].

To allow a meaningful comparison of $\mathrm{SNR}_{DW}(\vec{f})$ and $\mathrm{SNR}_Q(\vec{f})$ a common set of wavelength, light level, and seeing conditions must be established. For the results which follow the following assumptions were made. The mean wavelength of operation for the wave front sensor λ_W was chosen to be $\lambda_W = 600$ nm for DWFS, and the mean wavelength for image measurement λ_I for both DWFS and speckle interferometry was chosen to be $\lambda_I = 700$ nm.

TABLE 6.2
Input parameters for DWFS signal-to-noise ratio results.

Parameter	Value
Pupil diameter, D	1 m
Subaperture side length, d	0.1 m
Wave front sensor wavelength, λ_W	500 nm
Imaging wavelength, λ_I	600 nm
# Zernike polynomials	40
r_0	0.5 m, 0.1 m, 0.07 m at λ = 500 nm
SNR_W	∞, 5, 3, 2
$\overline{K}$	∞, 1963.5, 706.9, 314.2

Photon-limited detection was assumed for both the wave front sensor and the imaging camera. Integration times in the wave front sensor and the imaging camera were assumed to be identical. It was further assumed that equal photon flux densities, specified in units of photons/$(\text{m}^2 \cdot \text{s})$, were present at the telescope pupil at both λ_I and λ_W, and that the optical losses and detector quantum efficiencies in the imaging and wave front sensor legs were identical. Results for four wave front sensor signal-to-noise ratio values are presented, $\text{SNR}_W = \infty$, 5, 3, and 2, corresponding to wave front sensor light levels of $\overline{K}_W = \infty$, 25, 9, and 4, respectively. With these assumptions the mean number of photoevents per image per integration time $\overline{K}$ can be computed by noting that the telescope aperture is approximately 78.5 times larger than a wave front sensor subaperture. Hence, the appropriate average numbers of photoevents per image per integration time are $\overline{K} = \infty$ for $\text{SNR}_W = \infty$, $\overline{K} = 1963.5$ for $\text{SNR}_W = 5$, $\overline{K} = 706.9$ for $\text{SNR}_W = 3$, and $\overline{K} = 314.2$ for $\text{SNR}_W = 2$. Note that the $\text{SNR}_W = \infty$ case eliminates the effects of measurement noise in the wave front sensor and image plane, and hence this provides an upper bound on the performance of both DWFS and speckle interferometry. The Fried seeing parameter r_0 was specified at a wavelength of 500 nm. Specific values of r_0 used to obtain the results which follow are $r_0 = 50$ cm, 10 cm, and 7 cm.

Results are presented for two objects: a point source representing a single simulated star, and a computer generated rendering of a simulated satellite object shown in Fig. 6.5. $\text{SNR}_{DW}(\vec{f})$ results for the point source object are shown in Fig. 6.6, and $\text{SNR}_{DW}(\vec{f})$ for the computer generated satellite object are shown in Fig. 6.8. Corresponding results for $\text{SNR}_Q(\vec{f})$ are shown in Figs. 6.7 and 6.9. In Figs. 6.6 through 6.9 the plots labeled (a) are for $r_0 = 50$ cm, and the plots labeled (b) are for $r_0 = 10$ cm. The results presented in Figs. 6.6 through 6.9 were obtained by computing the full two-dimensional SNR array, and then averaging around circles of constant radius to express $\text{SNR}_{DW}(\vec{f})$ and $\text{SNR}_Q(\vec{f})$ as functions of a scalar spatial frequency.

Comparison of Figs. 6.6 and 6.7 shows that speckle interferometry has a signal-to-noise ratio advantage over DWFS for *point source objects* for all the conditions tested. In Section 4.3.2 the signal-to-noise ratio limited cutoff frequency f_{SNR} was defined as the normalized spatial frequency at which the relevant signal-to-noise ratio fell below some arbitrary threshold. If the threshold for computing f_{SNR} is chosen to be 0.1 and attention is restricted to the finite signal cases, then we see that for point source objects speckle interferometry provides $f_{\text{SNR}} \geq 0.95$, while DWFS provides $0.5 \leq f_{\text{SNR}} \leq 0.9$. Inspection of Figs. 6.8 and 6.9 shows that this signal-to-noise ratio advantage diminishes for the *extended object* case since $\text{SNR}_{DW}(\vec{f})$ depends upon $|O_n(\vec{f})|$, while $\text{SNR}_Q(\vec{f})$ depends on $|O_n(\vec{f})|^2$, and $|O_n(\vec{f})| < 1$ everywhere except $\vec{f} = 0$ for the extended object. For the extended object cases where finite signal level was used, inspection of Figs. 6.8 and 6.9 shows that for speckle interferometry $0.35 \leq f_{\text{SNR}} \leq 0.8$, while for DWFS $0.37 \leq f_{\text{SNR}} \leq 0.85$. These results allow us to draw the conclusion that speckle interferometry and DWFS can provide competitive performance on extended objects.

The conclusions drawn from Figs. 6.6 and 6.7 are based on a comparison of the SNR for the speckle interferometry power spectrum estimate to the SNR for the DWFS complex image spectrum estimate.

FIGURE 6.5
Computer generated rendering of a simulated satellite object.

We have also investigated SNR performance for the case in which the two SNRs being compared are for the same estimated quantity [18]. Since speckle interferometry estimates the image power spectrum, we considered comparing the DWFS SNR performance for estimating the image power spectrum to the speckle interferometry SNR. A theoretical investigation of the SNR expressions reveals that DWFS has a slight SNR advantage over speckle interferometry regardless of the object characteristics (points or extended source) for the case that performance is limited by image shot noise. In the case that the variance of the OTF limits performance we find that speckle interferometry has a slight SNR advantage over DWFS [18]. Due to the fact that these performance differences are slight we can generally conclude that DWFS and speckle interferometry give the *same* performance in terms of estimating the image power spectrum.

As a final comment, we note that these comparisons between speckle interferometry and DWFS may be misleading. The point to remember is that both of our SNR comparison approaches are lacking in that we have not considered the performance of the image phase spectrum estimate. In the case of speckle interferometry, the technique only gives an estimate of image power spectrum and another approach must be used to obtain the phase of the image spectrum. These other approaches involve higher order correlations of the complex object spectrum as discussed in Chapter 4. We have investigated the comparative accuracy of the phase estimate of DWFS and the phase estimate from Knox-Thompson or the bispectrum when being used in conjunction with speckle interferometry. We note that there is anecdotal evidence suggesting that the error in the phase estimate obtained from the bispectrum is closely tied to the speckle interferometry power spectrum SNR. Matson *et al.* [19] has shown that the point at which the speckle interferometry power spectrum SNR goes below unity corresponds roughly to the point at which the spectrum phase estimate error exceeds 1 rad. If we can assume the same association between the SNR of the complex image spectrum for DWFS and the error in DWFS's phase spectrum estimate, we conclude that speckle imaging will generally provide better performance for point like objects since the DWFS SNR will always be less than speckle interferometry SNR. Making the same statement for extended objects is less straightforward because of the differing effect of the object

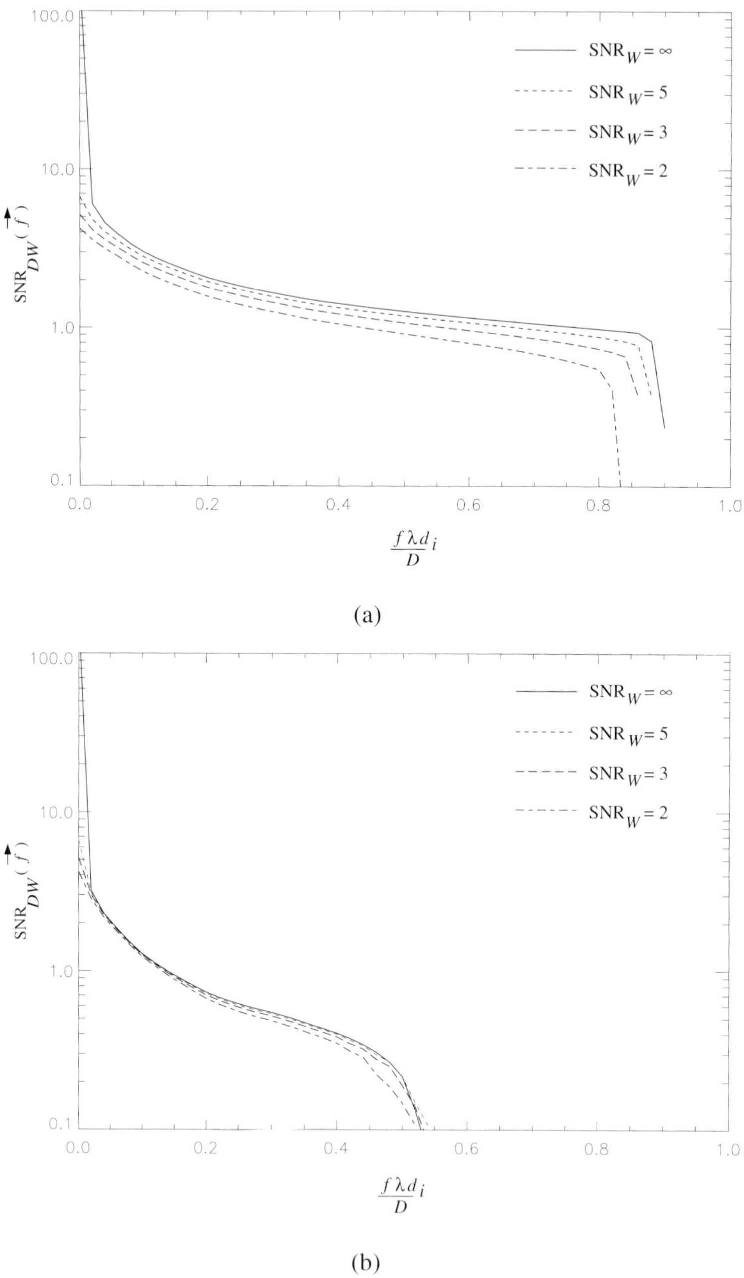

FIGURE 6.6
$SNR_{DW}(\vec{f})$ for a point source object for $SNR_W = \infty, 5, 3$, and 2 for: (a) r_0 = 50 cm; and (b) r_0 = 10 cm.

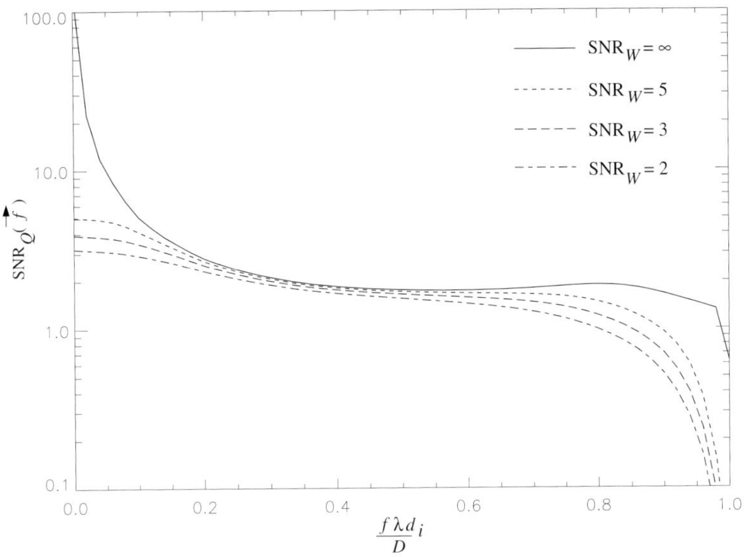

(a)

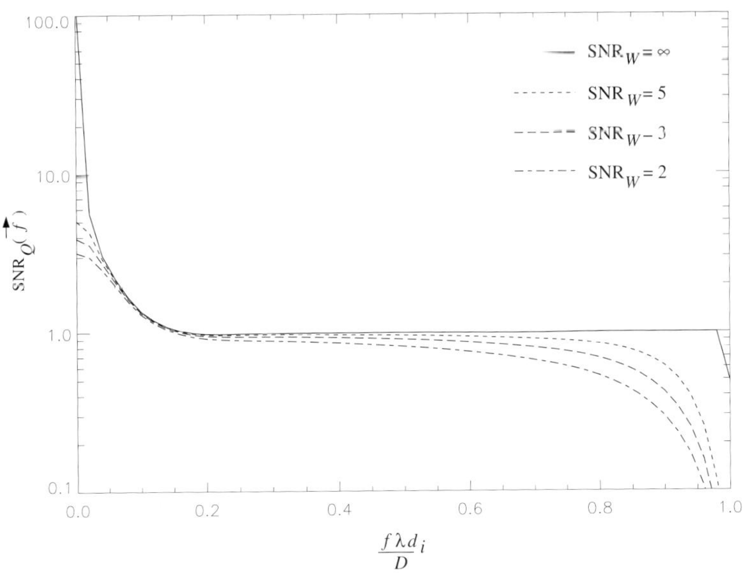

(b)

FIGURE 6.7
$\mathrm{SNR}_Q(\vec{f})$ for a point source object for $\mathrm{SNR}_W = \infty, 5, 3$, and 2 for: (a) $r_0 = 50$ cm; and (b) $r_0 = 10$ cm.

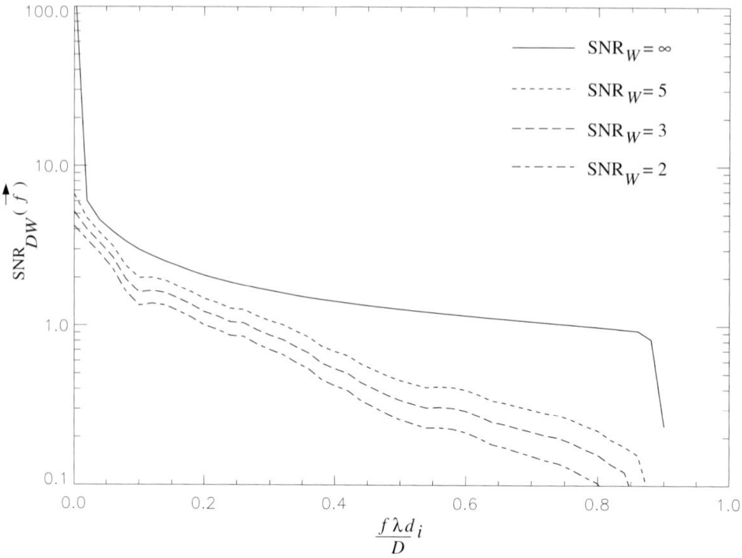

(a)

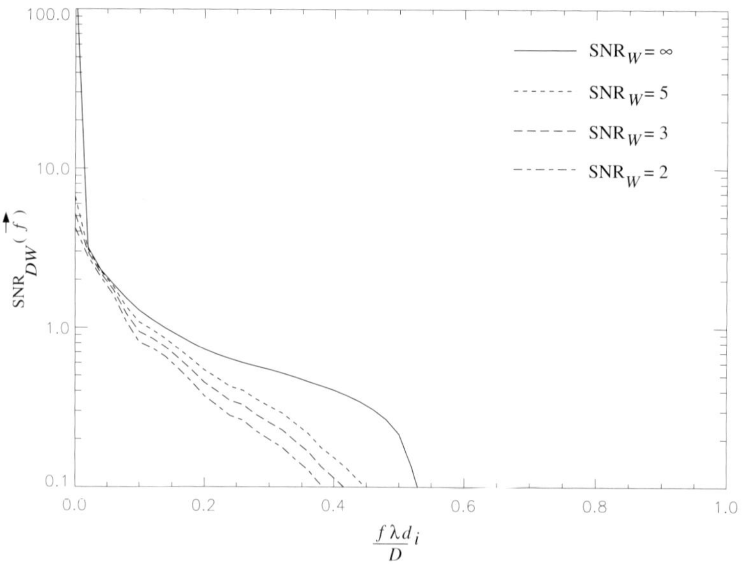

(b)

FIGURE 6.8
SNR$_{DW}(\vec{f})$ for computer generated satellite object for SNR$_W$ $= \infty,\ 5,\ 3,$ and 2 for: (a) r_0 = 50 cm; and (b) r_0 = 10 cm.

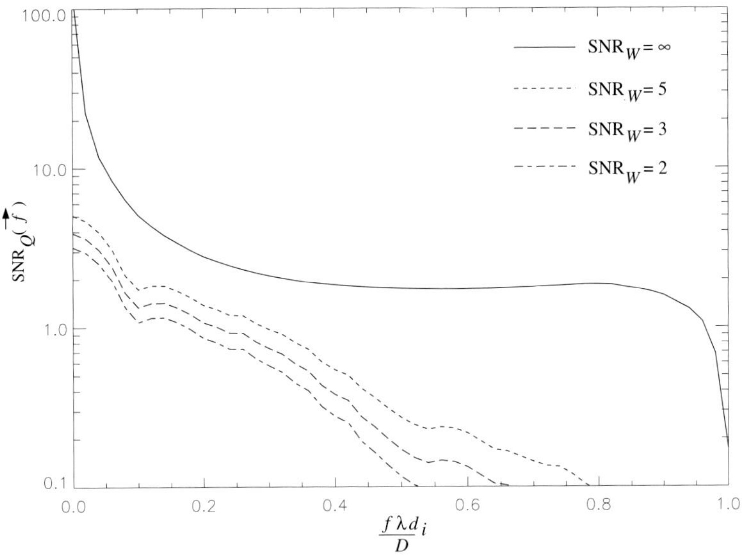

(a)

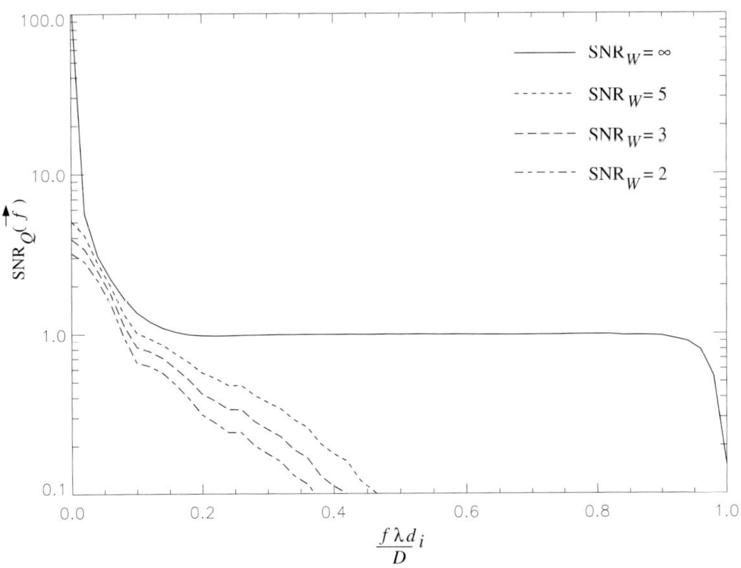

(b)

FIGURE 6.9
$SNR_Q(\vec{f})$ for computer generated satellite object for $SNR_W = \infty,\ 5,\ 3,$ and 2 for: (a) $r_0 = 50$ cm; and (b) $r_0 = 10$ cm.

TABLE 6.3
Input parameters for DWFS imaging results.

Parameter	Value
Pupil diameter, D	1 m
Subaperture side length, d	0.1 m
Wave front sensor wavelength, λ_W	600 nm
Imaging wavelength, λ_I	700 nm
# Zernike polynomials	40
Number of independent frames	200
r_0	0.1 m, 0.07 m at λ = 500 nm
$\overline{K}_W$	1,201, 76, 8
$\overline{K}$	746,928, 47,131, 2,974

spectrum, O_n, on the SNR expressions.

6.2.6 Examples of DWFS imaging performance

In this subsection we present DWFS image reconstruction examples for the simulated binary star object. The results were obtained using the simulation described in Section 6.2.3. The parameters used to obtain the results in Fig. 6.10 are summarized in Table 6.3. A telescope with diameter $D = 1$ m with an unobscured aperture was modeled. A Hartmann-type wave front sensor was used with square subapertures having side length of 10 cm. The first 40 Zernike polynomials, excluding the piston term, were used to reconstruct the wave front phase. The mean wavelength for the wave front sensor was selected at $\lambda_W = 600$ nm, and the imaging wavelength was selected as $\lambda_I = 700$ nm. A bandwidth of $\lambda_W = 600 \pm 60$ nm was used for the wave front sensor, and a bandwidth of $\lambda_I = 700 \pm 35$ nm was used for the imaging camera. The Fried seeing parameter was set at two different values: $r_0 =$ 10 cm and 7 cm. Three different visual magnitudes were used for the binary star, $m_v = 2$, $m_v = 5$, and $m_v = 8$, and integration times in both the wavefront sensor and the imaging camera were set at 10 ms. Transmission efficiency from the top of the atmosphere through the outputs of both the imaging camera and the wave front sensor were assumed to be identical, and equal to 10%. The resulting average photoevent rates in the wave front sensor were $\overline{K}_W = 19,030$ for $m_v = 2$; $\overline{K}_W = 1,201$ for $m_v =$ 5; and $\overline{K}_W = 76$ for $m_v = 8$. The corresponding average photoevent rates in the imaging camera were $\overline{K} = 746,928$ for $m_v = 2$; $\overline{K} = 47,131$ for $m_v = 5$; and $\overline{K} = 2,974$ for $m_v = 8$. With these choices the diffraction-limited smallest resolvable angle is given by $\lambda_I/D = 0.7\ \mu$rad, and the seeing-limited smallest resolvable angles are given by $\lambda_I/r_0 = 7\ \mu$rad for $r_0 = 10$ cm, and $\lambda_I/r_0 =$ 10 μrad for $r_0 = 7$ cm. The simulated binary star components had a brightness ratio of 2:1, and were separated by 2.8 μrad. A data set consisting of 200 independent frames of object and reference star data was used to obtain all results presented.

Image reconstruction results are presented in Fig. 6.10. Images with the following set of input parameters appear in Fig. 6.10: (a) $m_v = 2$, $r_0 = 10$ cm; (b) $m_v = 2$, $r_0 = 7$ cm; (c) $m_v = 5$, $r_0 =$ 10 cm; (d) $m_v = 5$, $r_0 = 7$ cm; (e) $m_v = 8$, $r_0 = 10$ cm; and (f) $m_v = 8$, $r_0 = 7$ cm. The object spectrum estimates $\tilde{O}(\vec{f})$ were reconstructed using Eq. (6.17), with $\gamma(\vec{f}) = 0$. The E$\{\tilde{O}(\vec{f})\}$ were then filtered to reduce high frequency noise effects with the conical filter defined in Eq. (4.130) as

$$H(\vec{f}) = \begin{cases} 1 - \frac{|\vec{f}|}{f_c}, & |\vec{f}| \leq f_c \\ 0, & |\vec{f}| > f_c \end{cases}, \tag{6.48}$$

where f_c is a user-selected cutoff frequency. The cutoff frequencies were chosen by radially averaging $\sqrt{M}\mathrm{SNR}_{DW}(\vec{f})$, where $M = 200$ is the number of independent frames averaged, for each case and selecting as f_c the spatial frequency at which $\mathrm{SNR}_{DW}(\vec{f})$ fell below 10. The resulting cutoff normalized spatial frequency for each case in Fig. 6.10 is as follows: (a), $f_c = 0.85D/(\lambda d_i)$; (b), $f_c = 0.42D/(\lambda d_i)$; (c), $f_c = 0.85D/(\lambda d_i)$; (d), $f_c = 0.37D/(\lambda d_i)$; (e), $f_c = 0.57D/(\lambda d_i)$; and (f), $f_c = 0.27D/(\lambda d_i)$. The general trend evident from the reconstructions in Fig. 6.10 is that for high signal levels in the wavefront sensor and imaging camera, and wave front sensor size set at approximately $r_0 \approx d$, where d is the subaperture side length, good image reconstructions are obtained using DWFS. However, at lower wave front sensor signal levels and when $r_0 < d$, the image quality is severely degraded.

This concludes our discussion of DWFS. We now turn our attention to the topic of hybrid imaging techniques which require adaptive optics. As discussed in the introduction to this chapter, hybrid imaging techniques which require adaptive optics offer a tradeoff between doing full compensation for turbulence effects using adaptive optics, and using only post processing techniques, such as speckle imaging or DWFS to overcome turbulence effects.

6.3 Methods involving adaptive optics

Hybrid imaging techniques which require the use of adaptive optics have been developed with two primary motivations: (1) relieving some of the hardware requirements for "fully compensating" for atmospheric turbulence effects by performing some correction in real time with adaptive optics, and some correction using post detection image processing; and (2) extracting greater performance from existing adaptive optics systems under suboptimal operational conditions, such as poor seeing or high Greenwood frequency conditions. All of the hybrid imaging techniques presented here have been analyzed either theoretically or using simulations [3, 4, 5, 6, 7, 20, 21, 22]. Some of the hybrid imaging techniques have been tested experimentally [23, 24].

The functional block diagram for hybrid imaging systems was introduced in Fig. 1.6, and is repeated in Fig. 6.11 for reference. The hybrid imaging paradigm is as follows. Images of the object of interest, and possibly simultaneous wave front sensor outputs, are measured using an adaptive optics system. This data is typically recorded for later processing. Calibration data from a reference star is also recorded to use in the post detection image processing steps. The image, wave front sensor, and reference star measurements are then processed according to the appropriate algorithm to create a single high resolution estimate of the object irradiance distribution.

Four hybrid imaging techniques are presented in the following subsections. In the next subsection deconvolution of compensated images is presented. This technique relies on the fact that high spatial frequency information with good signal-to-noise ratio is present in images measured with partially compensated adaptive optics, even though the raw image measurements look blurred because the atmosphere-telescope OTF is highly attenuated. The deconvolution process boosts the attenuated high spatial frequency components to their appropriate values, resulting in sharper images. The performance of speckle imaging post processing is examined in Section 6.3.2 for the case of adaptive optics imaging. In Section 6.3.3 the DWFS concept is extended to compensated images. The application of DWFS to adaptive optics imaging systems is referred to as compensated DWFS (CDWFS). CDWFS has been shown to be useful for compensating some of the effects of time delays in adaptive optics systems, and for overcoming some of the effects of anisoplanatism. Finally, in Section 6.3.4 the concept of *frame selection* is examined. In frame selection a large number of short exposure images of the object and a reference star are measured. These images are then sorted using an image sharpness criterion, and only the sharpest images are used to compute the average image. The image spectrum resulting from the frame selection and averaging process has been shown to have higher signal-to-noise ratio than the

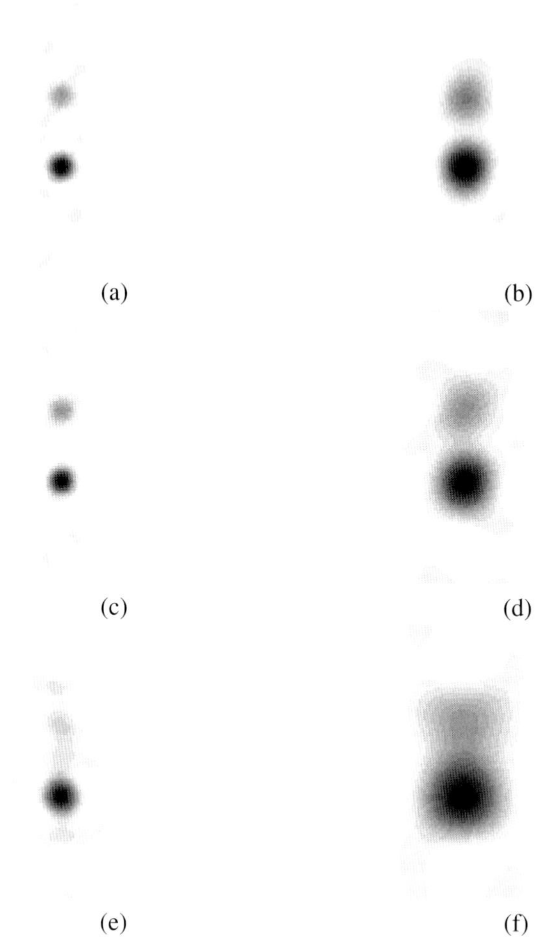

FIGURE 6.10
Binary star images reconstructed using DWFS: (a) $r_0 = 10$ cm, $m_v = 2$; (b) $r_0 = 7$ cm, $m_v = 2$; (c) $r_0 = 10$ cm, $m_v = 5$; (d) $r_0 = 7$ cm, $m_v = 5$; (e) $r_0 = 10$ cm, $m_v = 8$; (f) $r_0 = 7$ cm, $m_v = 8$. Angular separation of the binary components = 2.8 μrad, telescope diameter = 1.0 m, and ratio of brightnesses = 2:1. The central 100×100 pixels of 256×256 pixel images are presented. Results are presented for 200 independent frames of data. Negative images are displayed for clarity.

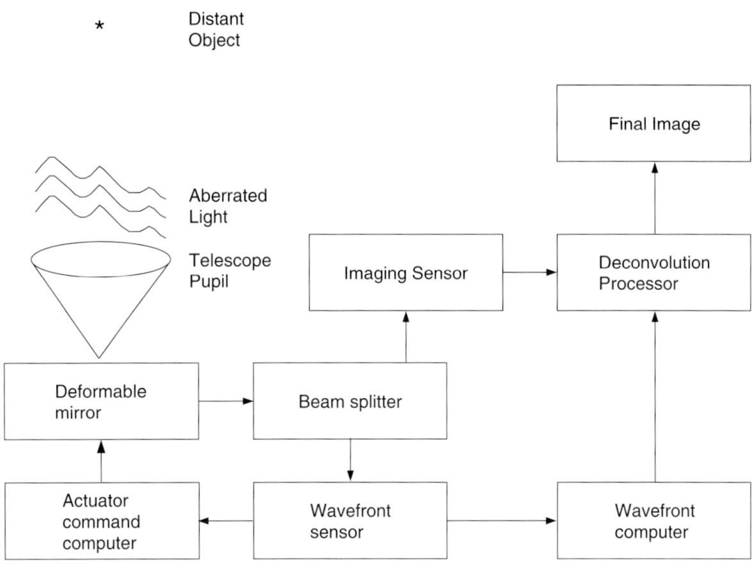

FIGURE 6.11
Block diagram of a hybrid imaging system.

average image spectrum resulting from averaging all the frames.

6.3.1 Deconvolution of compensated images

In adaptive optics imaging systems the OTF varies randomly from instant-to-instant. Any analysis of adaptive optics system imaging performance must account for this source of randomness. Both fully and partially compensated adaptive optics imaging systems provide good signal-to-noise ratio at high spatial frequencies in the measured images [3, 5]. However, the average OTF of the atmosphere-telescope system may be highly attenuated in a partially compensated system, leading to blurred raw image measurements. Deconvolution of these blurred measured images provides sharper, higher resolution images. There are, however, signal-to-noise ratio-imposed limits on deconvolution which must be understood to avoid amplifying noise effects in the deconvolution process. In this subsection we address the spatial frequency domain signal-to-noise ratio of the detected image in adaptive optics imaging systems, present a linear deconvolution measurement and processing algorithm, and discuss the signal-to-noise ratio imposed limits on deconvolution of adaptive optics images.

The theoretical basis for deconvolving compensated images derives from Eq. (2.126), which gave the expression for the signal-to-noise ratio $\mathrm{SNR}_D(\vec{f})$ of the detected image spectrum $D(\vec{f})$ under conditions of a random OTF as

$$\mathrm{SNR}_D(\vec{f}) = \frac{\overline{K}|\mathrm{E}\{\mathcal{H}(\vec{f})\}||O_n(\vec{f})|}{\sqrt{\overline{K} + (\overline{K})^2|O_n(\vec{f})|^2\mathrm{var}\{\mathcal{H}(\vec{f})\} + P\sigma_n^2}}, \tag{6.49}$$

where $\overline{K}$ is the average number of photoevents per image, $O_n(\vec{f})$ is the object spectrum normalized to have a value of unity at $\vec{f} = 0$, P is the number of pixels in the image, and σ_n is the RMS value of the additive noise in the detector pixels expressed in units of photoevents. Recall that the $\overline{K} \to \infty$ limit of

TABLE 6.4
Input parameters for adaptive optics imaging results.

Parameter	Value
Pupil diameter, D	1 m
Subaperture side length, d	0.1 m
Actuator spacing	0.12 m
Wave front sensor wavelength, λ_W	600 nm
Imaging wavelength, λ_I	700 nm
Number of independent frames	200
r_0	0.2 m, 0.1 m, 0.07 m, 0.05 m at λ = 500 nm
$\overline{K}_W$	120
$\overline{K}$	5,358

Eq. (6.49) is

$$\lim_{\overline{K}\to\infty} \{\text{SNR}_D(\vec{f})\} = \frac{|\text{E}\{\mathcal{H}(\vec{f})\}|}{\sqrt{\text{var}\{\mathcal{H}(\vec{f})\}}} = \text{SNR}_{\mathcal{H}}(\vec{f}), \tag{6.50}$$

where $\text{SNR}_{\mathcal{H}}(\vec{f})$ is referred to as the SNR of the random OTF. It should also be noted that if the exposure time is sufficiently long so that the average OTF during the exposure approaches the true average OTF during the duration of the exposure, then the term $\text{var}\{\mathcal{H}(\vec{f})\}$ in the denominator of Eq. (6.49) approaches zero so that the expression for $\text{SNR}_D(\vec{f})$ becomes

$$\text{SNR}_D(\vec{f}) = \frac{\overline{K}|\text{E}\{\mathcal{H}(\vec{f})\}||O_n(\vec{f})|}{\sqrt{\overline{K} + P\sigma_n^2}}. \tag{6.51}$$

The key point enabling deconvolution is that it can be shown that a sufficiently high signal-to-noise ratio exists at sufficiently high frequencies to allow deconvolution of images measured with highly attenuated OTFs.

Of course, the form of $\text{SNR}_{\mathcal{H}}(\vec{f})$ is critical. Examples of $\text{E}\{\mathcal{H}(\vec{f})\}$, and $(\text{var}\{\mathcal{H}(\vec{f})\})^{1/2}$, are shown in Fig. 6.12. The associated plot of $\text{SNR}_{\mathcal{H}}(\vec{f})$ is shown in Fig. 6.13. The adaptive optics simulation described in Section 5.4.6 was used to obtain all of the results presented in Fig. 6.12 using input parameters shown in Table 6.4. A one meter diameter unobscured telescope aperture was modeled. A Hartmann-type wave front sensor with 10 cm subapertures was used. The mean imaging wavelength was set at $\lambda_I = 700$ nm, and the mean wavelength of the wavefront sensor was set at $\lambda_W = 600$ nm. The deformable mirror had actuators placed on a Cartesian grid with grid spacing of 12 cm. The influence functions of the actuators were modeled as two-dimensional triangle functions [25] with base width equal to two times the actuator grid spacing. Least squares reconstruction, as described in Section 5.3.3, was used to map wave front sensor measurements to actuator commands. It was assumed that no delay between sensing and correcting the turbulence-induced aberration was present in the adaptive optics system, and that the beacon for the wave front sensor was infinitely bright. Note that the assumptions of an infinitely bright beacon and no time delays make the results presented in Fig. 6.12 upper bounds on performance The Fried seeing parameter r_0 was specified at the wavelength of 500 nm. Results for four different r_0 conditions are presented: $r_0 = 20$ cm, $r_0 = 10$ cm, $r_0 = 7.5$ cm, and $r_0 = 5$ cm. The statistical quantities presented in Fig. 6.12 were obtained for 200 frame data sets. In all cases the appropriate two-dimensional function was created with the simulation, and the two-dimensional arrays were then averaged around circles of constant radius to obtain results which are a function of a scalar spatial frequency variable. The horizontal axes of Fig. 6.12 are normalized to the diffraction-limited cutoff frequency, $D/(\lambda d_i)$.

Observe from Fig. 6.12 that the $\text{E}\{\mathcal{H}(\vec{f})\}$ plots for $r_0 = 7.5$ cm and $r_0 = 5$ cm are significantly lower

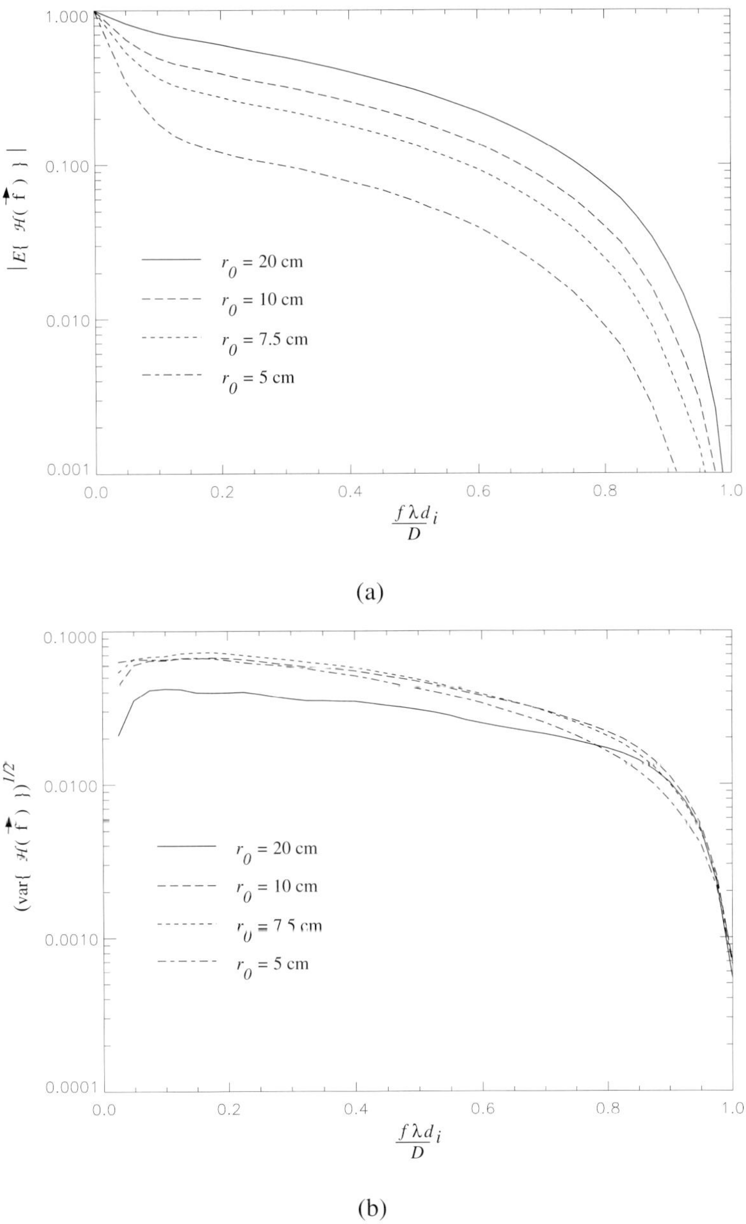

FIGURE 6.12
Adaptive optics performance for $r_0 = 20$ cm, $r_0 = 10$ cm, $r_0 = 7.5$ cm, and $r_0 = 5$ cm: (a) average OTF, $E\{\mathcal{H}(\vec{f})\}$, and (b) $\left(\text{var}\{\mathcal{H}(\vec{f})\}\right)^{1/2}$ for 1 meter diameter telescope.

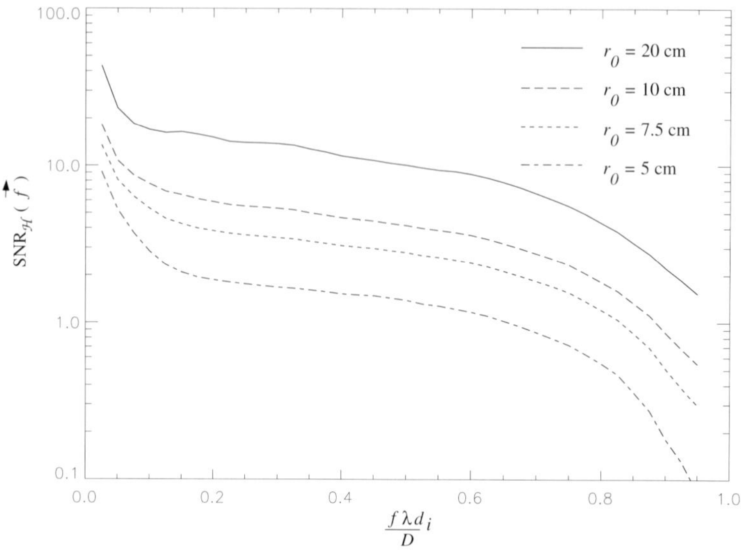

FIGURE 6.13
$\mathrm{SNR}_{\mathcal{H}}(\vec{f})$ for $r_0 = 20$ cm, $r_0 = 10$ cm, $r_0 = 7.5$ cm, and $r_0 = 5$ cm for a 1 meter diameter telescope.

than for the $r_0 = 20$ cm and $r_0 = 10$ cm cases at all spatial frequencies $|\vec{f}| > 0$. Thus, it is expected that the images measured in the $r_0 = 5$ cm, 7.5 cm, and 10 cm cases will be more blurred than the image measured in the $r_0 = 20$ cm case. Figure 6.14 shows this expectation to be correct. In Fig. 6.14 average simulated images of a binary star with components separated by 1.2 μrad and brightness ratio of 2:1 are shown for the $r_0 = 20$ cm, 10 cm, 7.5 cm, and 5 cm cases. The adaptive optics configuration used to generate these figures was identical to the configuration used to generate Fig. 6.12. However, the images shown in Fig. 6.14 contain the effects of finite signal levels in the wave front sensor and in the image plane. The binary star was assumed to have visual magnitude $m_v = 6$ with the same spectral distribution as the sun. The wave front sensor was assumed to operate in the wavelength range $\lambda_W = 600 \pm 60$ nm, and the imaging camera was assumed to operate in the wavelength range $\lambda_I = 700 \pm 70$ nm. Integration time of 1 ms was assumed in both the wavefront sensor and the imaging camera, and the transmission efficiency from the top of the atmosphere to the output of both detectors was assumed to be 10%. The result of these assumptions is that the average number of photoevents per integration time per subaperture in the wave front sensor is $\overline{K}_W = 120$, and the average number of photoevents per image is $\overline{K} = 5358$. Data sets of 200 images were averaged to obtain Fig. 6.14. Observe that as r_0 decreases the images of the binary star components become broader and the individual binary star components become more strongly overlapping, consistent with a loss of resolution.

The image spectrum signal-to-noise ratios $\mathrm{SNR}_D(\vec{f})$ for the binary star images shown in Fig. 6.14 are shown in Fig. 6.15. The plots in Fig. 6.15 were created by averaging the two dimensional $\mathrm{SNR}_D(\vec{f})$ arrays along circles of constant radius to create plots which are a function of a scalar spatial frequency. The horizontal axis of Fig. 6.15 is normalized by the diffraction-limited cutoff frequency for the OTF, $D/(\lambda d_i)$ for the aperture diameter $D = 1$ m, and the imaging wavelength $\lambda_I = 700$ nm. Choosing the threshold of 0.1 on $\mathrm{SNR}_D(\vec{f})$ to define the signal-to-noise ratio limited effective cutoff frequency f_{SNR}, which corresponds to $\sqrt{M}\mathrm{SNR}_D(\vec{f}) \approx 1.4$ for the case of $M = 200$ used here, we observe that f_{SNR} lies in the range $0.89 \le f_{\mathrm{SNR}}\lambda d_i/D \le 0.97$. Hence, even though the images shown in Fig. 6.14 are blurred, high signal-to-noise ratio information is present in the images at high spatial frequencies.

The blurred images shown in Fig. 6.14 can be sharpened using deconvolution [3, 4, 26, 27]. The goal of deconvolution is to remove, or at least reduce the effects of the average OTF on images. Two simple

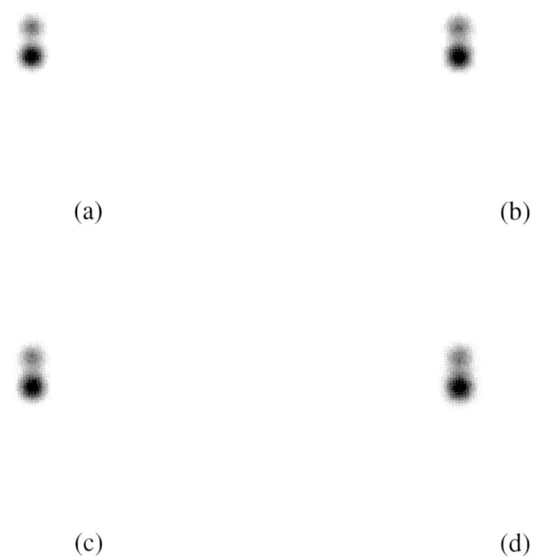

FIGURE 6.14
Average simulated binary star images using adaptive optics: (a) $r_0 = 20$ cm; (b) $r_0 = 10$; (c) $r_0 = 7.5$ cm; (d) $r_0 = 5$ cm. Angular separation of the binary components = 1.2 μrad, telescope diameter = 1.0 m, and ratio of brightnesses = 2:1. Negative images are displayed for clarity.

and widely used deconvolution techniques are the inverse filter and the Wiener filter [26]. Both of these filters are based on the relationship between the object spectrum $O(\vec{f})$, the image spectrum $I(\vec{f})$, and the OTF $\mathcal{H}(\vec{f})$, first stated in Eq. (2.41) for the case of infinite signal, as

$$I(\vec{f}) = O(\vec{f})\mathcal{H}(\vec{f}). \tag{6.52}$$

An estimate of the object spectrum $\tilde{O}(\vec{f})$ within the pass band of $\mathcal{H}(\vec{f})$ could easily be obtained from Eq. (6.52) if the OTF $\mathcal{H}(\vec{f})$ is known by using

$$\tilde{O}(\vec{f}) = \frac{I(\vec{f})}{\mathcal{H}(\vec{f})} = \frac{O(\vec{f})\mathcal{H}(\vec{f})}{\mathcal{H}(\vec{f})} \tag{6.53}$$

which is the classic inverse filter [26]. In the context of adaptive optics imaging of space objects it must be noted that both the OTF and the detected image are random. Hence, it is necessary to work with the average detected image spectrum $\mathrm{E}\{D(\vec{f})\}$ and an estimate of the average OTF obtained from a reference star $\mathrm{E}\{\mathcal{H}^{\mathrm{ref}}(\vec{f})\}$. The estimate of the average OTF $\mathrm{E}\{\mathcal{H}^{\mathrm{ref}}(\vec{f})\}$ is obtained by measuring images of a nearby reference star $d^{\mathrm{ref}}(\vec{x})$, and then computing the average OTF using

$$\mathrm{E}\{\mathcal{H}^{\mathrm{ref}}(\vec{f})\} = \mathrm{E}\left\{\frac{D^{\mathrm{ref}}(\vec{f})}{D^{\mathrm{ref}}(0)}\right\}, \tag{6.54}$$

where $D^{\mathrm{ref}}(\vec{f})$ has been used to represent the Fourier transform of $d^{\mathrm{ref}}(\vec{x})$. Note that expectation in Eq. (6.54) is obtained in practice using the sample-based estimate of the average.

The appropriate form of the object spectrum estimate $\tilde{O}(\vec{f})$ obtained from the detected image spectrum

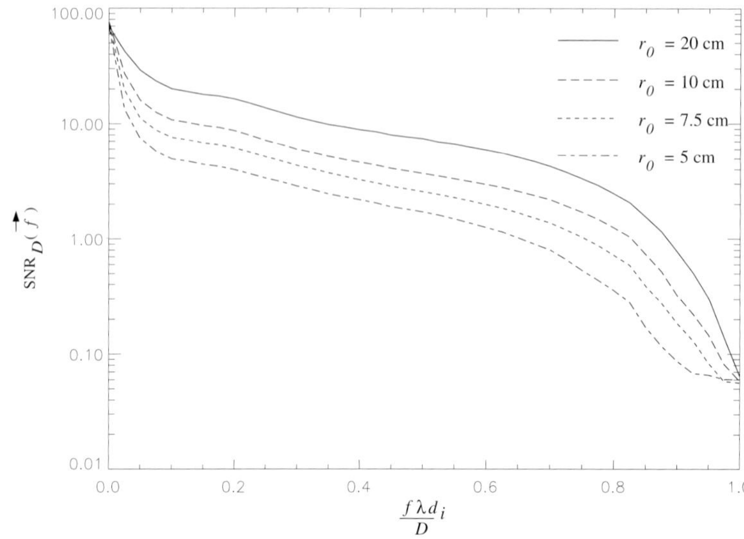

FIGURE 6.15
Image spectrum signal-to-noise ratio $\mathrm{SNR}_D(\vec{f})$ for the binary star images shown in Fig. 6.14.

$D(\vec{f})$ using the classic inverse filter is given by

$$\tilde{O}(\vec{f}) = \frac{\mathrm{E}\{D(\vec{f})\}}{\mathrm{E}\{\mathcal{H}^{\mathrm{ref}}(\vec{f})\}}, \tag{6.55}$$

where $\mathrm{E}\{\mathcal{H}^{\mathrm{ref}}(\vec{f})\} \neq 0$, and

$$\tilde{O}(\vec{f}) = 0, \tag{6.56}$$

where $\mathrm{E}\{\mathcal{H}^{\mathrm{ref}}(\vec{f})\} = 0$. The effect of dividing $\mathrm{E}\{D(\vec{f})\}$ by $\mathrm{E}\{\mathcal{H}^{\mathrm{ref}}(\vec{f})\}$ is to boost the spatial frequency components of $\mathrm{E}\{D(\vec{f})\}$ to approximate the object spectrum $O(\vec{f})$, thereby correcting for the attenuation of the object spectrum caused by $\mathrm{E}\{\mathcal{H}(\vec{f})\}$. To the extent that $\mathrm{E}\{\mathcal{H}^{\mathrm{ref}}(\vec{f})\}$ accurately approximates $\mathrm{E}\{\mathcal{H}(\vec{f})\}$, an accurate object spectrum estimate is obtained using Eqs. (6.55) and (6.56) in the absence of measurement noise.

Unfortunately, the classic inverse filter cannot be directly applied to realistic astronomical imaging problems because it neglects measurement noise effects. To see how the object spectrum estimate $\tilde{O}(\vec{f})$ obtained from the classic inverse filter is affected by noise, consider the simplified case of an image spectrum $I_C(\vec{f})$ corrupted with additive noise $N(\vec{f})$ given by

$$I_C(\vec{f}) = O(\vec{f})\mathcal{H}(\vec{f}) + N(\vec{f}), \tag{6.57}$$

where we will assume that the OTF $\mathcal{H}(\vec{f})$ is perfectly known. Let us now apply the classic inverse filter to $I_C(\vec{f})$ to obtain an estimate of the object spectrum $\tilde{O}_C(\vec{f})$ given by

$$\tilde{O}_C(\vec{f}) = \frac{I_C(\vec{f})}{\mathcal{H}(\vec{f})} = \frac{O(\vec{f})\mathcal{H}(\vec{f})}{\mathcal{H}(\vec{f})} + \frac{N(\vec{f})}{\mathcal{H}(\vec{f})}. \tag{6.58}$$

The first term of Eq. (6.58) is the desired result. The second term of Eq. (6.58) is not desired, since it contains the effects of additive noise. Recall that $\mathcal{H}(\vec{f})$ takes its maximum value of unity at $\vec{f} = 0$, and is less than unity everywhere else. Hence, we conclude from Eq. (6.58) that the effect of inverse filtering on images corrupted with additive noise is to both sharpen the image and *amplify* the noise.

Photon noise effects in $\mathrm{E}\{D(\vec{f})\}$ are also amplified by the classic inverse filter, even though this noise cannot be modeled as additive in nature.

Noise amplification by the classic inverse filter is not a serious problem in spatial frequency regions where $\mathrm{E}\{\mathcal{H}^{\mathrm{ref}}(\vec{f})\}$ and $\mathrm{SNR}_D(\vec{f})$ are significantly greater than zero. However, when $\mathrm{E}\{\mathcal{H}^{\mathrm{ref}}(\vec{f})\}$ or $\mathrm{SNR}_D(\vec{f})$ are small, noise amplification effects can dominate the object irradiance distribution estimate $\tilde{o}(\vec{x})$ [26, 27]. The general character of $\tilde{o}(\vec{x})$ obtained using the classic inverse filter directly on noisy images is that $\tilde{o}(\vec{x})$ is also very noisy, such that in some cases the actual object irradiance distribution may be completely obscured by the noise effects [27].

One technique for overcoming some of the noise effects associated with the classic inverse filter is to pass the $\tilde{O}(\vec{f})$ obtained from the inverse filter through an additional filtering step. A new object spectrum estimate $\tilde{O}_1(\vec{f})$ is obtained from this additional filtering step given by

$$\tilde{O}_1(\vec{f}) = \tilde{O}(\vec{f})H(\vec{f}), \tag{6.59}$$

where $H(\vec{f})$ represents the filter applied to $\tilde{O}(\vec{f})$. Useful forms of $H(\vec{f})$ include, but are not limited to, the cone filter given in Eq. (6.48) and the tophat filter given in Eq. (4.131) as

$$H(\vec{f}) = \begin{cases} 1, & |\vec{f}| \leq f_c \\ 0, & |\vec{f}| > f_c \end{cases}, \tag{6.60}$$

where $f_c \leq D/(\lambda d_i)$ is a user-selected cutoff frequency. The new object spectrum estimate $\tilde{o}_1(\vec{x})$ is obtained by applying the inverse Fourier transform operation to $\tilde{O}_1(\vec{f})$. Note that the object spectrum reconstruction given in Eq. (6.59) is not a minimum error filter [27]. However, reconstructions obtained with Eq. (6.59) have been shown to be visually appealing [23].

Qualitatively, the function of f_c is to implement a tradeoff between the amount of noise present in $\tilde{o}_1(\vec{x})$ and the resolution of $\tilde{o}_1(\vec{x})$. This tradeoff arises because the resolution of $\tilde{o}_1(\vec{x})$ increases with increasing f_c, until $f_c = D/(\lambda d_i)$, the diffraction-limited cutoff frequency. However, $\mathrm{SNR}_D(\vec{f})$ is generally a monotonically decreasing function of $|\vec{f}|$ so that increasing f_c also allows more noise into $\tilde{o}_1(\vec{x})$. In practice, several $\tilde{o}_1(\vec{x})$ are reconstructed using different values of f_c, and the subjectively assessed "best" reconstruction is selected as the final image.

The results of reconstructing the images shown in Fig. 6.14 using Eqs. (6.55) and (6.59) are shown in Fig. 6.16. The cone filter given in Eq. (6.48) was used for $H(\vec{f})$ in Eq. (6.59) with $f_c = f_{\mathrm{SNR}}$, and the threshold of $\sqrt{200}\mathrm{SNR}_D(\vec{f}) = 1.4$ was used to select f_{SNR}. Comparison of Figs. 6.14 and 6.16 shows that the outcome of the deconvolution process is a sharper image, that is, an image where the components of the binary star are more clearly defined, and the halo surrounding the binary star components has been reduced. However, consistent with the earlier discussion of noise effects in inverse filtering, the deconvolved images are somewhat noisier than the measured images. Noise effects are particularly evident in Fig. 6.16(d), where some noise-induced artifacts in the reconstructed image are present.

Another technique for overcoming noise effects in deconvolution is the so-called Wiener filter [26, 27]. The appropriate form of the Wiener filter $H_W(\vec{f})$ in the present context is

$$H_W(\vec{f}) = \frac{\mathrm{E}\{\mathcal{H}^{\mathrm{ref}}(\vec{f})\}^*}{\left|\mathrm{E}\{\mathcal{H}^{\mathrm{ref}}(\vec{f})\}\right|^2 + \mathrm{E}_N(\vec{f})/\mathrm{E}_O(\vec{f})}, \tag{6.61}$$

where $\mathrm{E}_N(\vec{f})$ is the power spectral density of the noise, and $\mathrm{E}_O(\vec{f})$ is the power spectral density of the object irradiance distribution. The object spectrum estimate obtained using the Wiener filter is given by

$$\tilde{O}(\vec{f}) = \mathrm{E}\{D(\vec{f})\}H_W(\vec{f}) = \frac{\mathrm{E}\{D(\vec{f})\}\mathrm{E}\{\mathcal{H}^{\mathrm{ref}}(\vec{f})\}^*}{\left|\mathrm{E}\{\mathcal{H}^{\mathrm{ref}}(\vec{f})\}\right|^2 + \mathrm{E}_N(\vec{f})/\mathrm{E}_O(\vec{f})}. \tag{6.62}$$

While the Wiener filter is a minimum error filter, the form of the Wiener filter given in Eq. (6.61) is rarely used in practice because of the requirement to know the power spectral density of the object

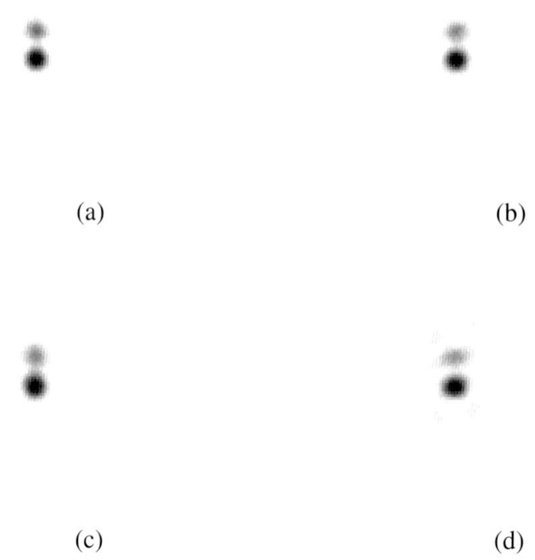

FIGURE 6.16
Simulated binary star images measured using adaptive optics and reconstructed using the classic inverse filter followed by the cone filter: (a) $r_0 = 20$ cm; (b) $r_0 = 10$ cm; (c) $r_0 = 7.5$ cm; (d) $r_0 = 5$ cm. Angular separation of the binary components = 1.2 μrad, telescope diameter = 1.0 m, and ratio of brightnesses = 2:1. Negative images are displayed for clarity.

irradiance distribution $\mathrm{E}_O(\vec{f})$ [27]. Rather, modified, suboptimal versions of the Wiener filter are used. One useful form for a modified Wiener filter is given by

$$H_W^{\text{mod}}(\vec{f}) = \frac{\mathrm{E}\{\mathcal{H}^{\text{ref}}(\vec{f})\}^*}{\left|\mathrm{E}\{\mathcal{H}^{\text{ref}}(\vec{f})\}\right|^2 + \alpha/\mathrm{SNR}_D(\vec{f})}, \tag{6.63}$$

where $0 \leq \alpha \leq 1$ is a user-selected parameter. The motivation for the second term in the denominator of Eq. (6.63) derives from the interpretation of the $\mathrm{E}_N(\vec{f})/\mathrm{E}_O(\vec{f})$ term in Eq. (6.61) as a "noise-to-signal" ratio. Qualitatively, the purpose of the $\alpha/\mathrm{SNR}_D(\vec{f})$ term in Eq. (6.63) is to reduce the response of $H_W^{\text{mod}}(\vec{f})$ in regions were $\mathrm{SNR}_D(\vec{f})$ is small, thereby mitigating some of the noise amplification effects which would be present in the classic inverse filter. The parameter α can be varied in Eq. (6.63) to obtain a visually pleasing $\tilde{o}(\vec{x})$. As a final comment, we note that while the Wiener filter defined in Eq. (6.61) minimizes a measure of mean squared error in image space, the reconstructions obtained with the Wiener filter have been criticized as "overly smooth" for human visual system interpretation [27].

This concludes the presentation of linear deconvolution of adaptive optics images. In the next subsection the effects of applying speckle imaging techniques to adaptive optics images are discussed.

6.3.2 Compensated speckle imaging

Compensated speckle imaging is the term used to describe the application of speckle imaging techniques, introduced in Chapter 4, to images measured with adaptive optical imaging systems. The compensated speckle imaging paradigm is as follows. An adaptive optical imaging system is used to measure a data set of the object of interest. A similar data set is obtained for a nearby reference star. The object spectrum is then estimated from the measurements using speckle imaging algorithms, and an estimate of the object irradiance distribution is obtained by applying the inverse Fourier transform operation to the estimated spectral data. Motivation for compensated speckle imaging arises from the fact that it has been shown that limited or partial wave front phase correction improves the residual wave front phase statistics, improving the statistical performance of the speckle imaging estimators [4].

Historically, Nisenson and Barakat [20] were the first to note that limited wave front correction significantly improves the performance of speckle imaging estimators. Roggemann and Matson [4] later extended this idea by deriving a rigorous expression for the signal-to-noise ratio of the unbiased speckle interferometry estimator $Q(\vec{f})$. Roggemann and Matson also showed that even very limited predetection wave front correction results in improvements in both the signal-to-noise ratio of $Q(\vec{f})$, $\text{SNR}_Q(\vec{f})$, and the phase spectrum estimate. One key result of Ref. [4] is that when adaptive optics correction is applied $\text{SNR}_Q(\vec{f})$ is not bounded from above by unity for spatial frequencies in the range $|\vec{f}| > r_0/(\lambda d_i)$, as is the case when no adaptive optics compensation is present [28]. Subsequent experimental results confirmed the key predictions of earlier theoretical work in compensated speckle imaging [24].

Compensated speckle imaging is most easily studied using a simulation. The simulation used to obtain the results that follow implements predetection wave front correction using the adaptive optics simulation described in Section 5.4.6. Speckle imaging post processing was then applied to the simulated images. Labeyrie's speckle interferometry technique, presented in Section 4.3, was used to estimate the modulus squared of the object spectrum $|\tilde{O}_n(\vec{f})|^2$. The bispectrum phase reconstruction technique, which was presented in Section 4.4.2, was used to obtain an estimate of the phase of the object spectrum $\tilde{\phi}(\vec{f})$.

The specific telescope and adaptive optics configuration, seeing, and light level conditions for which results are presented are now described. The telescope was modeled as a one meter diameter unobscured aperture. A Hartmann-type wave front sensor was modeled which had square subapertures having side length of 10 cm. The deformable mirror actuator influence functions were modeled as two-dimensional triangle functions [25] with base width equal to two times the actuator grid spacing. The deformable mirror actuators were placed on a Cartesian grid with grid spacing of 12 cm. Least squares reconstruction, described in Section 5.3.3, was used to map wave front sensor measurements to actuator commands. It was assumed that no delay between sensing and correcting for the turbulence-induced aberration was present in the adaptive optics system.

To illustrate the effects of adaptive optics compensation on $\text{SNR}_Q(\vec{f})$, adaptive optics imaging of a single star was simulated. The star was assumed to have visual magnitude $m_v = 6$ with the same spectral distribution as the sun. The wave front sensor was assumed to operate in the wavelength range $\lambda_W = 600 \pm 60$ nm, and the imaging camera was assumed to operate in the wavelength range $\lambda_I = 700 \pm 70$ nm. The Fried seeing parameter r_0 was specified at the wavelength of 500 nm, and four different values of r_0 were used: $r_0 = 20$ cm, 10 cm, 7.5 cm, and 5 cm. Integration times of 1 ms were assumed in both the wavefront sensor and the imaging camera, and the transmission efficiency from the top of the atmosphere to the output of both detectors was assumed to be 10%. The result of these assumptions is that the average number of photoevents per integration time per subaperture in the wave front sensor is $\overline{K}_W = 120$, and the average number of photoevents per image is $\overline{K} = 5358$. Data sets of 200 images were used to obtain sample-based estimates of $\text{E}\{Q(\vec{f})\}$, $\text{SNR}_Q(\vec{f})$, and the average bispectrum $\text{E}\{B(\vec{f}_1, \vec{f}_2)\}$. These input parameters are identical to those used for the adaptive optics results in Section 6.3.1, and are summarized in Table 6.4.

$\text{SNR}_Q(\vec{f})$ results for the single star images are presented in Fig. 6.17. The plots in Fig. 6.17 were created by averaging the two dimensional $\text{SNR}_Q(\vec{f})$ arrays along circles of constant radius to create plots which are a function of a scalar spatial frequency. The horizontal axis of Fig. 6.17 is normalized by the diffraction-limited cutoff frequency of the OTF, $D/(\lambda d_i)$, for the aperture diameter $D = 1$ m, and the imaging wavelength $\lambda_I = 700$ nm. Choosing the threshold of 0.1 on $\text{SNR}_Q(\vec{f})$ to define f_{SNR}, which corresponds to $\sqrt{M}\text{SNR}_Q(\vec{f}) \approx 1.4$ for the case of $M = 200$ used here, we observe that the

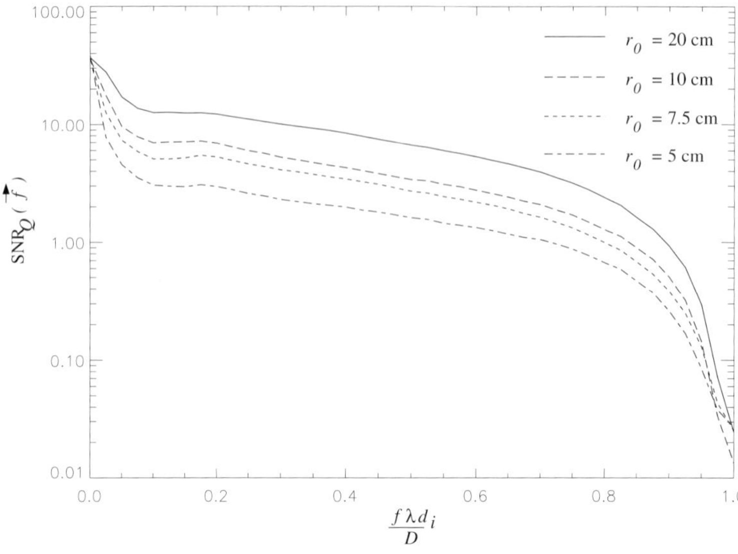

FIGURE 6.17
$\mathrm{SNR}_Q(\vec{f})$ for adaptive optics imaging case, single star with $m_v = 6$, and $r_0 = 20$ cm, 10 cm, 7.5 cm, and 5.0 cm.

signal-to-noise ratio limited cutoff frequency f_{SNR} lies in the range $f_{\mathrm{SNR}}\lambda d_i/D \geq 0.95$ for all cases shown in Fig. 6.17. Note also from Fig. 6.17 that, in contrast to the uncompensated speckle imaging case, illustrated in Fig. 4.5, $\mathrm{SNR}_Q(\vec{f})$ is not bounded from above by unity for spatial frequencies in the range $|\vec{f}| \geq r_0/(\lambda d_i)$ when adaptive optics are used. Hence, in compensated speckle imaging fewer frames must be averaged to obtain a desired level of $\mathrm{SNR}_Q(\vec{f})$, or a higher value of $\mathrm{SNR}_Q(\vec{f})$ is obtained from a fixed number of compensated frames.

The quality of the object phase spectrum estimate obtained using the bispectrum technique and the quality of the reconstructed images are also key issues in compensated speckle imaging. To address these issues the binary star system simulated in Section 6.3.1 was used in the compensated speckle imaging simulation. Recall that the binary star components were separated by 1.2 μrad, and the brightness ratio of 2:1 was assumed for the components. The binary star was assumed to be of visual magnitude $m_v = 6$. All other adaptive optics and imaging parameters were assumed to be the same as for the binary star system simulated in Section 6.3.1, and the input parameters are summarized in Table 6.4. The reference stars were simulated under the same seeing and light level conditions as the binary stars. Data sets of 200 frames were used to obtain all of the results presented. $\mathrm{SNR}_Q(\vec{f})$ results for the binary star example are shown in Fig. 6.18. Choosing a threshold of 0.1 on $\mathrm{SNR}_Q(\vec{f})$, which corresponds to $\sqrt{M}\mathrm{SNR}_Q(\vec{f}) \approx 1.4$ for the $M = 200$ case studied here, yields $f_{\mathrm{SNR}} > 0.93D/(\lambda d_i)$ for every case. The estimate of the modulus squared of the object spectrum $|\tilde{O}_n(\vec{f})|^2$ was obtained using Eq. (4.129) with $\alpha = 0.001$. The resulting estimate $|\tilde{O}_n(\vec{f})|^2$ was then passed through the tophat filter of Eq. (4.131), with $f_c = f_{\mathrm{SNR}}$. The square root of $|\tilde{O}_n(\vec{f})|^2$ was then computed to obtain the estimated modulus of the object spectrum $|\tilde{O}_n(\vec{f})|$. The estimate of the phase of the object spectrum $\tilde{\phi}(\vec{f})$ was obtained using the bispectrum technique described in Section 4.4.2, using the recursive reconstructor. The estimated modulus and phase of the object spectrum were then combined to obtain the estimated object spectrum $\tilde{O}_n(\vec{f})$ given by

$$\tilde{O}_n(\vec{f}) = |\tilde{O}_n(\vec{f})| \exp\{\tilde{\phi}(\vec{f})\}. \tag{6.64}$$

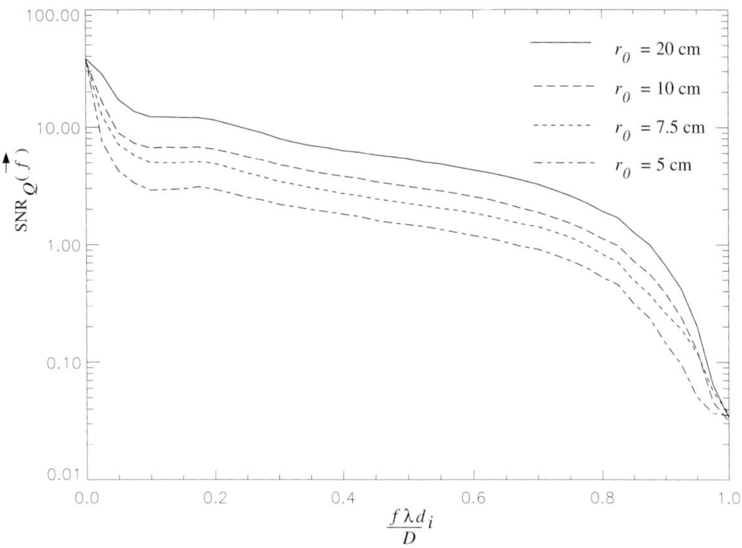

FIGURE 6.18
Radially averaged $\mathrm{SNR}_Q(\vec{f})$ results for compensated speckle imaging applied to binary star with $m_v = 6$, for $r_0 = 20$ cm, 10 cm, 7.5 cm, and 5 cm.

Finally, the estimated object irradiance distribution $\tilde{o}(\vec{x})$ was obtained by applying the inverse Fourier transform operation to $\tilde{O}_n(\vec{f})$.

Reconstructed binary star images obtained from compensated speckle imaging are shown in Fig. 6.19. Note that these images are highly resolved. Comparison of the reconstructed images shown in Fig. 6.19 to the reconstructed images obtained by deconvolution of the average compensated images shown in Fig. 6.16 shows that similar reconstructions are obtained for the $r_0 = 20$ cm, 10 cm, and 7.5 cm cases. However, for the poor seeing case of $r_0 = 5$ cm the compensated speckle imaging reconstruction is corrupted with fewer artifacts caused by noise. We conclude that compensated speckle imaging can provide high resolution images which have fewer noise-induced effects than deconvolution of the average image obtained with a partially compensated system. When turbulence effects are well-compensated by the adaptive optics, compensated speckle imaging provides performance roughly equivalent to that obtained by simply deconvolving the compensated images.

The quality of the bispectrum phase reconstruction estimate can be evaluated by computing the squared error between the object spectrum phase and the estimated object spectrum phase. This squared error is denoted $\phi_\epsilon^2(\vec{f})$, and is defined in Eq. (4.128) as

$$\phi_\epsilon^2(\vec{f}) = \left(\phi_o(\vec{f}) - \tilde{\phi}_o(\vec{f})\right)^2, \tag{6.65}$$

where $\phi_o(\vec{f})$ is the true object spectrum phase. Two dimensional arrays of $\phi_\epsilon^2(\vec{f})$ were computed for each binary star case. These two dimensional arrays were then averaged along circles of constant radius to obtain the plots of $\phi_\epsilon^2(\vec{f})$ as a function of a scalar spatial frequency variable shown in Fig. 6.20. Inspection of Fig. 6.20 shows that $\phi_\epsilon^2(\vec{f}) \leq 1$ rad^2 is obtained for $|\vec{f}| \leq 0.75D/(\lambda d_i)$ for all cases analyzed. These results are consistent with excellent phase spectrum estimation [4], leading to the high quality reconstructed images presented in Fig. 6.19.

As a final comment, we note that while only single and binary star results are presented in this subsection, the application of compensated speckle imaging to truly extended objects has been studied. In Ref. [4] it is shown that similar results to those presented here are obtained when compensated speckle

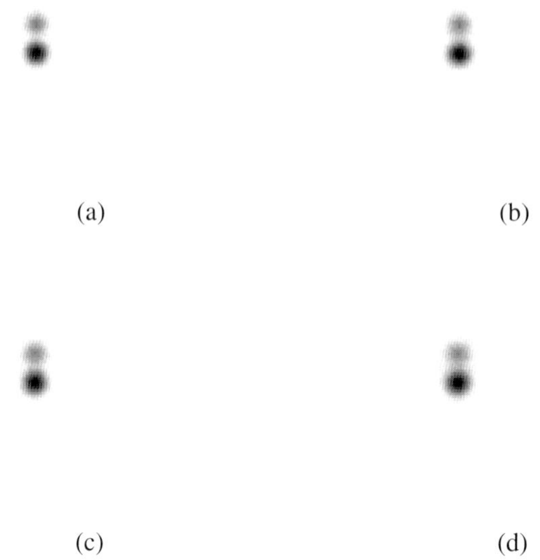

FIGURE 6.19
Simulated binary star images measured using adaptive optics and reconstructed using speckle imaging: (a) $r_0 = 20$ cm; (b) $r_0 = 10$ cm; (c) $r_0 = 7.5$ cm; (d) $r_0 = 5$ cm. Angular separation of the binary components = 1.2 μrad, telescope diameter = 1.0 m, and ratio of brightnesses = 2:1. Negative images are displayed for clarity.

imaging is applied to extended objects. This concludes the treatment of compensated speckle imaging. In the next subsection the topic of compensated deconvolution from wave front sensing is addressed as a means for overcoming some of the effects of time delays and anisoplanatism in adaptive optics systems.

6.3.3 Compensated deconvolution from wavefront sensing

The method of compensated deconvolution from wave front sensing (CDWFS) uses wave front sensor information in closed loop adaptive optics systems to reconstruct images [6, 7]. The CDWFS paradigm of operation is similar to the DWFS paradigm introduced in Section 6.2. However, there are two key technical distinctions between CDWFS and DWFS: (1) the wave front information in CDWFS comes from a wave front sensor measuring the residual aberration inside a closed-loop adaptive optics system; and (2) a different object spectrum estimator has been used in CDWFS than has been used in DWFS. CDWFS also has a different range of applications than DWFS. For example, CDWFS has been shown to be useful for overcoming some of the effects of time delays between sensing and correcting for the turbulence-induced aberration in an adaptive optics imaging system, and for overcoming some of the effects of anisoplanatism in adaptive optics imaging systems. The same image reconstruction estimator is used in CDWFS for either application. Hence, a generalized development of the CDWFS technique which encompasses both CDWFS applications is presented in the discussion which follows.

The functional block diagram for the two CDWFS applications is shown in Fig. 6.21. All of the elements of a conventional adaptive optics system are present in Fig. 6.21. Solid lines are used in Fig. 6.21 to represent connections to elements required for an adaptive optics system, and broken lines are used to represent connections to elements unique to CDWFS. A beacon coincident with the optical

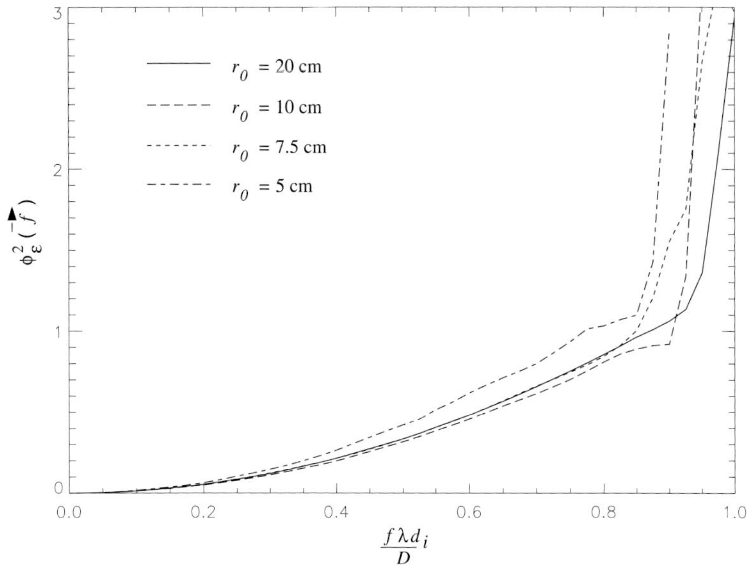

FIGURE 6.20
Radially averaged object spectrum squared phase error $\phi^2_\epsilon(\vec{f})$ for compensated speckle imaging applied to binary star with $m_v = 6$, for $r_0 = 20$ cm, 10 cm, 7.5 cm, and 5 cm.

axis of the imaging system, corresponding to a field angle $\vec{\theta}_f = 0$, provides signal to wave front sensor A. The residual wave front error in the $\vec{\theta}_f = 0$ direction is sensed by wave front sensor A, and is denoted $\epsilon(\vec{x}, \vec{\theta}_f = 0)$, where $\vec{x}$ is a two dimensional pupil location. Measurements made by wave front sensor A are input to the actuator command computer to drive the deformable mirror, optimizing optical performance in the $\vec{\theta}_f = 0$ direction. Wave front sensor A outputs may also be input to a wave front computer, indicated by a broken line in Fig. 6.21, which is used to compute an estimate of $\epsilon(\vec{x}, \vec{\theta}_f = 0)$, denoted $\tilde{\epsilon}(\vec{x}, \vec{\theta}_f = 0)$. The wave front estimate $\tilde{\epsilon}(\vec{x}, \vec{\theta}_f = 0)$ is used to compute an estimate of the instantaneous OTF in the $\vec{\theta}_f = 0$ direction, denoted $\tilde{\mathcal{H}}(\vec{f}, \vec{\theta}_f = 0)$, which can be used in post processing to reconstruct high resolution images of the object of interest.

As noted in Section 5.3.3, the temporal sampling rate of the wave front sensor is generally at least ten times the closed loop bandwidth. Hence, wave front phase information is available at a much higher rate than the deformable mirror can cause the residual aberration to change significantly. If the turbulence is evolving faster than the closed loop bandwidth of the adaptive optics system, then the closed loop wave front sensor measurements contain information about *uncorrected aberrations* arising from the relatively slow temporal response of the adaptive optics system. Uncorrected turbulence effects due to time delays in the adaptive optics system can be a major source of reduced optical performance. CDWFS has been shown to offer a means of recovering some of the optical performance lost due to time delays in the adaptive optics system [6].

Figure 6.21 also contains a beam splitter relaying light to a second wave front sensor, denoted wave front sensor B, which is receiving signal from an off-axis beacon at field angle $\vec{\theta}_f > 0$. Wave front sensor B and the associated processing blocks are unique to CDWFS for correcting some of the effects of anisoplanatism, and broken lines are used in Fig. 6.21 to indicate connections between these blocks. Recall from Section 5.2.1 that light arriving at the aperture from a direction $\vec{\theta}_f > 0$ experiences a different path through the atmosphere than light arriving from the $\vec{\theta}_f = 0$ direction, and acquires a different aberration, denoted $\psi(\vec{x}, \vec{\theta}_f > 0)$. The wave front aberration $\psi(\vec{x}, \vec{\theta}_f > 0)$ is altered by the deformable mirror, and the resulting residual aberration in the $\vec{\theta}_f > 0$ direction is denoted $\epsilon(\vec{x}, \vec{\theta}_f)$.

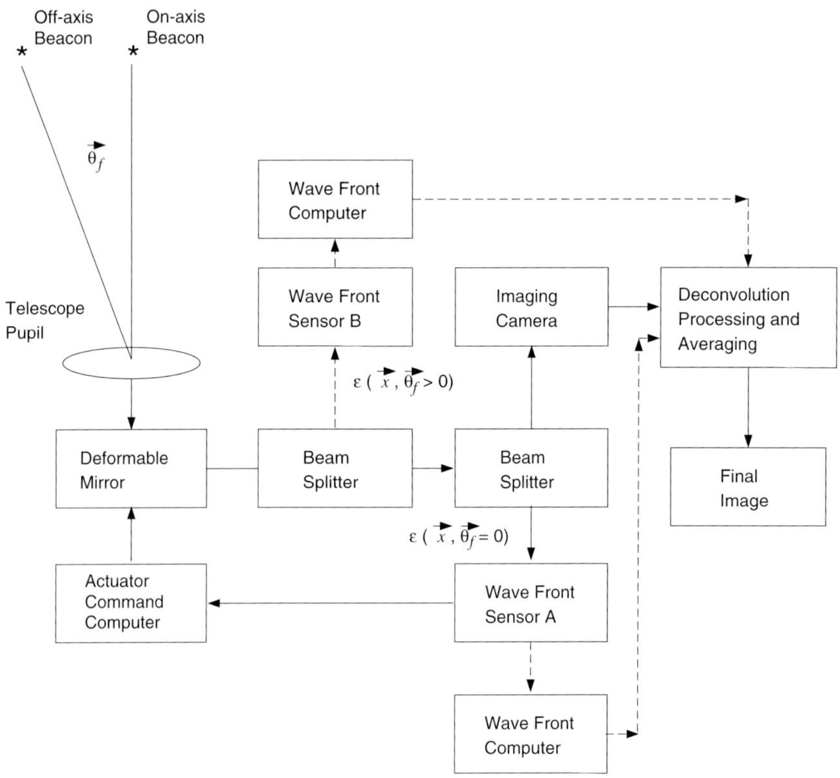

FIGURE 6.21
Functional block diagram for CDWFS. Solid lines indicate connections to elements required for an adaptive optics system. Broken lines indicate elements unique to CDWFS.

Wave front sensor B measures $\epsilon(\vec{x}, \vec{\theta}_f > 0)$, and provides its outputs to a wave front computer. The wave front computer calculates an estimate of the residual aberration in the $\theta_f > 0$ direction, denoted $\tilde{\epsilon}(\vec{x}, \vec{\theta}_f)$, which is used to compute an estimate of the associated OTF denoted $\tilde{\mathcal{H}}(\vec{f}, \vec{\theta}_f > 0)$. The OTF estimate $\tilde{\mathcal{H}}(\vec{f}, \vec{\theta}_f > 0)$ is input to the deconvolution processor to reconstruct an image of an object in the $\vec{\theta}_f > 0$ direction.

A major limitation to the wide application of adaptive optics to astronomy is the lack of natural sources to act as beacons for the wave front sensor. As mentioned in Section 5.2.1, the usual measure of the correctable field of view of an adaptive optics telescope is the isoplanatic angle θ_0 given by [29]

$$\theta_0 = 58.1 \times 10^{-3} \lambda^{6/5} \left[\int_0^L dz C_n^2(z) z^{5/3} \right]^{-3/5}, \tag{6.66}$$

where $C_n^2(z)$ is the structure constant of the turbulence induced index of refraction fluctuations, L is the path length through turbulence, and λ is the mean wave length of operation. The isoplanatic angle may be considered to be the largest field angle over which the optical path length through the turbulence does not differ significantly from the on-axis optical path length through the turbulence. The isoplanatic angle is known to be quite small, being on the order of 5-10 μrad at visible wavelengths. As the field angle $|\vec{\theta}_f|$ is increased beyond θ_0 the average OTF in the $\vec{\theta}_f$ direction rapidly narrows so that at field angles of several times θ_0 essentially uncompensated optical performance is obtained. Thus, to effectively use conventional adaptive optics the object of interest must lie within θ_0 of the beacon used to drive the wave front sensor. Unfortunately, less than 1% of the sky is within one isoplanatic angle θ_0 of a star bright enough to act as a wave front sensor beacon [30].

It has been shown that CDWFS gives significant improvements in object spectrum estimation over the detected image for field angles in the range $|\vec{\theta}_f| \leq 6\theta_0$ [7]. Hence, the area of the sky which can be covered with adaptive optics imaging can be increased by CDWFS. Note that, as discussed in Section 5.2.1, the principle of reciprocity in optical propagation prevents an artificial beacon from providing tilt information about the turbulence-corrupted wave front. Thus, CDWFS requires a natural beacon in the $\vec{\theta}_f = 0$ direction to provide tilt and higher order wave front aberration information to wave front sensor A. An artificial beacon could, in principle, be used to provide signal to wave front sensor B in Fig. 6.21. Such use of a natural and an artificial beacon removes much of the concern about the brightness of the object in the $|\vec{\theta}_f| > 0$ direction, since light from that object can be used entirely for the purpose of creating an image.

The CDWFS estimator is presented in the next subsection, and the performance of this estimator is discussed in the subsection which follows. A single estimator is used for both CDWFS applications. Hence, a generalized notation is adopted to present the CDWFS estimator. In this generalized notation the size of the field angle $\vec{\theta}_f$ is not made explicit. Thus, the closed loop residual aberration for the entire range of possible values $\vec{\theta}_f \geq 0$ is denoted by $\epsilon(\vec{x}, \vec{\theta}_f)$. Similarly, the estimate of $\epsilon(\vec{x}, \vec{\theta}_f)$ is denoted $\tilde{\epsilon}(\vec{x}, \vec{\theta}_f)$, the OTF is denoted by $\mathcal{H}(\vec{f}, \vec{\theta}_f)$, and the estimate of the OTF is represented by $\tilde{\mathcal{H}}(\vec{f}, \vec{\theta}_f)$.

CDWFS object spectrum estimation

In this subsection the CDWFS object spectrum estimator is presented. A single realization of the CDWFS object spectrum estimate $\tilde{O}(\vec{f}, \vec{\theta}_f)$ is obtained from the noise free image spectrum measurement $I(\vec{f}, \vec{\theta}_f)$ using

$$\tilde{O}(\vec{f}, \vec{\theta}_f) = I(\vec{f}, \vec{\theta}_f) \frac{\tilde{\mathcal{H}}^*(\vec{f}, \vec{\theta}_f)}{|\tilde{\mathcal{H}}(\vec{f}, \vec{\theta}_f)|}. \tag{6.67}$$

Note that the quantity $\tilde{\mathcal{H}}^*(\vec{f}, \vec{\theta}_f)/|\tilde{\mathcal{H}}(\vec{f}, \vec{\theta}_f)|$ is a unit phasor which, in the limit of perfect estimation of the residual aberration $\epsilon(\vec{x}, \vec{\theta}_f)$, has the conjugate of the phase of the OTF. The relationship between the noise free image spectrum $I(\vec{f},\vec{\theta}_f)$, the object spectrum $O(\vec{f},\vec{\theta}_f)$, and the OTF $\mathcal{H}(\vec{f},\vec{\theta}_f)$ can be used to rewrite Eq. (6.67) as

$$\tilde{O}(\vec{f},\vec{\theta}_f) = O(\vec{f},\vec{\theta}_f)\frac{\mathcal{H}(\vec{f},\vec{\theta}_f)\tilde{\mathcal{H}}^*(\vec{f},\vec{\theta}_f)}{|\tilde{\mathcal{H}}(\vec{f},\vec{\theta}_f)|}. \tag{6.68}$$

With reference to Eq. (6.68), the CDWFS transfer function $S(\vec{f},\vec{\theta}_f)$, which maps the object spectrum to the noise free image spectrum, is defined as

$$S(\vec{f},\vec{\theta}_f) = \frac{\mathcal{H}(\vec{f},\vec{\theta}_f)\tilde{\mathcal{H}}^*(\vec{f},\vec{\theta}_f)}{|\tilde{\mathcal{H}}(\vec{f},\vec{\theta}_f)|}. \tag{6.69}$$

Many realizations of $\tilde{O}(\vec{f},\vec{\theta}_f)$ are averaged in CDWFS to reduce noise effects. The average CDWFS transfer function is given by

$$\mathrm{E}\{S(\vec{f},\vec{\theta}_f)\} = \mathrm{E}\left\{\frac{\mathcal{H}(\vec{f},\vec{\theta}_f)\tilde{\mathcal{H}}^*(\vec{f},\vec{\theta}_f)}{|\tilde{\mathcal{H}}(\vec{f},\vec{\theta}_f)|}\right\}. \tag{6.70}$$

Note that the average CDWFS transfer function given in Eq. (6.70) differs from the DWFS transfer function given in Eq. (6.14) by the presence of estimated OTF $\tilde{\mathcal{H}}(\vec{f},\vec{\theta}_f)$ in the denominator of Eq. (6.70), rather than a product of the actual OTF and the estimated OTF present in the denominator of Eq. (6.14). Recall that the primary motivation for the form of the DWFS transfer function given in Eq. (6.14) is that the uncompensated average OTF is approximately zero for spatial frequencies in the range $|\vec{f}| \geq r_0/(\lambda d_i)$, while the second moment of the OTF is non-zero out to spatial frequencies approaching the diffraction-limited cutoff frequency. However, when adaptive optics compensation is present the mean OTF has significant value at spatial frequencies approaching the diffraction-limited cutoff frequency, and there is no need to resort to second moment methods to estimate the object spectrum.

The purpose of the CDWFS estimator is to improve the overall system transfer function by correcting the phase of the OTF from instant to instant so that $\mathrm{E}\{S(\vec{f},\vec{\theta}_f)\} > \mathrm{E}\{\mathcal{H}(\vec{f},\vec{\theta}_f)\}$. The effect of the CDWFS process can be understood by writing the OTF and the estimated OTF in phasor form so that

$$\mathcal{H}(\vec{f},\vec{\theta}_f) = |\mathcal{H}(\vec{f},\vec{\theta}_f)| \exp\{j\phi_{\mathcal{H}}(\vec{f},\vec{\theta}_f)\}, \tag{6.71}$$

and

$$\tilde{\mathcal{H}}(\vec{f},\vec{\theta}_f) = |\tilde{\mathcal{H}}(\vec{f},\vec{\theta}_f)| \exp\{j\phi_{\tilde{\mathcal{H}}}(\vec{f},\vec{\theta}_f)\}, \tag{6.72}$$

where $\phi_{\mathcal{H}}(\vec{f},\vec{\theta}_f)$ is the phase of the OTF, and $\phi_{\tilde{\mathcal{H}}}(\vec{f},\vec{\theta}_f)$ is the phase of the estimated OTF. Hence, the CDWFS transfer function is given by

$$S(\vec{f},\vec{\theta}_f) = |\mathcal{H}(\vec{f},\vec{\theta}_f)| \exp\{j[\phi_{\mathcal{H}}(\vec{f},\vec{\theta}_f) - \phi_{\tilde{\mathcal{H}}}(\vec{f},\vec{\theta}_f)]\}, \tag{6.73}$$

which differs from the OTF only by virtue of having a different phase from instant to instant. To the extent that $\phi_{\tilde{\mathcal{H}}}(\vec{f},\vec{\theta}_f)$ is a good approximation of $\phi_{\mathcal{H}}(\vec{f},\vec{\theta}_f)$ the phase of $S(\vec{f},\vec{\theta}_f)$ will be closer to its mean value of zero than the phase of the OTF. Note that the length of the phasor describing the instantaneous OTF is unchanged by CDWFS. Thus, it is easy to qualitatively understand that the average effect of correcting the phase of the OTF from instant to instant is to *increase* the mean value of the CDWFS transfer function over the mean value of the OTF.

The theoretical conditions required for $\mathrm{E}\{S(\vec{f},\vec{\theta}_f)\} > \mathrm{E}\{\mathcal{H}(\vec{f},\vec{\theta}_f)\}$ have been addressed in the literature [6], and are of interest. It is reasonable to assume that both $\phi_{\mathcal{H}}(\vec{f},\vec{\theta}_f)$ and $\phi_{\tilde{\mathcal{H}}}(\vec{f},\vec{\theta}_f)$ are symmetrically distributed about zero. It follows that the real and imaginary parts of $\mathcal{H}(\vec{f},\vec{\theta}_f)$ and $\tilde{\mathcal{H}}(\vec{f},\vec{\theta}_f)$ are uncorrelated. If it is further assumed that the real and imaginary parts of $\mathcal{H}(\vec{f},\vec{\theta}_f)$ and $\tilde{\mathcal{H}}(\vec{f},\vec{\theta}_f)$ are Gaussian distributed then the real and imaginary parts of $\mathcal{H}(\vec{f},\vec{\theta}_f)$ are also independent. A similar argument can be used to conclude that the real and imaginary parts of $\tilde{\mathcal{H}}(\vec{f},\vec{\theta}_f)$ are independent. Note that at low and mid spatial frequencies the mean compensated OTF is much larger than the standard deviation of the compensated OTF [3, 5, 6]. Hence, at low and mid spatial frequencies the compensated OTF can reasonably be modeled as a strong constant phasor, represented by the mean OTF, which is entirely real, plus a weak random phasor [6, 31]. Under the statistical conditions described above the modulus and phase of the OTF are approximately independent so that we may write [6, 31]

$$\mathrm{E}\{\mathcal{H}(\vec{f},\vec{\theta}_f)\} = \mathrm{E}\{|\mathcal{H}(\vec{f},\vec{\theta}_f)|\}\mathrm{E}\left\{\exp\{j\phi_{\mathcal{H}}(\vec{f},\vec{\theta}_f)\}\right\}, \tag{6.74}$$

and

$$\mathrm{E}\{\tilde{\mathcal{H}}(\vec{f},\vec{\theta}_f)\} = \mathrm{E}\{|\tilde{\mathcal{H}}(\vec{f},\vec{\theta}_f)|\}\mathrm{E}\left\{\exp\{j\phi_{\tilde{\mathcal{H}}}(\vec{f},\vec{\theta}_f)\}\right\}. \tag{6.75}$$

Equations. (6.73), (6.74), and (6.75) can be used to write the mean CDWFS transfer function as

$$\mathrm{E}\{S(\vec{f},\vec{\theta}_f)\} = \mathrm{E}\{|\mathcal{H}(\vec{f},\vec{\theta}_f)|\}\mathrm{E}\left\{\exp\{j[\phi_{\mathcal{H}}(\vec{f},\vec{\theta}_f) - \phi_{\tilde{\mathcal{H}}}(\vec{f},\vec{\theta}_f)]\}\right\}. \tag{6.76}$$

If it is now assumed that $[\phi_{\mathcal{H}}(\vec{f},\vec{\theta}_f) - \phi_{\tilde{\mathcal{H}}}(\vec{f},\vec{\theta}_f)]$ is Gaussian distributed the second expectation on the right hand side of Eq. (6.76) can be evaluated using Eq. (3.46) as

$$\mathrm{E}\left\{\exp\left\{j\left[\phi_{\mathcal{H}}(\vec{f},\vec{\theta}_f) - \phi_{\tilde{\mathcal{H}}}(\vec{f},\vec{\theta}_f)\right]\right\}\right\} =$$

$$\exp\left\{-\frac{1}{2}\left[\mathrm{E}\{\phi^2_{\mathcal{H}}(\vec{f},\vec{\theta}_f)\} + \mathrm{E}\{\phi^2_{\tilde{\mathcal{H}}}(\vec{f},\vec{\theta}_f)\} - 2\mathrm{E}\{\phi_{\mathcal{H}}(\vec{f},\vec{\theta}_f)\phi_{\tilde{\mathcal{H}}}(\vec{f},\vec{\theta}_f)\}\right]\right\}. \tag{6.77}$$

Inspection of Eq. (6.77) shows that if $\phi_{\mathcal{H}}(\vec{f},\vec{\theta}_f)$ and $\phi_{\tilde{\mathcal{H}}}(\vec{f},\vec{\theta}_f)$ are highly correlated then $\mathrm{E}\{S(\vec{f},\vec{\theta}_f)\} > \mathrm{E}\{\mathcal{H}(\vec{f},\vec{\theta}_f)\}$. It is reasonable to expect $\phi_{\mathcal{H}}(\vec{f},\vec{\theta}_f)$ and $\phi_{\tilde{\mathcal{H}}}(\vec{f},\vec{\theta}_f)$ to be highly correlated when the residual wave front is highly sampled in space and time, and when sufficient signal-to-noise ratio is present in the wave front sensor measurement.

The signal-to-noise ratio of the CDWFS estimator $\mathrm{SNR}_{CD}(\vec{f},\vec{\theta}_f)$ is easily obtained from the derivation of the signal-to-noise ratio of the DWFS method presented in Section 6.2.5. However, to use the derivation in Section 6.2.5 the quantity $\exp\{-j\phi_{\tilde{\mathcal{H}}}(\vec{f},\vec{\theta}_f)\}$ must be substituted for $\tilde{\mathcal{H}}(\vec{f})$ in Eq. (6.39), and this substitution must be carried through the derivation to obtain an expression analogous to Eq. (6.47). The resulting expression for the signal-to-noise ratio of the CDWFS estimator is

$$\mathrm{SNR}_{CD}(\vec{f},\vec{\theta}_f) = \frac{\overline{K}|O_n(\vec{f},\vec{\theta}_f)\mathrm{E}\{S(\vec{f},\vec{\theta}_f)\}|}{\left(\overline{K} + (\overline{K})^2|O_n(\vec{f},\vec{\theta}_f)|^2\mathrm{var}\{S(\vec{f},\vec{\theta}_f)\}\right)^{1/2}}, \tag{6.78}$$

where $O_n(\vec{f},\vec{\theta}_f)$ is the object spectrum normalized to have the value of unity at $\vec{f} = 0$, and $\overline{K}$ is the average number of photoevents per image. Note that in the limit of infinite light level $\overline{K} \to \infty$, and Eq. (6.78) takes a form similar to Eq. (6.50)

$$\lim_{\overline{K}\to\infty}\{\mathrm{SNR}_{CD}(\vec{f},\vec{\theta}_f)\} = \frac{|\mathrm{E}\{S(\vec{f},\vec{\theta}_f)\}|}{\sqrt{\mathrm{var}\{S(\vec{f},\vec{\theta}_f)\}}} = \mathrm{SNR}_S(\vec{f},\vec{\theta}_f), \tag{6.79}$$

where $\mathrm{SNR}_S(\vec{f})$ is referred to as the SNR of the random CDWFS transfer function $S(\vec{f},\vec{\theta}_f)$.

This concludes the discussion of the CDWFS estimator. In the next subsection we discuss performance measures for CDWFS, and present representative results.

Simulating CDWFS

The results in the next subsection were created using a CDWFS simulation. Two separate simulations were used: one unique simulation was used to simulate the performance of CDWFS to overcome time delays in an adaptive optics system, and a different simulation was used to simulate the performance of CDWFS to overcome anisoplanatic effects in adaptive optics systems. Both of these simulations are described in this subsection.

A simulation suitable for studying the performance of CDWFS for overcoming time delays in adaptive optics systems is shown in Fig. 6.22. Fig. 6.22 shows an open loop adaptive optics simulation with an additional wave front sensor and the associated processing blocks needed to simulate CDWFS. This open loop model for simulating closed loop CDWFS was first presented in Ref. [6]. The discussion of simulating CDWFS borrows heavily from the adaptive optics simulation presented in Section 5 and the DWFS simulation presented in Section 6.2.3. Here we dwell on describing the unique aspects simulating CDWFS.

The setup stage is identical to that described in Section 5.4.6, where the topic of simulating an adaptive optics system was presented. The CDWFS simulation begins by creating a random phase screen $\psi(\vec{x}, \vec{\theta}_f = 0)$. The phase screen $\psi(\vec{x}, \vec{\theta}_f = 0)$ is input to the wave front sensor used to obtain measurements for computing deformable mirror actuator commands, denoted wave front sensor 1 in Fig. 6.22. A vector of wave front slopes $\mathbf{s}_1$, defined in Eq. (6.24), is output by wave front sensor 1, providing the input to the actuator command computer. The deformable mirror actuator weight vector $\mathbf{c}$, defined in Eq. (5.42), is computed by the actuator command computer and passed to the deformable mirror model. The phase of the deformable mirror $\phi_{DM}(\vec{x}, \vec{\theta}_f = 0)$ is computed using Eq. (5.41). To simulate the effects of a time delay between sensing and correcting for $\psi(\vec{x}, \vec{\theta}_f = 0)$, the phase $\psi(\vec{x}, \vec{\theta}_f = 0)$ is shifted by the amount $\vec{v}t$, where $\vec{v}$ is the wind velocity moving the screen, and t is the time delay between sensing and correcting for the wave front. The shifted phase screen is denoted $\psi(\vec{x} - \vec{v}t, \vec{\theta}_f = 0)$. The residual wave front error $\epsilon(\vec{x}, \vec{\theta}_f = 0)$ is computed using

$$\epsilon(\vec{x}, \vec{\theta}_f = 0) = \psi(\vec{x} - \vec{v}t, \vec{\theta}_f = 0) - \phi_{DM}(\vec{x}, \vec{\theta}_f = 0), \tag{6.80}$$

which is the error due to applying actuator commands obtained from an earlier instant in time to a wave front which has shifted by $\vec{v}t$.

The residual wave front phase $\epsilon(\vec{x}, \vec{\theta}_f = 0)$ is used for two purposes: (1) a realization of the instantaneous OTF $\mathcal{H}(\vec{f}, \vec{\theta}_f = 0)$ is computed using $\epsilon(\vec{x}, \vec{\theta}_f = 0)$ as the phase of the generalized pupil function; and (2) $\epsilon(\vec{x}, \vec{\theta}_f = 0)$ is applied to the input of a second wave front sensor, denoted wave front sensor 2 in Fig. 6.22. Note that wave front sensor 2 is exposed to the residual wave front error, similar to the situation that would exist in a closed loop adaptive optics system, but without the complexity of simulating a closed loop adaptive optics system. The vector of wave front slopes output by wave front sensor 2, denoted by $\mathbf{s}_2$, is input to a wave front computer which computes an estimate of $\epsilon(\vec{x}, \vec{\theta}_f = 0)$, denoted by $\tilde{\epsilon}(\vec{x}, \vec{\theta}_f = 0)$, using the method described in Section 6.2.2. An estimate of the OTF $\tilde{\mathcal{H}}(\vec{f}, \vec{\theta}_f = 0)$ is calculated using $\tilde{\epsilon}(\vec{x}, \vec{\theta}_f = 0)$ as the phase of the generalized pupil function.

Finally, the OTF $\mathcal{H}(\vec{f}, \vec{\theta}_f = 0)$ and the estimated OTF $\tilde{\mathcal{H}}(\vec{f}, \vec{\theta}_f = 0)$ are used to compute a single realization of the CDWFS transfer function $S(\vec{f}, \vec{\theta}_f = 0)$. This entire process is repeated a sufficiently large number of times to insure that good sample based estimates of the quantities of interest are obtained. Statistical quantities required to compute the desired statistics of $S(\vec{f}, \vec{\theta}_f = 0)$ are accumulated at each iteration. Sample-based estimates of $\mathrm{E}\{S(\vec{f}, \vec{\theta}_f = 0)\}$ and $\mathrm{var}\{S(\vec{f}, \vec{\theta}_f = 0)\}$ are computed after the last iteration of the simulation.

A functional block diagram of a simulation used to study the performance of CDWFS for overcoming anisoplanatic effects in adaptive optics systems is shown in Fig. 6.23. Use of the simulation shown in Fig. 6.23 was first presented in Ref. [7]. The simulation shown in Fig. 6.23 differs from the simulation shown in Fig. 6.22 primarily in the nature of the phase screen generator used.

To study CDWFS for overcoming anisoplanatic effects it is necessary to simulate the effects of an

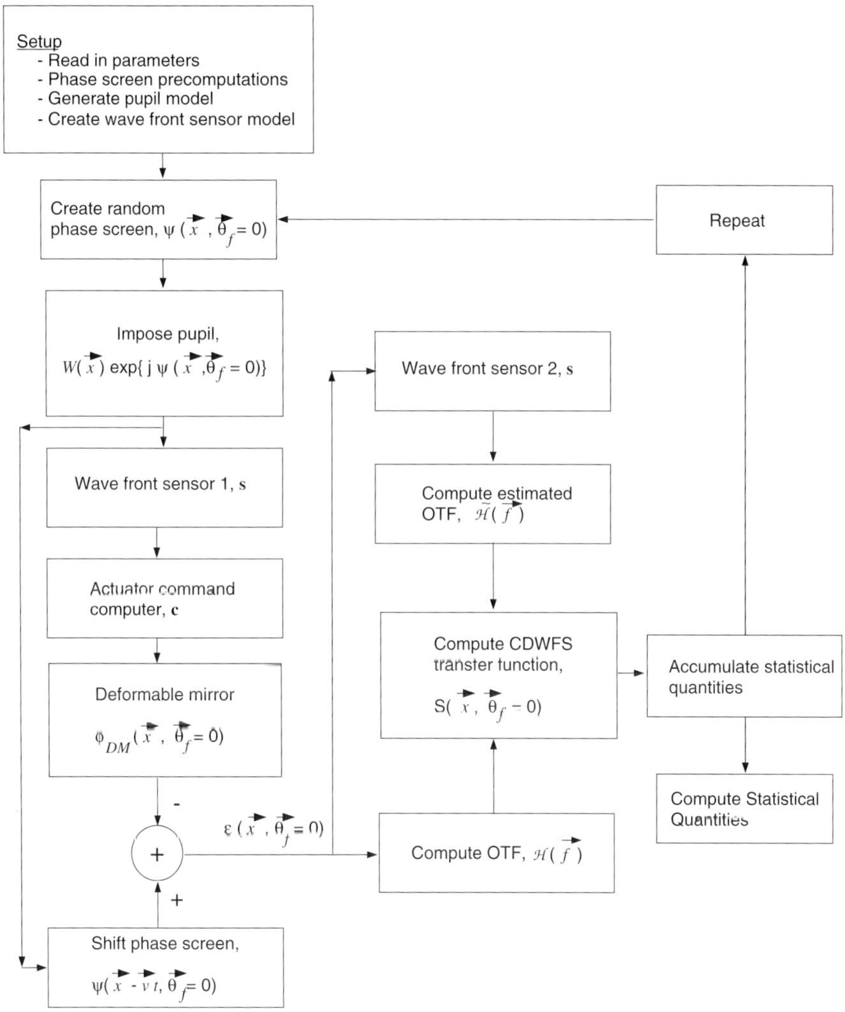

FIGURE 6.22
Block diagram for a simulation to study the performance of CDWFS for overcoming time delay effects in adaptive optics systems.

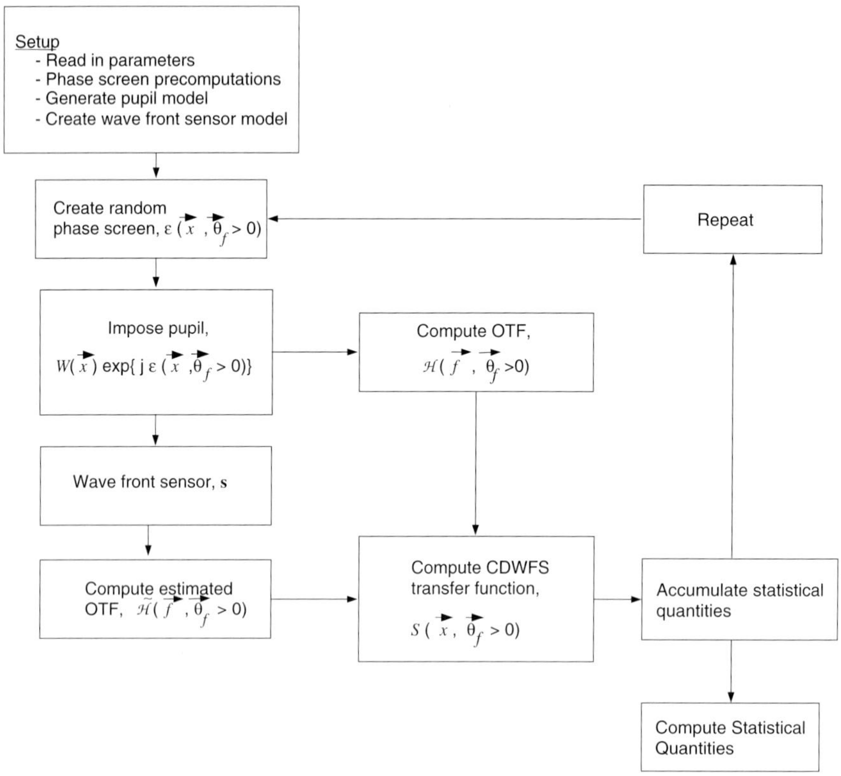

FIGURE 6.23
Block diagram for a simulation to study the performance of CDWFS for overcoming anisoplanatic effects in adaptive optics systems.

adaptive optics system on a wave front arriving from the $\vec{\theta}_f > 0$ direction. The approach taken in Ref. [7] was to perform an analytic calculation of the covariance of the closed loop residual phase error of a wave front arriving from the $\vec{\theta}_f > 0$ direction, denoted $\epsilon(\vec{x}, \vec{\theta}_f > 0)$, using the method presented in Ref. [32]. A phase screen generator based on the concepts presented in Section 3.7 was implemented to make random draws of $\epsilon(\vec{x}, \vec{\theta}_f > 0)$. Because random draws of $\epsilon(\vec{x}, \vec{\theta}_f > 0)$ were created directly by the phase screen generator there was no need to simulate the components of an adaptive optical system.

The wave front phase $\epsilon(\vec{x}, \vec{\theta}_f > 0)$ is used for two purposes in the simulation: (1) $\epsilon(\vec{x}, \vec{\theta}_f > 0)$ is used to create a realization of the OTF in the $\vec{\theta}_f > 0$ direction denoted $\mathcal{H}(\vec{f}, \vec{\theta}_f > 0)$; and (2) $\epsilon(\vec{x}, \vec{\theta}_f > 0)$ is input to a wave front sensor model, corresponding to wave front sensor B in Fig. 6.21. The wave front sensor in Fig. 6.23 computes the closed loop wave front slope vector **s** due to $\epsilon(\vec{x}, \vec{\theta}_f > 0)$. The closed loop wave front slope vector **s** is input to a wave front computer which computes an estimate of $\epsilon(\vec{x}, \vec{\theta}_f > 0)$, denoted $\tilde{\epsilon}(\vec{x}, \vec{\theta}_f > 0)$, using the method described in Section 6.2.2. An estimate of the OTF, denoted $\tilde{\mathcal{H}}(\vec{f}, \vec{\theta}_f > 0)$, is computed using $\tilde{\epsilon}(\vec{x}, \vec{\theta}_f > 0)$ as the phase of the generalized pupil function. Corresponding realizations of $\mathcal{H}(\vec{f}, \vec{\theta}_f > 0)$ and $\tilde{\mathcal{H}}(\vec{f}, \vec{\theta}_f > 0)$ are then used to compute a realization of $S(\vec{f}, \vec{\theta}_f > 0)$. Quantities needed to compute the statistics of $S(\vec{f}, \vec{\theta}_f > 0)$ are accumulated at each iteration, and the entire process is repeated. Sample-based estimates of $\mathrm{E}\{S(\vec{f}, \vec{\theta}_f > 0)\}$ and $\mathrm{var}\{S(\vec{f}, \vec{\theta}_f > 0)\}$ are computed after the last iteration.

This concludes the discussion of techniques for simulating the performance of CDWFS. In the next subsection examples of CDWFS performance are presented.

CDWFS performance

In this subsection the performance of the CDWFS technique is discussed. Since the CDWFS transfer function $S(\vec{f}, \vec{\theta}_f)$ is closely related to the compensated OTF $\mathcal{H}(\vec{f}, \vec{\theta}_f)$, it is natural to compare performance measures based on $S(\vec{f}, \vec{\theta}_f)$ and $\mathcal{H}(\vec{f}, \vec{\theta}_f)$. Two measures of performance are compared here: (1) $\mathrm{E}\{S(\vec{f}, \vec{\theta}_f)\}$ is compared to $\mathrm{E}\{\mathcal{H}(\vec{f}, \vec{\theta}_f)\}$; and (2) the signal-to-noise ratio of the CDWFS transfer function $\mathrm{SNR}_S(\vec{f}, \vec{\theta}_f)$ is compared to the signal-to-noise ratio of the OTF $\mathrm{SNR}_{\mathcal{H}}(\vec{f}, \vec{\theta}_f)$. The results presented in this subsection were obtained with the simulations described above. In this subsection the specific details of both of the simulations are described. Representative CDWFS performance results are then presented.

Results are presented for a 1.6 meter diameter unobscured telescope aperture. The wave front sensors used to obtain the CDWFS time delay results were identical, and consisted of Hartmann sensors with square subapertures having side length of d = 10 cm. This arrangement provided for 164 fully illuminated subapertures in the telescope pupil. The wave front sensor modeled for the CDWFS anisoplanatism results was also a Hartmann sensor with square subapertures. However, in the CDWFS anisoplanatism case the subapertures had side length of d = 9.37 cm, yielding a total of 208 fully illuminated subapertures in the telescope pupil. In both the CDWFS time delay and anisoplanatism cases the wave front sensor and the imaging camera were assumed to operate at the wavelength of $\lambda_I = \lambda_W = 500$ nm.

The deformable mirrors modeled in both the time delay and anisoplanatism cases had influence functions defined by two dimensional triangle functions [25] with base width given by two times the subaperture side length. Actuators were placed on a Cartesian grid of points with a grid spacing of 11 cm in the CDWFS time delay case. This arrangement of actuators provided for 169 actuators in the telescope pupil in the CDWFS time delay case. Actuators were also placed on a Cartesian grid of points for the CDWFS anisoplanatism case, but with a grid spacing of 9.37 cm. This arrangement of actuators provided for 241 actuators in the telescope pupil for the CDWFS anisoplanatism case. Deformable mirror actuator commands were obtained using the least squares reconstructor described in Section 5.3.3. All input parameters are summarized in Table 6.5 for the CDWFS time delay results, and in Table

TABLE 6.5
Input parameters for CDWFS time delay results.

Parameter	Value
Pupil diameter, D	1.6 m
Subaperture side length, d	0.1 m
Actuator spacing	0.11 m
Wave front sensor wavelength, λ_W	500 nm
Imaging wavelength, λ_I	500 nm
Elementary functions for phase reconstruction	First 40 Zernike polynomials
Number of independent frames	200
r_0	0.16 m, 0.1 m λ = 500 nm
SNR$_W$	10
Greenwood frequency, f_G	20, 30, 40, 50, 60 Hz

TABLE 6.6
Input parameters for CDWFS anisoplanatism results.

Parameter	Value
Pupil diameter, D	1.6 m
Subaperture side length, d	0.0937 m
Actuator spacing	0.0937 m
Wave front sensor wavelength, λ_W	500 nm
Imaging wavelength, λ_I	500 nm
Elementary functions for phase reconstruction	2-D triangle functions
Number of independent frames	200
r_0	0.133 m, 0.1 m λ = 550 nm
θ_0	7 μrad at λ = 550 nm
θ	$2\theta_0, 4\theta_0, 6\theta_0$
SNR$_W$	∞, 10, 5, 2

6.6 for the CDWFS anisoplanatism results.

The wave front phase estimate $\tilde{\epsilon}(\vec{x}, \vec{\theta}_f)$ was obtained using the least squares wave front reconstruction technique described in Section 6.2.2. Zernike polynomials, described in Section 3.6, were used as elementary functions for phase reconstruction for the CDWFS time delay results. All results presented here were created using the first 40 Zernike polynomials, excluding the piston term, for phase reconstruction. Two dimensional triangle functions [25] were used as the elementary functions for the CDWFS anisoplanatism case results. In the CDWFS anisoplanatism case the two dimensional triangle functions were placed on a Cartesian grid with grid spacing on 13 cm, yielding a total of 121 elementary functions used for wave front phase reconstruction.

Results specific to the CDWFS time delay case are now presented. To simulate the temporal effects of turbulence the phase screen was shifted by $\vec{v}t$ between being measured by wave front sensor 1 in Fig. 6.22, and being corrected by the deformable mirror. The time delay t between sensing and corrected for a given realization of turbulence was set at 0.01 s, consistent with an adaptive optics bandwidth $f_{AO} = 100$ Hz. The velocity $\vec{v}$ had direction parallel to the x axis of the pupil. The magnitude of the velocity was chosen to explore a range of Greenwood frequencies f_G. Recall from Eq. (5.146) that the relationship between the Fried seeing parameter r_0, $|\vec{v}|$, and f_G is

$$|\vec{v}| = \frac{f_G r_0}{0.428}. \tag{6.81}$$

The distance the phase screen was shifted in each case was chosen to provide Greenwood frequencies of $f_G = 20$ Hz, 30 Hz, 40 Hz, 50 Hz, and 60 Hz. Two values of the Fried seeing parameter were used: $r_0 = 10$ cm, and 16 cm, which yielded $D/r_0 = 16$ and 10, respectively. The wave front sensor signal-to-noise ratio SNR_W was set at $\text{SNR}_W = 10$ for all results presented here.

Comparisons of $E\{S(\vec{f}, \vec{\theta}_f = 0)\}$ and $E\{\mathcal{H}(\vec{f}, \vec{\theta}_f = 0)\}$ for $r_0 = 16$ cm are presented in Fig. 6.24. A similar set of plots for the $r_0 = 10$ cm case is presented in Fig. 6.25. Comparative plots of $\text{SNR}_{\mathcal{H}}(\vec{f}, \vec{\theta}_f)$ and $\text{SNR}_S(\vec{f}, \vec{\theta}_f)$ are shown in Figs. 6.26 and 6.27. Figures 6.24 through 6.27 were generated by using the simulation to compute the entire two dimensional function of interest. The two dimensional functions were then averaged along circles of constant radius to create the functions of a scalar spatial frequency displayed in Figs. 6.24 through 6.27. The horizontal axes in Figs. 6.24 through 6.27 have been normalized to the diffraction-limited cutoff spatial frequency for the aperture, $D/(\lambda d_i)$, where D is the diameter of the aperture, λ is the wavelength of operation, and d_i is the distance from the exit pupil to the image plane.

Figure 6.24 corresponds to a case where good wave front sampling by the wave front sensor is provided, since the ratio of the wave front sensor side length d to r_0, $d/r_0 = 0.625$, is less than unity. Figure 6.24 shows that for the case of good wave front sampling and temporal effects characterized by $f_G/f_{AO} \leq 0.3$ the situation $E\{S(\vec{f}, \vec{\theta}_f = 0)\} \approx E\{\mathcal{H}(\vec{f}, \vec{\theta}_f = 0)\}$ is obtained using CDWFS. However, as the ratio f_G/f_{AO} increases to $f_G/f_{AO} > 0.3$ for the case of good wave front sampling we see that $E\{S(\vec{f}, \vec{\theta}_f = 0)\} > E\{\mathcal{H}(\vec{f}, \vec{\theta}_f = 0)\}$ is obtained. In fact, in the range $f_G/f_{AO} \geq 0.4$ in the good wave front sampling case Fig. 6.24 shows that $E\{S(\vec{f}, \vec{\theta}_f = 0)\}$ is a factor two to ten times greater than $E\{\mathcal{H}(\vec{f}, \vec{\theta}_f = 0)\}$ at mid and high spatial frequencies.

Figure 6.25 corresponds to a case where the wave front is less well sampled by the wave front sensor than the case presented in Fig. 6.24, since in Fig. 6.25 $d/r_0 = 1$. Figure 6.25 shows that for the case of diminished wave front sampling and temporal effects characterized by $f_G/f_{AO} \leq 0.2$ the situation $E\{S(\vec{f}, \vec{\theta}_f = 0)\} \approx E\{\mathcal{H}(\vec{f}, \vec{\theta}_f = 0)\}$ is obtained using CDWFS. Figure 6.25 also shows that in the range $0.3 \leq f_G/f_{AO} \leq 0.5$ the situation $E\{S(\vec{f}, \vec{\theta}_f = 0)\} > E\{\mathcal{H}(\vec{f}, \vec{\theta}_f = 0)\}$ is obtained. However, the degree to which CDWFS improves the overall transfer function of the system is less than that for the good wave front sampling case presented in Fig. 6.24. For $f_G/f_{AO} \geq 0.6$ Fig. 6.25 shows that $E\{S(\vec{f}, \vec{\theta}_f = 0)\} \approx E\{\mathcal{H}(\vec{f}, \vec{\theta}_f = 0)\}$.

The signal-to-noise ratio performance of the CDWFS transfer function is compared to the signal-to-noise ratio performance of the OTF in Figs. 6.26 and 6.27. Figure 6.26 shows $\text{SNR}_S(\vec{f}, \vec{\theta}_f = 0)$ and $\text{SNR}_{\mathcal{H}}(\vec{f}, \vec{\theta}_f = 0)$ corresponding to the good wave front sampling case of $r_0 = 16$ cm and $d/r_0 = 0.625$. Inspection of Fig. 6.26 shows that for $f_G/f_{AO} \leq 0.1$, $\text{SNR}_S(\vec{f}, \vec{\theta}_f = 0) \approx \text{SNR}_{\mathcal{H}}(\vec{f}, \vec{\theta}_f = 0)$. However, for $f_G/f_{AO} \geq 0.2$ the we obtain $\text{SNR}_S(\vec{f}, \vec{\theta}_f = 0) > \text{SNR}_{\mathcal{H}}(\vec{f}, \vec{\theta}_f = 0)$ at mid and high spatial frequencies. In fact, Fig. 6.26 shows that for $f_G/f_{AO} \geq 0.4$, $\text{SNR}_S(\vec{f}, \vec{\theta}_f = 0)$ is from two to ten times larger than $\text{SNR}_{\mathcal{H}}(\vec{f}, \vec{\theta}_f = 0)$ at mid and high spatial frequencies.

Figure 6.27 shows $\text{SNR}_S(\vec{f}, \vec{\theta}_f = 0)$ and $\text{SNR}_{\mathcal{H}}(\vec{f}, \vec{\theta}_f = 0)$ corresponding to the diminished wave front sampling case of $r_0 = 10$ cm and $d/r_0 = 1$. Inspection of Fig. 6.27 shows that for $f_G/f_{AO} \leq 0.1$, $\text{SNR}_S(\vec{f}, \vec{\theta}_f = 0) \approx \text{SNR}_{\mathcal{H}}(\vec{f}, \vec{\theta}_f = 0)$. For $f_G/f_{AO} \geq 0.2$ the we again obtain $\text{SNR}_S(\vec{f}, \vec{\theta}_f = 0) > \text{SNR}_{\mathcal{H}}(\vec{f}, \vec{\theta}_f = 0)$ at mid and high spatial frequencies. However, in the range $f_G/f_{AO} \geq 0.2$ Fig. 6.27 shows that the degree of improvement in signal-to-noise ratio provided by CDWFS is smaller than for the $d/r_0 = 0.625$ case shown in Fig. 6.26. Figure 6.27 shows that at mid and high spatial frequencies $\text{SNR}_S(\vec{f}, \vec{\theta}_f = 0)$ is on the order of two times larger than $\text{SNR}_{\mathcal{H}}(\vec{f}, \vec{\theta}_f = 0)$ for the case $d/r_0 = 1$.

Several conclusions can be drawn from this study of using CDWFS to mitigate temporal effects in adaptive optics systems. CDWFS does not provide an improvement in either the mean or signal-to-noise ratio performance of the adaptive optics if the Greenwood frequency is low compared to the adaptive optics frequency. However, in the range $0.2 \leq f_G/f_{AO} \leq 0.6$ we find that CDWFS can provide significant improvements to both the effective mean transfer function of the optical system and to the signal-to-noise ratio performance of the system. Good wave front sampling by the wave front sensor is required to maximize the performance of CDWFS, with $d/r_0 \leq 0.625$ being desired. The requirement for good wave front sampling derives from the need to reconstruct an accurate estimate of the residual wave front phase error used to compute the estimate of the OTF. Thus, CDWFS offers a means of

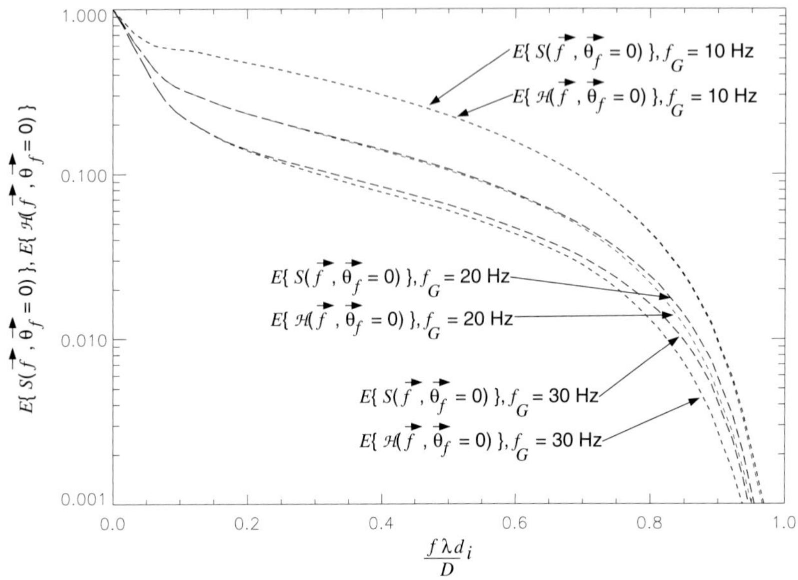

(a)

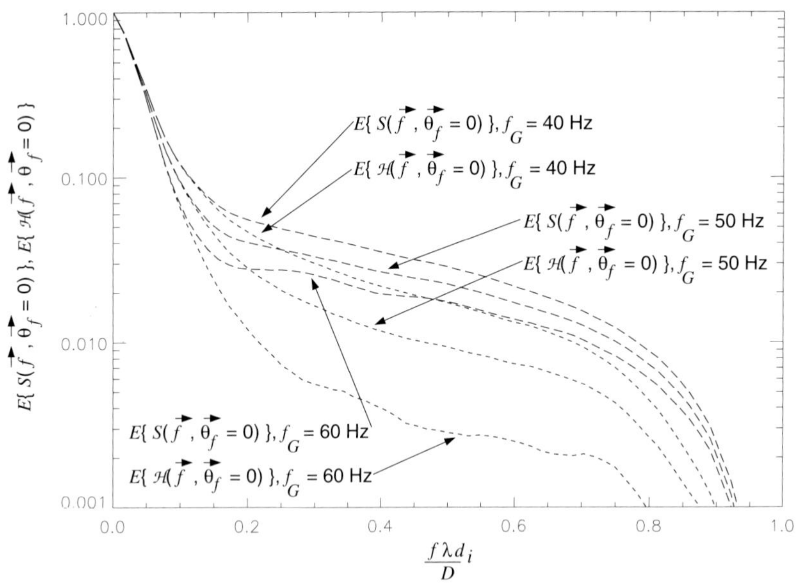

(b)

FIGURE 6.24
$E\{S(\vec{f}, \vec{\theta}_f = 0)\}$ and $E\{\mathcal{H}(\vec{f}, \vec{\theta}_f = 0)\}$ for $r_0 = 16$ cm: (a) $f_G = 10$ Hz, 20 Hz, and 30 Hz; (b) $f_G = 40$ Hz, 50 Hz, and 60 Hz. Adaptive optics bandwidth for all cases of $f_{AO} = 100$ Hz.

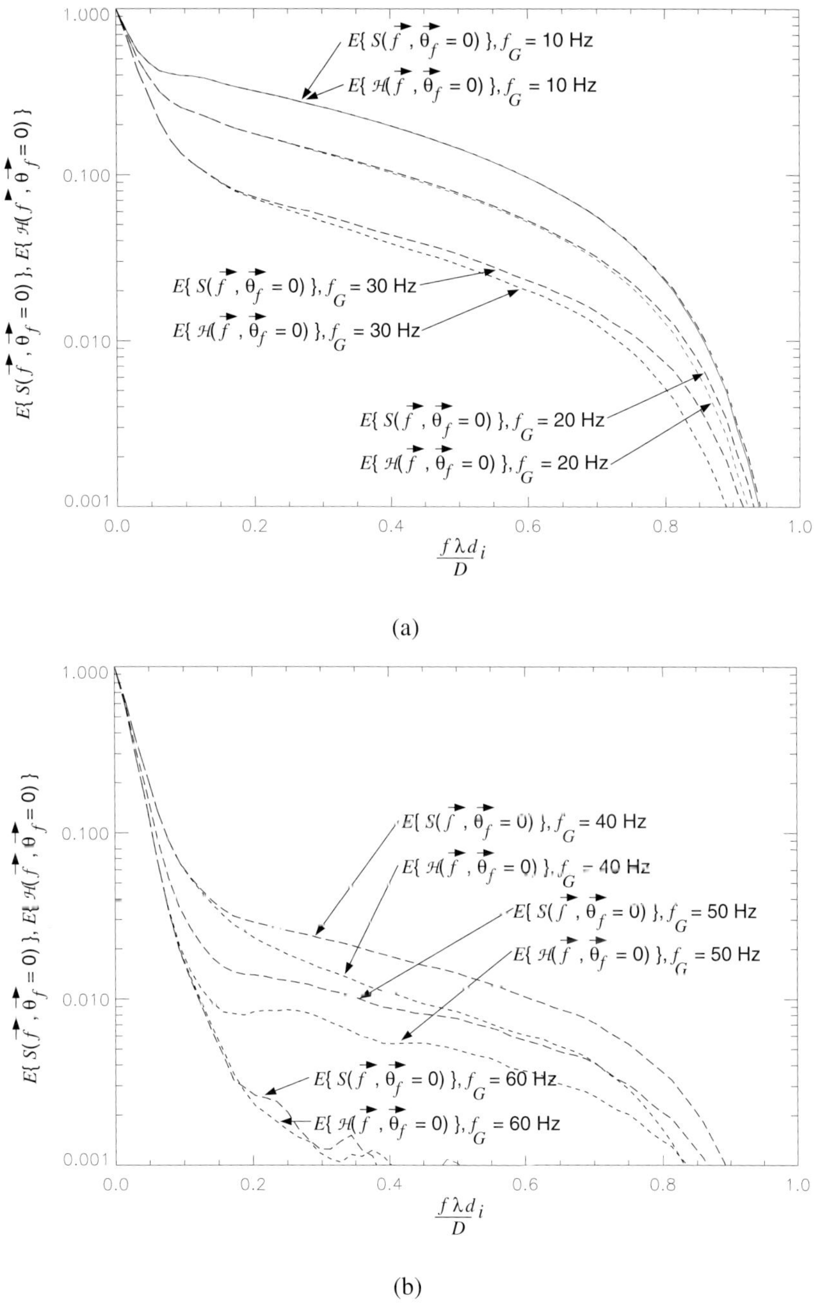

FIGURE 6.25
E$\{S(\vec{f},\vec{\theta}_f = 0)\}$ and E$\{\mathcal{H}(\vec{f},\vec{\theta}_f = 0)\}$ for r_0 = 10 cm: (a) f_G = 10 Hz, 20 Hz, and 30 Hz; (b) f_G = 40 Hz, 50 Hz, and 60 Hz. Adaptive optics bandwidth for all cases of f_{AO} = 100 Hz.

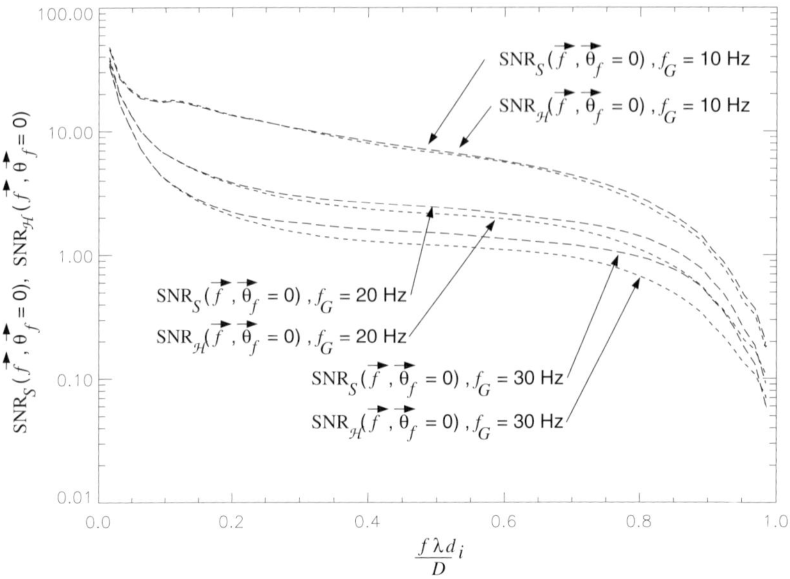

(a)

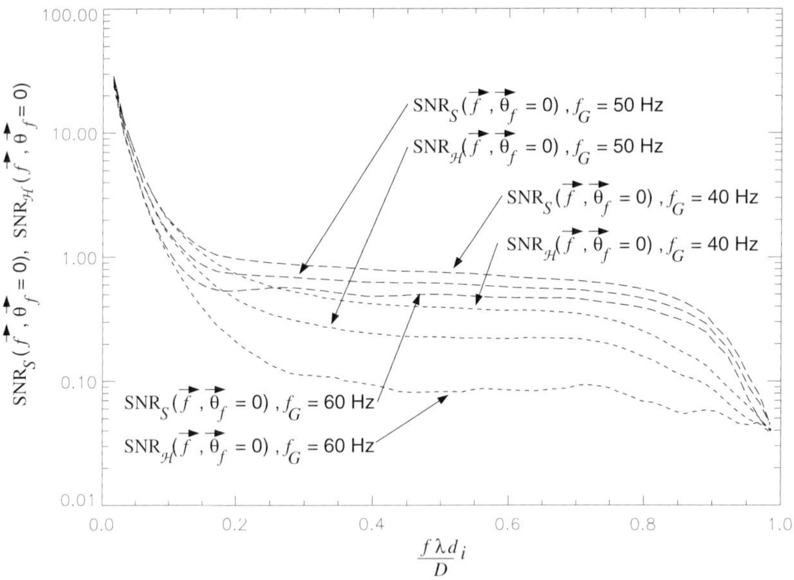

(b)

FIGURE 6.26
SNR$_{\mathcal{S}}(\vec{f}, \vec{\theta}_f = 0)$ and SNR$_{\mathcal{H}}(\vec{f}, \vec{\theta}_f = 0)$ for $r_0 = 16$ cm: (a) $f_G = 10$ Hz, 20 Hz, and 30 Hz; (b) $f_G = 40$ Hz, 50 Hz, and 60 Hz. Adaptive optics bandwidth for all cases of $f_{AO} = 100$ Hz.

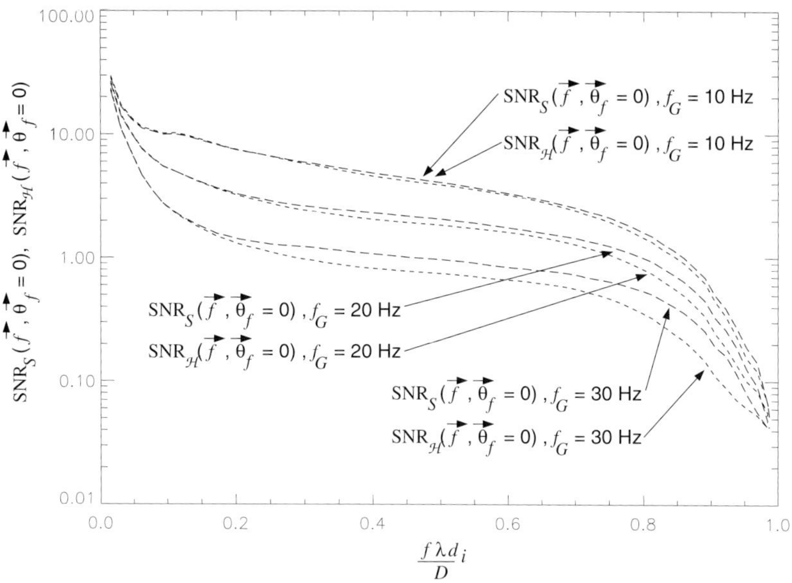

(a)

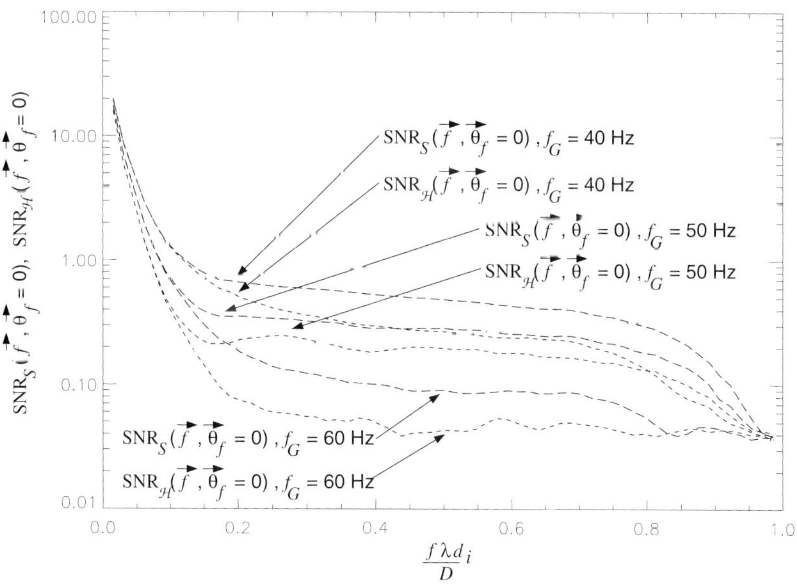

(b)

FIGURE 6.27
$\mathrm{SNR}_S(\vec{f}, \vec{\theta}_f = 0)$ and $\mathrm{SNR}_{\mathcal{H}}(\vec{f}, \vec{\theta}_f = 0)$ for $r_0 = 10$ cm: (a) $f_G = 10$ Hz, 20 Hz, and 30 Hz; (b) $f_G = 40$ Hz, 50 Hz, and 60 Hz. Adaptive optics bandwidth for all cases of $f_{AO} = 100$ Hz.

overcoming some of the effects of time delays in adaptive optics systems.

Performance results for using CDWFS to overcome anisoplanatic effects of turbulence are presented in Figs. 6.28 through 6.31. Results in these figures are presented in terms of the field angle $\vec{\theta}_f$ expressed as multiples of the isoplanatic angle θ_0 defined in Eq. (6.66). The results presented in Figs. 6.28 through 6.31 were computed for the Hufnagel-Valley $C_n^2(z)$ profile defined in Eq. (3.16), with the low altitude turbulence strength parameter A and the high altitude wind parameter v adjusted to provide $r_0 = 5$ cm and $\theta_0 = 7$ μrad at $\lambda = 550$ nm [7]. To vary r_0 the entire $C_n^2(z)$ was multiplied by a constant factor. This method of varying r_0 causes a proportional increase in θ_0. The values of $|\vec{\theta}_f|$ were adjusted in the results presented below to maintain the stated ratios of $|\vec{\theta}_f|/\theta_0$.

It was assumed that the off-axis beacon was displaced from the optical axis parallel to the x axis in the pupil. It was also assumed that both beacons were at infinite altitude, consistent with having natural sources for both beacons. In practice, the likelihood of having natural beacons available for both wave front sensors is quite small. A sodium layer artificial laser beacon in the off-axis direction is the most likely source of signal for the wave front sensor looking in the $\vec{\theta}_f$ direction. Sodium layer beacons are approximately 90 km in altitude. Hence, the results presented here are upper bounds on the performance of the CDWFS technique for correcting anisoplanatic effects, though the degradation in performance arising from the use of a sodium layer artificial laser beacon for the off-axis wave front sensor is expected to be quite small [7]. Full two dimensional realizations of $\mathcal{H}(\vec{f},\vec{\theta})$ and $S(\vec{f},\vec{\theta})$ were created in the simulation. Results are presented for slices through $\mathrm{E}\{\mathcal{H}(\vec{f},\vec{\theta}_f)\}$, $\mathrm{E}\{S(\vec{f},\vec{\theta})\}$, $\mathrm{SNR}_{\mathcal{H}}(\vec{f},\vec{\theta}_f)$, and $\mathrm{SNR}_S(\vec{f},\vec{\theta}_f)$ along the f_x axis, where the spatial frequency variable f_x corresponds to the Fourier transform of the x axis. The horizontal axes in Figs. 6.28 through 6.31 have been normalized to the diffraction-limited cutoff frequency for the aperture, $D/(\lambda d_i)$, where D is the diameter of the aperture, λ is the wavelength of operation, and d_i is the distance from the exit pupil to the image plane.

Figure 6.28 shows the comparative performance of $\mathrm{E}\{S(\vec{f},\vec{\theta}_f > 0)\}$ and $\mathrm{E}\{\mathcal{H}(\vec{f},\vec{\theta}_f > 0)\}$ for the cases $\mathrm{SNR}_W = \infty$, $|\vec{\theta}_f| = 2\theta_0$, $4\theta_0$, and $6\theta_0$ and $r_0 = 13.3$ cm, and 10 cm. Inspection of Fig. 6.28 shows that $\mathrm{E}\{S(\vec{f},\vec{\theta}_f > 0)\} > \mathrm{E}\{\mathcal{H}(\vec{f},\vec{\theta}_f > 0)\}$ in every case at mid and high spatial frequencies. The degree of improvement in overall transfer function provided by CDWFS is a function of seeing conditions and the field angle $\vec{\theta}_f$. The most striking improvements provided by CDWFS occur at the largest field angles, where $\mathrm{E}\{\mathcal{H}(\vec{f},\vec{\theta}_f > 0)\} \approx 0$, while $\mathrm{E}\{S(\vec{f},\vec{\theta}_f > 0)\}$ has significant value. For example, in Fig. 6.28(b), $\mathrm{E}\{\mathcal{H}(\vec{f},\vec{\theta}_f = 6\theta_0)\} \approx 0.01$ occurs at approximately $0.35 f_x \lambda d_i/D$. while $\mathrm{E}\{S(\vec{f},\vec{\theta}_f = 6\theta_0)\} = 0.01$ does not occur until approximately $0.5 f_x \lambda d_i/D$.

Figure 6.29 shows the effects of varying the wave front sensor signal-to-noise ratio SNR_W on $\mathrm{E}\{S(\vec{f},\vec{\theta}_f > 0)\}$ and $\mathrm{E}\{\mathcal{H}(\vec{f},\vec{\theta}_f > 0)\}$ for a single field angle, $\vec{\theta}_f = 4\theta_0$. Results for three different values of SNR_W are presented, $\mathrm{SNR}_W = 10$, 5, and 2; and two different values of r_0, $r_0 = 13.3$ cm, and 10 cm. Inspection of Fig. 6.29 shows that $\mathrm{E}\{S(\vec{f},\vec{\theta}_f = 4\theta_0)\} > \mathrm{E}\{\mathcal{H}(\vec{f},\vec{\theta}_f = 4\theta_0)\}$ for every case at mid and high spatial frequencies. Figure 6.29 also shows that both $\mathrm{E}\{S(\vec{f},\vec{\theta}_f = 4\theta_0)\}$ and $\mathrm{E}\{\mathcal{H}(\vec{f},\vec{\theta}_f = 4\theta_0)\}$ degrade gracefully as SNR_W decreases.

The transfer function signal-to-noise ratios $\mathrm{SNR}_S(\vec{f},\vec{\theta}_f > 0)$ and $\mathrm{SNR}_{\mathcal{H}}(\vec{f},\vec{\theta}_f > 0)$ are presented in Fig. 6.30 for the cases of $\mathrm{SNR}_W = \infty$, $|\vec{\theta}_f| = 2\theta_0$, $4\theta_0$, and $6\theta_0$, and $r_0 = 13.3$ cm, and 10 cm. Note that $\mathrm{SNR}_S(\vec{f},\vec{\theta}_f > 0)$ and $\mathrm{SNR}_{\mathcal{H}}(\vec{f},\vec{\theta}_f > 0)$ are shown at $\vec{\theta}_f = 0$ since $\mathrm{SNR}_S(\vec{f},\vec{\theta}_f > 0)$ and $\mathrm{SNR}_{\mathcal{H}}(\vec{f},\vec{\theta}_f > 0)$ take the deterministic value of unity, and hence have infinite signal-to-noise ratio at $\vec{\theta}_f = 0$. Inspection of Fig. 6.30 shows that $\mathrm{SNR}_S(\vec{f},\vec{\theta}_f > 0) > \mathrm{SNR}_{\mathcal{H}}(\vec{f},\vec{\theta}_f > 0)$ at mid and high spatial frequencies. Hence, CDWFS will allow spatial frequencies to be measured with higher signal-to-noise ratio than could be obtained using adaptive optics alone.

Finally, Fig. 6.31 shows $\mathrm{SNR}_S(\vec{f},\vec{\theta}_f = 4\theta_0)$ and $\mathrm{SNR}_{\mathcal{H}}(\vec{f},\vec{\theta}_f = 4\theta_0)$ for the cases $\mathrm{SNR}_W = 10$, 5, and 2, and $r_0 = 13.3$ cm, and 10 cm. Figure 6.31 shows that both $\mathrm{SNR}_S(\vec{f},\vec{\theta}_f = 4\theta_0)$ and $\mathrm{SNR}_{\mathcal{H}}(\vec{f},\vec{\theta}_f = 4\theta_0)$ degrade gracefully as SNR_W decreases. Further, $\mathrm{SNR}_S(\vec{f},\vec{\theta}_f = 4\theta_0) > \mathrm{SNR}_{\mathcal{H}}(\vec{f},\vec{\theta}_f = 4\theta_0)$ is obtained at mid and high spatial frequencies in every case.

We conclude from the results presented here that the CDWFS measurement and post processing technique can be used to improve the performance of adaptive optics imaging systems over that obtained

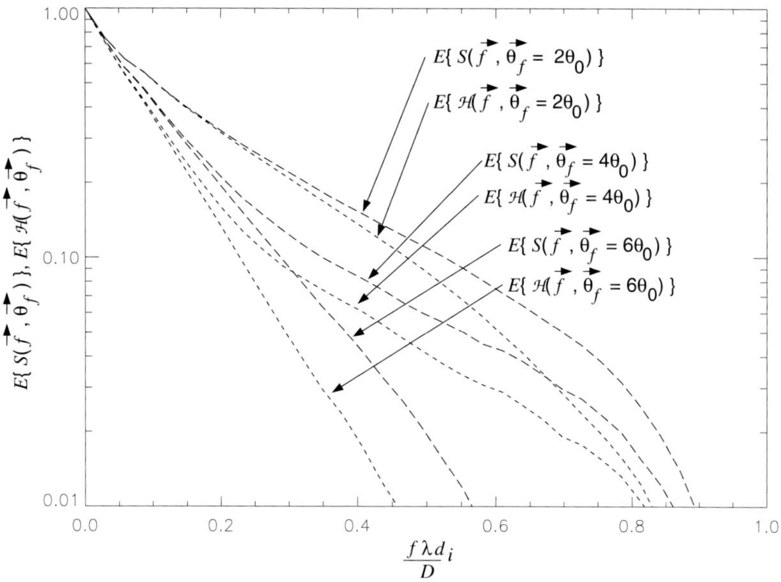

(a)

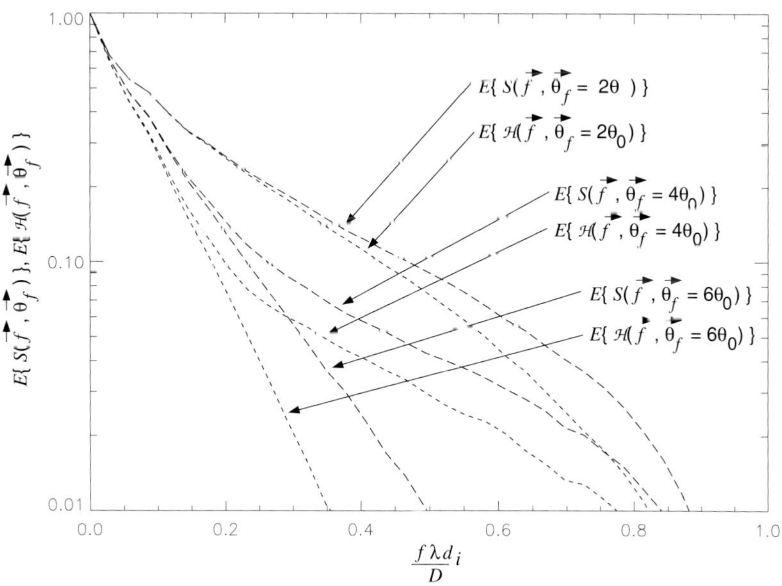

(b)

FIGURE 6.28
E$\{S(\vec{f}, \vec{\theta}_f > 0)\}$ and E$\{\mathcal{H}(\vec{f}, \vec{\theta}_f > 0)\}$ for SNR$_W = \infty$ and $|\vec{\theta}_f| = 2\theta_0, 4\theta_0$, and $6\theta_0$: (a) $r_0 =$ 13.3 cm; and (b) $r_0 =$ 10 cm.

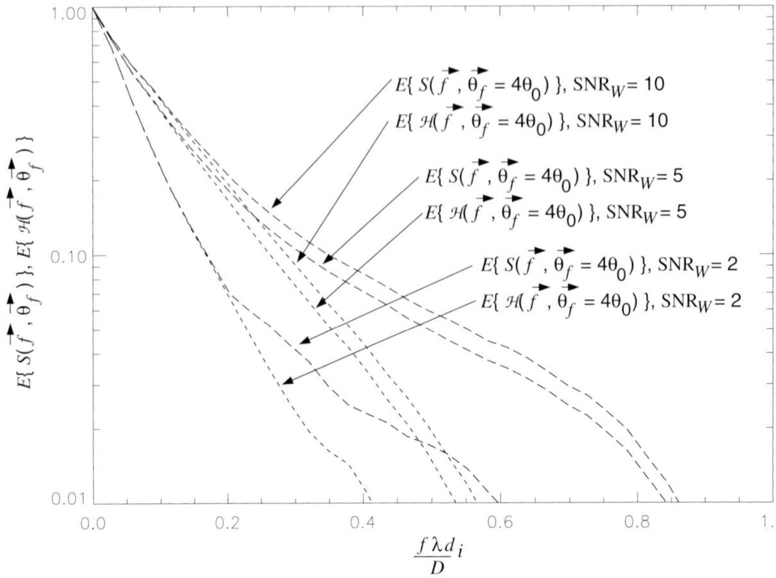

(a)

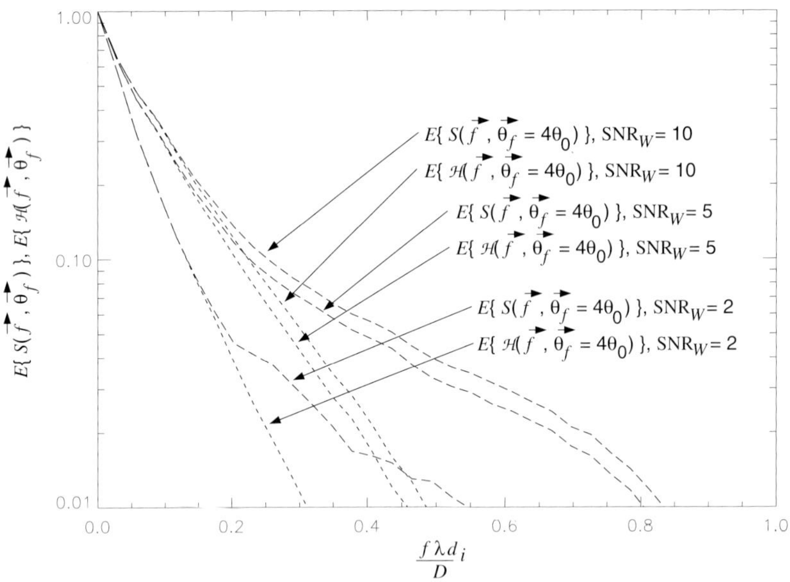

(b)

FIGURE 6.29
E$\{S(\vec{f}, \vec{\theta}_f = 4\theta_0)\}$ and E$\{\mathcal{H}(\vec{f}, \vec{\theta}_f = 4\theta_0)\}$ for SNR$_W$ = 10, 5, and 2: (a) r_0 = 13.3 cm; and (b) r_0 = 10 cm.

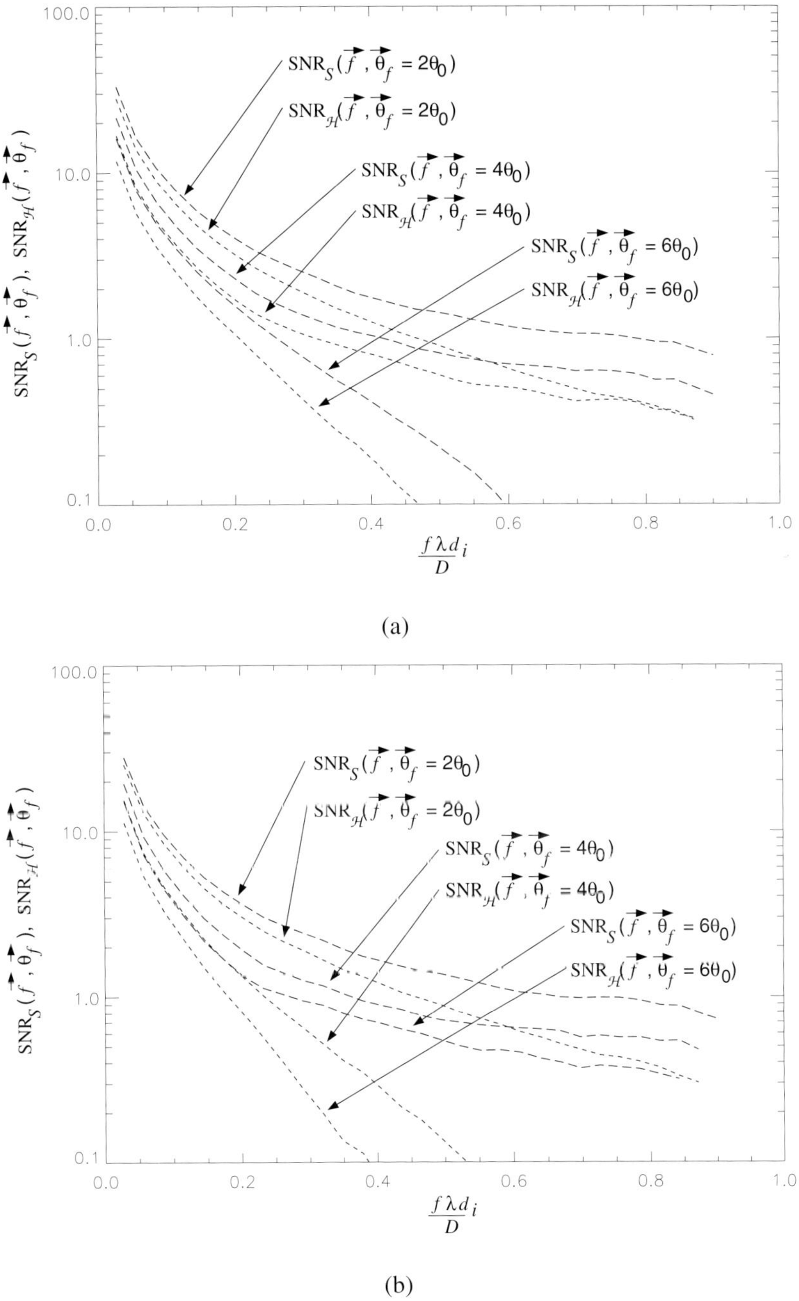

FIGURE 6.30
$\mathrm{SNR}_S(\vec{f}, \vec{\theta}_f > 0)$ and $\mathrm{SNR}_{\mathcal{H}}(\vec{f}, \vec{\theta}_f > 0)$ for $\mathrm{SNR}_W = \infty$ and $|\vec{\theta}_f| = 2\theta_0, 4\theta_0$; and $6\theta_0$: (a) $r_0 = 13.3$ cm, and (b) $r_0 = 10$ cm.

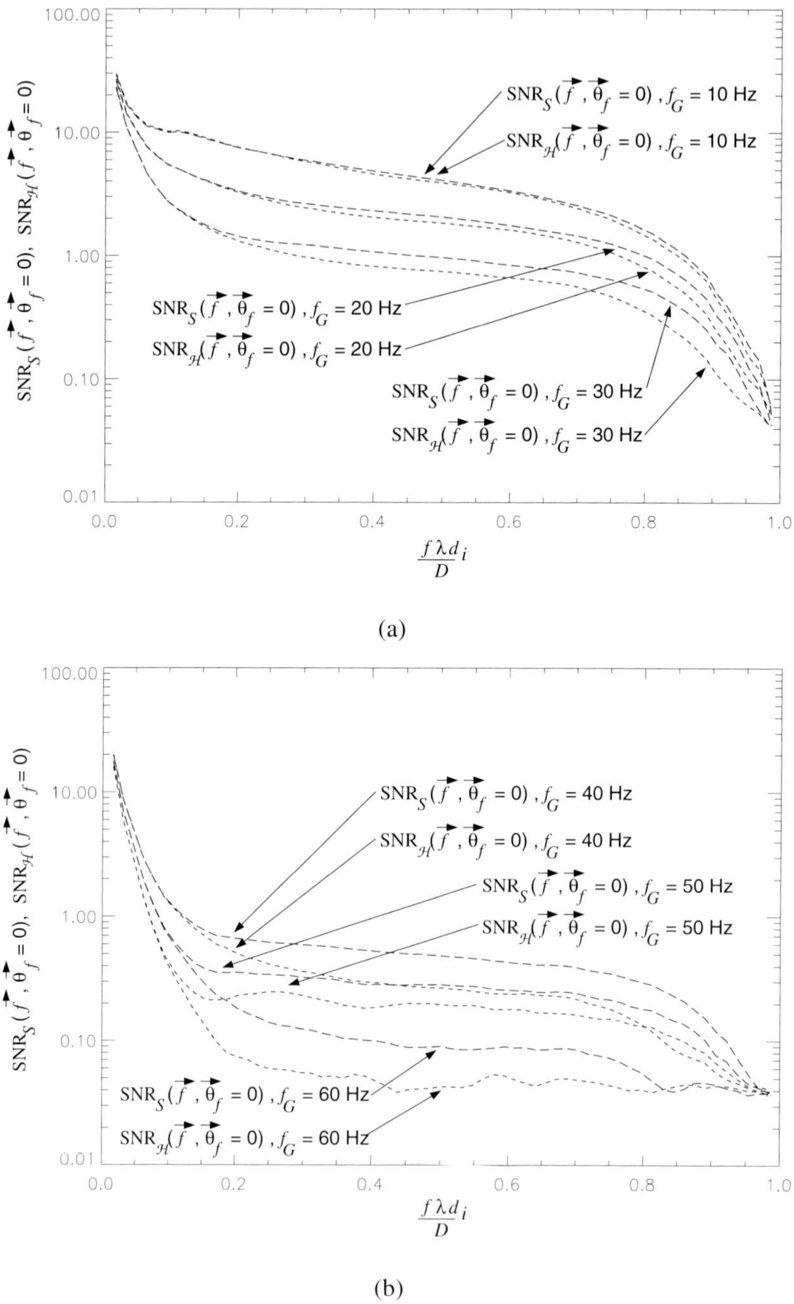

FIGURE 6.31
SNR$_S(\vec{f}, \vec{\theta}_f = 4\theta_0)$ and SNR$_{\mathcal{H}}(\vec{f}, \vec{\theta}_f = 4\theta_0)$ for SNR$_W$ = 10, 5, and 2: (a) r_0 = 13.3 cm; and (b) r_0 = 10 cm.

with the adaptive optics alone. In fact, it has been shown that $E\{S(\vec{f},\vec{\theta}_f)\} > E\{\mathcal{H}(\vec{f},\vec{\theta}_f)\}$ and $\text{SNR}_S(\vec{f},\vec{\theta}_f) > \text{SNR}_{\mathcal{H}}(\vec{f},\vec{\theta}_f)$ at mid and high spatial frequencies for using CDWFS to overcome both time delay and anisoplanatic effects. Thus, CDWFS allows spectral components to be accurately measured which cannot be measured using adaptive optics alone, increasing the effective resolution of the overall imaging system. In the next section we examine another hybrid imaging technique referred to as frame selection.

6.3.4 Frame selection

In this section we address the hybrid imaging technique referred to as frame selection. Frame selection involves using image quality metrics to select the best subset of a random data set of images, and then averaging only the selected images to create the final image [33, 34]. Motivation for the frame selection technique arises from the fact that experience with both simulations [3, 4] and experimental data [23, 24] has shown that some members of a set of short exposure images are better than others in the sense that the best images appear to contain sharper details of the object - that is, the best images appear to have higher resolution. This is a natural and expected consequence of the statistical nature of short exposure image measurement. In this section we examine the statistical consequences of selecting and using only the best images in a finite data set of short exposure images. Clearly, a tradeoff exists: by selecting the best images to process, the underlying statistics of the subset used to create the final image are more favorable for object estimation than the statistics of the entire data set, but the amount of data available for averaging has been reduced. This tradeoff is addressed here using the signal-to-noise ratio of the detected image spectrum, $\text{SNR}_D(\vec{f})$, first introduced in Eq. (2.107) [5, 35] as the metric for comparing the performance of the frame selection method to the performance obtained using the entire data set.

In this section we restrict attention to the use of frame selection as applied to images measured with adaptive optics imaging systems. However, Fried has shown that there is a finite probability that an uncompensated short exposure image will be nearly diffraction limited because the wavefront distortion over the aperture is negligible during the exposure [36]. The term short exposure image is used in this context to describe the image resulting from instantaneous removal of the tilt component of the aberrated wave in conjunction with exposure times sufficiently short to freeze the turbulence-induced aberration during the exposure. Fried defined the criterion for a good image as one for which the pupil averaged mean square phase distortion, c^2, is 1 rad^2 or less. Based on this criterion, the probability of getting a good short exposure, Pr, was derived to be

$$Pr \cong 5.6 \exp\left[-0.1557\left(\frac{D}{r_0}\right)^2\right], \quad \left(\frac{D}{r_0} \geq 3.5\right), \tag{6.82}$$

where r_0 is the atmospheric coherence diameter, and D is the diameter of the telescope [2]. This result demonstrates that the probability of capturing a single, nearly diffraction limited short exposure image decreases as the aperture diameter increases. For example, to obtain a probability of getting a nearly diffraction-limited image on the order of 10^{-3} dictates use of a telescope diameter no greater than $8r_0$; a probability on the order of 10^{-6} dictates use of a diameter no greater than $10r_0$. Thus, at a site with a one meter diameter telescope aperture and $10\text{cm} \leq r_0 \leq 12.5\text{cm}$, one would have to collect 1,000-1,000,000 short exposure images to have a reasonable probability of capturing a single nearly diffraction-limited image.

The limitation on the aperture size resulting from using frame selection on short exposure images is unacceptable when imaging dim objects, and the requirement to collect such a large number of short exposure images imposes severe data collection and storage requirements. Further, it was necessary in the derivation of Eq. (6.82) to assume that r_0 remains constant during the period of data collection. This assumption is only reasonable over short periods of time. Other factors in image quality, such as shot noise in the case of photon-limited detection, are not accounted for in Eq. (6.82). Further, the derivations of Eq. (6.82) do not allow for the presence of an adaptive optics system which would partially compensate for the effects of atmospheric turbulence. Also, the derivation which yields Eq. (6.82) does not suggest a means for selecting the best frames from a data set; rather, the derivation demonstrates the finite probability of the existence of such frames. Finally, given the statistical nature of the image formation and detection process, it is necessary to examine the SNR implications of frame selection.

In this section we examine the use of frame selection for photon-limited images measured through atmospheric turbulence using adaptive optics to correct for some of the turbulence effects before the image is measured. Image sharpness metrics [37], described later, were used to select the best subset of frames from the simulated data set of images. The functional block diagram for the frame selection method is shown in Fig. 6.32. Light from a distant object passes through the turbulent atmosphere and is intercepted by the telescope pupil. An adaptive optics system partially corrects for the effects of the turbulence-induced phase aberration, and an image is formed. The residual aberration is random, and as a consequence, the telescope-atmosphere point spread function varies from instant to instant. The presence of photon noise in the image detection process is an additional source of randomness corrupting the measured images. A number, N, of images are measured and stored in the frame selection technique. Image quality metrics are applied to each image, and the images are then rank ordered based on these metrics to choose a subset of the M best images within the data set of N images, $M < N$. The frame selection rate, f_{sr}, is defined as the ratio:

$$f_{sr} = \frac{M}{N}. \tag{6.83}$$

To study the benefits and range of applicability of frame selection, statistical quantities were accumulated to calculate the image spectrum SNR for both the selected subset, containing M elements, and the full set, containing N elements. The image spectrum SNRs for both data sets are then compared. The statistical information accumulated can also be used to compute the average point spread function of the adaptive optics imaging system and the average *effective* point spread function of the frame selection method. A set of representative results of frame selection performance are provided here.

The remainder of this section is organized as follows. In the next subsection the image sharpness metrics used to rank order images are presented. The discussion of image sharpness metrics is followed by a brief discussion of comparative performance measures which are used to evaluate the effectiveness of frame selection. In the final subsection the performance of frame selection as applied to point sources is presented.

6.3.5 Image sharpness metrics

To implement frame selection it is necessary to have an automated means of selecting the best frames from a data set. In Ref. [37], Muller and Buffington proposed a set of eight image quality measures which they referred to as image sharpness metrics. The S_1, S_4, and S_8 image sharpness metrics of Muller and Buffington [37], along with another function developed in Ref. [33] and referred to as S_{S1}, were evaluated as candidate quality metrics for the frame selection technique. The image sharpness metrics of Muller and Buffington are of interest in frame selection because it is known that they take their maximum values when the diffraction-limited image is used in the calculation of the metric [37]. The S_1, S_4, and S_8 image sharpness metrics are defined as [37]

$$S_1 = \int \mathrm{d}\vec{x} d^2 \left(\vec{x}\right) , \tag{6.84}$$

$$S_4 = \int\int \mathrm{d}x\mathrm{d}y \left| \frac{\partial^{m+n} d\left(x,y\right)}{\partial x^m \partial y^n} \right|^2 , \text{ and} \tag{6.85}$$

$$S_8 = -\int \mathrm{d}x \left| d\left(\vec{x}\right) - i_0\left(\vec{x}\right) \right|^2 . \tag{6.86}$$

In Eqs. (6.84) through (6.86) the quantity $d(\vec{x})$ is the detected image. The model for a photon-limited

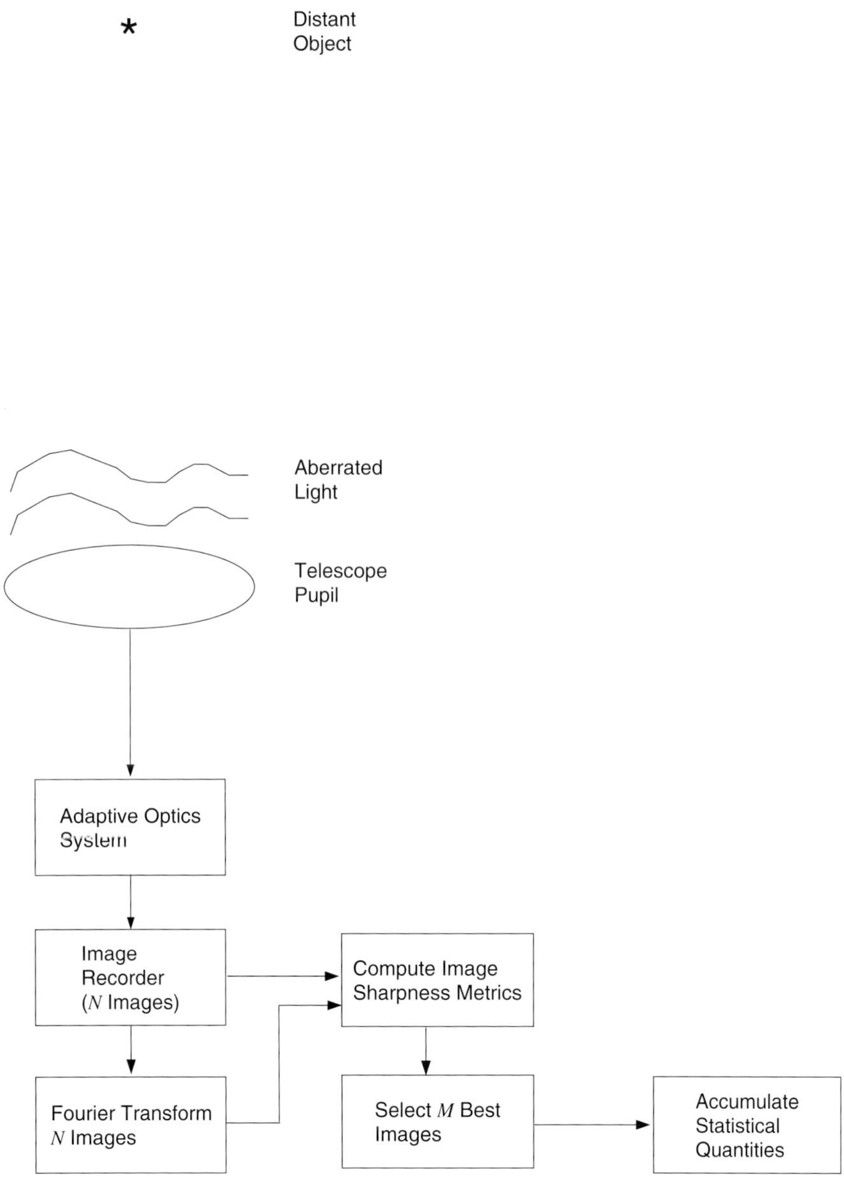

FIGURE 6.32
Block diagram of approach used to study signal-to-noise ratio effects of frame selection.

detected image $d(\vec{x})$ was given in Eq. (2.83) as:

$$d(\vec{x}) = \sum_{n=1}^{K} \delta(\vec{x} - \vec{x}_n), \tag{6.87}$$

where K is the total number of photo-events in the image, and $(\vec{x}_n)$ is the location of the nth photo-event. The metrics S_1, S_4, and S_8 are defined by integration over the image plane coordinate $\vec{x}$. In Eq. (6.85) the image plane coordinate $\vec{x}$ is written in terms of its orthogonal components, $\vec{x} = (x, y)$, to make explicit the fact that $m \neq n$ is possible. In S_8 the function $i_0(\vec{x})$ is the image of the object which would be formed by the optical system in the absence of the atmosphere. S_8 is a measure of the error between the actual image and the best possible image, but S_8 requires prior knowledge of the object. Thus, S_8 is not a useful metric in field experiments where the object is not known prior to the experiment. However, S_8 was implemented to provide a standard for comparing the performance of S_1, S_4, and S_{S1}, which is described now.

A modified, frequency domain version of S_1, denoted S_{S1}, was also used in frame selection [33]. By Parseval's theorem, the energy in the image and its Fourier transform must be equal [31]:

$$\int \mathrm{d}\vec{x}|i(\vec{x})|^2 = \int \mathrm{d}\vec{f}|I(\vec{f})|^2. \tag{6.88}$$

However, imaging systems are band limited, with cutoff frequency, $|\vec{f_c}| = D/(\lambda d_i)$, where D is the diameter of the aperture, λ is the mean wavelength of operation, and d_i is the distance from the exit pupil to the image plane [1]. Any energy in the image at frequencies higher than $|\vec{f_c}|$ is due solely to photon noise, motivating the use of a metric which excludes energy outside the cutoff frequency. One such metric, referred to here as S_{S1}, is defined as:

$$S_{S1} = \int \mathrm{d}\vec{f}\, W_{OTF}(\vec{f})|D(\vec{f})|^2, \tag{6.89}$$

where $D(\vec{f})$ is the detected image spectrum, and $W_{OTF}(\vec{f})$ is the support of the diffraction-limited OTF of the imaging system, defined by

$$W_{OTF}(\vec{f}) = \begin{cases} 1, & |\vec{f}| \leq |\vec{f_c}|, \\ 0, & |\vec{f}| > |\vec{f_c}| \end{cases}. \tag{6.90}$$

6.3.6 Performance metrics

In this section we introduce various performance metrics which have been used to evaluate the performance of frame selection. Performance metrics used for comparing the performance of the selected subset of images to the performance obtained with the full data set include the mean image spectrum $\mathrm{E}\{D(\vec{f})\}$, the variance of the image spectrum $\mathrm{var}\{D(\vec{f})\}$, and the average point spread function $\mathrm{E}\{s(\vec{x})\}$ [33, 34]. An additional performance metric is based on comparing the detected image spectrum signal-to-noise ratio for the total data set and for the selected subset of images. Hence, the detected image spectrum signal-to-noise ratio is reviewed here, and a comparative performance measure based on the detected image spectrum signal-to-noise ratio is presented.

The detected image spectrum $D(\vec{f})$ is defined as the Fourier transform of $d(\vec{x})$, and was given in Eq. (2.85). The image formed with an adaptive optics system contains two sources of randomness: (1) randomness arising from the random arrival times and locations of photo-events; and (2) randomness arising from fluctuations in the instantaneous OTF. Such processes are called doubly stochastic Poisson random processes [31]. $\mathrm{SNR}_D(\vec{f})$ for a doubly stochastic Poisson random process was derived in Section 2.3.2, and is stated in Eq. (2.125) as

$$\mathrm{SNR}_D(\vec{f}) = \frac{|\mathrm{E}\{D(\vec{f})\}|}{\mathrm{var}\{D(\vec{f})\}^{1/2}}$$

$$= \frac{\overline{K}|O_n(\vec{f})||\mathrm{E}\{\mathcal{H}(\vec{f})\}|}{\sqrt{\overline{K} + \overline{K}^2|O_n(\vec{f})|^2\mathrm{var}\{\mathcal{H}(\vec{f})\}}}, \tag{6.91}$$

where var$\{D(\vec{f})\}$ is the variance of the detected image spectrum, $\overline{K}$ is the mean number of photo-events per image, $O_n(\vec{f})$ is the image spectrum normalized to have a value of unity at $\vec{f} = 0$, E$\{\mathcal{H}(\vec{f})\}$ is the mean OTF, and var$\{\mathcal{H}(\vec{f})\}$ is the variance of the OTF. For statistically independent realizations of the detected image the effect of averaging P frames into a single image is to increase the image spectrum signal-to-noise ratio of the averaged image, denoted by $\mathrm{SNR}_D^P(\vec{f})$, according to

$$\mathrm{SNR}_D^P(\vec{f}) = \sqrt{P}\mathrm{SNR}_D(\vec{f}). \tag{6.92}$$

Applying frame selection to a data set of N images results in selection of M images, $M < N$, which are averaged to create a single image. $\mathrm{SNR}_D^M(\vec{f})$ and $\mathrm{SNR}_D^N(\vec{f})$ are functions of the number of short exposure images used in the averaging process. Clearly, applying the selection rule will result in the use of a smaller number of short exposure frames in the averaging process for the selected subset of images. However, the act of selecting the best frames changes the underlying statistics of the OTF used to form the averaged image. To compare $\mathrm{SNR}_D^N(\vec{f})$, obtained from the entire set of N frames, to $\mathrm{SNR}_D^M(\vec{f})$, obtained from the selected subset consisting of M frames, the signal-to-noise ratio gain function $G(\vec{f})$ is defined:

$$G(\vec{f}) = \frac{\mathrm{SNR}_D^M(\vec{f})}{\mathrm{SNR}_D^N(\vec{f})}. \tag{6.93}$$

$G(\vec{f})$ is a comparison of the relative gain or loss of signal-to-noise ratio that results from applying frame selection. $G(\vec{f}) > 1.0$ at a particular spatial frequency constitutes a gain in signal-to-noise ratio while any value of $G(\vec{f})$ below 1.0 constitutes a loss of SNR. We now examine the performance of frame selection.

6.3.7 Frame selection performance

In this section we study the performance of the frame selection technique. All of the performance results for frame selection presented here were obtained using an accurate simulation [33, 34]. Simulating frame selection follows directly from the adaptive optics imaging simulation described in Section 5.4.6. Two additional elements beyond the standard adaptive optics simulation described in Section 5.4.6 are required to simulate frame selection. As shown in Fig. 6.32, after the N simulated images are formed image sharpness metrics are computed for each image. After the image sharpness metrics are computed the images are rank ordered based on their image sharpness metric value, and the M best images, $M < N$, are selected and averaged. To compare performance between the selected subset of data and the results obtained from processing the whole data set statistical quantities of interest are computed for both data sets.

All of the input parameters for the results presented here are summarized in Table 6.7. A $D = 1$ meter diameter pupil was modeled. A Hartmann-type wave front sensor with square subapertures with side length of 10 cm was assumed, yielding a total of 64 wave front sensor subapertures completely contained in the telescope pupil. Photon-limited detection was assumed in the wave front sensor detector plane. The light levels available to both the wave front sensor and the image plane derive from object radiometric and integration time considerations. The variance of the additive noise in the wave front sensor slope measurement is given by Eqs. (5.26) through (5.33). Least-squares wavefront reconstruction was used to map wave front sensor measurements to deformable mirror actuator commands as described in Section 5.3.3 [3, 14]. The actuators in the deformable mirror were modeled as two-dimensional triangle functions [25], with centers located on a Cartesian grid with the origin at the center of the pupil, and with grid spacing equal to 11 cm. This configuration yields 69 actuators within the telescope pupil. The actuator influence function base width was equal to two times the grid spacing. All results presented

TABLE 6.7
Input parameters for frame selection results.

Parameter	Value
Pupil diameter, D	1 m
Subaperture side length, d	0.1 m
Actuator spacing	0.11 m
Wave front sensor wavelength, λ_W	500 nm
Imaging wavelength, λ_I	500 nm
Number of independent frames	500
r_0	0.1 m at λ = 500 nm
$\overline{K}_W$	20,215, 518
$\overline{K}$	1,040,430, 26, 635

here are derived from data sets with $N = 500$ frames.

Light levels were chosen in the following manner. Both the WFS and the imaging sensor were modeled as operating at a mean wavelength of 500 nm with a fractional bandwidth of 10%. An intensity splitter was modeled which sent 40% of the incident photons to the imaging camera and 60% of the incident photons to the WFS. Integration times of 9.1 ms were assumed in both the wavefront sensor and the imaging sensor. An overall transmission efficiency of 50% was assumed from the top of the atmosphere to the output of both detectors. Results are presented for objects with visual magnitude $m_v = 2$ and 6. In all cases, the objects were assumed to have the same spectral signature as the sun.

A simple experiment was used to establish the existence of a best subset of M frames within the larger data set of N frames. In this experiment the simulation was run for various selection rates f_{sr} using the S_8 metric, and $\mathrm{E}\{D(\vec{f})\}$, $\mathrm{var}\{D(\vec{f})\}$, and $\mathrm{E}\{s(\vec{x})\}$ were computed [33, 34]. The results of this experiment are shown in Fig. 6.33, where $\mathrm{E}\{D(\vec{f})\}$ and $\mathrm{var}\{D(\vec{f})\}$ are plotted for the case of a star object with $m_v = 2$ and $r_0 = 10$ cm. The plots shown in Fig. 6.33 were obtained by using the simulation to compute the two dimensional $\mathrm{E}\{D(\vec{f})\}$ and $\mathrm{var}\{D(\vec{f})\}$ arrays. The $\mathrm{E}\{D(\vec{f})\}$ and $\mathrm{var}\{D(\vec{f})\}$ arrays were then averaged along circles of constant radius to obtain functions of a scalar spatial frequency. The horizontal axes in Fig. 6.33, and all subsequent figures where appropriate, have been normalized to the diffraction-limited cutoff frequency for the aperture $D/(\lambda d_i)$, where λ is the mean wavelength of operation and d_i is the distance from the exit pupil to the image plane [1]. Inspection of Fig. 6.33 shows that as the selection rate is decreased $\mathrm{E}\{D(\vec{f})\}$ increases and $\mathrm{var}\{D(\vec{f})\}$ decreases, strongly suggesting that $\mathrm{SNR}_D^M(\vec{f}) > \mathrm{SNR}_D^N(\vec{f})$ could arise at some spatial frequencies. The average point spread functions $\mathrm{E}\{s(\vec{x})\}$ for the seeing and light level cases illustrated in Fig. 6.33 are shown in Fig. 6.34. Observe in Fig. 6.34 that as f_{sr} decreases the point spread function approaches the diffraction-limited point spread function.

The signal-to-noise ratio gain $G(\vec{f})$ for the seeing and light level cases presented in Figs. 6.33 and Fig. 6.34 is shown in Fig. 6.35. In Fig. 6.35, and all subsequent $G(\vec{f})$ plots, a horizontal line at $G(\vec{f}) = 1$ is provided to illustrate the regions of the spatial frequency domain where improved signal-to-noise ratio is obtained. Specifically, if $G(\vec{f}) > 1$ then $\mathrm{SNR}_D^M(\vec{f}) > \mathrm{SNR}_D^N(\vec{f})$, indicating that frame selection has resulted in improved estimation of the spatial frequency components where $G(\vec{f}) > 1$. Figure 6.35 clearly shows that frame selection improves the image spectrum signal-to-noise ratio, providing a maximum $G(\vec{f}) \approx 1.3$ for $f_{sr} = 60\%$. Note the fact that $G(\vec{f}) < 1$ at low spatial frequencies is not a serious problem in astronomy because $\mathrm{SNR}_D^N(\vec{f})$ is generally quite high at low spatial frequencies, typically on the order of many tens to a few hundred [23]. Similarly, the fact that $G(\vec{f}) < 1$ at high spatial frequencies is not a serious problem in astronomy because the various sources of noise in the image measurement force $\mathrm{SNR}_D^N(\vec{f})$ to be quite low at the highest spatial frequencies. Hence, the loss in signal-to-noise ratio caused by frame selection at the high spatial frequencies does not generally result in a serious loss of information in the image [33].

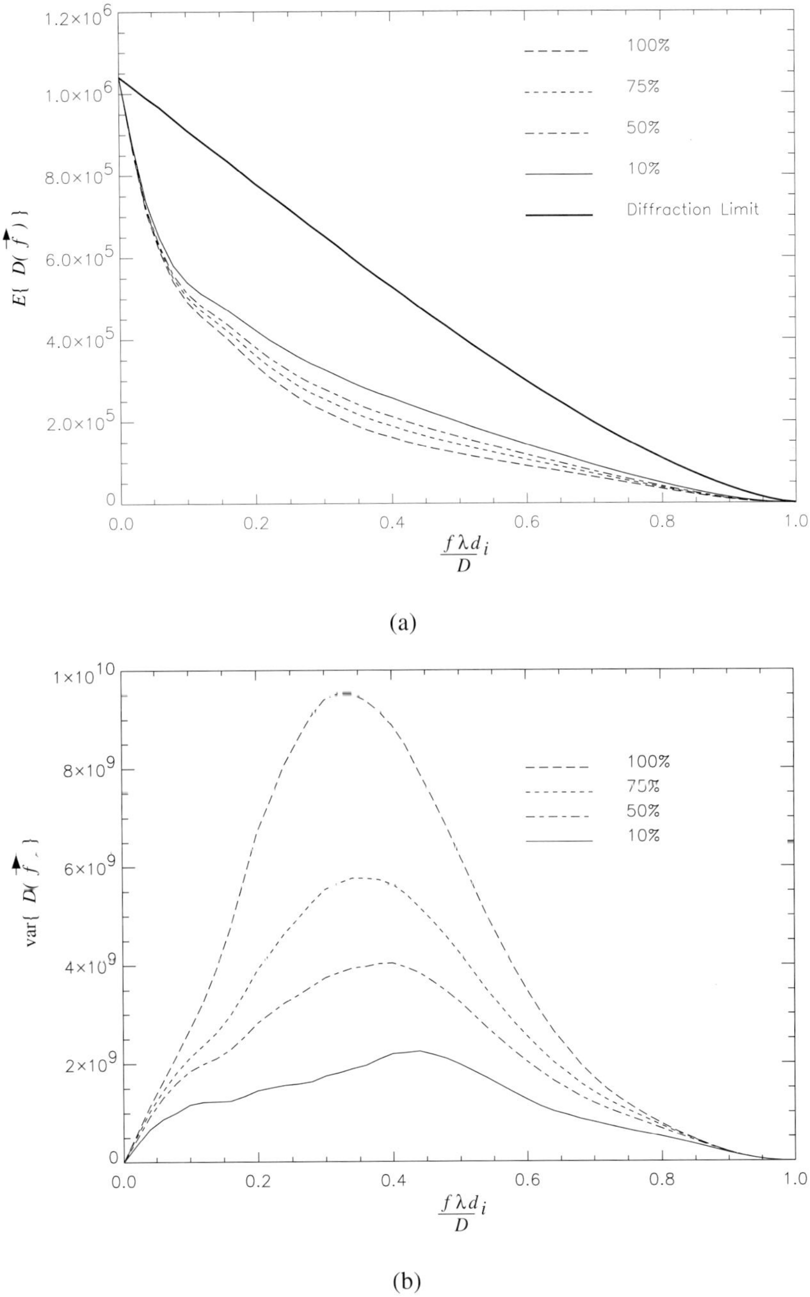

FIGURE 6.33
(a) E$\{D(\vec{f})\}$; and (b) var$\{D(\vec{f})\}$ for various selection rates; $m_v = 2, r_0 =$ 10 cm, and S_8 metric used.

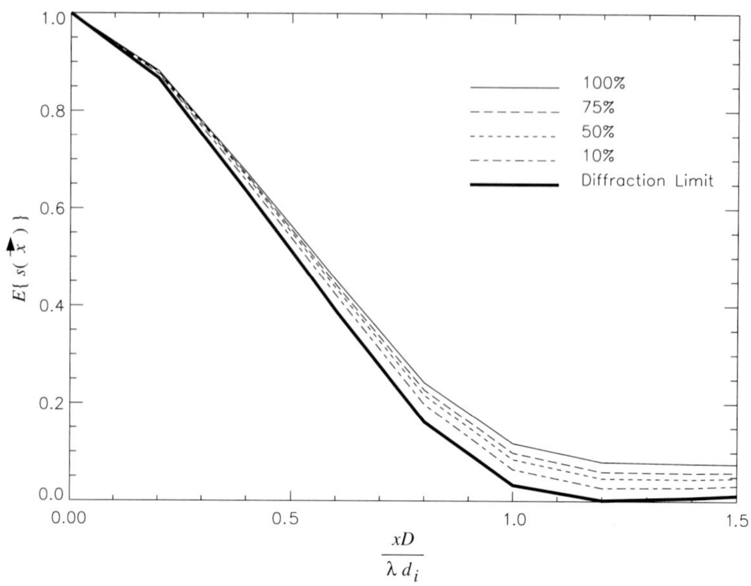

FIGURE 6.34
E$\{s(\vec{x})\}$ for various selection rates; $m_v = 2$, $r_0 = 10$ cm, and S_8 metric used.

The results presented in Figs. 6.33 through 6.35 establish that a best subset of M images exists within a larger data set of N images. However, the S_8 metric, which requires prior knowledge of the diffraction-limited image, was used to obtain these results. The S_8 metric cannot be used in real experiments because the diffraction-limited image is not generally known prior to the experiment. Figure 6.36 shows the results of a study of the comparative performance four image sharpness metrics, S_1, S_4, S_8, and S_{S1}, for the case of a star object with $f_{sr} = 75\%$, $m_v = 2$, and $r_0 = 10$ cm. Note from Fig. 6.36 that, as expected, $G(\vec{f})$ performance is maximized using the S_8 metric. However, Fig. 6.36 also shows that the S_{S1} metric provides $G(\vec{f})$ performance which closely approximates the performance obtained with the S_8 metric. Since the $G(\vec{f})$ performance obtained using the S_{S1} metric closely approximates the performance obtained with the S_8 metric, and because the S_{S1} metric does not require prior knowledge of the diffraction-limited image, the S_{S1} metric was adopted for additional study in Refs. [33] and [34].

A final question regarding the performance of frame selection derives from the observation that the plots of E$\{D(\vec{f})\}$ shown in Fig. 6.33 are well below what would be obtained with a diffraction-limited imaging system. Hence, it may be desirable to apply deconvolution, as presented in Section 6.3.1, to images of extended objects [34]. The paradigm of operation for a system combining frame selection and deconvolution is as follows. A data set of images of both the object of interest and a nearby reference star would be collected and stored. Frame selection using the same image sharpness metric and the same selection rate f_{sr} would then be applied to both the object and the reference star data sets. The outcome of applying frame selection to the reference star data is taken as the effective overall transfer function, as described in Eq. (6.54), and would be used to deconvolve the image of the extended object to increase the resolution of that image. This process will provide for small errors in the deconvolved image only if the effective average point spread function of frame selection applied to the extended object and the reference star. Since the frame selection process is not linear, there is no guarantee that application of frame selection to images of extended objects and star objects will yield the same effective average point spread function. The results of an investigation of the effective average point spread function for frame selection applied to the extended object shown in Fig. 6.5 and to a star object are presented in Fig. 6.37. Note in Fig. 6.37 that the point spread functions arising from applying frame selection to the extended

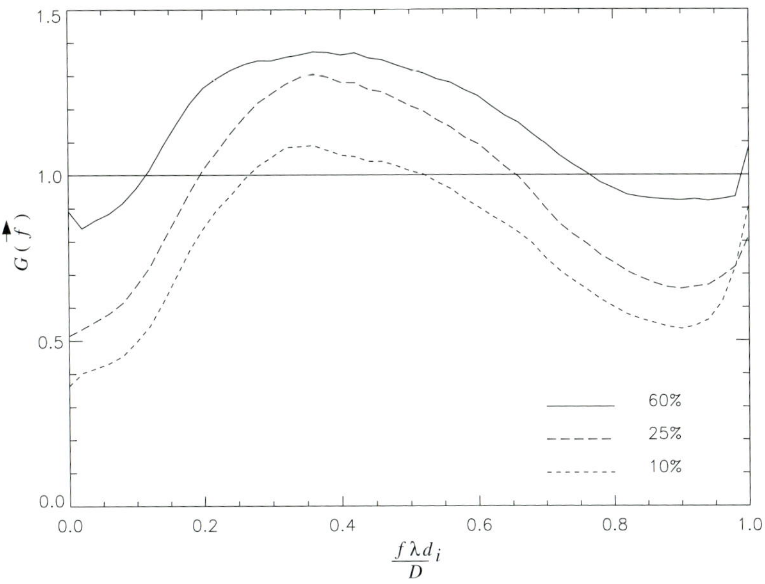

FIGURE 6.35
$G(\vec{f})$ for star object using various selection rates; $m_v = 2$, $r_0 = $ 10 cm, and S_8 metric used.

object and the star object are nearly indistinguishable, leading to the conclusion that deconvolution can be used in conjunction with frame selection under some circumstances.

We conclude that frame selection offers a means of improving the image spectrum signal-to-noise ratio, leading to images with lower noise and higher resolution. The S_8 metric does not require prior knowledge of the diffraction-limited image, and hence provides a experimentally useful metric from ordering the measured images. Finally, deconvolution can be used to sharpen images obtained with frame selection because the effective point spread functions of the frame selection technique applied to both point sources and extended objects are nearly identical.

6.4 Conclusion

In this chapter a number of hybrid imaging techniques, which combine elements of adaptive optics and post detection image processing, have been presented. Hybrid imaging techniques offer a tradeoff between complexity in adaptive optics and image post processing. We have seen that hybrid imaging techniques studied here provide improved performance, both in the mean sense and in the signal-to-noise ratio sense, over adaptive optics alone. There are two primary roles for hybrid imaging: (1) to allow simpler, partially compensated adaptive optics implementations with lower degrees of turbulence compensation, with the remaining imaging performance improvements obtained by post processing; and (2) to provide an imaging performance improvement to systems which are "fully compensated" under some nominal seeing condition when seeing conditions fall below the design parameters. We have seen that in some cases hybrid techniques can work with image data alone, such as deconvolution of compensated images, compensated speckle imaging, and frame selection. However, in the deconvolution from wave front sensing hybrid imaging paradigm additional information in the form of wave front sensor data is required.

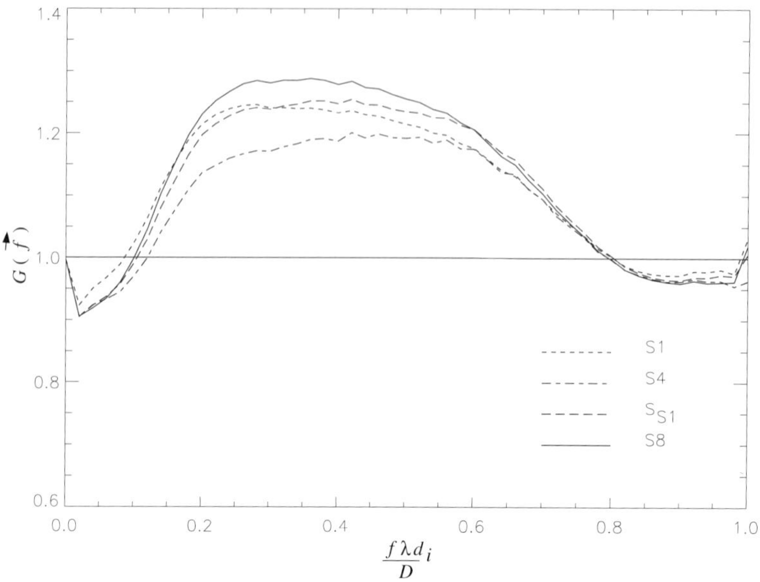

FIGURE 6.36
$G(\vec{f})$ **for star object using various image sharpness metrics;** $f_{sr} = 75\%$, $m_v = 2$, **and** $r_0 = 10$ **cm.**

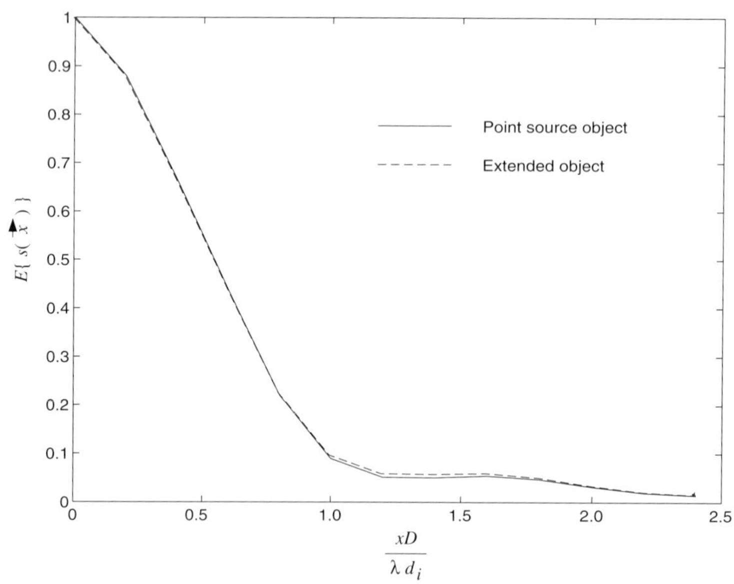

FIGURE 6.37
Effective average point spread functions $E\{s(\vec{x})\}$ **for star object and extended object obtained using the** S_8 **metric;** $f_{sr} = 70\%$, $m_v = 2$, **and** $r_0 = 10$ **cm.**

Bibliography

[1] J. W. Goodman, *Introduction to Fourier Optics*. New York: McGraw-Hill Book Co., 1968.

[2] D. L. Fried, "Optical resolution through a randomly inhomogeneous medium for very long and very short exposures," *J. Opt. Soc. Am.*, vol. 56, pp. 1372–1379, 1966.

[3] M. C. Roggemann, "Optical performance of fully and partially compensated adaptive optics systems using least-squares and minimum variance phase reconstruction," *Comp. & Elec. Eng.*, vol. 18, pp. 451–466, 1992.

[4] M. C. Roggemann and C. L. Matson, "Power spectrum and fourier phase spectrum estimation by using fully and partially compensating adaptive optics and bispectrum postprocessing," *J. Opt. Soc. Am. A*, vol. 9, pp. 1525–1535, 1992.

[5] M. C. Roggemann, "Limited degree-of-freedom adaptive optics and image reconstruction," *Appl. Opt.*, vol. 30, pp. 4227–4233, 1991.

[6] M. C. Roggemann and J. A. Meinhardt, "Image reconstruction by means of wave-front sensor measurements in closed-loop adaptive-optics systems," *J. Opt. Soc. Am. A*, vol. 10, pp. 1996–2007, 1993.

[7] M. C. Roggemann, B. L. Ellerbroek, and T. A. Rhoadarmer, "Widening the effective field-of-view of adaptive optics telescopes using deconvolution from wavefront sensing: average and signal-to-noise ratio performance," *Appl. Opt.*, vol. 34, pp. 1432–1444, 1995.

[8] D. L. Fried, "Post detection wavefront compensation," in *SPIE Proceedings on Digital Image Recovery and Synthesis*, vol. 828, pp. 127–133, 1987.

[9] J. Primot, G. Rousset, and J. C. Fontanella, "Deconvolution from wave-front sensing: a new technique for compensating turbulence-degraded images," *J. Opt. Soc. Am. A*, vol. 7, pp. 1589–1608, 1990.

[10] J. D. Gonglewski, D. G. Voelz, J. S. Fender, D. C. Dayton, B. K. Spielbusch, and R. E. Pierson, "First astronomical application of postdetection turbulence compensation: images of α aurigae, ν ursae majoris, and α geminorum using self referenced speckle holography," *Appl. Opt.*, vol. 29, pp. 4527–4529, 1990.

[11] B. M. Welsh and R. N. Von Niederhausern, "Performance analysis of the self-referenced speckle holography image reconstruction technique," *Appl. Opt.*, vol. 32, pp. 5071–5078, 1993.

[12] M. C. Roggemann, B. M. Welsh, and J. Devey, "Biased estimators and object-spectrum estimation in the method of deconvolution from wavefront sensing," *Appl. Opt.*, vol. 33, pp. 5754–5763, 1994.

[13] M. C. Roggemann and B. M. Welsh, "Signal-to-noise ratio for astronomical imaging by deconvolution from wavefront sensing," *Appl. Opt.*, vol. 33, pp. 5400–5414, 1994.

[14] E. P. Wallner, "Optimal wave front correction using slope measurements," *J. Opt. Soc. Am.*, vol. 73, pp. 1771–1776, 1983.

[15] D. L. Fried, "Atmospheric turbulence optical effects: understanding the adaptive-optics implications," in *Adaptive Optics for Astronomy* (D. Alloin and J.-M. Mariotti, eds.), vol. C423 of *NATO Advanced Study Institute Series*, pp. 25–57, Kluwer Academic Publishers, 1994.

[16] G. Rousset, "Wavefront sensing," in *Adaptive Optics for Astronomy* (D. Alloin and J.-M. Mariotti, eds.), vol. C423 of *NATO Advanced Study Institute Series*, pp. 115–137, Kluwer Academic Publishers, 1994.

[17] B. M. Welsh and C. S. Gardner, "Performance analysis of adaptive optics systems using slope sensors," *J. Opt. Soc. Am. A*, vol. 6, pp. 1913–1923, 1989.

[18] B. M. Welsh and M. C. Roggemann, "Signal-to-noise comparison of deconvolution from wave front

sensing and speckle imaging," in *SPIE Proceedings on Image Reconstruction and Restoration*, vol. 2302, pp. 281–292, July 1994.

[19] C. L. Matson, I. A. DeLarue, T. M. Gray, and I. E. Drunzer, "Optimal fourier spectrum estimation from the bispectrum," *Comp. & Elec. Eng.*, vol. 18, pp. 485–497, 1992.

[20] P. Nisenson and R. Barakat, "Partial atmospheric correction with adaptive optics," *J. Opt. Soc. Am. A*, vol. 4, pp. 2249–2253, 1987.

[21] R. C. Smithson and M. L. Peri, "Partial correction of astronomical images with active mirrors," *J. Opt. Soc. Am. A*, vol. 6, pp. 92–97, 1989.

[22] F. Roddier, M. Northcott, and J. E. Graves, "A simple low-order adaptive optics system for near-infrared applications," *Pub. Astron. Soc. Pac.*, vol. 103, pp. 131–149, 1991.

[23] M. C. Roggemann, D. W. Tyler, and M. F. Bilmont, "Linear reconstruction of compensated images: theory and experimental results," *Appl. Opt.*, vol. 31, pp. 7429–7441, 1992.

[24] M. C. Roggemann, E. L. Caudill, D. W. Tyler, M. J. Fox, M. A. Von Bokern, and C. L. Matson, "Compensated speckle imaging: theory and experimental results," *Appl. Opt.*, vol. 33, pp. 3099–3110, 1994.

[25] J. D. Gaskill, *Linear Systems, Fourier Transforms, and Optics*. New York: John Wiley & Sons, 1978.

[26] R. C. Gonzalez and R. E. Woods, *Digital Image Processing*. Reading, Massachusetts: Addison-Wesley, 1993.

[27] J. Biemond, R. L. Lagendijk, and R. M. Mersereau, "Iterative methods for image deblurring," *Proc IEEE*, vol. 78, pp. 856–883, 1990.

[28] J. C. Dainty and A. H. Greenaway, "Estimation of spatial power spectra in speckle interferometry," *J. Opt. Soc. Am.*, vol. 5, pp. 786–790, 1979.

[29] D. L. Fried, "Anisoplanatism in adaptive optics," *J. Opt. Soc. Am.*, vol. 72, pp. 52–61, 1982.

[30] F. Rigaut and E. Gendron, "Laser guide star adaptive optics: the tilt determination problem," *Astron. Astrophys.*, vol. 261, pp. 677–684, 1993.

[31] J. W. Goodman, *Statistical Optics*. New York: John Wiley & Sons, 1985.

[32] B. L. Ellerbroek, "First-order performance evaluation of adaptive optics systems for atmospheric turbulence compensation in extended field-of-view astronomical telescopes," *J. Opt. Soc. Am. A*, vol. 11, pp. 783–805, 1994.

[33] M. C. Roggemann, C. A. Stoudt, and B. M. Welsh, "Image spectrum signal-to-noise ratio improvements by statistical frame selection for adaptive optics imaging through atmospheric turbulence," *Opt. Eng.*, vol. 33, pp. 3254–3264, October 1994.

[34] M. C. Roggemann, B. M. Welsh, S. D. Ford, and C. A. Stoudt, "Frame selection for adaptive optics imaging through atmospheric turbulence," in *SPIE Proceedings on Image Reconstruction and Restoration*, vol. 2302, pp. 42–53, 1994.

[35] P. S. Idell and A. Webster, "Resolution limits for coherent optical imaging: signal-to-noise analysis in the spatial-frequency domain," *J. Opt. Soc. Am. A*, vol. 9, pp. 43–56, 1992.

[36] D. L. Fried, "Probability of getting a lucky short-exposure image through turbulence," *J. Opt. Soc. Am.*, vol. 68, pp. 1651–1658, 1978.

[37] R. A. Muller and A. Buffington, "Real-time correction of atmospheric turbulence degraded telescope images through image sharpening," *J. Opt. Soc. Am.*, vol. 64, pp. 1200–1210, 1974.

Index